Praise for

WHEN THE WORLD SEEMED NEW

"At one of the great turning points in modern history, America and the Free World (the phrase meant more then) were fortunate to be led by a president, George H. W. Bush, who is only now receiving his due. In this epic book, Jeffrey A. Engel explains how Bush presided over the momentous conclusion of the Cold War. With searching scholarship, a gift for the telling human detail, and an appreciation of how the personal and the political interact in often subtle ways, Engel has written a landmark account of a president, a nation, and a global order at a crossroads. This is a terrific work of history."

— Jon Meacham, author of
*Destiny and Power: The American Odyssey
of George Herbert Walker Bush*

"*When the World Seemed New* is a fine, often stirring account of these times . . . An absorbing book." — *Wall Street Journal*

"Usually, when empires fall, war and chaos follow. Amazingly, when the Soviet Union — with its twenty thousand or so nuclear weapons — collapsed, the world became (for a time, anyway) more peaceful and prosperous, thanks in no small part to the wisdom, vision, and restraint of President George H. W. Bush. Jeffrey Engel has written a rich, marvelous narrative history stocked with lessons for our own dangerous times."

— Evan Thomas, author of
Ike's Bluff and *Being Nixon*

"Engel's excellent history forms a standing — if unspoken — rebuke to the retrograde nationalism espoused by Donald J. Trump."

— *New York Times Book Review*

"The Cold War's end offered peril and promise, and Jeffrey A. Engel's revealing and deeply researched new history demonstrates that George H. W. Bush navigated it brilliantly. Instability loomed. Yet Bush's personal diplomacy ensured Germany's successful unification, the Soviet Union's peaceful collapse, victory in the Persian Gulf, and preservation of Sino-American relations after the horror of Tiananmen Square. His reputation as a statesman has rightly grown over time. Peering into the halls of power on both sides of the Iron Curtain, and in the Oval Office in particular, Engel's gripping account shows us why."

— **Stephen Hadley, former national security adviser**

WHEN THE WORLD SEEMED NEW

WHEN THE WORLD SEEMED NEW

George H. W. Bush and the
End of the Cold War

JEFFREY A. ENGEL

Mariner Books
Houghton Mifflin Harcourt
Boston   New York

First Mariner Books edition 2018

Copyright © 2017 by Jeffrey A. Engel

For information about permission to reproduce selections from
this book, write to trade.permissions@hmhco.com or to
Permissions, Houghton Mifflin Harcourt Publishing Company,
3 Park Avenue, 19th Floor, New York, New York 10016.

hmhco.com

Library of Congress Cataloging-in-Publication Data

Names: Engel, Jeffrey A., author.
Title: When the world seemed new : George H. W. Bush and the end of
the Cold War / Jeffrey A. Engel.
Description: New York, New York : Houghton Mifflin Harcourt, 2017. |
Includes bibliographical references and index.
Identifiers: LCCN 2017044499 (print) | LCCN 2017050301 (ebook) |
ISBN 9780544931848 (ebook) | ISBN 9780547423067 (hardback) |
ISBN 9781328511652 (pbk.)
Subjects: LCSH: United States — Foreign relations — 1989–1993. |
Bush, George, 1924– Influence. | Cold War — Diplomatic history. | United States — Foreign
relations — Soviet Union. | Soviet Union — Foreign relations — United States. | National
Security Council (U.S.) — History. | Germany — History — Unification, 1990. | Soviet
Union — History — 1985–1991. |
Persian Gulf War, 1991. | BISAC: HISTORY / United States / 20th Century. | POLITICAL
SCIENCE / International Relations / Diplomacy. | BIOGRAPHY & AUTOBIOGRAPHY /
Presidents & Heads of State.
Classification: LCC E840 .E54 2017 (print) | LCC E840 (ebook) |
DDC 327.73009/04—dc23
LC record available at https://lccn.loc.gov/2017044499

Book design by Chloe Foster

This book is dedicated to my mentors:

Walter LaFeber
who taught me history

Tom McCormick
who taught me to think

Richard Immerman
who taught me to care

and Lory and Toby Engel
who taught above all else by loving example

CONTENTS

We have it in our power to begin the world over again. A situation, similar to the present, hath not happened since the days of Noah until now. The birthday of a new world is at hand, and a race of men, perhaps as numerous as all Europe contains, are to receive their portion of freedom from the events of a few months.

— Thomas Paine, 1776

I come back to the word of prudent — managing of what we do and what we say — and resist flamboyant actions. Things are moving our way . . . Democracy? Freedom? They are moving our way. And so, we don't need to be out there trying to micromanage the desire for change.

— George H. W. Bush, 1991

INTRODUCTION

HE HAD NEVER SEEN anything quite like it. Onlookers cheered the armored motorcade as it barreled through the streets of Washington, D.C. They waved flags and rhythmically chanted his name. The energy was electric. George Bush had seen crowds before. Vice president for almost seven full years by 1987, he had witnessed inaugurations and national political conventions, military parades and Super Bowls. He had thrown out more first pitches than he could count. He had heard cheering his entire life.

But not like this. And never for him. The masses were screaming for the other man in the backseat of their limousine: Mikhail Gorbachev, general secretary of the Communist Party of the Soviet Union, arguably the most popular man in the world. Americans were cheering the communist as if he were a rock star. Perhaps they were right to be so enthusiastic. In little more than two years in office, Gorbachev had transformed his country, and prospects for global peace, fundamentally changing the tone and tenor of the Cold War between the United States and the Soviet Union that had raged for four decades, more than once bringing humanity to the brink of nuclear annihilation.

Those days now seemed past because of him. Gorbachev had started nothing less than a revolution throughout the communist world. He called for *perestroika:* a general "reconstruction" of Soviet communism complete with increased political and economic freedom for his people. He offered *glasnost:* a new "openness" for a Soviet society long governed by secrecy and fear. Each catalyzed latent democracy movements throughout Moscow's empire, prompting even hard-line anticommunists like President Ronald Reagan to accept that

Gorbachev was a new kind of leader. The whole world seemed new again, and poised to enter a new democratic future.

Gorbachev may have been the Prometheus of his age, but he was also perpetually behind schedule. This morning was no different. Tasked with delivering their guest from the Soviet embassy to the White House, Bush waited impatiently. Gorbachev's handlers first asked him to sit in an anteroom while their boss finished his morning meetings. When those sessions stretched on, they suggested perhaps the vice president might prefer simply to proceed without him. Bush refused. Vice presidents were rarely made to wait. Neither were presidents, and Reagan was waiting too. More important, Bush had lobbied for weeks for this opportunity to be seen at Gorbachev's side, and he was not about to let the chance for that ideal photo op evaporate merely because he had to cool his heels for a few moments more. Bush wanted to be president. He had been preparing for that job his entire life. His campaign to succeed Reagan was about to kick off, and he needed voters to see him not merely as a loyal underling but as a global leader in his own right. So he waited.

Plus, he had a message to deliver. Election season would soon be heating up, he told Gorbachev when the pair finally made their way into the back of the Soviet limousine for the short drive across downtown Washington. Only a translator rode with them. They had met two years before, though only briefly, during the funeral for Gorbachev's predecessor. Bush had been to many Soviet funerals over the years, and they were always hectic affairs. This was their first time alone. "The campaign would be filled with strong, hardline statements from all kinds of sources about US-Soviet relations," he noted. Gorbachev should not pay too close attention.

The Soviet leader didn't seem to be paying attention now, either. Instead, he was fixated on the teeming crowd chanting his name. Bush plowed on. Re-calling one of Chinese leader Mao Zedong's favorite phrases — and one of his own after serving as Washington's chief diplomat in Beijing more than a decade before — deriding "empty cannons of rhetoric," Bush warned Gorbachev not to take his campaign statements "too seriously." He would have to say and do all sorts of things to get elected. This was how things were done in America. Politicians played to the crowd.

Gorbachev nodded but continued to stare silently into the masses of humanity just outside the car. Bush wondered if perhaps the translator had delivered the message incorrectly. He wondered if Gorbachev was even listening at all.

Suddenly the Russian leaned forward. "Stop the car!" Gorbachev ordered his driver. Tires screeched. The limousine's doors flew open, and Gorbachev threw himself directly into the nearest throng. Soviet and American security personnel screamed into their handsets and scrambled to catch up. Before Bush truly knew what was happening, his companion was engulfed by the ecstatic mob. Gorbachev smiled and shook hands like a man who had been kissing babies and wooing voters his entire life.

In a flash he was at Gorbachev's side, lest he fail to appear in photos sure to run in the next day's newspapers. "Do you do this a lot?" Bush gasped when the two finally tumbled back into the safe confines of their car. Flush with adrenaline, Gorbachev nodded. "Leaders should be equal to the people," he said. His twinkling gray eyes told Bush that Soviet leaders could be unpredictable. And that they too could deliver messages.

It was not an auspicious beginning to their relationship. Neither man truly heard the other. Each instead seemed primarily interested in proving his own importance. Like new rivals in a schoolyard, the world's largest, they seemed interested most of all in intimidation.

Their relationship would in time help ease the world through one of the twentieth century's most perilous periods. Within a year Bush would be president. Within two the Cold War would effectively be over. Within three the Soviet Union would be gone, and European communism with it. No one was predicting these events in 1987, and it was unthinkable that they might be accomplished with little bloodshed. Empires rarely ceded power gracefully. Collapsing great powers typically leave wars in their wake. Plus, the Soviets were different. Empires had risen and fallen for millennia. Never before in human history had a great power broken apart with more than twenty thousand nuclear weapons in its midst, enough firepower to end all of human life. There were no helpful precedents for what Bush eventually faced as president, and no guarantees it would all end well.

He governed during remarkable times. It is hard to believe that so much of consequence could occur in a single four-year presidential term. The Berlin Wall fell. Democracy took root behind the Iron Curtain. Germany, divided since World War II, reunited. The Soviet Union surrendered, revolted, and then ceased to exist. Bulgaria, Czechoslovakia, Hungary, East Germany, and Poland all threw out socialism and their Soviet overlords. Romanians did so violently. Yugoslavians descended into outright civil war. Some of the others came perilously close to violent eruptions. China did erupt, roiled by a democratic surge and its violent suppression. During this same period, Bush ordered American forces into Panama (to remove a dictator), into Somalia (to prevent a famine), and into the Middle East at the head of a massive international coalition formed to repel Iraq's invasion of Kuwait. It was the largest American military expedition since Vietnam, which remains true at this writing. Closer to home, his administration largely concluded negotiations for a North American Free Trade Association, joining the continent economically as never before, while on the other side of the Atlantic, preliminary talks for expansion of the North Atlantic Treaty Organization began to shape a new and contentious vision for Europe.

One or perhaps two such events would have been enough to occupy any White House's attention. That they all occurred during the same four years makes Bush's presidency the most internationally complex since that of Franklin D. Roosevelt, who led the country through World War II and to newfound heights of global influence and power. By the time he left office, Bush was even more powerful, in a global sense, than FDR. The United States he left to his successors appeared hegemonic as never before in the Cold War's wake. It was popular, too. Newfound allies clamored to join its side. Democracy, which American leaders had long championed, and no one more vigorously than Bush himself, was the global rage. If one ranks the American empire as the world's most powerful, rivaled only by imperial Rome in its heyday, then for a brief moment, by the close of his time in office George H. W. Bush was the most powerful man in human history.

He could foresee none of these things as he contemplated his run for the Oval Office in 1987. But he knew something big was under way within the communist world. His biggest challenge, aside from getting elected, was embodied by the other man in the back of that Soviet limousine. Gorbachev had started a revolution. He had also unleashed forces well outside his control. Popular beyond description abroad, he was increasingly under siege at home, where his reforms had brought economic hardship, instability, and most important of all a sense that the Soviet people were about to lose everything they had worked for over generations, including their status as a global superpower. Their vassal states and republics, won at the cost of more than 20 million lives during World War II, yearned for independence from Moscow and a chance to join the American-led West, and more narrowly the Western European economic and political union forming just beyond the Iron Curtain. Millions found his words inspiring. Gorbachev spoke of change and of self-determination. He even spoke of freedom. Throughout the communist world, people took to the streets to demand precisely those things, with no certainty that their own governments were willing to cede them. Time and again crowds confronted troops. Each confrontation could have turned bloody. The threat of civil war loomed, as did ethnic and religious conflicts that could easily have leaped national boundaries. Global wars had started over less. Communism's collapse was a remarkably dangerous prospect.

Bush was the most powerful man in the world once seated in the Oval Office, yet he was also remarkably helpless when the revolution of ideas Gorbachev inspired flowered on the ground. Any soldier with a gun, or any protester with a rock, could do more in any given standoff between regimes and their people to determine the course of events than the president of the United States thousands of miles away. The Cold War's end is often told as a story of crowds. Crowds marched, regimes fell. Throughout Eastern Europe and ultimately even in the Soviet Union itself, crowds toppled dictators in a series of democratic revolutions, many so smooth they were soon termed "velvet." Historian Will Hitchcock nicely summed up this strain of thinking when he wrote: "Gorbachev did not give Eastern Europeans their freedom in 1989. They took it."

It might have happened very differently. Not every regime ceded power willingly. In China, the crowds marched and were destroyed. In Romania, crowds marched and the regime fled, ultimately ripped to shreds. In Yugoslavia, freedom quickly became license to slaughter. Sometimes the violence began with an official order. Sometimes it was spontaneous. When faced with chanting crowds and a changing world, Soviet bloc leaders had to choose sides, deciding whether to paddle with or against what Germany's great nineteenth-century statesman Otto von Bismarck termed the "stream of time." They needed to decide if they would go with the flow of history or against it. Whether to march with their people toward the future or mow them down to preserve the past.

Western leaders, and Bush in particular, had choices to make as well. The wrong move or gesture in response to the turmoil throughout the communist world could have scuttled the entire democratic movement. Too quick an embrace of reformers might have rallied conservative and reactionary forces against them, leading to their demise. Move too slow, and Western leaders could have equally doomed Eastern Europe's democratic reformers by denying them the legitimacy and aid (moral and material) they needed to flourish. To call this a tightrope understates the stakes. Falling meant not merely one death but potentially tens of millions.

Bush's challenge was to keep moving forward along the stream of history, influencing both crowds and leaders to go in the right direction, even as he could never hope to control their actions from half a world away. His challenge was in part to work with Gorbachev, but not so close as to hand Gorbachev's enemies a propaganda cudgel, keeping *perestroika* and *glasnost* alive long enough to ensure that the democratic surge behind the Iron Curtain survived. He also had his own country to worry about. Indeed, given the pervasive talk that America's empire might pass into history soon after the Soviets', maintaining American security and influence throughout the world was more than just what he pledged to do when taking the oath of office. It was also to Bush's mind the one certain way to ensure global peace.

This is the story of the Cold War's end, and the real story of American leadership skillfully employed to aid its safe passage into history: the way Bush cajoled, talked, bribed, guided, and advised, when he could, and how he displayed self-restraint during a perilous moment. Where others might have found the powers of his office too intoxicating to ignore, unleashing forces they could not control by exerting American influence and force too vigorously, Bush succeeded through what I call his style of "Hippocratic diplomacy." That is, he first strove to do no harm.

Others might have rushed to respond to miraculous changes. Not only is power intoxicating, but also its deployment frequently makes leaders, and democratically elected politicians in particular, popular — for a while at least. Bush typically waited. Revolutions are inherently dangerous, he and those around him believed. Violence can erupt even at the moment of most hopeful optimism. Resisting the opportunity for short-term gain, and ignoring cries from critics demanding more public exultation at all the apparent victories of their democratic era, Bush tempered the hopes and fears of his age with caution, and in large measure his caution helped carry the world safely through this tumultuous time.

To be clear, Bush did not end the Cold War. No single person did. But he could have ruined its outcome. A leader with a different temperament, with more of a hair-trigger penchant for action and less respect for the danger of chaos, catalyzed by the apparent victory for democracy, free markets, and the loose conglomeration of Western ideals seemingly demonstrated every time protesters marched for freedom, might have intervened in ways that backfired — as many interventions, even American ones, have done. "Things are going our way," he instead frequently said during his time in office. His job was to keep them going in that direction. And not to lose. "Democracy? Freedom?" he said. "They are moving our way."

Gorbachev hated boats but liked nautical analogies. "One might say I had to assume the role of the captain of a ship, riding out the storm . . . occasionally turning so sharply that I thought the wheel would be ripped from my hands," he concluded long after his political career had come to an end. Bush had a different job. His responsibility as captain of the world's largest vessel, with a rapidly growing flotilla following its course, was avoiding rocky shoals, tacking along with the current toward the new and better world that seemed to beckon. This is his story, the story of his partnership with Gorbachev, and the story of the Cold War's surprisingly peaceful end.

This is also a story of international policy making far beyond merely the Kremlin and the White House. Any history of such a turbulent period must take into account voices and actors throughout the world if we are to gain a full picture of what happened and why. This book reveals much that was until recently locked away in American archives. For several years, declassification requests submitted over my signature — so archivists told me — exceeded the combined rolls of all other such requests submitted at all the nation's presidential libraries. It draws upon these newly available materials, and private interviews with many of the key American actors at the time. So too does it draw upon the wealth of similar research conducted by scholars and interest groups in other countries. Soviet, Chinese, British, Canadian, French, German (East and West), Polish, Hungarian, Czechoslovak, Japanese, and even more new sources have been tapped for this book, making it in many ways a product not possible until the twenty-first century, when the interconnected web of scholars is matched by the plethora of new documentary resources published en masse and in heretofore unimaginable quantity on the Internet. A study of an American president in an international age, it is at heart an international story.

It is also the story of an underappreciated president. He was a one-term extension of Reagan to some, merely the father of a seemingly more impactful president to others. He was transitional but not transformational. The last World War II veteran to hold the office, he was during his time as president and immediately afterward considered unimaginative, beholden to the past even as he struggled, in his own words, to grasp "the vision thing" others described for the future.

Solely with regard to his foreign policy, this view demands readjustment. Evaluating a strategist's success is no easy thing. Some fail by being too brash, others by not trying hard enough. Some can conceive of new worlds beyond their reach; others fail to see the need for change. Results are difficult to gauge, being wholly dependent on an observer's perspective. Is a peace that lasts a generation proof of a successful strategy? Or is the fact that it lasts only a generation a sign of failure? If conflict erupts anew, should it have been foreseen, or is it perhaps the fruit of seeds planted by leaders who, up until their peaceful world's collapse, celebrated their own achievement?

Having taught and written about grand strategists for nearly two decades, asking students to evaluate leaders from ancient times to our own day, I have found one metric that is meaningful, if not the final word: Did a strategist accomplish all he or she desired? That question does not cover the quality or the long-term consequences of that strategist's work, but it offers a starting point for judgment. Put simply, did he or she have a plan (shockingly few strategists do) and then make it happen?

For Bush, the answer clearly is yes to both elements of that basic query. New evidence reveals that the man said to lack vision had one indeed, even if rooted in familiar tropes drawn from his generation's reading of history. Determined to maintain America's position as the world's dominant power during a time of great transition, determined to ensure the survival of international institutions (many designed by American strategists after World War II), and determined to make certain that the stream of history continued to flow democracy's way, he achieved his goals. In doing so, he secured Germany's reunification and with it the survival of NATO, which he considered the essential element of the post-1945 American-led peace in Europe — his single greatest diplomatic achievement — and kept China engaged in the world even after the horrors of a crackdown on protesters that changed Tiananmen Square in Western minds from a place to an event. Believing engagement to be the root cause of China's transformation to date, and its isolation inherently dangerous, Bush ensured at the least that Beijing and Washington could continue to communicate in its aftermath. Neither liked what the other had said or had done. But they continued to talk.

That was not all. Subsequently determined to demonstrate that the Cold War's end would not produce international anarchy, Bush led an international coalition in the Middle East to repel a tyrant's invasion, setting a precedent for respecting sovereignty, and the international community's will, in what he dubbed (unoriginally) the "new world order." Not every ensuing president followed his lead or agreed with his strategies. His decisions are frequently questioned and criticized in these pages as well. Many clearly sowed the seeds of subsequent conflicts. What cannot be disputed is Bush's success in attaining what he designed and desired. To the surprise of many who considered him lacking in vision, he passes the first test of a truly successful strategist with ease. He had a plan and achieved it.

Bush's diplomacy also offers a useful example, demonstrating that leadership, experience, and patience matter. The wording of a letter; the tenor of a conversation; knowing when to listen and when to yell. Knowing most of all that success demands hard work and the investment of time. Like all powerful lessons, this one appears obvious at first blush. But too often history has turned a dangerous corner because such prudent leadership was lacking when needed most. "These are the times that try men's souls," Thomas Paine wrote in 1776. His writing also provides this book's title: "We have it in our power to begin the world over again," he wrote as the American Revolution began.

The world that emerged from the ashes of the Cold War was indeed new. It was not perfect. But neither was it designed to be, at least according to its principal architect. As Bush baldly stated, the goal — his goal — was simply to improve upon the past. A world "freer from the threat of terror, stronger in the pursuit of justice, and more secure in the quest for peace ... A hundred generations have searched for this elusive path to peace, while a thousand wars raged across the span of human endeavor." Humanity's ills would not be solved within a single generation's lifetime, but civilization could, step by step, continue its path toward a better world. His close friend and national security adviser summed up the administration's diplomatic ethos by offering that "the world could be a better place . . . but don't get carried away." In perilous times, that is no small vision. Would that other American leaders in Bush's wake had followed his lead, going with rather than against the stream of history, content to ride its current rather than speed recklessly at a faster clip, accomplishing more by changing course less.

1

SWAN SONG AND SURPRISE

REAGAN WAS READY. Thanksgiving was but days away. After eight years in Washington, there was one final turkey to pardon, one more White House Christmas tree to light, one more New Year's celebration. He looked forward to a long break, leisurely trail rides with his wife, and retirement. The oldest man ever elected president, he'd once been full of energy to match his legendary optimism. Now he was tired. Never one for details, he'd been forgetting more and more. "He's not really working at the job and not in touch with reality," his chief of staff complained. Other top aides had the same experience: "I can't tell whether I'm really helping him or not because he listens and I don't get a sense that he disagrees with me or agrees with me or what," the CIA director confessed. "I'm in the same boat," his national security adviser commiserated.

Reagan, however, had few regrets. The economy was rolling. Unemployment was down, the stock market up. Nearly three quarters of Americans polled in 1988 considered their personal prospects on the rise, the highest rate of his presidency. A majority believed that the country was headed in the right direction. To his eyes, Americans seemed proud again, and the world largely at peace. Even if he'd failed to eradicate nuclear weapons, as he'd long dreamed and nearly achieved, he'd helped bring the superpowers closer than ever before. He'd even forged a personal friendship of sorts with the energetic new Soviet leader, Mikhail Gorbachev. For over forty years, Soviet and American leaders had waged a Cold War for global dominance, holding the world hostage to their nuclear-charged competition. Those days appeared over. Never before had the men who ran the White House and the Kremlin seemed so close.

Some even whispered that the Cold War itself might be coming to an end.

Powerful forces were calling for reform throughout the communist world. In Moscow the talk was of political and economic change, *perestroika* and *glasnost*, restructuring and openness. In hotspots like Poland and Hungary, deep inside the Soviet empire, reformers promising open markets and open elections had taken their protests from the streets into the government. Even the Kremlin appeared less menacing. Words like "peace" and "new thinking" had replaced ominous threats of war and promises of perpetual class conflict. Soviet leaders once boasted of burying the West, and of churning out nuclear-tipped missiles like sausages. Moscow's current head talked instead about building a "common European home" where all could prosper and live in peace.

Times had indeed changed over the course of Reagan's eight years. Even if his administration had nearly foundered a few years back over the revelation of convoluted arms deals with Iran designed to help anticommunist forces in Central America, a deal Reagan still did not—nor want to—fully comprehend, the communist menace he'd long despised finally appeared to be in retreat. He felt vindicated and proud. Voters had just elected his vice president, George Bush, to succeed him in office, the first time an incumbent had pulled off that trick since 1831. As far as Reagan was concerned, the American people had just voted him a third term.

The world would soon be Bush's problem. As the White House buzzed with the fervor of transition, there remained a few final things Reagan wanted to wrap up before he left Washington for good. High on his list was giving proper thanks to Margaret Thatcher, Britain's Iron Lady and prime minister since 1979. The two shared much. Both were deeply conservative, favoring blanket endorsements of the market and the overriding power of unbridled "freedom." Both loathed authoritarianism in all forms and were quick to identify totalitarian streaks in political opponents. For eight years they had weathered international storms together, confiding in each other, offering support when times got tough. They didn't agree on everything. The world could look quite different when viewed from the Potomac or the Thames. But when Iran-contra seemed ready to scuttle his entire presidency in late 1986, with impeachment or resignation not wholly unthinkable, it was Thatcher's handwritten letter that most buoyed his spirits. "The press and media are always so ready to criticize," she'd consoled him, "but your achievements in restoring America's pride and confidence and in giving the West the confidence it needs are far too substantial to suffer any lasting damage." Reagan called the prime minister right away to thank her. Now as his presidency was coming to an end, he planned a grand valedictory dinner to thank her anew.

Thatcher's role in Reagan's presidency went beyond mere confidant and cheerleader. The pair frequently seemed to operate in tandem. His lofty words softened her sharper edges, while pundits often assumed she voiced his true beliefs. He also listened to her, taking to heart her assessment in the mid-1980s that the new Soviet leader's fresh ideas stood in stark contrast to the staid mold of traditional Kremlin oligarchs. Gorbachev was someone "with whom we could do business," she said. He appeared to be "an unusual Russian in that he was much less constrained, more charming, open to discussion and debate, and did not stick to prepared notes," she told the president as they chatted beside a roaring fire at Camp David in 1985. Even as she cautioned that "the more charming the adversary, the more dangerous," she implored Reagan to meet him, to listen to him.

Only someone he truly trusted could penetrate Reagan's lifelong contempt for a political ideology he so despised. "Communism is neither an economic or political system," he'd told radio audiences in 1975. "It is a form of insanity . . . contrary to human nature." Reagan rode similar rhetoric to California's governor's mansion, then the White House. Anticommunism formed the bedrock of his entire philosophy, including the belief that the Soviet Union's leaders were scheming to conquer the world by force. "I know of no Soviet leader since the revolution, and including the present leadership," Reagan had declared at his first presidential press conference in 1981, "that has not more than once repeated in the various Communist congresses they hold their determination that their goal must be the promotion of world revolution and a one-world Socialist or Communist state." To achieve world domination, he warned, Soviet leaders "have openly and publicly declared that the only morality they recognize is what will further their cause, meaning they reserve unto themselves the right to commit any crime, to lie, to cheat, in order to attain that." Until they changed, in effect surrendered, he vowed, the Soviets would find in him an implacable enemy.

Once in power, Reagan sought out evidence confirming his long-held belief that the Kremlin's military was stronger than his own. CIA director Stansfield Turner told him otherwise, briefing him after the election that the best available intelligence suggested no real Soviet strategic advantage after all, in particular because of the technological superiority of America's nuclear arsenal. Reagan responded with silence. Turner had hoped to retain his job despite the change of administration. The president-elect's stone-faced look made it clear his "days as a passenger in an official CIA armored limousine were over."

He'd be retired within weeks. Reagan knew what he believed. "The Soviets have spoken as plainly as Hitler in "Mein Kampf,"" he warned his National Security Council. "They have spoken world domination." For American policy makers of his generation, invocation of Hitler demanded a singularly forceful response. Communists could not be appeased, he warned; they would never be satisfied. Reagan planned to stop them. Inheriting a renewed arms race begun by his predecessor following the 1979 Soviet invasion of Afghanistan, he authorized the largest military buildup in American peacetime history. "The United States will invest as much as is necessary in the arms race until we are in first place," he declared. Pentagon spending doubled. Soviet-American relations crumbled in turn. The Cold War was not merely the result of some giant international misunderstanding, Reagan charged in 1983, nor traditional great power politics played out on a modern stage. It was instead a "struggle between right and wrong and good and evil." The Kremlin, he said, was not simply immoral. It ruled an "evil empire." Its sins required eradication through a "crusade of freedom," which he would lead, to purge the world by fire, if need be, of communism's stain.

As Reagan later admitted, his speechwriters coined the term "evil empire." But the sentiment was his, and he proudly gave the speech, and several more like it, "with malice aforethought." It embodied his fundamentally Manichaean sense of a world divided into light and dark, right and wrong, good and evil. He thought it obvious which side each superpower represented. "We're enjoined by Scripture and the Lord Jesus to oppose it [evil] with all our might," Reagan sermonized, and so long as Kremlin leaders "preach the supremacy of the state, declare its omnipotence over individual man, and predict its eventual domination of all peoples on earth, they are the focus of evil in the modern world." He called Soviet leaders "terrorists" after their air force mistakenly downed a Korean jetliner carrying 269 passengers and crew, including one U.S. congressman. He also frequently deployed the modifier "godless" alongside the term "communists." He could think of nothing more damning.

Soviet leaders heard Reagan's harsh words—and, more important, measured his rapid military buildup—and feared the worst. "The beginning of the Reagan presidency calls to mind the fascist seizure of power," Moscow's high-

est-ranking soldier warned, recalling the Nazi invasion and the 20 million Soviet deaths that ensued. The Soviets had been caught unawares in 1941. Determined not to make that mistake again, Soviet leaders forty years later ordered
their global intelligence network to ferret out confirmation of Washington's
intention to attack. Confirmation was not hard to find, especially after Reagan
followed up his "evil empire" speech with announcement of his Strategic Defense Initiative (SDI), designed to shield the United States and its allies from
nuclear attack. Using Reagan's own words, the Kremlin warned its Warsaw Pact
allies that Washington "has declared a 'crusade' against socialism as a social
system. Those who have now ordered to deploy new nuclear weapons on our
threshold link their practical policies to this reckless undertaking."

Politburo fears led to jittery fingers on the nation's nuclear button. In late
September 1983, satellite sensors erupted with warnings of a massive and unexpected American assault. Soviet nuclear crews swung into action, warming
up their missiles and scrambling their planes aloft. It seemed Reagan's military
crusade had begun. Fortunately, frazzled Kremlin policy makers realized it
was a false alarm mere moments before ordering a retaliatory strike. In calmer
times, commanders might have suspected faulty equipment when told of an
unprovoked full-scale nuclear attack. These were not calm times.

This would not be the world's only close brush with annihilation during
Reagan's first term. By November 1983, Soviet nerves proved so frayed that
an American-led military exercise in Europe nearly prompted another preemptive nuclear strike. NATO maneuvers appeared a bit too realistic to Soviet
strategists, who once more debated firing before American missiles could land.
Cooler heads thankfully prevailed, and the tense peace returned. Previous crises over Berlin or Cuba had unfolded under the white-hot lights of international scrutiny. In late 1983, only the world's most powerful leaders knew how
close to the abyss they'd come.

Reagan's missile defense plan, quickly dubbed "Star Wars" by critics and supporters alike, exacerbated Soviet anxieties nearly beyond the tipping point. The
Kremlin's top scientists predicted little chance of such a technologically audacious plan ever working. Most American scientists concurred, though Soviet
military strategists took a far warier view. Trained to consider worst-case scenarios, they warned that the system, if it ever became operational, would upset
the delicate balance of nuclear deterrence that had governed the entire nuclear
age. Mutually assured destruction had kept crises from slipping into hot war

for more than a generation. So long as each superpower retained the ability to annihilate the other, the theory ran, neither would ever dare to attack. Perhaps no longer. When coupled with the short-range missiles Washington hoped to deploy in Europe, designed to effectively eliminate Soviet response time, Reagan's defensive shield seemed the ultimate offensive weapon, affording Washington the opportunity to launch a debilitating surprise attack while parrying any retaliatory strike. "All this confirms our worst fears," Soviet general secretary Yuri Andropov lamented from his hospital bed, his strength sapped by the kidney failure that would eventually claim his life. "The U.S. ruling circles have embarked on a sudden application of a nuclear attack on the Soviet Union." Andropov had already made international headlines by calling his American counterpart an insane liar, privately telling the head of East Germany's security service that "Reagan's vulgar speeches show the true face of the military industrial complex." His farewell address to the Soviet people thus read more like a warning: "If anybody ever had any illusions about the possibility of an evolution to the better in the policy of the present American administration, these illusions are completely dispelled now."

A sad and dangerous situation had developed by mid-decade, awful in its irony, darkly comical save for the stakes. American and Soviet leaders trusted their own promises of peace. Recalling their shared Nazi foe, and believing their adversaries similarly driven more by fanaticism than by reason, they also believed that their current enemy intended to strike. Each side's ensuing defensive reactions, seemingly reasonable in the face of such fervent opposition, appeared menacing in its opponent's eyes, and thus only further raised tensions. The dangerous cycle repeated, and Soviet-American relations crumbled in turn. "It's coming, Bud," Reagan had earlier confessed to deputy national security adviser Bud McFarlane. "This inexorable building of nuclear weapons on our side, and the Russians' side, can only lead to Armageddon."

Andropov soon died, as quickly did his successor, leaving the relatively youthful Gorbachev in charge by 1985. Within a few short years, Reagan then reversed a lifetime of distrust and disdain of communism by learning to work with the new Soviet leader, just as Thatcher promised. "I bet the hardliners in both our countries are bleeding when we shake hands," Reagan quipped to Russia's leader in 1985. Gorbachev nodded his agreement. A year later in Reykjavik, in one of history's most astonishing might-have-beens, the two men nearly concluded an agreement to scrap all nuclear weapons by the twenty-first century.

Nearly, but not quite. The deal was on the table, but neither would sign. "If I go back to Moscow and say that despite our agreement on deep reductions of nuclear weapons, despite our agreement on the 10-year period, we have given the United States the right to test SDI in space so that the U.S. is ready to deploy it at the end of that period, they will call me a fool and an irresponsible leader," Gorbachev told him. He'd not remain long in power, or perhaps even breathing, after that. Nevertheless, he virtually pleaded as the negotiations neared collapse, "if you will agree to banning tests in space, we will sign the document in two minutes."

Reagan wouldn't hear of it. Star Wars would force the Soviets to the negotiating table, he'd earlier promised. Star Wars then proved the one thing he would not concede once there. "It's too bad we have to part this way," Reagan replied. "We were so close to an agreement. I think you didn't want to achieve an agreement anyway. I'm very sorry."

The pair had at least made progress. By 1987 they signed a treaty limiting their respective nuclear arsenals. A year later, standing together in the middle of Moscow's Red Square, Reagan publicly recanted his earlier condemnations. His description of the Soviet Union as an evil empire embodied "another time," he said, and "another era." The following day he again noted that times had changed, adding, "I think that a great deal of it is due to the General Secretary, who I have found different from previous Soviet leaders." Reagan had loathed Andropov, refusing to "honor that prick" by attending his funeral, and barely knew his short-term successor, Konstantin Chernenko, who lasted barely a year in office. "They keep dying on me," he complained before authorizing Vice President Bush to attend yet another Soviet state funeral in his stead.

What made Gorbachev different from previous Soviet leaders, aside from the fact that he survived, was less his recognition of his country's problems than his refusal to sit idly by in response. Kremlin predecessors understood their society's economic sclerosis. Hard-liners like Andropov or Chernenko nonetheless refused to undertake the fundamental restructuring required if socialism were to remain competitive into the 1990s and beyond. Gorbachev accepted the challenge. Indeed, he did not think he had a choice. Educated beyond communist dogma and privileged with the opportunity to travel abroad, he could appreciate what his country lacked. The West was rushing forward into a new century, filled with potential technological marvels unthinkable even a decade before. Soviets citizens of the early 1980s lived in much the same closed world as their parents, their eyes fixed less on the future than on securing Moscow's approval of the present, and of the acceptable status quo. "We can't go on living like this," he confided to his wife the night he assumed power. "The system was dying away, its sluggish senile blood no longer contained any vital juices."

Walking with her in the cool nighttime air, he vowed to reform socialism in order to save it. It was a vow, it should be noted, that he only felt fully comfortable discussing outdoors and on the move, away from the prying eyes and ears of political rivals who'd deem such words treasonous.

Gorbachev would in time go even further than "merely" changing his own state, instead skillfully exploiting his nation's economic crisis in order to reboot an entire international system he believed had gone awry, changing it in ways he'd never imagined or desired. An angrier ideologue in the Kremlin might have lashed out in defiance against their disintegration, choosing war rather than change, national suicide over submission. Not everyone caught in the stream of history is capable of discerning its direction. Some simply refuse to accept even currents they can feel. "History punishes those who act too late," Gorbachev warned the doddering old men who ruled Moscow's Eastern European authoritarian satellites. Many appeared more willing to die than change. The forces he unleashed would soon bring protesters into the streets throughout the Soviet empire. Similar forces pervaded communist China as well. Would those old men embrace those crowds or order their soldiers to open fire?

Even Gorbachev, this modern-day Prometheus of change, needed partners willing to reform their decrepit system before it was too late. Troglodytes rarely ceded power peacefully, nor did powerful empires typically recede without first lashing out. To put it bluntly, Gorbachev faced an unprecedented challenge. He could not stand pat if he hoped for long-term Soviet success. Neither could he ratchet down Cold War tensions alone, or likely survive for all his efforts. Gorbachev would need Bush's support to face the coming chaos, the American's steady reaction to unforeseen crises behind the Iron Curtain would determine Europe's ensuing strategic map. More than any other foreign leader, Bush kept the dangerous crisis of communism's collapse from exploding. Many thousands can take credit for helping the Cold War end, Gorbachev most of all. As leader of the world's most powerful country while that conflict concluded with a speed no one anticipated, however, Bush would matter most of all in determining the world's response. Nothing of consequence happened on the international stage without his input, while he confronted an even larger problem than merely the potentially dangerous erosion of his country's greatest adversary: he inherited a world in which the United States, the globe's most dominant power since imperial Rome in its heyday, appeared suddenly deprived of its primary mission. He believed peace required American leadership, which had to his mind kept a general war and historic conflicts at bay for a generation. The anticommunist struggle begun in the 1940s provided policy makers reason to remain engaged; indeed, there was no real choice even for those who longed to retreat behind the nation's ocean moats after World War II. The Cold War's end augured something new: eradication of the collective purpose behind American foreign policy, and with it the end of the stabilizing American presence throughout the world. Hundreds of thousands of Americans, military and civilian alike, stood guard around the globe against communism as Bush prepared to take office. He considered their presence the vital condition for global peace. Yet without the communist threat, what reason would they have to be overseas at all? Managing the Cold War's collapse, while simultaneously ensuring conditions for a sustainable peace in its wake, thus became both Bush's greatest challenge and his greatest accomplishment.

This is why all of Washington, indeed all the world, watched the new president-elect with particular interest as Reagan's final weeks in the White House wound down. These would soon be his problems. All of them. Before his final trip into retirement, however, Reagan planned one more meeting with Gorbachev, with Bush at his side, following the Soviet leader's much-ballyhooed speech to the United Nations in mid-December. "This will be our last such meeting," Reagan noted in his weekly radio address. "And I must admit that I would not have predicted after first taking office that someday I would be waxing nostalgic about my meetings with Soviet leaders."

He also desired one more chance to toast his special relationship with Thatcher. Reagan authorized a massive state dinner in her honor a fortnight after Bush's election in November 1988, a celebration as elaborate as any his staff had yet arranged. She'd been the first foreign guest welcomed to the White House after his inauguration in 1981, so it seemed only right she should be honored as the last. A nineteen-gun salute greeted the prime minister, adding sparkle despite the London-like drizzle that chilled the air, as did the army's

Old Guard in their Revolutionary-era uniforms. It was an ironic choice, given their guest of honor, but it was Reagan's favorite military outfit. Dinner featured lobster, veal, and wines to accompany every course and toast. No expense was spared. Music and dancing beckoned soon after. Joining the usual cast of Washington's political elite, actor Tom Selleck swept his mother across the dance floor, adding that extra bit of Hollywood glamour Reagan particularly appreciated.

It was a night for celebration, but also for self-congratulation. "We set out to change a nation," the president toasted the prime minister, and "instead, we changed the world." Needless to say, Thatcher agreed. They began their association, she recalled, back when their nations were gripped by "dark days" of recession, inflation, and fear of communism worldwide. They stood jointly convinced that "together we could get our countries back on their feet, restore their values, and create a better yet safer world." Working together, she said while raising her glass, they had done just that.

Neither Reagan nor Thatcher was truly the center of attention, however. All of Washington was eyeing George Bush instead, wondering how the man who had just won one of the most contentious elections in recent memory might actually lead. Thatcher met privately with both men in the Oval Office before the dinner, but like everyone else, she knew Reagan's opinion was no longer the one that mattered, and that Bush's assumption of power meant the introduction of an unknown variable into the world's most powerful office during a moment of potentially historic international change. "The jury is still out" on Soviet reforms, Bush had said during one presidential debate. She wanted to know what he meant. Until she did, it was to her mind still out as well on the new American leader.

After a working breakfast with Bush and top aides, Thatcher wrapped up two days of meetings and festivities with an afternoon reserved for the press, who were eager for some headlines beyond the nostalgia. She did not disappoint. Indeed, she spoke with great clarity. The world had turned a fundamental page, Thatcher reported. The Cold War, the very foundation of international relations that had governed the world system for nearly a half century, was over. "We are not in a Cold War now," she said, but instead East and West enjoyed "a new relationship much wider than the Cold War ever was."

This was no simple statement. It is little exaggeration to state that every international decision since 1946 engaged in some way the bipolar struggle at the core of the world system. Untold billions had been spent on this fundamental battle between capitalism and communism. Untold millions of lives had been touched, and too frequently lost. It surely colored the lives of every single person of influence and power by the mid-1980s, many of whom had spent their entire career in its political universe. The Cold War was all they really knew, and if Thatcher was to be believed, all they knew for certain was now past.

A new wind of change was sweeping the communist world, Thatcher continued, adding, "I expect Mr. Gorbachev to do everything he can to continue his reforms." The Western powers, London and Washington especially, must be ready to help him whenever the need arose, "both verbally and in practice." Bush, in other words, should carry on Reagan's work — the work Reagan and Thatcher had accomplished together — and embrace change the way Reagan had. He should, in turn, embrace Mikhail Gorbachev. Those words still hanging in the air, she left for the airport and home, leaving pundits and policy makers in Washington, London, Moscow, and beyond wondering if she once more had given voice to an American president's desires.

Bush, however, had different plans.

2

BUSH'S RISE

HE WAS WHOLLY KNOWN, yet a mystery. George Bush had been a congress-man, a businessman, a war hero, America's ambassador to the United Nations and its chief representative to the People's Republic of China, chair of the Re-publican National Committee, director of the Central Intelligence Agency, and, most recently, for eight years Reagan's loyal vice president. He was arguably the most qualified person ever elected to the White House, and loyalty defined him. He was loyal to Reagan; loyal to the Republican presidents he'd served before and to their party's orthodoxies; loyal to his friends and business part-ners, oftentimes one and the same; and loyal to his family. Finally and most im-portant, Bush loyally accepted everything he considered part of the American system. This was an unquestioning loyalty indeed, beyond doubt and beyond introspection. He was a company man for what to his mind seemed the great-est corporation of all. "I will never apologize for the United States," he'd once boasted from the campaign trail. "I don't care what the facts are."

It was the kind of spontaneous statement that opens a window into a poli-tician's soul. Bush's entire life celebrated his vision of America, making him a true conservative in the traditional sense of the word, one who believed that future success required holding fast to the best features of the past, and in par-ticular to the freedom and free markets America represented. Both combined in Bush's mind to make the United States the most powerful and prosperous nation the world had ever seen, and, crucially, he believed both that power and that prosperity to be exportable. "The world looks to America for lead-ership not just because we're militarily strong, not just because we have the world's largest economy, but because the ideas we have championed are now dominant," he explained soon after taking office. "Freedom and democracy, openness, and the prosperity that derives from individual initiatives in the free marketplace — these ideas, once thought to be strictly American, have now become the goals of mankind." On the brink of tectonic change wrought in large measure by the United States, he believed to his core that the entire world need only become more like the United States to succeed.

Such beliefs long predated his presidency. The Chinese people "undoubtedly are freedom loving," he'd written while in Beijing in 1975. He never doubted that their lives would be improved by embracing the American system, discarding their own politics along the way. Presidents typically say much the same thing, placing American society on a pedestal for others to emulate. From George Washington to Ronald Reagan, each of Bush's predecessors had in some form or fashion praised his land as a model for freedom and its champion to the world, the precise ratio between example and liberator dependent on each man's particular circumstance and worldview. Frequently cheap applause lines for domestic crowds, such sentiments are also typically heartfelt. The new United States enjoys a peculiarly "detached and distant situation," Washington had declared in his farewell address, and a consequent freedom of action that "invites us to pursue a different course." Abraham Lincoln called the country freedom's "last best hope of earth." Reagan made the same point just as plainly, albeit from a position of global power that Washington and Lincoln could scarcely have fathomed. America remained the same "shining city on a hill" that had inspired Pilgrim settlers, Reagan said in 1984. In his farewell address he declared, "After 200 years, two centuries, she still stands strong and true on the granite ridge . . . and she's still a beacon, still a magnet for all who love freedom, for all the pilgrims from all the lost places who are hurtling toward the darkness, toward home." The trope even had a name: "American exceptionalism," the notion that the United States was special, ideal, at once unique yet full of answers for all who would but listen.

Bush believed this trope with particular passion, standing apart from his predecessors not so much in his faith in the American system but in his utter lack of interest in exploring alternatives. "We don't need radical new directions," he insisted when opening his formal campaign for the White House in late 1987. "We don't need to remake society — we just need to remember who we are." It all seemed obvious to him. "I like what's real," he said. "I'm not much for the airy and abstract. I like what works. I am not a mystic, and I do not yearn to lead a crusade."

For a man born into a world created by American success and largely successful in everything he had ever tried, it was natural to believe his country held all the answers. Words like "freedom" and "liberty" were easily interchangeable in his mind with "democracy" and "free markets." Each was divinely inspired and unquestionably good. Scholars could debate the nuances of positive versus negative rights. Lawyers and sophists could parse details of natural law. Such questions simply did not interest him, or seem worth his time. Americans "do have principles, and it is time we stood up for them without being contentious," he'd written in his diary in the early 1970s, even as others questioned their continued viability in an era of scandals (Watergate) and defeat (Vietnam). Yet if asked to explain precisely what they were, he conceded, "I'd be damned if I know how to define it."

Over time he had stopped trying. "He does not talk in a philosophical way," his national security adviser Brent Scowcroft observed. "He's very reticent to talk in those terms. He's uncomfortable talking in that way." Bush didn't disagree. "I didn't quote Shelley and Kant," he later confessed. "I didn't remember exactly what Thucydides had meant to me when I was only twelve." Besides, he reasoned, true expressions of principles, whether for a man or a people, came from within. "It doesn't have to be done with the most rhetorical flourish. It has to be your inner self. It's got to drive you . . . It can be your set of values. Your vision can be 'I want to live [according] to this code of behavior.'"

Critics had long taken umbrage at the image of a blue-blooded American aristocrat, the son of a United States senator and scion of two prosperous families, educated at the finest schools and reared in privilege; a man who'd never known the sting of discrimination or the pain of hunger, telling others that their society required neither tinkering nor improvement. George Bush was "born with a silver foot in his mouth," Texas governor Ann Richards once joked, mocking at the same time his frequently garbled syntax and his upbringing. It was the critique of another Texas politician, however, that hit closest to the mark. Bush was a man "born on third base," Jim Hightower quipped, "who thought he'd hit a triple."

The jest was more revealing than Highly intended. Runners rounding third and headed for home do not typically think that the ideal moment to question the rules of the game. Nor is the phenomenon limited to sports. Winners typically approve of the rules. The impulse for change, conversely, comes from those who think the rules don't work in their favor. Revolutionaries by definition are discontented, and George Bush was no one's idea of a revolutionary. Born on third, and living his life on the brink of scoring, he'd built a remarkable résumé by the late 1980s in both business and government, always operating within the innermost circles of American power, and always playing by the rules.

He also took office at a moment when the United States itself appeared to be rounding third and heading for home in the long contest of the Cold War. Bush was not among those who so optimistically believed that the great superpower game for global domination was at its end as Reagan's tenure came to a close, but by no means did he doubt America's inevitable victory. In turning toward democracy, the communist world appeared at long last to be accepting what his country — and he in particular — already knew to be true: the American system worked, and history itself flowed in democracy's direction. It would continue to flow America's way, because the freedom and free markets it exemplified and supported were simply right.

Put squarely in terms of baseball, a game Bush played in college despite epitomizing the old scouting adage "good field, no hit," having rounded third and with a clear path to the plate as the 1980s came to a close, the American-led free world needed only to keep from stumbling. Victory might not come tomorrow, or even soon. But so long as Americans held fast, it was inevitable and seemingly closer every day. Bush kept his oversized first baseman's mitt in the bottom drawer of his Oval Office desk for when times got toughest, placing his hand inside the well-oiled leather whenever he needed a reminder that the truest values were also the most well worn.

One of the most qualified men ever elected president, George Bush was not in truth one of the most accomplished. Reliable rather than revolutionary, personable and proper rather than a firebrand, he typically served with a steadying hand rather than an innovative one, climbing through positions of greater and greater responsibility, albeit without changing any of them along the way. He authored no significant legislation during his political career, developed no doctrine that bore his name, and transformed neither the CIA, the Republican Party, nor the mission to the UN. His skill — and a skill it most certainly was

—lay instead in his ability to provide stability and order, in particular in times of crisis, building loyalty as well as exemplifying it. "What are they trying to do to us?" he asked at his first senior staff meeting at the CIA, in a period when the agency was suffering unprecedented public scrutiny, immediately declaring himself an insider and including himself in the agency's fate.

An insider from the start, he also worked his political skills to the fullest. Blessed with a remarkable memory and energized in the company of others, he invariably learned the names of everyone he encountered (and their children's and grandchildren's as well), keeping in touch across the years and the miles by compulsively dashing off quick notes to friends and those he hoped soon would be. The Bush family Christmas card list eventually topped 25,000, each a potential political ally. Asked in 1980 during his first bid for the White House why he was running, and why he thought he could win, he answered in all seriousness, "I've got a big family, and lots of friends."

The answer made complete sense within Bush's world, though as one national magazine noted in 1987, a lifetime built on personal relationships and loyal but not innovative service left voters to wonder "what does Bush really stand for?" The charge dogged his career. No matter the depth of his conviction, voters heard nothing new in his words, nothing they'd not heard before. Boasting "I like what works," he promised implementation rather than innovation when running for the White House in 1980. He lost to a visionary little concerned with details. Bush promised competence. Reagan promised crusades—against communism, big government, and taxes. His last real rival for the Republican nomination, Bush trashed Reagan's ideas, questioned his ability to lead, and ridiculed his paper-thin proposals as "voodoo economics." The barb stuck as a critique of Reagan's financial plans, surviving long after most forgot who first uttered it.

Yet Bush then disavowed every one of those statements for the chance to join Reagan on the national ticket. "I'm following Mr. Reagan, blindly," he told reporters at the 1980 Republican convention, pledging the same in the pages of his diary. "I don't believe a President should have to be looking over his shoulder wondering if the Vice President was out there carving him up or undermining his programs in one way or another," he wrote. Naturally ambitious, if unable to secure the nation's top job, he nonetheless gladly accepted a promotion. Naturally competitive, he was consequently willing to do and to say whatever was necessary to win, and then to do whatever it took to remain in

Reagan's good graces. If being this close to the pinnacle of American politics and power meant subsuming his own opinions to his president's, so be it. He'd done so before.

Unlike nearly everyone else in the upper ranks of Reagan's White House, Bush never gave Reagan cause to question his loyalty, though that devotion did little to foster his own distinct public persona. One national publication dubbed him a "wimp" halfway through the decade, questioning if he had the courage of his convictions, or any convictions at all. Political lampoonist Garry Trudeau found it impossible even to depict Bush in his widely read *Doonesbury* comic strip. He was merely an ethereal voice, lacking in definition and substance. "He's becoming his own man!" reporters in the strip exclaimed in 1988 when it appeared Bush might finally contradict something Reagan had said, only to be disappointed when his contours disappeared once more. Trudeau's caricature of Bush ultimately placed his "manhood in a blind trust" for Reagan to use as he saw fit. "Fairly or unfairly," one pollster concluded by late 1987, "voters have a deep-rooted perception of him as a guy who takes direction, who's not a leader." The question plagued him particularly in the Republican primaries as the fight to succeed Reagan heated up. He could rebuke those charges by criticizing some of the president's policies, staking out an independent position on some key issue solely for the purpose of demonstrating his independence, Bush admitted during a late 1987 debate. "But I won't. I'm not going to start what I've not been doing for seven years. In my family, loyalty is a strength. Not a character flaw."

The incongruity of his New England manner and Texas mailing address didn't help. Born in Massachusetts in 1924 and raised in Maine and Connecticut, he looked west and south for a career, and an identity. "Where are you from?" a well-meaning lunch companion once asked his wife, Barbara. "We live in Houston," she answered. "Yes, but where are you from?" came the query once more. "We live in Houston," she answered anew, as though the Bushes could by force of will forge any identity they chose. They were Texan through and through, Bush boasted, but even then he never failed (save only once, during the war) to spend summers at his family home in Maine. In truth, he spent more time in Washington than anywhere else from the late 1960s until his presidency, living in Houston a mere four years between 1966 and 1992. As Congressman Jim Wright once teased, Bush was "the only Texan I know who eats lobster with his chili."

He had an identity, to be sure; it just wasn't found west of the Pecos. His paternal grandfather was one of the great industrialists of his day, tied by friendship and business to paragons of New York money and power. His father, Prescott Bush, was a Wall Street banker and a United States senator. His mother's side boasted an equally prestigious name and Wall Street connections of their own. Bush never rejected his background, but he displayed from an early age a desire to prove himself worthy of his heritage. Infuriated by the Japanese attack on Pearl Harbor during his senior year in prep school, he joined the navy upon graduation, against the advice of his father and of their family friend Secretary of War Henry Stimson, who thought young men of proper breeding and upbringing would make better officers after a bit of college seasoning. Unable to stop him from enlisting, they at least secured the young man a coveted spot in the navy's pilot-training program. Stationed in the Pacific, one of the youngest aviators in the entire fleet, he subsequently hatched a scheme to forgo college altogether after the war and jump into business immediately upon his return. "Next fall I'll have saved $3000," he wrote his parents in mid-1944. "We [he and Barbara, the daughter of another distinguished family whom he'd met and courted while on leave in 1943] can make that last a while." After the war, he continued, "I can get a job, a modest one at first, irregardless [sic] of my lack of college education. I wonder if you agree with me on this point?"

His father did not, enrolling his son at Yale instead at war's end. It was what the Bushes did, the young couple accepting the obligations of their social rank but also its benefits. Many veterans struggled to find affordable housing as campuses boomed as a result of the G.I. Bill. George and Barbara rented an apartment next door to Yale's president on one of the nation's most prestigious tree-lined avenues. A new baby arrived soon after, leaving their neighbor to ask only that the well-heeled young couple refrain from hanging little George's diapers out to dry during important university functions. In the culture of the university at the time, the young man outranked the old. The latter merely presided over the campus; Bush was a legacy. Administrators came and went; Yale's families endured.

Success followed. He graduated in three years with a degree in economics and a Phi Beta Kappa key, captained the university's baseball team to the College World Series, and won induction into Yale's most prestigious secret society, Skull and Bones, reserved for the campus elite. Prescott had been a Bonesman himself a generation before, rumored to have pilfered Geronimo's skull for the club's secretive chambers. Enthusiastic, optimistic, and energetic in the extreme, George Bush presented the classic portrait of a young man in a hurry, and his young family the picture-perfect ideal of postwar contentment and success.

He then left for the newly developed oil fields of west Texas, departing leafy Connecticut for a part of the world where scorpions and rattlesnakes outnumbered people and brush qualified as a tree. It was an adventure, albeit one with built-in safety measures, including generous lines of credit from his family's Wall Street contacts, along with the promise of a job back home should wildcatting for oil prove a bust. Though riskier than if he had taken a Wall Street job straight out of Yale, his was not the pioneer story of old. Bush drove west in a brand-new Studebaker rather than a Conestoga wagon, never worrying that his family would starve if he failed to strike it rich. But neither did he think he could be happy unless, at least in his own mind, he'd made it on his own. "George lives his life in chapters, like a lot of the Bushes do," his sister later observed. "The war was over and he left it behind. Some people had trouble doing that. George didn't."

He had good reason to want to forget. He'd known the horror of a flak-riddled bombing run, and the pain of having shared breakfast with shipmates who were gone before lunch. Sometimes their last moments involved a fiery plunge into the sea for all to witness. Sometimes they simply disappeared on patrol. By the time he was twenty, he'd written his fair share of condolence letters to families back home, and crash-landed more than once himself.

That was not the worst of it. Shot down by enemy fire on September 2, 1944, he lost the two crewmen under his command. One was a family friend, a classmate of his brother's, who'd volunteered to join the attack on the Japanese stronghold of Chichijima for the adventure of it all. They'd been bombing the island for days with little effect, targeting a radio tower set amidst volcanic cliffs. Feeling a jolt, "as if a massive fist had crunched into the belly of the plane," Bush held his smoke-filled bomber aloft long enough to complete his attack, eventually bailing out once back over the ocean. His crewmates were not so fortunate. "The fact that our planes didn't seem to be searching anymore showed me pretty clearly that they had not gotten out," he wrote his parents after his rescue. "I'm afraid I was pretty much of a sissy about it cause I sat in my raft and sobbed for a while . . . I feel so terribly responsible for their fate." He had just turned twenty-one.

Guilt plagued him for decades. He wondered what else he might have done to save them, and where his bombs might have inadvertently landed. He clung to news, passed along years later, that Japanese defenders counted two parachutes that fateful 1944 day. Two chutes meant he had held his plane steady long enough for at least one of his crewmen to eject, and that he had not abandoned them to a fiery end. "I think about those two all the time," he conceded on one of the rare occasions when he openly discussed their deaths. Most of the time he actively avoided talking about the incident, or the war years at all. I tried to raise the subject once as an interview session neared its end. Bush offered me a cookie instead (a Double Stuf Oreo), asking, "What else ya got?" It was clearly not a topic he'd choose to discuss with a historian one third his age.

The war deepened his faith, though this too was something rarely discussed. An infrequent churchgoer at best, Reagan nonetheless spoke frequently of Jesus's prescriptions and of Americans as the Lord's chosen people. This was simply not Bush's style, nor how he'd been raised. He attended church weekly, but not for show. "You know us Episcopalians," he told a meeting of Southern Baptists. "For me, prayer has always been important but quite personal." A true Yankee did not discuss such things. Yet if there was a moment in his life that offered spiritual purpose, it was that awful day spent adrift in the Pacific. "People talk about foxhole Christianity, where you're in trouble and think you are going to die, and so you want to make everything right with God," he later remarked. "But this was just the opposite of that. I had already faced death, and God spared me. I had this very deep and profound gratitude and a sense of wonder . . . In my own view there's got to be some kind of destiny and I was being spared for something on earth."

He attacked life with a real sense of purpose after the war, as though determined to leave no mental space for terrible memories to creep in. "If George Bush has a free day, he manufactures things that involve people," an aide later said. "The idea of spending a day alone I don't think would ever occur to him. It would be punishment of some kind." Sadness couldn't find you, he seemed to reason, if you kept moving fast enough.

Unfortunately there was still more pain to come. Moving west after college because, in the bank robber Willie Sutton's famous words, "that's where the money" was, he found a state in the throes of a population boom. Dusty Midland quickly filled with similar East Coast transplants drawn to the promise of fortunes buried deep in the ground. They formed clubs, populated golf courses best described as eighteen-hole sand traps, and built houses on streets named "Harvard" and "Dartmouth" and "Yale" as reminders of home, and to remind others from whence they came. They at least had houses. Midland and towns like it throughout the broad oil patch swelled beyond capacity, as they always did during flush times before the inevitable bust. Tent villages sprang up overnight, quickly followed by bars for sinning, then churches for repenting.

Bush found success while others similarly endowed retreated back east in defeat, and in time he received offers to head back east himself, including an invitation in 1950 to join his father's old Wall Street firm. "This suggestion did not originate with your father and, in fact, he has nothing to do with it," the offer from Brown Brothers Harriman insisted. Tempted, he stayed put instead. It was one thing to work with family friends and to invest family money. It would be another to work *for* them. George Bush wanted to be like Prescott, and if possible best him. Not be him.

Tall, athletic, handsome, and well on his way to making his first million, he seemed the picture of midcentury success as the 1950s commenced. "We like Texas," he reported to an old friend on New Year's Day 1951. "The kids have been well. Robin [his and Barbara's daughter, born December 20, 1949] is now walking around and Georgie has grown to be a near-man, talks dirty once in a while and occasionally swears, age 4½. He lives in his cowboy boots." In short, he reported, "Midland is a fine town . . . we really love it," and he was happy "selling oilfield equipment and supplies, a task which is great as far as gaining experience goes."

His comfortable bubble soon burst. Robin fell sick in 1953. Doctors in Midland diagnosed leukemia, advising her parents to take her home, make her comfortable, and await the inevitable. They instead flew her back to New York in hope of trying experimental treatments more advanced than could be found in west Texas. An uncle on staff at Sloan Kettering provided the connections needed to secure a bed at the hospital. For a time the girl rallied. "She is still full of fun and we hope that she will have many more months of active life," Bush wrote a trusted friend in August. His little girl passed away only weeks later. "We are settling back into our normal life," he wrote his father by late October. "At times I think we are just beginning to miss Robin."

It was only the beginning indeed. Still aching five years later, "Bar and I wonder how long this will go on," he confessed to his mother in 1958 in a late-night letter composed, drink in hand, at his kitchen table. Awkward and uncomfortable discussing his emotions in person, like any proper New Englander, he opened up to the blank page. "I sometimes wonder whether it is fair to our boys and to our friends to 'fly-high' that portrait of Robin which I love so much," he wrote, "but here selfishness takes over because every time I sit at our table with just our candlelight, I somehow can't help but glance at this picture you gave us and enjoy a renewed physical sense of closeness to a loved one." There is "about our house a need," he continued. "We need some soft blond hair to offset those crew cuts. We need a doll house to stand firm against our forts and rackets and thousand baseball cards . . . We need someone who's afraid of frogs. We need someone to cry when I get mad — not argue. We need a little one who can kiss without leaving egg or jam or gum. We need a girl. We had one once."

Part of him stayed with his lost child laid to rest in the family plot in Greenwich, but Bush once more pushed forward. The family prospered in Midland (adding four more children after Robin's passing, including finally another girl), but having tired of its paucity of cosmopolitan charms, he moved the clan to Houston in 1959 in pursuit of offshore drilling opportunities — and political ones. He found there as well one of the most important friendships of his career when he teamed up with another young Ivy League graduate in pursuit of their tennis club championship. It was a good fit, their styles on the court matching their personalities. Bush preferred the teamwork of doubles, and the challenge of extending rallies long enough for James Baker to make the kill shot. Partner in a prestigious Houston law firm, Baker was descended from a long line of the city's professional and country club elite. His true talent lay less in legal prowess or litigation skills than as a cold-blooded negotiator. Bush frequently appeared jovial to the point of downright goofiness in public, pointing out friends in crowds and making faces in search of a personal connection, seemingly unaware despite his decades in the public eye that others were watching and the cameras always rolling. Baker's professional mode was taciturn. He knew silence could be a lawyer's best friend, and thus frequently appeared dour, abrupt, and cutting when at work or in public, his private face well hidden from view. Bush had a salesman's constant patter. Though Baker could be wickedly funny in private, his joviality rarely made it into public view. He was a man built for control — of other people, of negotiations, and most important of himself.

More than Bush's complement, Baker also was in his debt. Several years younger, Baker had missed out on World War II and returned from his own military service to Houston in order to follow in his father's legal footsteps. He lived what might be considered an idyllic establishment life of his own, building a respected practice, marrying well and fathering five children, and alongside Bush winning championships with his racquet. Largely apolitical as he sought clients from both sides of the aisle, Baker strove whenever possible to follow his grandfather's advice to "work hard, study . . . and keep out of politics." (The phrase ultimately became his autobiography's purposefully ironic title.)

His wife, Mary Stuart, then found a lump in her breast. Baker would soon be widowed, their five children motherless, his life without its rudder. She'd managed the household and their social lives. He'd largely worked, her absence leaving him alone with his office, sad children, and a bottle to grieve with after they'd been put to bed. Mourning turned to despair, and then frequently to another open bottle.

Bush salvaged Baker's personal life and introduced him to politics in a single blow. Come help my congressional campaign, he suggested. It will take your mind off things. It will be good to be busy, he offered from sad experience, at night especially, when dark thoughts encroached. Baker initially mounted a weak protest: "Number one, I don't know anything about politics, and number two, I'm a Democrat."

"We can take care of that second problem" right away, Bush retorted, having made a habit of converting lifelong Texas Democrats to the political possibilities of the New South. Baker's own political career was thus formed less by personal ambition for office than by acting as the coordinator of another's plans. The role suited him. And he was good at it. Properly a Republican by the late 1960s, when establishment southerners began migrating to the party in droves, he helped orchestrate Richard Nixon's 1972 reelection effort in Texas, garnering attention from national officials, including President Ford's young chief of staff, a thirty-two-year-old Yale dropout from Wyoming named Dick Cheney, who selected Baker to organize Ford's national campaign. His rapid ascent continued, with Reagan tapping him for the pivotal position of chief of staff in 1980 despite his having led Bush's campaign. Baker orchestrated Reagan's White House through the 1984 election, ultimately enjoying the longest tenure of any of Reagan's appointees to the position. He moved from the White House to the top slot in the Treasury Department in 1985. Both Iran-contra and Reagan's transformation toward the Soviets occurred only after Baker left his side.

But Baker was always at heart a Bush man, even to the point of shaping his advice during a 1988 National Security Council meeting on Panama's Manuel Noriega in order to best position his friend's presidential bid. He'd already decided to leave Reagan's cabinet to become Bush's national campaign chair. "For the first time," Baker recalled, "Vice President Bush really disagreed with the administration's policies," noting, "I knew where I was going to end up." After a "spirited discussion" — often a diplomatic euphemism for a knockdown argument — "we carved out a little bit of a separate agenda and identity."

The "we" in this admission, it should be emphasized, was not the White House or those sitting around Reagan's conference table; it was Baker and Bush. More like brothers than mere friends, they displayed their fair share of sibling rivalry as well. Baker knew he had Bush's full confidence, and thus frequently made diplomatic offers without his boss's explicit consent, more than once overstepping. "If you're so smart, how come I'm the President?" Bush angrily quipped when Baker's critiques hit too close to home, or when his negotiating skills outstripped White House policy. In truth, Baker was eyeing the Oval Office for himself, thinking if all went well he might succeed his friend after two terms. In the meantime, he could privately voice to Bush objections or critiques others less favorably placed might have shied away from mentioning. He possessed the president's ear, and also what famed Secretary of State Dean Acheson once proclaimed the most important attribute for success in the position: "a killer instinct."

The pair were still a long way from the White House when Bush finally turned to the political career he'd always considered inevitable, becoming chairman of the Harris County Republican Party in 1960 and then running for the U.S. Senate four years later. He stressed consensus, hoping to build bridges, and a winning electoral coalition, out of the party's two main wings: its traditional pro-business and internationalist faction, from which he hailed, and the proponents of a more virulent right-wing brand of isolationist conservatism then coming to the fore throughout the West. The latter group, exemplified by the John Birch Society and by Arizona senator Barry Goldwater, the GOP presidential candidate in 1964, preached opposition: to communists, big government, and international commitments.

Many of Bush's political ilk wanted to cast the Birchers aside, thinking their radical brand unlikely ever to appeal to the wider electorate. Bush would have none of it. "We're all Republicans," he told constituents. Their state's electoral minority in 1964, Texas Republicans needed all the help they could muster, and thus "we're not going to divide ourselves, calling anyone 'crazies' or nuts." Theirs would be a big tent, open to anyone eager to secure a Republican victory, including his own.

Bush found it easy to accept divergent views during his first Senate bid because his own political identity was far from fully formed. Much like his father, a progressive and internationalist Republican known in the Senate as a man who could forge consensus but rarely offered original legislation, George Bush was willing to compromise. It was his principal conviction. He wanted passionately to hold office, believing the lesson preached in his youth that the privileged had a duty to seek public service. Yet he was not particularly sure how he might govern if elected, and thus found it uncomfortable to expose too much of himself to voters, a fear exacerbated as well by memories of his mother's constant chiding against the sin of self-promotion. Perceiving her son's natural charisma and athletic prowess, she warned throughout his childhood of the danger of the word "I." Put the team first and others at the fore, she'd instructed. "Listen, don't talk."

Her son took the lesson to heart in 1964, offering speeches devoted more to explaining what Republicans could offer than why he should be their choice. "You're talking about yourself too much, George," his mother nonetheless complained after one such rally. Voters wanted to hear what a candidate believed, he argued. "Well, I understand that," she replied, "but try to restrain yourself." She'd be offering the same critique years later, even as he ran for the White House.

Without a specific political agenda and uncomfortable selling himself to voters, Bush offered himself in 1964 as a hodgepodge of East and West, both pro-business and a frontier conservative. It was an inchoate blend at best, directed more at attacking his opponent's liberalism than anything else, in which adoption of a right-wing position led invariably to a caveat. America should "get the hell out of the UN," he told voters, sounding like a Bircher, before cautiously adding, "if that organization seated the Red Chinese." Washington should also "cut foreign aid," he proposed, another Bircher position, before quietly suggesting merely cutting out aid to nations shown to be "soft" on communism. America should be in Vietnam to win, he said, but should use only the military force the Pentagon requested. One can almost hear the pauses and see the ellipses when reading Bush's statements from this period, denoting a man more eager to hold office than certain why.

Nineteen sixty-four was a bad year to be running as a Republican, no matter the stripe. Lyndon Johnson won the White House in a landslide. Bush ran ahead of the rest of his ticket yet lost his own race by more than thirteen percentage points. It was never really close. Pundits interpreted the statewide and national results as the voters' full-throated repudiation of Goldwater's brand of conservatism. Republicans would never win unless they shed their dangerous right wing, the party's more liberal factions warned, leaving Bush finally willing to agree in defeat. "I am anxious to stay active in the Party," he confided to former vice president Richard Nixon, cultivating their long-standing relationship even when Nixon's own political future appeared unpromising, but he "felt that the immediate job would be to get rid of some of the people in the Party who [permit] no difference, who through their overly dedicated conservatism are going to always keep the party small." They were, in fact, "a bunch of 'nuts,'" Bush conceded, putting aside his previous prohibition on such labels, and "responsible people are going to have to stand up and do something about it." Two years later, lowering his sights from the Senate to the House of Representatives, he won a seat representing an affluent Houston district comfortable with his brand of liberal Republicanism. "I took some of the far right positions," he confessed to his minister after his failed 1964 Senate bid. "I hope I never have to do it again. I regret it."

Ambition drove him toward another Senate bid in 1970. Elected president two years before, Nixon promised the White House's full support, and a job if things turned sour. Bush reasoned that victory required only positioning himself just to the right of his expected opponent, the incumbent Ralph Yarborough, a Democrat whose liberal agenda voters had come to distrust as Johnson's Great Society agenda cracked beneath the weight of Vietnam and strained race relations. Lloyd Bentsen foiled those plans, his victory over Yarborough in the Democratic primary giving Bush an opponent whose conservatism equaled if not exceeded his own. It was hard for voters to distinguish any real difference between them, one Austin magazine noted, and even with party registration rolls turning over, "given a choice between Phillie Winkle and Winkle Pop, Texan'll take the dude with the Democratic label."

Forty-six, unemployed, a two-time loser in bids for statewide office, and with a large family to support, Bush turned to Nixon for his promised help, lobbying in particular for the open position of treasury secretary. Nixon barely considered the request. "Every cabinet should have at least one potential president in it," he concluded, and Nixon did not think Bush had the experience or the gravitas for the job, giving the position instead to another Texan, former governor John Connally, a Democrat, "much to George's disappointment," James Baker recalled.

If inexperience was the problem at Treasury, Bush subsequently lobbied, then inexperience was his best qualification for the open slot as Nixon's ambassador to the United Nations. The White House could use an ally in New York, he argued, where, given his family connections, he could "spell out [Nixon's] program with some style." Knowing absolutely nothing about diplomacy, he'd also make the perfect ambassador, being unable to do anything but follow orders without question or deviation. This was an appealing argument for a White House famous for its jealous hoarding of power, and in particular for Secretary of State Henry Kissinger, who yearned to control every aspect of Nixon's foreign policy. "He takes our line beautifully," Nixon's chief of staff agreed, and seasoned Foreign Service officers handled most of the office's day-to-day work in any event. He'd follow orders, Bush promised, adding that "even if someone who took the job didn't understand" his role, "Kissinger would give him a twenty-four-hour crash course in the subject."

Critics howled when the White House announced the news. "The appointment of a political loser," the *Washington Star* editorialized, and "a lame-duck congressman with little experience in foreign affairs and less in diplomacy — would seem a major downgrading of the U.N. by the Nixon Administration. And the senators who sit in judgment of the nomination are certain to question the appointment of a conservative Republican Texas oil millionaire to the nation's highest ambassadorial post." The *New York Times* was blunter: "There seems to be nothing in his record that qualifies him for this highly important position." Even Bush's closest associates expressed their bewilderment. "George," one longtime friend exclaimed when learning of his posting, "what the fuck do you know about foreign affairs?"

"You ask me that in ten days," Bush responded. "They laid down a challenge," he later wrote of his critics, and "got my competitive instincts flowing." Moving into the ambassador's residence at the Waldorf Astoria, Bush began one of the most enjoyable periods of his life, thrilling to the new work while relishing his prominent position in New York society. He entertained fellow Republicans in the salons of the city's well-heeled, which he'd frequented his entire life, and took groups to his uncle's private seats at Shea Stadium — the man was a part-owner of the New York Mets — introducing many foreign diplomats to base-ball. He even hosted China's entire United Nations delegation for brunch at his mother's home in nearby Greenwich. "Everybody said, 'Well, he doesn't know anything about diplomacy,' and everybody was right, I didn't," Bush recalled decades later. "But I learned, and worked hard at it." He strolled the halls of the United Nations, soaking in the atmosphere and, ever eager to win new friends, chatting up diplomats and staffers from around the world. It also did not escape him that his new job would be a final tweak to those right-wingers from home who despised international organizations. "It will be interesting to see what the Texas reaction is," he deadpanned in his diary.

More seasoned American diplomats balked at his behavior. "Let me give you a piece of advice," former UN ambassador Arthur Goldberg told the new man. "You represent the United States of America," the most powerful nation on earth. "They should come to see you . . . We are the United States. They should come see you." Bush saw things differently. He was a salesman, and he believed the same tactics that won him contracts in west Texas, votes in Houston, or popularity in school or within his flight squadron could curry favor in Manhat-tan as well. "I got there," he later recalled. "I go call on the Burundi ambassador, [and] I thought the woman [the office secretary] was going to have a heart attack." The visit, and others like it, had a broader purpose. "I knew that word would get all around the United Nations that we recognized and respected the sovereignty and the vote of every country there."

"You can't do this job if you don't like people," he continued, explaining him-self in the diary he'd begun keeping after the election, using it as a means of working through his disappointment but also in time as a sketchbook for work-ing through new ideas. His predecessors were too proud for their own good, and too rooted in old ways. "I think the best policy around here is to demon-strate your willingness to 'go to others,'" he therefore recorded, "to ask advice, to be grateful, and to get here earlier and leave later than the rest of the people."

His was "just a difference in approach" from that of previous American am-bassadors, Bush said, though adding, "I'm convinced mine was right." Indeed, he ultimately considered his UN experience the basis of his entire diplomatic approach, which to his mind "doesn't have anything to do with diplomacy; it has to do with life. Treat people with respect and recognize in diplomatic terms that the sovereignty of Burundi is as important to them as our sovereignty is. Slightly different scale, I might add. But nevertheless, this is just a value thing. This isn't any great diplomatic study from the Fletcher School or something. This is just the way you react to things."

Bush thus laid the foundation for his subsequent personal style of diplomacy nearly a full generation before he'd assume the presidency. Just because Bush chose a more open diplomatic style than his predecessors did not mean, however, that he disagreed with their overall worldview, including in particular their ingrained sense of America's privileged position in the world. He simply thought the time long past when Americans should expect others to accept their leadership without question. His country was right, but twenty years into the Cold War it also needed to convince and cajole rather than merely demand obedience. He was surrounded by "little wiener nations," he quipped soon after arriving in New York, but was too polite to say such things to his fellow ambassadors. Yet his ensuing conversations with them, and especially the experience of listening and thus learning their sense of the world, in time softened his views, in turn opening his mind to the variety of voices found within the international system. His time at the UN "had a profound effect on him in terms of understanding just how different the world is," Scowcroft concluded. "I think that gave him an appreciation of how to deal with the world, because [other] people don't always think the way you do." They could be convinced, but only over time, and only through effort.

His UN service also reinforced his faith in something he'd learned long before: the value of friendship. Personal ties could grease the cogs of diplomacy just as easily as for business and politics, making crises easier to manage, but only if developed before they were truly needed. One couldn't ask a foreign leader to trust one's promises at a critical moment if that was the first time they'd spoken. Because one never knew when a critical moment would arise, he thus reasoned that men and women he knew beforehand, and who knew him, would be much more likely to extend that critical though unquantifiable quality. It was a lesson his father had long preached, one Baker reinforced from his own legal experience, deeming trust so vital he ultimately gave it a formal name. In difficult negotiations he was after not merely trust but rather "*the* trust," because "if you have the trust, you can get a hell of a lot more done than if you don't. If you and I are trying to negotiate something and I know I can trust your word, we've got a better chance of getting there than we don't." When asked how one knew trust existed, Baker answered simply, "You test it."

Bush saw firsthand during the 1970s how the changing nature of international power would test America's leadership, and the trust of its allies, as never before since 1945. Serving amidst a great period of decolonization, he witnessed the building swell with new members. Newfound sovereignty also fueled a sense of solidarity, as leaders throughout the developing world increasingly argued that their strength would count only if counted together, in forums such as the United Nations General Assembly in particular, where even the smallest country wielded the same vote as a superpower.

China's fate offered a case in point, its continued representation in the UN in the early 1970s by the pro-Western Nationalist regime in Taiwan seeming increasingly out of date to the body's newest members, who largely believed that the communist regime in Beijing, in power on the mainland since 1949, deserved the country's General Assembly seat and voice on the Security Council. Like every other president since the communists took power, Nixon pledged to support the Nationalists, though their continued presence in the General Assembly appeared ever more unlikely by the fall of 1971. "All the UN people feel the ballgame is over, Peking is in and Taiwan is out," Bush confided to his diary. Yet charged with defending Taiwan's seat, he lobbied during the day and counted votes at night, his sense of responsibility to the longtime American ally fusing with his own competitive zeal. "It is all encompassing," he wrote a friend only days before a critical vote. "Night and day — at every meal — first thing in the morning, last thing at night. I think we can win for a policy I believe strongly in, but it's going to be terribly close."

Kissinger scuttled their hopes, appearing in Beijing only hours before the final vote as part of his overall wooing of the communist regime. Pictures of Nixon's personal envoy shaking hands with his hosts made plain which Chinese government the White House really valued. The expected close vote turned into a rout. Bush was even booed in the General Assembly. "There is no question that the U.N. will be a more realistic and vital place with Peking in here," he confided to a friend afterward, employing the old-fashioned name for the mainland capital, "but I had my heart and soul wrapped up in the policy of keeping Taiwan from being ejected." It was a telling remark, followed by a revealing gesture. When the final vote was tallied and cheers erupted from the winning side, the sight of Taiwan's ambassador exiting the hall brought jeers.

Bush rushed to his side and escorted the man from the hall, his arm around his shoulders, lest a friend be forced to suffer alone.

The entire debate over Taiwan left its mark on Bush's thinking, reinforcing in particular his sense that America's changing global role demanded a renewal of its traditional values. The Americans would thereafter have to be more responsive to global opinion than ever before, but could lead only by holding fast to their commitments, and to the notion of leadership itself. "I mean, we're the United States," as Bush later put it. Whenever a vote came up, "we could have said" to a foreign representative, "Look, you little bastard, you can go do it your way, but we are the higher power, [and] that's not the way it ought to be." Washington could have acted this way, but it wouldn't have worked for long, if at all. The final lesson he took from the UN was this: "In my view, diplomacy was best built on a foundation of respect rather than bullying."

He would have stayed longer, loving the work and his life in the capital of the world, but Nixon once more had other plans. Reshuffling his cabinet after winning reelection in 1972, he ordered his chief of staff to "eliminate the politicians." All, that is, "except George Bush. He'd do anything for the cause." True to the president's word, Bush reluctantly accepted Nixon's reassignment, becoming head of the Republican National Committee.

It was a decision he'd regret. Washington soon after erupted in a scandal unlike any in recent memory, as a break-in at the Democratic Party headquarters at the Watergate complex eventually led investigators to the White House. While never implicated in any of the political shenanigans, Bush as head of the party was forced to offer daily defenses of an increasingly beleaguered president, never once revealing the growing suspicions of Nixon's guilt he confessed to his diary. Living "in the eye of the storm" simply exhausted him. "Heaven knows," he told Baker in November 1973, "I wish we were moving back to Houston today . . . but I must stay here." Poll numbers suggested Bush's reputation in Texas was not faring well enough to support another bid for statewide office in any event. By the following March, the relentless controversy had taken its toll. "This job is no fun at all," he admitted to a friend. He "longed for an escape," he said, "and an escape in my fantasy usually takes the form of running around in the boat in Maine — no telephone" especially.

Nixon's resignation left Gerald Ford in charge, and Bush with a more dramatic escape than he had ever imagined. Thinking that Bush deserved a safe landing after so faithfully serving the party over the previous months, and hoping to rid Washington of anyone even vaguely associated with Watergate, Ford offered to name him ambassador to either France or the United Kingdom. Bush instead asked to be posted to China, the country whose rise, or at least whose recognition, he'd earlier worked tirelessly to forestall. Beijing would be exotic, and his time at the UN had convinced him that Asia's future prospects could no longer be ignored. Moreover, unlike serving in Europe, where ambassadors were expected to supplement their embassy's meager social budget, the Beijing post would prove downright cheap. After a decade of public service (and with three more children still to put through college), Bush relished its affordability. Even champagne and caviar were inexpensive, he soon realized to his delight, filling his diary with detailed accounts of sumptuous banquets as well as recollections of the political players he met along the way. "What the hell," he told one reporter who asked if a posting so far away would hurt his future political prospects. When he was done in Beijing, he'd still be only fifty. Think of it like a sabbatical, he explained. "It won't hurt anything."

Many questioned his choice, including Barbara, who learned of her husband's plans only after he'd accepted the job. The Beijing of 1974 was not the vibrant (if smog-choked) global capital of today, but instead a dark place still reeling from the Cultural Revolution sporadically fought over the previous decade, as hard-liners and reformers battled for control of the Communist Party. The city's populace, and to a greater extent its bureaucrats, seemed in Bush's eyes subdued, afraid of standing out and of befriending an American in particular. Requests to meet with government officials often went unheeded for months, frustrating his hope of smoothing Sino-American relations, at least in some small way, by cultivating relationships with the country's leaders. Kissinger thought Bush foolish even to try. "It doesn't matter if they [the Chinese] like you or not," he told Bush during one Beijing stopover. Strategic power was all that really mattered when it came time for states and statesmen to make decisions, and he didn't think that Bush qualified as a statesman in any event. Bush disagreed on both counts. Kissinger "seems to put no faith in individual relationships," he wrote in his diary in early 1975, adding years later: "I do think it [personal diplomacy] is important . . . I suppose there is a danger that one can be naively lulled into complacency if one expects friendships will cause the other party to do things your way, but I thought the danger was remote. For me, personal diplomacy and leadership went hand in hand." Their benefits did not merely happen, he noted. "You can't develop or earn this mutual trust and respect unless you deliberately work at it."

Though never granted the access to China's leadership he desired, Bush enjoyed his sabbatical immensely. He read, slept late, and played tennis at the local club and table tennis with the embassy staff and their families. He ate well, toured the country, and spent considerable time with the rest of the city's diplomatic community. Most important of all, he wrote a political ally, "my phone doesn't ring — after many years of incessant ringing, it's rather weird." He said the same to a close friend only weeks into his stay: "It's the new me, meditative, no phones ringing, little mail, time to think but plenty to do. How awesome!"

China also provided time to think, in particular about the changing world in which American power seemed on the wane. It had been a constant in Bush's life, imprinted during his days serving in the most powerful armada the world had ever produced. American power, to Bush's thinking, had liberated the Pacific, rebuilt Japan and saved Korea, and, even more impressively to his mind, prevented Europeans from going to war for a third time in the twentieth century. American power had been the only thing capable of standing up to the Soviets, he reasoned, and was thus the only thing standing in the way of total communist domination. American power, like the ideas of freedom and free markets he'd so fully internalized that they required no further examination, was for Bush more than just the greatest force for good on earth. It was what kept the free world safe. After China fell to the communists in 1949 and Eastern Europe to the Soviets in the wake of World War II, like so many other Americans, Bush believed that the United States had drawn impenetrable lines around the rest of the world, saying, "No more." And to their minds, it was a promise long and largely kept.

Largely, but not everywhere. The Americans had been unable to save South Vietnam, which fell while Bush was living in Beijing. Cambodia descended into anarchy as well, as did Laos, their fate further weakening the resolve of longtime American allies in the region. More accurately, communist gains in the early 1970s despite U.S. opposition further weakened the faith Asian leaders placed in American promises. When coupled with Washington's apparent rejection of the economic system built by a previous generation of American policy makers with the express purpose of ensuring international stability, a move embodied for conservatives by Nixon's rejection of the gold standard in

1971, the United States appeared weaker in their eyes, and less reliable, than at any time since the end of World War II.

Bush saw all this from Asia in a way he never could have if he had remained in Washington or taken a more traditional route and accepted Ford's offer of Paris or London. He was instead in Beijing the day the last American helicopters evacuated the final holdouts from the nation's besieged embassy in Saigon, and then witness to the Chinese capital's massive victory celebration. Along with his staff, he also marked the lengthy procession of Asian leaders, including longtime American allies, who subsequently journeyed to Beijing to pay homage to the continent's new great power. It only made sense for them to hedge their bets and explore their prospects with communist China, Bush reasoned. "Clearly as the United States has reneged on commitments and pulled back," he confided in his diary, "and is unwilling to support recommendations of the president, the free countries, the Asian countries in Southeast Asia, are concerned" about the worth of future allegiance.

Bush feared that any reduction in American power would lead invariably to communist gains, which in turn would lead to even greater losses for the free world. He believed, in other words, in the domino theory, though not in a simplistic version of this powerful Cold War thesis that saw countries infected with communism topple one another like so many dominoes. He instead held the more nuanced view that communist gains despite massive American support infected an entire region, indeed the entire world, with skepticism about the reliability of American promises. The dominoes wavered and might topple, in this view, all at once rather than singly or in order. "China continues to support revolutions in all these countries," Bush noted after Saigon fell, "and yet many of the countries like the Philippines, Malaysia and others keep trying to get closer to China." Their tack away from Washington and toward Beijing was logical, even if it saddened him. "They have to [turn away] because they don't see in the US the firm kind of interventionist support that they have been able to count on in the past." In short, he wrote after reviewing the aftermath of Saigon's fall, "many of our allies are compelled to move toward the PRC [the People's Republic of China]. The domino theory is alive and well."

It was the first time he'd seen his country fail, the blow on the international stage coming on the heels of Watergate and the general disillusionment with American domestic institutions it caused even for true believers like Bush. Together they signaled a dangerous decline in American power. Long supportive of the war in Vietnam — when a president asked for support, a company man said yes — he began in its aftermath to question the wisdom of American intervention in the first place. "I am wondering what I would have recommended" to the president, he told his diary, "if I had been in a position of major authority," because "I don't know all the facts, but from here [Beijing] it appears to be a losing cause." His country needed to back its allies, he reasoned. It needed to stand up for freedom and democracy. Yet it also could no longer afford to fight quixotic crusades where vital interests were not involved. "We have got to be realistic. We have to have our eyes open." Most important to his mind in 1975, Washington needed to determine with its allies the issues that truly mattered, so that "we are not committed in wars we shouldn't be involved in, where we'd have no support from the American people." They could no longer do it all, he reasoned, at least not until Americans renewed their strength, and their sense of civic purpose, by reinforcing their first principles, rebuilding the trust in the American system he still refused to question.

If they'd faltered, it was not because their tenets were flawed. Freedom still worked. The flaw was in the execution. They'd faltered when a president left office mired in scandal. They'd faltered when sending such mixed signals to the world that they allowed a UN ambassador to support one China in New York while a secretary of state simultaneously courted the other in Beijing. They'd faltered when they'd given up explicit leadership of the global economic system they'd created after 1945. "Institutions that have served to preserve freedom, or to hopefully prop it up in some places, are being dismantled," Bush wrote. What they needed now most of all was not something new to stem the erosion, but rather renewal. "As soon as America doesn't stand for something in the world," he told his diary, lamenting his country's fate yet at the same time vowing to change it, "there is going to be a tremendous erosion of freedom." More deeply internalized than ever as a result of his time in China, this fundamental faith in his country and its role in the world would ultimately guide his presidency.

A surprise phone call soon brought him home. Ford asked him to head the Central Intelligence Agency. It was perhaps Washington's most difficult job. The agency was suffering from unprecedented scrutiny in 1975, and criticism as well, following revelations of a long litany of potentially illegal, immoral, or at the least controversial acts. Whoever took over its leadership would need to walk an impossibly fine line between improving transparency and ensuring morale.

It was also considered a career killer. General wisdom held that no one associated with the secret world of espionage could ever expect to win elected office again. At least, that was the hope of some in the White House, where Don Rumsfeld, Ford's chief of staff, schemed with his young deputy, Dick Cheney, to scuttle the prospects of any potential rival. Friends and colleagues reported Bush "depressed" at the idea of returning home to a job that could end his political ambitions, but he nonetheless accepted Ford's offer, believing there was only one acceptable answer whenever a president called. Truth be told, Beijing had begun to grow tiresome as well. He was ready to come home.

He found he loved the job and the drama of being the country's "head spook," and even asked Jimmy Carter if he might continue in the post after Ford's defeat in the 1976 election. The new president's desire to name his own CIA head left Bush little choice but to return to Texas to rebuild his oil business and plot his next political move. He visited forty-two states in 1978 alone, officially on business, but always with time to meet with local Republican officials. There was always time to make new friends, too, especially those who might prove helpful when he took the inevitable plunge. "I'll go about anywhere and do about anything that's legal to try and win this thing," he told an aide, "but please don't put me in a single-engine aircraft anymore!"

He announced for the presidency in 1979, campaigning as a more moderate alternative to Reagan, winning an upset victory at the Iowa caucuses by visiting every county in the state while his front-running opponent made plans for bigger primaries down the road. Enthusiasm can take a candidate only so far, however, and the former actor turned politician went on to capture twenty-nine of the next thirty-three contests. Bush ultimately conceded defeat by May.

Yet for another chance at the vice presidency, he accepted Reagan's eleventh-hour offer to join him on the Republican ticket. Bush was not his first choice. Reagan initially discussed an unprecedented "co-presidency" with Gerald Ford, but negotiations eventually made plain its unworkability. Bush subsequently accepted without hesitation, vowing despite their campaign differences to follow Reagan's lead in every instance. "I am not going to be building my own constituency or doing things like background conferences to show that I am doing a good job," he wrote at the outset of his vice presidency. His diary subsequently showed the strain of holding his tongue while Reagan implemented a far more conservative agenda than Bush would have liked, but he never once broke stride with the administration. Even as old friends advised him to distance himself from Reagan for the sake of his own political identity, and to avoid any stain from Iran-contra in particular, Bush continued to remind himself that "the president must know that he can have the vice president for him and he must not think that he has to look over his shoulder."

Eight years of loyalty carried a price, deepening the reputation he'd fought since the early 1960s: that he had no real political backbone. He'd been steady and useful as vice president, but *Newsweek*'s editors wondered as Bush prepared his next White House bid if "beneath such surface qualms lie deeper doubts." Reagan's unlikely turn from hawk to peacemaker in his second term, more specifically his embrace of this unproven Soviet reformer, a man with little track record and perhaps even less sincerity, gave Bush the perfect opportunity to show he could indeed step out of the president's shadow. Plus, he believed Reagan was wrong. Dangerously so.

3

GORBACHEV AT THE UN

THOUGH UNWILLING TO SAY SO publicly while serving as his vice president, Bush disagreed with Reagan on the most prominent foreign policy issue of their day: how to deal with Gorbachev. He thought Reagan too trusting, and painfully uninterested in evaluating the consequences of *perestroika*'s success or failure. He also feared Reagan, in striving for a new era of détente, had ignored the real meaning behind communism's apparent change of heart: that because of their rapidly globalizing world, Moscow's decline might be but a mere step ahead of Washington's.

Reagan had put his faith in Gorbachev's revolution, but George Bush was no revolutionary. Stability seemed safer, and change, even when necessary, required care. "All these revolutions present unparalleled opportunity," Bush said in the fall of 1988, speaking of the growing calls for democracy throughout the Soviet empire, "and risk." Revolutions are destabilizing by their very nature, he feared, even if begun from the inside. As Alexis de Tocqueville wrote of his native France before 1789, "The most dangerous time for a bad government is when it starts to reform itself." France's revolution started out with high ideals, initially catalyzed by the royal court's concessions. Decades of war and strife resulted. Two hundred years later, the world could not afford, or perhaps even hope to survive, similar turmoil in the nuclear age. The Soviet empire spanned nearly a dozen time zones, featuring dozens of languages and ethnicities, each with its own history (and historical grievances), so it was foolish in the extreme to believe all that had held Moscow's vast vassalage together could simply erode without consequences.

Having sown a whirlwind, could Gorbachev possibly hope to control it?

Would he even survive? *Perestroika* had powerful enemies. Not everyone within the Soviet military-industrial complex relished amicable relations with the United States, their enemy for four-plus decades. True believers in socialism's promise remained as well, as did those who feared that Gorbachev's remedies were worse than what plagued them. Together they formed a powerful obstacle to his revolutionary plans, with a powerful legacy. To date, only one Soviet leader since Lenin had departed the Kremlin alive, and even then Nikita Khrushchev survived office only through the good graces of the coup plotters who removed him from power. Most left in coffins, their successors changing course the moment they seized power.

With enemies at home, Gorbachev needed allies abroad. Most specifically, he needed Bush on his side, because without a willing partner in the White House — without another Reagan — he could never slow down the spiraling defense expenses that doomed all hope of Soviet prosperity. Gorbachev needed a peace dividend, which ultimately only Bush could provide. Their relationship, as much as anything, would determine if the Cold War could end after all, or if Reagan's final years in office would prove to have been merely a temporary and quickly forgotten warming, a blip of optimistic relations within an otherwise longer narrative of competition and distrust.

Their dance began in New York, a month after Bush won election yet six weeks before he'd fully take the reins of power, where Gorbachev traveled to address the United Nations. He'd told his Kremlin staff in October to prepare a speech capable of keeping *perestroika*'s international momentum rolling no matter which candidate won, Bush or his opponent, Massachusetts governor Michael Dukakis. Both presented unknowns. Bush loyally backed his president, but few could say for certain what he might do once finally in charge in his own right. Dukakis was even more of a question mark, offering no foreign policy record whatsoever. I want to deliver an "anti-Fulton," seizing the diplomatic initiative no matter November's outcome, Gorbachev told his aides, a "Fulton in reverse." His staff immediately understood the reference to Winston Churchill's famous 1946 speech in Fulton, Missouri, where he proclaimed that an "Iron Curtain" had descended between Eastern and Western Europe. Featuring no date that would live in infamy or formal declaration of war, the Cold War lacks an official starting date. Absent those traditional markers, Churchill's Fulton address represents the Cold War's most widely recognized commencement.

Gorbachev wanted to declare its end. It was time for *perestroika* to go global, he thus told his speechwriters, and in New York they could launch a new offensive in the battle for world opinion. "We should present our worldview philosophy based on the results of the last three years," he instructed. "We should stress the process of demilitarization of our thinking, and the humanization of our thinking," making the case amidst Reagan's departure that it was time for the Soviets to assume a more prominent role as a leader not just of the communist world but of the entire world.

Gorbachev had changed since taking power. His entire cadre had, realizing over time the power that derived from being able to open every aspect of their society for reevaluation. He'd already argued that *perestroika* meant a more peaceful Soviet Union, though this was a move made as much of necessity as of desire. Whatever he could save in military spending he hoped to invest in his country's overall economic restructuring. Yet he'd also come to believe that the spirit of *perestroika* could set the tone for a new era of transparency, trust, peace, and ultimately a reformulation of international relations with himself its one true prophet.

He consequently longed to give a speech so eloquent and profound that the Americans, and their incoming president in particular, would have no choice but to recognize not only the power of his ideas but also their sincerity, especially as Soviet analysts predicted superpower relations would likely take a step backwards following Reagan's departure no matter which candidate prevailed. Reagan's willingness to cooperate delivered "a blow at the most sensitive spot of the entire 'cold war' structure," influential Kremlin adviser Georgy Arbatov counseled, including the West's "starting assumption that Soviet communism was ingrained with hostility and bent on confrontation." But not everyone was Reagan. Washington's "more influential circles" would most likely demand a period of retrenchment after so many months of reform. "They [the American 'political elite'] will try to impose on us a more and more slow tempo in the real development of relations," Arbatov warned, all done "in the name of 'caution' and 'realism.'"

Gorbachev faced similar calls for realism and retrenchment of his own. "What is this 'new thinking'?" Boris Ponomarev had argued during a Communist Party meeting in 1986, not long after Gorbachev first assumed power. Head of the party's International Department, and thus responsible for coordinating the global communist movement, the conservative Ponomarev was well positioned to stymie the new general secretary. "Our thinking is already correct," he insisted, citing party-approved Marxist-Leninism. "Let the Americans change their thinking."

Gorbachev's increasingly influential foreign policy aide Anatoly Chernyaev had been quick with a retort. "Look at what Gorbachev's been saying" with his speeches and briefs, Chernyaev implored. "It's quite clear that he's referring to our thinking." *Perestroika* meant changing everything for the sake of survival, including how their country engaged its empire and beyond. Previous Soviet leaders had periodically spoken of coexistence with the West, Chernyaev argued. "Nobody believed them." But "they [Western leaders] trust Gorbachev because he's begun to make our deeds match our words."

Ponomarev could only sputter in response. Gorbachev demanded nothing less than rejection of everything he held dear, including their country's most unifying idea of all: the powerful belief that Soviet citizens were collectively building a better society not just for themselves but for the world. Why did their country even exist if not to lead socialism's inevitable triumph? Communism and capitalism were dialectically opposed. Marx had said so. Lenin confirmed this point. Successive generations of Soviet citizens imbibed this foundational dogma. To suggest the real possibility of coexistence with the West, as Gorbachev demanded with his newfangled ideas, was akin to the pope denouncing God's Trinity, or the Dalai Lama discarding reincarnation. Some dogmas could not be questioned lest the entire edifice collapse, and at the heart of *perestroika* was an idea so powerful it could undermine the Soviet Union's entire reason for being and thus too its justification for leadership over its empire. There was more than one path to the truth, Gorbachev preached. Central authorities didn't always wield the right answer. "What are you trying to do to our foreign policy?" Ponomarev ranted incredulously.

Even loyal followers had difficulty accepting the full implications of what he proposed. For a while, Gorbachev had trouble with it himself. "We were at Zavidovo [a government retreat] working," Gorbachev later recalled of time spent with his inner circle in 1986 trying to square their ideological circles. "We really quarreled—for a day and a half we even stopped speaking to each other. What was the argument about? About . . . the fact that we live in an interdependent, contradictory but ultimately integral world. No, the new thinking wasn't just some policy shift, it required a major conceptual breakthrough." Interconnectivity was the key. So too conceding their system's imperfections. Communism was clearly not working on its own. It needed the wider world, just as the wider world could learn a thing or two from socialism, in particular a means of tempering capitalism's rougher edges with greater concern for the common good. The first step was forsaking the certainty that any one system had exclusive claim to the truth, and the next forgoing the idea that any people had the right to tell another how to live their lives. "Non-violence must become the basis of human co-existence," he said, and with it a fundamental respect for self-determination and sovereignty; yet to be believable, their own actions needed to better match their words. "The military doctrine we announced differs from what we are actually doing in military building," Gorbachev confessed to his closest advisers. But "if we publish how the matters stand, that we spend over twice as much as the US on military needs," he complained in exasperation, "if we let the scope of our expenses be known, all our new thinking and our new foreign policy will go to hell."

What they needed to make *perestroika* really work, Chernyaev argued, to bring the Americans along and their own Soviet military-industrial complex too, was to change the Cold War's "algebra" rather than merely its "arithmetic." Superpower security had for decades derived from addition, with each side adding destructive force until both were satisfied that neither had the ability to survive war with the other. But what if instead of worrying about making sure forces balanced, East and West found themselves on the same side of the equation, their cooperation adding up to something bigger? Something like peace?

They were hardly the first generation of Soviet leaders to recognize their system's imperfections. They were instead merely the first fully committed to finding a fix. Indeed, it is fashionable in American circles in the twenty-first century to argue that the United States and its allies won the Cold War, in large measure by demonstrating beyond doubt their economic and thus military superiority, and in turn the futility of the Soviet Union's hoping to compete. In the most extreme version of this argument, Reagan won the Cold War by ramping up military spending — and corresponding rhetoric — beyond anything the Soviets could ever hope to match. Realizing their inability to keep pace, this reading of history argues, Gorbachev's cadre wisely, if grudgingly, surrendered.

This America-centric reading of history is both wrong and dangerous, implying both that American action alone determines global affairs, and that any adversary can simply be steamrolled by America's economic might. Neither is true. The United States did not win the Cold War so much as the Soviet Union gave up. Soviet leaders could have continued indefinitely, consistently lowering their standard of living, but never fully ceding power. Their nuclear weapons alone ensured no outside force would ever dare attempt conquest; their internal security apparatus had for decades prevented any legitimate opposition from forming, and none did until *perestroika* made real dissent possible for the first time in the nation's history. Their system need not have ended at all, and certainly not beginning in the mid-1980s, save for the conviction of its leaders that the longer they delayed in implementing real reforms, the less likely their success.

If one must give the United States credit for communism's death, it was simply in the same way that a physician providing a lethal palliative ends a terminal patient's life. The syringe may speed the final moment, but the true cause of death is the sickness itself. In this case, communism's own flaws ensured its death, with policy makers of the time — first Reagan with his buildup, and then, more important, Gorbachev with *perestroika* — administering the last lethal dose, though it should be noted that the American did not anticipate his euthanasia taking effect with such speed, and that the Soviet hoped his medicine would save the patient in the end.

Either way, communism's decline was hardly unanticipated, and was predicted well before either Reagan or Gorbachev took office. In the late 1940s, George Kennan, the American diplomat whose "containment" doctrine offered a framework for the entire Western Cold War strategy for defeating communism without outright war, predicted that Soviet economic and political contradictions would doom the regime. Pen the Soviets in, he effectively argued, and wait for their collapse.

It took decades, as Kennan also predicted, but by the 1970s, Western analysts repeatedly observed fatal cracks within the Soviet system. Indeed, owing to ingrained secrecy at the Kremlin, even and especially among its own people, Washington's CIA knew it before Moscow's KGB, leading top Soviet policy makers to harvest American intelligence reports on the Soviet bloc's economy out of recognition that the quality of the American intelligence community's examinations of communism's problems far exceeded that of their own. In one infamous example of the lengths to which Kremlin officials were forced to go in order to gain information about their own country, spy planes designed to ferret out NATO capabilities were requisitioned to assess Central Asian cotton fields, as local reports on the annual crop proved wholly unreliable (especially as much of the crop was destined for the black market beyond Moscow's control). It was the only way Kremlin planners could know for certain what was going on within their own country.

This is more than just an amusing anecdote. It illuminates one key reason why the Soviets failed to keep pace with the West. To succeed, a planned economy requires quality information, something Soviet policy makers wholly failed to cultivate, in large measure because in their closed society, information could not be safely distributed. In Stalin's day it had been a crime even to own a typewriter, lest ideas spread too quickly beyond the state's direct control. Times had changed as the 1980s dawned, but the Soviet Union's penchant for secrecy remained. Travel was restricted and news heavily regulated. Even to place a phone call between Moscow and Leningrad, the nation's two largest cities, required an appointment at a telephone center, where on-site operators monitored every conversation.

Information, as much as any natural resource, proved the West's key commodity and the critical ingredient lacking from Soviet calculations. Even Kremlin officials found information impossibly curtailed. "On taking office as General Secretary," Gorbachev recounted, "I was immediately faced with an avalanche of problems." Worse yet, "the problem could not even be analyzed," because "all statistics concerning the military-industrial complex were top secret, inaccessible even to members of the Politburo." No wonder leaders like Leonid Brezhnev, in power since 1964, failed to address their society's sclerosis. They had little idea.

Precocious, ambitious, and as a consequence largely lacking in deference, Gorbachev had walked into Brezhnev's office soon after taking up his first position within the Kremlin hierarchy. He had no appointment. Junior staffers never simply arrived unannounced, but Gorbachev felt a need to "share his ideas" for improving the country's agricultural sector with the aging leader. Brezhnev's office manager let him in. He must have an appointment, the man thought. Who would show up without one? What happened next was instructive. Gorbachev talked, but Brezhnev just stared into space. The younger man wanted to make changes in a hurry; the older man could only respond to reports of mismanagement with the question "Is it really that bad?"

It hadn't always been this way. Always secretive, Soviet society had once been productive. Indeed, for a brief moment it appeared superior to the capitalist alternative. During the 1950s and early 1960s, postwar reconstruction fueled admirable growth rates beyond anything the United States could consistently achieve. "History is on our side," Khrushchev boasted. Soviet living standards skyrocketed, as quite literally did Soviet prestige following initial advances in the space race. The Soviets launched the world's first manmade satellite, and the first person into space. Literacy rates and the percentage of citizens receiving health care soared for the generation born after 1945. Crime plummeted, as did rates of infant mortality and, for a time at least, even the state's authoritarian aura. Khrushchev promised "socialism with a human face," pledging economic gains alongside increased respect for social justice, greatly influencing a young Gorbachev, then in school in Moscow. By the early 1960s, Soviet officials even felt sufficiently confident to promise, at a party congress an impressionable Gorbachev attended, the prospective transition from socialism to communism by 1980. Soviet society would enter history's final phase, a utopia of justice, equality, and cooperative prosperity. The end of history was nigh, and what Gorbachev's generation considered their due.

Their promised utopian era never arrived, their state artificially propped up first by the stimulus of postwar rebuilding and then later by oil. More accurately, oil masked Soviet decline. Massive petroleum discoveries in Siberia throughout the 1960s transformed the country into an energy exporter, accounting for 80 percent of Soviet hard currency earnings by the ensuing decade. Arms exports made up most of the rest, in particular to oil-rich states in the Middle East. Prosperity built on a nonrenewable resource is fleeting. Soviet economic expansion slowed to a crawl in the fifteen years preceding Gorbachev's installation, plummeting from an enviable rate of nearly 5 percent annual growth at the end of the 1960s to an anemic 1 percent by 1985, as the generation of factories hastily erected from the rubble of World War II reached industrial obsolescence in one big wave. Because employment mattered as much to central policy makers as productivity, inefficient or unnecessary factories were rarely if ever shuttered.

Soviet centralization promoted additional unexpected inefficiencies. Steel production, for example (always a favorite metric of central planners), was evaluated by technocrats and apparatchiks in Moscow on the basis of a factory's overall output, measured in gross weight. The factory that produced the most by total weight was most valued. Yet as the twentieth century progressed, innovations in metal design, in particular for high-tech products such as aircraft and automobiles, placed a premium on lighter yet stronger materials. Thus the determination by central authorities to assess metrics across an entire system actually encouraged the manufacture of inferior products so long as they could be measured in accordance with Moscow's grand design.

The comparison with the American experience is instructive, and important as a way of understanding precisely what each state faced as the 1980s came to a close. As historian Stephen Kotkin has sagely noted, foreign competition led to massive layoffs within American manufacturing throughout the 1970s, and in turn to widespread unemployment and misery. Employment in steel production alone dropped by half from 1980 to 1988. Bethlehem Steel laid off whole cadres of workers throughout the decade, while its hometown in eastern Pennsylvania nearly crumbled as a result in a story repeatedly played out throughout the American Rust Belt. "Well we're living here in Allentown," songwriter Billy Joel lamented in a sad but successful chart-topper, "and they're closing all the factories down." The pain was real, though the result was trimmed-down and teched-up industries, and the freed-up capital to support them, capable of competing with the world's best in new arenas, or of recognizing the efficiency of not competing at all, leaving less remunerative industries to developing parts of the globe.

The Soviets never went through this pain, even as the West moved (both metaphorically and in a real sense) from the age of steel to the age of computers. Soviet planners continued to tally and assess production in much the same way they had since the 1950s. Abacuses dominated shops and offices alike. Conversely, the United States military alone boasted twenty thousand computers by the start of the 1980s, with thousands more coming online every year. The American private sector deployed an even greater number of high-speed processors. So too its universities. Nowhere near that number of computers existed throughout the entire Soviet Union.

The price of oil then dropped. Elevated energy exports buttressed Soviet wealth during the decade's first years, but the ensuing drop in the cost of petroleum eroded the total value of Soviet exports by the mid-1980s. World oil prices plummeted 69 percent in 1986 alone, shredding Soviet profits just as Gorbachev hoped to implement his first reforms. At the same time, production from Siberian wells temporarily faltered.

The timing could not have been worse. The Kremlin officially touted the aforementioned 1 percent growth rate for the early 1980s, but by some estimates the Soviet economy in fact shrank during that time. Under Brezhnev,

Soviet planners devoted more than 30 percent of their total state investment to agriculture, yet that sector progressed by a mere 1.2 percent, less than population growth. This meant that by Gorbachev's tenure, the state was spending more per capita on food than in Stalin's day, yet its people remained hungrier. Even in privileged Moscow, food lines grew. A country once envied for its agricultural potential turned with great reluctance and shame to importing food from abroad. The problem was magnified by Moscow's client states, whose economic failings forced the Kremlin to subsidize its empire's subsistence. Rather than an asset, the Soviet bloc was a drain.

Military expenses sucked up whatever was left. With the Soviet Union largely on a wartime footing since Stalin's day, arms consumed upward of one third of the state's budget by the 1980s. When indirect costs were included, total defense-related expenses may have accounted for nearly 40 percent. The unexpectedly long-lasting Soviet occupation of Afghanistan, which began in 1979, pushed these percentages even higher. By comparison, even at the height of Reagan's military buildup, American officials routinely spent less than a third that amount on defense from their far larger gross national product.

Two sudden events demonstrated the true state of Soviet long-term problems by the mid-1980s. The first was Chernobyl. In the early morning of April 26, 1986, a huge explosion decimated the city's nuclear reactor deep within Ukraine's heartland. The worst nuclear accident in history ensued. Thousands perished from acute radiation poisoning. Hundreds of thousands more were affected over time, as ten times as much radioactivity was released as from the Hiroshima or Nagasaki bombs. The reactors lacked the heavy containment structures Western governments demanded of their nuclear facilities, ensuring that any accidental meltdown resulted in immediate exposure of the surrounding areas. Soviet engineers knew how North American, Japanese, and Western European plants mitigated nuclear risks; but they simply deemed such measures an unnecessary expense for machinery designed to work, not to fail. Indeed, the Kremlin acknowledged the problem only after Western scientists reported dangerous clouds of radiation drifting beyond Soviet borders. "The Chernobyl disaster tells us about the deficiencies of the Soviet political and administrative system," Nobel laureate in physics Hans Bethe concluded, "rather than about problems with nuclear power."

Chernobyl exemplified to Gorbachev the Soviet Union's two fundamental problems: first, technological incompetence had led to an unconscionable crisis (in other words, the system was broken), and second, the state had proved unable to acknowledge a troubling truth even for the sake of saving lives (and thus could never undertake the frank debates necessary to save itself). His reform agenda featured two planks intended as remedies. The Soviets had to become more responsive to market forces and more democratic, he reasoned. The first step was admitting they had a problem. "Our work is now transparent to the whole people, to the whole world," Gorbachev sarcastically told the Politburo weeks after the disaster, in yet another plea for change. "There are no interests that could force us to hide the truth," he said, arguing that the greatest reform Soviet leaders could endorse was that of their own relationship with reality.

Gorbachev made a similar argument — and made use of a second embarrassing episode — a year later when Soviet airspace was penetrated not by the latest high-tech British or American jet but instead by a West German teenager piloting a small recreational plane. Mathias Rust not only eluded Soviet defenses but also landed smack-dab in the middle of Red Square. Gorbachev seized the opportunity to purge the Red Army's top levels of opponents and old guard conservatives. "Transparency and candor" would henceforth mark Soviet defense policy, he vowed. As one of his closest foreign policy aides recalled, Gorbachev pledged by the summer of 1987 to overhaul "the whole system — from economy to mentality." Domestically this meant encouraging open questioning of the state's infallibility. Internationally this meant projecting a more peaceful Soviet image abroad, designed not only to reduce tensions for their own sake but also because the East-West rivalry placed such a heavy military strain upon the economy. "It was clear to both of us," Foreign Minister Eduard Shevardnadze later wrote of himself and Gorbachev, "that if we did not change our foreign policy by removing the main sources of distrust — the use of force and rigid ideology — we would never create a zone of security around our country."

Gorbachev's reforms spurred by necessity and decline left Western analysts, and Bush especially, to wonder, did the Soviets genuinely want peace, or merely time to renew their strength? If Gorbachev acted only so his country could catch its breath before resuming its struggle for global dominance, any reduction in American military preparedness would be a mistake. Perhaps Gorbachev had a more Machiavellian plan in mind. Being unable to keep pace with the Americans, he may have hoped they might be persuaded to keep pace with his reductions, cutting their defenses and, in a best-case scenario, departing from Europe believing the Cold War over, thus leaving an undefended continent ripe in time for eventual Soviet conquest. He might mean what he said; or it might all be a ruse, and no one in Washington could say for certain which was the case.

American leaders faced questions of their own as the 1990s drew near, beyond merely wondering about Moscow's ultimate intentions, as their own post–World War II economy aged and transformed. The Soviets remained a military threat, and the superpowers still controlled most of the world's sophisticated weaponry, yet an equal number of Americans queried in 1988 ranked economic rivals as the nation's primary future competition. Asia seemed on the rise, led by Japan, as did a nascent European Union with Germany at its head. The developing world was catching up fast as well. By sparring for so many decades against each other, had both the United States and the USSR exhausted their resources, opening the door to their successors, like two prizefighters never quite the same after an epic bout? While they were still dominant, was their power in decline because the very metrics of power in the modern age had shifted beneath them?

It had happened before. The surprise runaway bestseller of the period, *The Rise and Fall of the Great Powers* by Yale historian Paul Kennedy, touched a sensitive nerve within the American psyche. Published in 1987, it predicted the ebbing of American power by "imperial overstretch." Just like the United States in the middle decades of the twentieth century, previous powers had dominated the international systems by controlling more than their share of global commerce. Wealth and power followed. Yet empires were their own worst enemies, hubristically overextending their reach and failing to reinvest for the long term while masking their decline with increased military spending. Imperial overstretch doomed the Spanish, Dutch, British, and Hapsburg empires, and seemed to be already unraveling the Soviet Union. If this historical pattern held, the United States would suffer the same fate.

By some measures its decline had already begun. America's share of global production tumbled in the 1980s, just as its deficits soared. Star Wars, lowered taxes, and renewed military spending did not come cheap. The United States had an international current accounts surplus of $6.9 billion in 1981, and a deficit of $8.7 billion in 1982. By 1986 and 1987, conversely, the nation's annual debt to the world rose to $140 billion and $160 billion, respectively. Within five years, therefore, the nation's annual trade debt had soared 2,000 percent.

Reagan left the country in the red to the tune of more than $400 billion, while the total value of American assets owned by foreigners doubled between 1982 and 1986 alone.

The country Bush was elected to lead, therefore, appeared according to Kennedy's quantifiable measures a classic example of an empire afflicted by overstretch: its military knew no immediate rival, yet its treasury was empty, and its factories faced stiff competition from abroad, in particular from those whose more recent investments had flourished under the very shadow of American protection. Deficit spending could hide the problem for a while, but competitors historically always appeared when overstretched empires inevitably faltered. "The only other example which comes to mind of a Great Power so increasing its indebtedness in peacetime is France in the 1780s," Kennedy warned, "where the fiscal crisis contributed to the domestic political crisis." As already noted, that crisis in French politics led first to revolution, then to the guillotine, then ultimately to decades of war.

James Baker knew enough to be worried, noting before leaving his post as treasury secretary in 1988 to run Bush's campaign that "the scope of the national-security debate has been broadened to include the economic dimension. And I think the reason for that is that the economic domination of the United States, which has long been taken for granted, is to some extent now being questioned." On the other side of the political aisle, former defense secretary Harold Brown simultaneously noted that "containment of the Soviet Union has succeeded," but "it's not the Soviet Union we have to watch, but Japan, to some extent Europe, and increasingly the newly industrializing countries," such as South Korea, or even in a more distant future China.

Concerns over America's place in the modern world frequently contained this Asian hue. It was one thing to lose market share to European rivals, but those same levels of decline in comparison to Asian countries rankled in a way few would openly discuss yet many privately felt. Marketers knew how to tap into this fear. The cover for the first edition of Kennedy's book, for example, depicted what the author would later describe as "medieval wheels of fortune or clocks of fate," featuring "Uncle Sam" at the pinnacle about to be overtaken by "an Oriental-looking gentleman bearing the flag of the rising sun." Uncle Sam himself had taken the top spot from Britain's John Bull decked out in topcoat and bowler hat, but more than merely national competition appeared at stake. Asians were supplanting Caucasians.

The image would not have seemed out of place as propaganda during World War II, and by the 1980s that conflict's losers were looking like its long-term winners, as newspapers and magazines filled throughout the decade with tales of the superhuman zeal and work ethic of Japanese workers, the focus of Japanese firms, and by extension the roll call of American factories bankrupted by insurmountable Japanese exports. German investment in the United States in 1987 alone was more than double Japan's. The French invested even more than the Germans, and the British still more. Yet American pundits anxiously singled out Asia's rising powers for scrutiny, and for concern. Japanese investors owned more than $42 billion worth of American real estate by the close of Reagan's second term. Two thirds had been acquired in the previous two years alone. Late in 1989 the Mitsubishi group would buy New York's famed Rockefeller Center, itself named for one of America's long-gone financial titans. The entire complex had been designated a national historic site only two years before, leading to charges that Japanese investors were scooping up not only prime American real estate but America's history as well. "It used to be we could say America should be moving into the future," a leading American trade negotiator concluded after years of dealing with the Japanese. "Now we are finding we have no future."

What prosperity remained in the United States also rested in fewer hands, another metric Kennedy warned boded ill for the future. White-collar employment, exemplified by finance and the service industries, boomed even as manufacturing declined. These trends formed the leading edge of broader economic currents as American industry evolved for the coming information age and the twenty-first century. More visible to most, however, the 1970s and 1980s offered those at the tail end of the post-1945 baby boom the unprecedented — for their lifetimes — realization that their generation's prosperity might not exceed their parents'. Of course, not everyone lamented such change. "Greed, for lack of a better word, is good," fictional financier and corporate raider Gordon Gekko famously told stockholders — and movie audiences — in the 1987 film *Wall Street*. The line became famous, a hallmark of the exuberance and excess of the time, and in some circles a statement of the real purpose of Reagan's trickle-down economics.

Less often remembered, though of equal import to audiences at the time, was the context in which the line was uttered: a debate over the downsizing of a company for the sake of short-term profit, but at the cost of hundreds of previously respectable working-class jobs. To truly sense the national uncertainty over America's economic future as Reagan departed and Bush took over, the uncertainty that gnawed away at the exuberance over potentially impending Cold War victory, one should recall not just Gekko's most memorable one-liner but his entire closing statement: "Greed, in all of its forms; greed for life, for money, for love, knowledge, has marked the upward surge of mankind, and greed, you mark my words, will not only save Teldar Paper but that other mal-functioning corporation called the U.S.A."

No one in the film, or perhaps in the audience, could deny the possibil-ity that their nation was in fact, for all its power and glamour, and for all its immediate prospects of triumph over communism, malfunctioning just as he claimed. How long could the system last? Years? Decades? No matter the time frame, the outcome seemed certain. "The United States has become a hegemon in decay," warned Johns Hopkins University scholar David Calleo, "set on a course that points to an ignominious end."

Bush could say none of this from the campaign trail, or as president-elect. He'd run as Reagan's successor and inheritor of his optimistic economic mes-sage, but also as the candidate most likely to keep Americans safe. "Our policies are working and I want to build on those policies," he told one crowd, during one of the few occasions when Reagan consented to campaign at his side, de-spite "all talk about this book out of Boston."

Bush knew that Kennedy taught at Yale, and his alma mater's Connecticut location. But Dukakis hailed from Boston, a city portrayed by Republicans as the bastion of effete liberalism. The opportunity to paint Dukakis as both weak and down on America's future by linking him to a bestseller whose thesis ques-tioned America's prospects was simply too tempting to forgo. "My opponent sees the task ahead as how to manage the decline of this country and I see it as how to open further the golden promise of opportunity that is America," Bush continued. Let the pessimists talk. "The American century has not drawn to a close. We are not in decline." Americans were special, he argued, once more refusing to question the foundations of the success that had served him and the nation so well for so long. They even had the power to overturn history, provided they had a man in charge strong enough to hold international un-certainties at bay. "I will keep America moving forward, always forward," Bush promised in accepting his party's nomination, "for a better America, for an endless enduring dream, and a thousand points of light."

He would also keep Americans safe. Two Bush campaign ads pushed voters to see their man as the candidate of strength and security, the other of weakness and uncertainty. The first has become legendary, displaying the Dukakis campaign's own staged footage of their pocket-sized candidate careening around an empty field atop a massive tank. They'd hoped to give Dukakis a martial air, though immediately attempted to quarantine the video once they recognized the comical result. Bush's team, seeing the same, replayed the footage incessantly. "Michael Dukakis has opposed virtually every defense system we developed," a narrator intones in the background, "and now he wants to be our commander in chief." As the camera zooms in on Dukakis's giddy grin, his smile nearly as wide as the oversized helmet bearing his name on a crooked strip of masking tape, looking more like Snoopy than Patton, a deep voice warns, "America can't take that risk." Voters worried about their country's long-term place in the world would not be reassured by the image.

A second commercial explicitly revealed the Bush campaign's effort to forge political gains from latent Cold War anxieties. It begins with a photo of Reagan, Bush, and Gorbachev shaking hands in front of the White House. Reagan slowly disappears from view as a somber voice says: "This is no time for uncertainty . . . Somebody is going to have to find out if he is real." The "he," of course, was Gorbachev. For all the hope and wonder the Soviet leader had inspired by 1988, his promises of change struck the majority of Americans as yet unproven. "It would be a grave error for any U.S. President to base peace with the Soviets on trust," Bush's handlers suggested he state from the campaign trail, because ultimately "peace must be based through strength." Bush followed their script precisely, adding, "His [Dukakis's] foreign policy views born in Harvard Yard's boutique would cut the muscle out of our defense." The killer words were not the name of the university where Dukakis attended law school, but rather its characterization as an effeminate "boutique," lacking "muscle." Bush, by contrast, promised old-fashioned manly protection.

And protection, most of all, meant staying both strong and attentive, allowing him as the campaign drew to a close to emphasize his skepticism when it came to Gorbachev without moving too far from Reagan's position. "What I think we ought to do is take a look at *perestroika* and *glasnost*, welcome them, but keep our eyes open," he said during one debate. "Be cautious. Because the Soviet change is not fully established yet. Yes, I think it's fine to do business with them. And, so, I'm encouraged . . . but can they pull it off?"

It was the question Gorbachev came to New York in December to answer, face-to-face with Reagan's successor. As for *perestroika*, "the world will see that it's not empty talk, these are policies," he told his aides in preparation for their trip. In New York, Gorbachev vowed, "we will advance the entire process" with the two audiences that mattered: global opinion and President-elect Bush. If he could convince the latter that his reforms were sincere, lasting, and likely to succeed, the new interdependent world he longed to develop really could come to pass. Americans could stop worrying about their decline, and masking it through increased military spending, and they could in turn both enjoy the dividends of peace.

The city welcomed him like a visiting hero. News anchors interrupted day-time soap operas to describe Gorbachev's arrival at Kennedy Airport. CBS's Dan Rather praised his wife's "light up the room" smile. Tom Brokaw of NBC complimented her luxurious fur coat. Tens of thousands of normally staid New Yorkers, typically blasé about celebrities, lined the motorcade, chanting "Gorby! Gorby!" Flags bearing the hammer and sickle outnumbered the Stars and Stripes. In Times Square the ticker read, "Welcome Comrade General Secretary Gorbachev." Even Wall Street reacted positively. The markets climbed during Gorbachev's visit.

He did not disappoint those who hoped to see history in the making. "We are witnessing the emergence of a new historic reality," Gorbachev told the packed United Nations General Assembly. "The history of the past centuries and millennia has been a history of almost ubiquitous wars, and sometimes desperate battles, leading to mutual destruction . . . However, parallel with the process of wars, hostility, and alienation of peoples and countries, another process, just as objectively conditioned, was in motion and gaining force: The process of the emergence of a mutually connected and integral world." Western Europe was even then in the final stages of a long-dreamed-of unification process begun in earnest in the aftermath of World War II. Proponents promised greater economic efficiency, political harmony, and, in the minds of the most optimistic, the kind of cooperation that eliminated any need for war. As the president of the European Commission had recently declared, "The watchword of post-1945 politics, 'We must never go to war with each other again,' buoyed the hopes of those who built the Community. That objective has been achieved."

Stop and consider that statement. By unifying, Europeans believed they had eradicated war. They stood poised to enter a new age, Gorbachev declared, but his vision transcended mere continental affairs, and certainly extended beyond just the European Union forming in the continent's western half. The entire globe demanded a "new world order," and the formula of development "at another's expense" was outdated. Competition was outdated. "In light of present realities, genuine progress by infringing upon the rights and liberties of man and peoples, or at the expense of nature, is impossible." East and West, North and South, developed and not, capitalist and communist: each dialectic divided humanity; each division could be pushed aside through integration.

Gorbachev offered tangible steps as proof of his nation's willingness to lead the way toward a new cooperative order. First, he pledged unprecedented support for national self-determination. "Force," he argued, "and the threat of force can no longer be and should not be instruments of foreign policy." These were grand statements for a man whose country boasted a long history of violent crackdowns on client states that dared pursue a different path. It was nothing less than public affirmation that he was prepared to let Eastern Europe take *perestroika* in its own direction, or in multiple directions. "They're sick of us and we're sick of them," he had already privately complained to Chernyaev. "Let's live in a new way, that's fine."

Gorbachev saved his most dramatic revelation for last. To prove their goodwill, the Soviets would unilaterally reduce troop levels by a half-million soldiers by 1991 and withdraw fifty thousand troops and five thousand tanks from Eastern Europe. Remaining Soviet units in the region would be reorganized on a purely defensive footing, further proving his desire for "the demilitarization of international relations." He concluded with a direct appeal. "The future U.S. administration headed by newly elected President George Bush will find in us a partner, ready—without long pauses and backward movements—to continue the dialogue in a spirit of realism, openness, and goodwill, and with a striving for concrete results."

Gorbachev finished his speech to silence. He had spoken for an hour. Brief remarks by his standard; perhaps the assembled delegates expected more. Slowly, the magnitude of his words crept over the crowd. They began applauding. Then standing. Then roaring approval. Veteran diplomats could recall no other UN speaker so enthusiastically embraced. Newspaper editorialists and pundits soon joined in the acclaim. Gorbachev had "unfurled a blueprint for saving the planet and democratizing the world," one writer declared in the *Washington Post*. "No 'thousand points of light', no 'I'm on your side', no one-liners, no sound bites. This was cosmic stuff, announcement of a new order, one in which the Soviet Union will march side by side with, although a step ahead of, other nations toward peace and reason on earth." Editors at the *New York Times* were no less effusive: "Perhaps not since Woodrow Wilson presented his fourteen points in 1918 or since Franklin Roosevelt and Winston Churchill promulgated the Atlantic Charter in 1941 has a world figure demonstrated the vision displayed yesterday at the United Nations." Longtime Washington insider (and frequent Bush confidant) Senator Daniel Patrick Moynihan termed it "the most astounding statement of surrender in the history of ideological struggle."

Even American intelligence analysts predicting dramatic moves were taken aback by the scale of Gorbachev's proposals. His speech overlapped with a closed-door hearing before the Senate Intelligence Committee on the future of Soviet reform, and officials took turns answering questions and jumping into the adjoining room to catch snippets of Gorbachev's remarks. "In all honesty, had we said a week ago that Gorbachev might come to the UN and offer a unilateral cut of 500,000" worldwide, the CIA's director of the Office of Soviet Analysis confessed, "we would have been told we were crazy."

Gorbachev would not have long to savor victory. He received devastating news upon entering his limousine for the short drive to lunch with Reagan and Bush. A massive earthquake had struck the Soviet province of Armenia. Measuring 6.9 on the Richter scale, it was sure to have caused widespread damage and untold casualties. Reports grew worse as the afternoon progressed. Casualty estimates soared past 100,000. Red Army units arrived quickly on the scene, but the soldiers could do little to help the region's thousands buried beneath the rubble, or the far more numerous cold and hungry survivors suddenly exposed to the elements. The scale of destruction was immense. Whole city blocks had tumbled like dominoes. It was later revealed that steel rods intended as reinforcement for concrete had instead been sold on the black market by local officials. It was, in short, a very Soviet disaster, coupling corruption with bureaucratic indifference, just like Chernobyl.

Unlike in 1986, however, this time communist officials did little to hide news of the suffering. Soviet journalists covered the tragedy and its aftermath as never before. Foreign reporters and relief agencies were on the scene as well. Even Bush's son and young grandson, Jeb and George P. Bush, traveled to Armenia as part of the international relief effort. Soviet society had not yet overcome its structural deficiencies, but at least it was far more open about its trials and hardships.

Cutting his trip short to attend to the disaster at home, Gorbachev had one opportunity to meet with Reagan and Bush before facing the crisis in Armenia. The threesome met on Governors Island, off the southern tip of Manhattan, shaking hands before a wall of flashbulbs and a sea of reporters. Yelling above the din, one member of the press managed to ask Reagan, "Has he [Gorbachev] taken a propaganda advantage with his major proposal today?" Gorbachev answered incredulously for the president. "This is not [a] serious question," he replied with real anger in his voice. "If we score any points, we can do it only together. If we try to score points alone, nothing will happen." Out of earshot of the press, he deadpanned to the two Americans that he "hoped what he said at the UN had not contained surprises."

Ever the gracious host, Reagan asked, "Have I ever told you about Lyndon Johnson's remark concerning the press?" Without waiting for a response, Reagan recounted Johnson's frequent lament that "if he ever walked from the White House to the Potomac and walked on top of the water, the press would report that the President could not swim!" Gorbachev laughed politely. Reagan had told him that story before. More than once. By the close of 1988 it was clear the aging president had fully exhausted his new material.

Gorbachev really cared only for Bush's opinion in any event. What was the president-elect's reaction to the UN speech?, one reporter yelled before the lot was ushered out of the room. "I support what the President says," Bush answered with a grin. Long used to playing the role of Reagan's junior partner, he knew his lines well. That's "one of the best answers of the year," Gorbachev said, laughing heartily this time.

Gorbachev was far less pleased when the party settled in for a private lunch. He "would need a little time to review the issues," Bush told him as Reagan looked on. His new administration planned a top-to-bottom review of all of Washington's policies. "What had been accomplished could not be reversed," Bush assured him, yet with a new team taking over, every aspect of American strategy would be reassessed.

Gorbachev could barely contain his incredulity. Wasn't Bush, as vice president, already up to speed on all the issues? Wasn't he ready to lead? Gorbachev obviously liked Washington's direction under Reagan and knew that reviews frequently led to new policies. They also took time, perhaps more than he could stomach, or even survive. "The name of the game is continuity," Gorbachev responded hastily, having only that morning warned in his speech against "long pauses and backward movements." He had no intention of "stalling things," Bush promised. He merely wanted "to formulate prudent national security policies."

The conversation grew tense, with Reagan sitting largely mute as the two younger men jousted. Press secretary Marlin Fitzwater later reported he'd never before seen a president treated as a mere piece of furniture. Trying to reset the tone, Reagan asked Gorbachev how his reforms were faring, which the increasingly frustrated Soviet only interpreted as a challenge. "Have you completed all the reforms you need to complete?" he shot back.

The situation threatened to deteriorate. Recognizing he could not rebuff Gorbachev's questions forever, and tired of playing defense, Bush offered a question of his own. He wanted to know if *perestroika* and *glasnost* were for real. "What assurance can you give me," he asked, "that I can pass to American businessmen who want to invest in the Soviet Union" that "reforms will succeed?" Gorbachev snapped at him yet again: "Not even Jesus Christ knows the answer to that question!"

The Russian continued to berate him: "I know what people are telling you now — that you've won the election, you've got to go slow, you've got to be careful, you've got to review, that you can't trust us, that we're doing all this for show. You'll see soon enough that I'm not doing this for show, and I'm not doing this to undermine you or surprise you or take advantage of you . . . I'm doing this because I need to. I'm doing this because there's a revolution taking place in my country. I started it. And they all applauded me when I started it in 1986, and now they don't like it so much. But it's going to be a revolution nonetheless . . . Don't misread me, Mr. Vice President."

Their meeting, much like the entire day, failed to live up to Gorbachev's expectations. His speech, designed to announce a tectonic shift in international affairs, was sadly overshadowed by an actual tectonic shift back home. He had hoped to warm his personal friendship with Bush, yet he left Governors Island more worried than before. His address, meant to propel Soviet-American relations forward, instead prompted the opposite reaction from Bush's inner circle. Real progress was won through deliberate negotiations rather than at a speaker's podium, Baker reminded Bush when the two found time to dissect all that had occurred. Trust came from deeds, not words. Soviet troop reductions were of course always welcome, but in the end would do little to change strategic calculations. Gorbachev wanted an immediate response, perhaps even a move in kind, but Baker advised that Bush "avoid rashness." In his experience, the greatest mistakes a president and an administration could make, in particular in their first months in power, were "those of commission, not omission."

Incoming national security adviser Brent Scowcroft seconded the motion for prudence. Gorbachev was a "clever bear," he told Bush, "potentially more dangerous than his predecessors." The Soviet leader was clearly "attempting to kill us with kindness," but Scowcroft's fear "was that Gorbachev could talk us into disarming without the Soviet Union having to do anything fundamental to its own military structure, and in a decade or so, we could face a more serious threat than before." With advisers like these two men, there was simply nothing Gorbachev could have done to convince Bush otherwise. The new president was determined to take his time.

There would be no early summit, nor any sudden American attempt to compete with the Soviets in the crucial arena of global public opinion. "The Soviets in general and Gorbachev in particular were masters at creating these enervating atmospheres," Scowcroft later recalled. "Gorbachev's UN speech had established, with a largely rhetorical flourish, a heady atmosphere of optimism." He was a masterly speaker, with charisma to burn, and an increasingly global following. Scowcroft's man had substance but could hardly match the eloquent revolutionary word for word and speech for speech. In a battle of dramatic gestures pitting Gorbachev against Bush, the Americans would lose. Gorbachev would say or do almost anything. Increasingly desperate and eager for revolutionary change, the Soviet leader was many things, but he was not prudent. The new American leader had prudence to spare, and experience, but was no match for the man he had just replaced, or the one he now faced, when it came to inspirational rhetoric.

Bush did reach out once more to Gorbachev before his inauguration, sending a personal letter to Moscow delivered by hand by Henry Kissinger, a man he neither liked nor trusted. Their mutual antagonism lingered from the early 1970s. But Kissinger had easy access to the Kremlin, and could thus deliver a private message without raising alarm.

"As I explained" in New York the prior month, Bush wrote to Gorbachev, "my new national security team and I will need time to reflect on the range of issues — particularly those relating to arms control — central to our bilateral relationship, and to formulate our own thoughts on how best to move that relationship forward . . . beyond the details of arms control proposals to the issues of the larger political relationship we should want to create." He was willing to work with Gorbachev; just not yet.

"Stable relations can only be built on the basis of the long-term interests of each side," Kissinger further elaborated to the Soviets, having of course taken the opportunity to read Bush's private letter before delivering it. Clearly the world was changing, and "it is impossible to stop history." Turning from professor to powerbroker, he offered a deal: Moscow would continue to allow Eastern Europe's transition to democracy, and in exchange the Americans would do nothing to undermine Soviet influence over the region. "We should not be trying to reform you," Kissinger elaborated, "and you agree to live in conditions of relative and not absolute security." Bush in turn "would be willing to work on ensuring conditions in which a political evolution could be possible but a political explosion would not be allowed." Bush desired stability above all else. It was his very nature. This was why he would take his time in deciding his next move.

Gorbachev heard Kissinger out but wanted more. There were "new forces at large in the world," he replied, yet despite transformational changes, in his view "the United States and the Soviet Union still had the principal responsibility for preserving the peace." Both countries should keep an eye on Germany, "and by that I mean both Germanies," Kissinger reported Gorbachev to have said. "We must not do anything to unsettle Europe into a crisis." Moreover, the Soviet leader continued, at least in Kissinger's recollection of their conversation, "as far as Eastern Europe his view was as follows: life brings certain changes which no one can stop and that applied as well to Eastern Europe. Both sides should be careful not to threaten each other's security. That was the spirit in which he would approach the dialogue."

"I lead a strange country," Gorbachev confessed as he led Kissinger out the door. "I am trying to take my people in a direction they do not understand and many do not want to go. When I became General Secretary I thought by now Perestroika would be completed. Instead the economic reform has only just begun. But one thing is sure . . . this country will never be the same again."

4

"WE KNOW WHAT WORKS"

GORBACHEV DID NOT LET UP. Overall Soviet military spending would drop an additional 14 percent, the Soviet leader promised early in the new year. Weapons procurement would plunge nearly one fifth. Moscow would withdraw three fourths of its troops deployed along the Chinese border, and pull its remaining soldiers from Afghanistan in mid-February. Foreign Minister Eduard Shevardnadze subsequently announced a 20 percent reduction in nuclear forces in Europe. Whether out of financial necessity or genuine pacific intent, the Soviets gave every public appearance of ramping down. Western Europe, in particular, breathed easier.

Yet Bush's anxiety rose with each of Moscow's pronouncements, which to his eyes appeared calculated not to ease tensions in Europe but to weaken the Atlantic alliance, and American leadership in particular. Traumatized by two world wars in as many generations during the first half of the twentieth century, and subsequently terrified for two generations by the prospect of Soviet invasion, Western Europeans had long looked west for security. In the oft-repeated quip of one British diplomat, NATO was designed to "keep the Russians out, the Americans in, and the Germans down."

Gorbachev's peaceful overtures, openly derided within Bush's inner circle as a "peace offensive," offered a different formula, an appealing one for Western Europeans fatigued by decades of Washington's perceived hawkishness. Europe's residents need not look west for security if they no longer had to fear Soviet incursions. They could instead look east as never before, toward a new Soviet Union eager to partner rather than conquer. Together with their new European Union, they might construct a peaceful twenty-first century for the

entire continent, even as the Americans continued to live by the tired milita-
rized rules of the twentieth. The implication was clear in Washington: Gorbachev's promises of peace threatened the end of transatlantic ties, and the end
of American influence on the European side. Gorbachev had already "eroded
U.S. leadership in Europe," Bush would complain to his staff soon after taking
office. "If we don't regain leadership, things are going to fall apart."

At least once inaugurated, he could finally begin making his own case. On
January 20, 1989, Bush stared out over a crowd that stretched halfway down
the National Mall. Rows of dignitaries behind him sat muffled against the cold.
Onlookers might have been forgiven for thinking he had just become leader
of an overcoat appreciation society. Reagan, well bundled, a broad white scarf
wrapped around his neck in deep contrast to his dark topcoat and hair, occupied a place of honor to his left. Bush's eighty-seven-year-old mother smiled
proudly from his right. His five grown children sat a row behind. Millions more
watched on TV as the ceremonies unfolded along the Capitol building's west
front, decked with federal-style bunting reminiscent of the 1790s. It was exactly two hundred years since George Washington first uttered the presidential
oath of office, leading Bush to swear his oath on Washington's personal Bible
(with a Bush family Bible added for good measure), and giving rise to endless
quips about the nation's journey from George to George. The crowd burst into
the traditional applause when he'd finished. Trumpets blared on cue. Cannons
boomed in the distance. After fifty inaugurations, Americans knew their roles.

After eight years in Reagan's shadow, and a lifetime of political subordination before that, George Herbert Walker Bush was finally in charge. Gazing
at the crowd, he thanked Reagan, asked for a moment of prayer, and then declared the start of a "new chapter" in the nation's long journey. Peoples around
the globe were "moving toward democracy through the door to freedom," he
declared. "Men and women of the world move toward free markets through the
door to prosperity. The people of the world agitate for free expression and free
thought through the door to the moral and intellectual satisfactions that only
liberty allows." Theirs was indeed the beginning of "a peaceful and prosperous
time." In sum, "the day of the dictator is over."

That was the most memorable line of the speech but not its most important.
Bush offered a direct rebuttal of Gorbachev's incessant calls for a new world
order by pleading for the world to stay its current course. The new breeze of
freedom was blowing, Bush said, because the communist world was rapidly

awakening to the realization that the West had been right all along. So right, in fact, that it didn't need radical new ideas pulled from the shards of socialism. It didn't need new options; indeed, the world didn't need anything new at all. "We know what works," he said: "Freedom works. We know what's right: Freedom is right. We know how to secure a more just and prosperous life for man on Earth: through free markets, free speech, free elections, and the exercise of free will unhampered by the state." The American system worked, Bush said. For all. "For the first time in this century, for the first time in perhaps all history, man does not have to invent a system by which to live," Bush declared. Democracy had won.

Bush's inaugural address stated plainly one of the era's most powerful arguments: that its Cold War triumph represented a fundamental victory for democracy over all other political systems. Indeed, he preceded by weeks the mid-level State Department analyst typically given credit for this insight. There were far fewer people in the room in February for a University of Chicago seminar than attended Bush's inaugural. The proceedings, under the provocative title "Are We Approaching the End of History?," were neither televised nor recorded. The speaker hoped merely to impress his mentors, and to add another line to his résumé. If luck held, he might find a publisher as well. By year's end he would instead be known throughout the world. "I don't understand it myself," Francis Fukuyama conceded. "I didn't write the article with any relevance to policy. It was just something I'd been thinking about."

Specifically, like every other international analyst, he had been thinking about what to make of communism's apparent surrender. Fukuyama believed the answer could be found deep in the writings of the German philosopher Georg Hegel and the French Hegel scholar Alexandre Kojève, which together revealed a far greater victory than merely bringing the Cold War to an end. It was instead the culminating moment of all history, which for millennia had featured an ongoing series of experiments all designed to identify the ideal government and societal structure. Kings, despots, oligarchs, collectives, and even a führer had been tried and found wanting. Better systems consistently prevailed. Weaker systems perished. Fascism lost earlier in the twentieth century to a combination of democracy and communism. It took only another forty years for communism to fail as well.

Just one system remained. Democracy had won. Moreover, with no other competitors on the horizon, its victory was likely to stick. All of history, according to Fukuyama, had led to this moment. The process would take years more to unfold, perhaps even centuries, before every society fully got the message. But just as Bush believed the stream of history flowed toward an inevitable democratic end, Fukuyama argued that the democratic destination was not only in sight but unalterable as well. History had revealed its final point.

The triumphalist argument took its author somewhere indeed, especially after *The National Interest* published his essay later that summer. "Fukuyama's bold and brilliant article," his mentor Allan Bloom concluded, "is the first word in a discussion imperative for us, we faithful defenders of the Western alliance . . . It is the ideas of freedom and equality that have animated the West and have won by convincing almost all nations that they are true." Irving Kristol, the journal's editor, similarly praised his "brilliant analysis," including its powerful conclusion. "We may have won the Cold War, which is nice," Kristol wrote, but as Fukuyama shows, "it's more than nice, it's wonderful."

Why such a fuss? Why were the magazine's editors and a list of celebrated commentators so enthused by, of all things, a labored discourse on Hegel? First, because Fukuyama said something powerful people liked to hear: that they had been right all along. Western-style democracy ran on a straight path from the Athenian polis to the Roman Republic and then through Magna Carta, conservative champions such as Bloom and Kristol had long argued, until ultimately landing in the United States. It need never migrate again. Recorded history would continue as before. Human passions would never allow for the complete eradication of violence, greed, competition, or conquest. But its overall democratic end point was no longer in doubt. "We close this millennium with the sure knowledge that in liberal, pluralistic, capitalist democracy," Fukuyama wrote, "we have found what we have been looking for." As the editor of the *New York Times Magazine* summed it up, "In other words, we win."

Like "imperial overstretch" a few years before, Fukuyama's pithy phrase "the end of history" quickly entered the national lexicon. Copies of *The National Interest* containing the essay flew off the shelves at a record pace. It was "outselling everything, including pornography!" one newsstand owner in Washington marveled. Leaders as far away as Margaret Thatcher wrote in search of the text. The influential French quarterly *Commentaire* announced a special issue devoted wholly to discussing Fukuyama's thesis. His face appeared in *Time* magazine and news crews clamored for interviews. Of all the statements in this book this one stands most clear: Fukuyama's essay easily ranks among the best-selling studies of Hegelian philosophy ever produced.

He said what Bush had already tried to articulate, albeit saying it better, and therefore in hindsight helps us understand a central paradox of the incoming administration's mindset, in which triumphalism and trepidation commingled. Bush and those around him believed they represented the right side of history; they also feared both that the end of American influence might well be at hand even as democracy triumphed, and that democracy's inevitable victory could still be forestalled, perhaps dangerously so, by unexpected events. "The quest for democracy," said James Baker in a major address in early May, "is the most vibrant political fact of our time." Yet just because democracy was destined to win, that did not mean victory would occur on their watch. Neither did democracy's triumph necessarily mean continuation of American power and influence. Freed of its need for American protection, democratic Europe might turn elsewhere. Asia too might find its own democratic way without American influence. To return to an earlier metaphor, the United States was rounding third but could still stumble before reaching home. Democracy would win in the end, but another might score the winning run, perhaps not for generations to come. "We are in a truly extraordinary period," deputy national security adviser Robert Gates told an audience at the beginning of April. "Capitalism and democracy are ascendant. Economic statism and political despotism are in retreat." Despite those positive trend lines, however, the way forward required further thought. Overstretch remained a concern, and so too the potential waning of American influence even in a new democratic age. Gates cautioned that Bush's new team was therefore "resisting the siren song of the quick fix and the big headline" while "trying to think about the 1990s."

If mismanaged, the potential end of history might yet go completely awry. "There is a new world out there," Scowcroft told colleagues in mid-February. They just "might stand at the door of a new era." Yet their charge as American leaders was to ensure not only international stability, he privately warned Bush, but also America's place within any new world to come. This was not mere selfishness; it was a matter of long-term survival. Marked and annotated in Bush's own hand, Scowcroft's note reminded the new president that whatever success the democratic world currently enjoyed derived first and foremost from American power. Americans maintained the peace in Europe after World

War II and won the Cold War through the long investments of time and resources by successive administrations, Democratic and Republican alike. Some
things transcend partisanship. The end of communism was not simply a global
victory, Scowcroft forcefully wrote. It was a validation of everything Americans believed and had sacrificed to achieve. "Soviet flexibility is, in large part,
a response to Western, especially U.S. policies," he noted. "Clearly, it would
be unwise to walk away from a successful strategy which has brought us to
this point. This is our agenda, not his [Gorbachev's], and we need to recapture
credit for it."

Enduring peace could be assured only through continued American vigilance, oversight, and strength. This reading of history, a coupling of democratic triumphalism with the centrality of American power, would drive almost
everything Bush's inner circle did in office. Nowhere was this historical lesson
clearer in their minds than in Europe, whose history seemed to offer an unend-
ing succession of wars and violence from the moment Rome collapsed until
the Americans showed up on the scene. More accurately, until the Americans
showed up and, in 1945 (unlike in 1918), refused to leave. There was simply
something innate to Europeans that made them unable to live peacefully with
their neighbors, Bush and those around him believed. The continent nearly
committed societal suicide twice in the twentieth century alone. Only American oversight had prevented a third try. Only American power deployed to the
far side of the ocean had beaten back fascism and kept communism at bay. If
the democratic end of history had truly arrived, the United States was its mid-
wife. It needed to remain its nurse and nanny.

American strategists had for generations quietly believed that their long-
term European cantonment, while publicly portrayed as defense against Soviet
invasion, had as much to do with restraining Europe's broader self-destructive
tendencies. This is a part of American history rarely taught in schools, yet it
was a central catechism taught to successive generations of postwar American
policy makers. "The history of the past two hundred years in Europe showed
that Western Europe would tear itself to pieces" without some form of outside
supervision, Secretary of State John Foster Dulles explained in the early 1950s.
Secretary of State Dean Rusk promoted the same line throughout the 1960s.
"Without the visible assurance of a sizeable American contingent," he warned,
"old frictions may revive, and Europe could become unstable once more." Richard Nixon's strategists believed the same in the 1970s. Beyond deterring Soviet

aggression, his National Security Council concluded, "the US commitment to Europe contributes to a stable relationship between those European countries into whose wars we have been drawn twice in this century." Translation: Americans were indeed in Europe to keep the Soviets out and the Germans down, but also to keep Europeans from one another's throats.

Bush and those around him believed with every fiber of their being that the Old World required oversight from the New, which is why they feared anything that threatened the Atlantic relationship. "The basic lesson of two world wars was that American power is essential to any stable equilibrium on the continent," Scowcroft reminded Bush in March. "The postwar era's success is founded on recognition of this fact." Baker was blunter when speaking publicly only weeks later. "Great sacrifices have been made," he noted in May 1989. "The burdens were — and indeed, the burdens still are sometimes — very difficult to bear." But in the end, because the United States stayed put after World War II, "we prevented for 40 years war in Europe."

The pronoun is what mattered. The "we" was not NATO. Neither was it a grand democratic alliance. It was the United States. Bush made this point as well only weeks into his administration. "We must never forget that twice in this century American blood has been shed over conflicts that began in Europe," he remarked with France's president at his side. "That is why the Atlantic alliance is so central to our foreign policy. And that's why America remains committed to the alliance and the strategy which has preserved freedom in Europe. We must never forget that to keep the peace in Europe is to keep the peace for America." The central lesson of the twentieth century, to Bush and those he gathered around him, was their own indispensability.

Perhaps needless to say, Europeans rarely considered themselves so wholly reliant on American oversight, and indeed European critics over time and with increasing frequency during the 1980s portrayed American militarism as the cause of, rather than the solution to, their ongoing security dilemma. Crowds numbering in the hundreds of thousands had only recently marched through Western Europe's great capitals protesting American plans to deploy a new generation of short- and intermediate-range nuclear missiles. Such weapons did not make them safer, they argued, but only exacerbated international tensions. Upward of 250,000 marched in London in 1981. An equal number joined across several Italian cities as well, and in Bonn in 1983. The Hague saw more than a million.

West German popular opinion in particular was running against Washington by the time Bush took office, precisely when Gorbachev was making his most impassioned plea for Europe to look anew to the east. Their country was only as wide as New York's Long Island, yet for decades it had been saturated with troops primed for battle. Only at the height of the Vietnam War were more American troops deployed to Southeast Asia than in Germany, leaving far more American veterans of the autobahn and the Fulda Gap than ever experienced the cities or jungles of Indochina.

Germans largely accepted the massive American presence both as a vestige of their penance for World War II and as a bulwark of their defense, but they increasingly balked at any enlargement of Washington's nuclear footprint, epitomized by plans to deploy updated versions of the Lance missile system, designed with a maximum range of seventy-five miles. Zealots even called for the Americans to remove the eighty-eight short-range missile launchers remaining on West German soil after the 1987 INF treaty took effect (even though the Soviets deployed more than 1,600 short-range missiles of their own in East Germany).

West German chancellor Helmut Kohl wavered under the pressure. Already struggling to hold together a loose governing coalition, he'd nearly lost his office during the early 1980s for supporting expanded nuclear deployment. Fearful of a similar backlash as the decade came to a close, Kohl was additionally influenced by the argument that the Lance's only real purpose would be in retaliation, in effect to devastate German towns and regions already overrun by Soviet tanks should the Cold War ever turn hot, leaving no chance for civilians to survive even if they somehow lived through the initial hostilities. "The shorter the missile," critics darkly noted, "the deader the German."

Recognizing an approaching political firestorm over the ongoing nuclear divide, Bush's incoming senior director for European and Soviet affairs on his National Security Council privately predicted at the start of the administration that the nuclear deployment in West Germany would be "the defining security issue of the next couple of years." The fate of NATO hung in the balance. It would rupture if the Germans, like the French a generation earlier, rejected American leadership and the reliance on nuclear arms that it entailed. It would rupture just as easily if the West Germans forced Bush to choose between his Pentagon's demand for a nuclear tripwire and his allies' desire for American troops stationed without their nuclear umbrella. "We need those wpn's [weapons] to defend our own troops," Baker wrote in his notes before a meeting with Germany's foreign and defense ministers that spring. The nuclear option allowed the "Pres[ident] of the U.S. to maintain U.S. forces in Europe." It was a point on which the new White House would tolerate no dissent.

The implications of a German withdrawal from NATO were nearly too horrible for Bush's policy makers to fathom. NATO would not long survive. Germany was the alliance's second-largest financial contributor and its largest manpower source, and home to some of its most vital installations. More fundamentally, without NATO the Americans would have no reason to stay in Europe, or at least no open-ended invitation, leaving the continent unsupervised and primed for Soviet domination should *perestroika* succeed, or for a resurgence of the dangerous nationalism of its past. Without NATO and without American oversight, in other words, Europe was destined to fall either under the Kremlin's control or into internecine warfare and chaos of a kind sure to drag Americans once more across the Atlantic to fight, and to die. The next time, they might not survive at all.

Germany was therefore central to American strategic planning as Bush took office, to democracy's advance, and to the retention of Washington's leadership role. As his National Security Council would put it in March, "Today, the top priority for American foreign policy in Europe should be the fate of the Federal Republic of Germany" because "the FRG's fidelity to Alliance security is uniquely indispensable to the success of that policy."

Bush inherited a fraying NATO relationship even as Gorbachev offered Western Europeans their first meaningful alternative to American leadership in generations. "Odd as it seems now," Washington insider Paul Wolfowitz recalled nearly twenty years later, thinking back to the time when he served as undersecretary of defense for policy, "in early 1989 the new administration feared that it was NATO — not the Soviet empire — that might collapse under the weight of Gorbachev's peace offensive." Gorbachev's weapons were words rather than tanks or planes, but Bush's inner circle reflexively believed that the Kremlin still desired the same dominant position within Europe that had animated his predecessors.

Bush's team was wary of Soviet designs, but not necessarily paranoid. Europe indeed remained Moscow's goal, just as it had been for the previous four decades, though *perestroika* prompted Soviet strategists to redefine their terms. They need not control Western Europe militarily in order to profit or to prosper, Gorbachev's inner circle came to realize. They could *join* Western Europe instead, ultimately profiting — perhaps even dominating — from within. Diminution of East-West tensions would save Moscow billions in military expenses, and there were billions more in profits and investments to be made on the far side of the Iron Curtain if barriers to East-West consolidation could somehow be overcome. Gorbachev would in time begin speaking of a "common European home" linking Eastern and Western Europe in not quite a confederation but a cooperative association, not fully socialist yet not imbued with cutthroat American-style capitalism either. It would be a Europe that would, at long last, accept the Russians as equals.

West Germany thus seemed to Gorbachev, no less than to Bush, the real key to continental influence, leading him to seek improved relations with Bonn even as he pondered the political transition in Washington. The Germans were the continent's most powerful state, too weak to dominate on their own, but too strong for any long-term stability without their consent. "First, there was the Federal Republic's key position in any future 'common European home,'" observed Gorbachev's aide Andrei Grachev, "which was gradually becoming one of the key elements of Gorbachev's foreign policy strategy in Europe. Secondly, West Germany could play an essential role in assisting the Soviet economy at a time when political perestroika started to face serious problems." In need of new markets and new sources of revenue, the same German industrial zones that once taunted Soviet military strategists now tempted Gorbachev's economic planners as well.

The feeling was mutual. Captivated by the prospects of untapped Soviet markets, West German bankers and businesses coordinated with their government to pledge billions in credits and future investments by 1988, capped by a personal reconciliation between Gorbachev and Kohl by the close of the year. The German leader had referred to Gorbachev as a latter-day fascist in 1986, likening him to Nazi propagandist Joseph Goebbels. Kohl's remarks "angered us to the depths of our souls," Foreign Minister Eduard Shevardnadze complained, while the official Soviet news organ lamented that Kohl's words had "poisoned" Soviet–West German relations "at a time when they were beginning to acquire a new dynamic."

Mutual opportunity helped patch over that wound two years later, leading Gorbachev to celebrate his West German counterpart during a visit to Moscow, little more than a month before he himself would travel to meet Bush and

Reagan in New York. "The ice has been broken," he told Kohl, who arrived at the head of a particularly large delegation including the chairman of Deutsche Bank and more than seventy top West German business leaders. Their nations shared a painful past, Gorbachev conceded, an understatement if ever there was one. But "we Europeans should, at last, behave in accordance with the logic of the new times: not to get ready for war, not to intimidate one another, not to compete in perfecting weaponry and not simply to try to prevent war, but to learn to make peace." These words were like music to the German electorate's ears, even as they stung in Washington. Gorbachev's own advisers simultaneously heard the clinking of valuable German deutsche marks, and the first notes of a new European order. They did not even critique Kohl's remarks in favor of a unified Germany and Berlin, a line that Soviet leaders of the past would have reflexively condemned. Let the Germans have their words and dreams of unity, Gorbachev's advisers reasoned. Everyone knew it would never happen.

WARY OF WEAKENING Atlantic ties and of the consequences of Gorbachev's warm words, the new administration made its first order of business the dampening of expectations, something Scowcroft went to work on the moment he arrived at his new West Wing office. He had a broader agenda in mind as well, which entailed nothing less than a restructuring of Washington's entire national security bureaucracy. Put simply, Brent Scowcroft came determined to clean up the mess left by Iran-contra, offering a more "mainstream" foreign policy rather than the fervor and ferment of Reagan's, a "tough, hard-headed, sort of power politics oriented — but with a relatively low ideological content." Their policies would be intentionally "unimaginative, perhaps," Scowcroft conceded in the ensuing weeks. But they would always tilt toward the preservation of stability.

He spoke from experience, having been Ford's national security adviser more than a decade earlier. No man before him had held that post twice, leading Scowcroft to hope for the top job at the Pentagon when Bush began filling his cabinet. He'd begun his government service in the military, having risen to the rank of lieutenant general in the air force despite being barred from flight training because of a severe back injury suffered during a crash soon after his graduation from West Point. "I prefer to call it a forced landing," Scowcroft subsequently insisted. Two years of hospitalization ensued, as did a career on the ground.

Together with Bush and Baker, Scowcroft formed the final leg of the administration's decision-making triad. One of these was not like the others, however. The two Texans were both tall and lean, brimming with the innate confidence a well-heeled upbringing can produce. Scowcroft, conversely, shied away from the spotlight. Elfin in stature, the son of a country grocer rather than a senator or big-city lawyer, he succeeded by working harder and being smarter than those born with better connections. He was the kind of man who saved time by running in the dark of night when the rest of the world slept, and who shopped, cooked, and laundered for his invalid and reclusive wife even after a long day in the White House. Bush and others in the West Wing who understood his stressful schedule largely gave him nothing more than a gentle ribbing about his resulting moments of "somnolent excellence" when meetings droned on.

They knew no one worked longer hours. "As Brent's wife became sicker during the administration," his principal deputy, Bob Gates, later recalled, "sometimes the President and I would conspire against Brent, and I would find a way during the day to let the President know that Jackie [Scowcroft] was in the hospital again and so the President would call Brent, maybe at 4:30 or 5:00 in the afternoon and tell him that he was going over to the residence, that he was done for the day. Then Brent would go to the hospital and I would call the President and the President would come back to the office and we'd do a couple more hours' work."

Grinding work had always mattered more to Scowcroft than flash or promise. Raised in scrubland at the foot of imposing mountains in Ogden, Utah, less a town than a railway junction, he earned first a coveted appointment to West Point, and then ultimately a Ph.D. in international relations from Columbia University, studying under Hans Morgenthau, one of the originators of the modern "realist" paradigm, whose work prioritized hard power over the power of ideas. This worldview fit Scowcroft's own, as did the idea that strategic thinkers and policy makers had something to learn from each other. "If I had been in either the Army or the Navy," both older services and thus more bureaucratically regimented, he recalled in retirement, "I could not have done what I did . . . But I was able to go from one job to another — challenging, interesting jobs, very unusual for a career pattern. And the Air Force did not hold it against me that I did not serve the normal number of times in operational units and so on." He spent much of the 1960s in the branch's long-range planning division and in the office of the secretary of defense before landing, in his words, "in a special small shop — which was the Air Force relationship to the Joint Staff — that worked the papers for the Joint Chiefs of Staff. It was [a] very key job." That role offered a view over the whole of the national security process to the military's top leaders. It was particularly well suited to a brain more comfortable with cold calculation than fiery bluster.

Fire lay just beneath the surface, however. Scowcroft's career took a dramatic turn in the early 1970s. While detailed to the White House, he stood up — literally — to H. R. Haldeman, Nixon's chief of staff, aboard Air Force One, inwardly fuming by some accounts, in others openly shouting. The topic of their dispute has largely been forgotten. Its impact was what mattered. Kissinger took note, struck by the image of a serving officer refusing to buckle under to political pressure even at the epicenter of American power. It was the kind of devotion to duty that Kissinger wanted on his own staff, which Scowcroft joined in 1973.

Bush and Kissinger in time grew to loathe each other, but Scowcroft bonded with each, first developing his friendship with the future president during Bush's time in Beijing. Scowcroft was his main contact with the White House, which, under Kissinger's Machiavellian reign, frequently preferred to keep the State Department out of the loop on sensitive relations with the Chinese. Determined to maintain an independent channel to President Ford, Bush employed the CIA's private communications network, orchestrated by its young chief of station in Beijing, a fellow Yale graduate named James Lilley, who decades later would play a crucial role as Bush's own ambassador.

Their back-channel correspondence helped engender a growing trust, which developed further during Ford's final year, when Bush headed up the CIA and Scowcroft took command of the National Security Council. The pair maintained their friendship through the 1970s and 1980s, when the latter worked first as a principal in Kissinger's lucrative consulting firm, and then later as a member of numerous blue-ribbon commissions on national security reform. Initially peeved when denied a formal position in Reagan's administration, Scowcroft came to appreciate this period outside the formal halls of power. "I had spent the '80s doing a lot of thinking about the process, government, how it ought to run," he recalled, "what the world was like, where it was going." He also avoided becoming enmeshed in Iran-contra. Instead, he co-chaired the commission formally charged with investigating the arms-for-hostages scandal.

Former senator John Tower of Texas lent his name to the wide-ranging investigation, but it was Scowcroft who authored the final report that harshly criticized the wholesale breakdown of order and judgment. "The arms transfers to Iran and the activities of the NSC staff in support for the Contras," the report concluded, "are case studies in the perils of policy pursued outside the constraints of an orderly process." Furthermore, "established procedures for making national security decisions were ignored. Reviews of the initiative by all the NSC principals were too infrequent. The initiatives were not adequately vetted below the cabinet level. Intelligence resources were underutilized. Applicable legal constraints were not addressed. The whole matter was handled too informally, without adequate written records of what had been considered, discussed, and decided."

Translation: the system had broken down. Scowcroft judged Reagan a dreamer, prone to big ideas and lofty visions. This did not, in his view, necessarily condemn the president to failure. Effective leadership demands vision. He could not forgive the chief executive's inattentiveness, however, nor the unprofessional work of his staff. Reagan was "an incurious President," he charged, "not interested in foreign affairs . . . not at all interested in details." Failing to care about details, he thus in Scowcroft's opinion too easily lost sight of principle when complex situations arose, and too frequently relied on professionals who let him down in favor of their own personal agendas.

Reagan's national security advisers — he ran through six in eight years — had taken it upon themselves to make rather than coordinate policy, and that was where the trouble began. That structure produced not only the scandals laid out in the Tower Commission report but (in Scowcroft's personal view) bad policies as well. Like so many others within Bush's inner circle, Scowcroft detested Reagan's embrace of Soviet reforms, believing that the president's personal dreams of disarmament outweighed any rational appraisal of Gorbachev's intentions, sincerity, or likelihood of success. The new Soviet leader talked a good game, Scowcroft charged in the late 1980s, but "atmospherics are not enough. We have to see Gorbachev not only saying that things are different . . . but actually see some actions."

Bush's presidency allowed Scowcroft the opportunity to turn his critique into actions. Speaking to reporters even before the last bits of confetti from the last inaugural ball had been swept away, Scowcroft delivered a careful, pessimistic message. Bush was right to be hopeful, Scowcroft said. The world was clearly moving in the right direction. But he also warned that this new administration intended to approach the open door to freedom cautiously. "There may be, in the saying, light at the end of the tunnel," Scowcroft allowed. But "I think it depends partly on how we behave whether the light is the sun, or an oncoming train."

He warned against putting too much stock in Soviet promises of reform, despite their frequency. Gorbachev "badly needs a period of stability," Scowcroft told the reporters gathered in his still unpacked office, "if not definite improvement in the [superpower] relationship so he can face the awesome problem he has at home of trying to restructure that economy." His next words were more controversial. More menacingly, Gorbachev seemed "interested in making trouble within the Western alliance," Scowcroft charged, hoping to divide NATO through a "peace offensive, rather than to bluster the way some of his predecessors have." All this added up to a singular conclusion, offered less than a week into the new administration by the man whose office connected to the new president's, and who would have more control over what passed through Bush's hands and across his desk than any other: "The Cold War is not over."

5

THE PAUSE

THE COMMUNIST WORLD SEETHED with unrest as Bush took office, though not because of him. Inspired and unleashed by Gorbachev's words, pro-democracy demonstrators in Czechoslovakia clashed with police. In Hungary the police watched warily as protesters strode by, but stood their ground. The Polish United Workers' Party meanwhile voted to legalize Solidarity, the first substantive opposition group in the country's postwar era, whose growing power had only a few short years before instigated imposition of martial law. In Washington, however, Bush hit the pause button, initiating a full-scale strategic review covering every aspect of American foreign policy. "I am interested in progress but I want to be prudent," he told reporters in December, revealing what he'd told Gorbachev during their meeting on Governors Island and would reiterate in January in a private letter. "My advisers responsible for national security and I will need some time to think through the entire range of issues," he wrote, before he could possibly meet the Soviet leader for face-to-face discussions. The process was expected to take months, which was its principal virtue, giving Bush nothing to do formally as he and his inner circle pondered their first crucial move. In truth, he did not fully know what to do, save that he wanted it done differently.

Baker's initial plans included cleaning house. "This is NOT a friendly takeover," he told Reagan holdovers hoping for continued employment following Bush's electoral victory, demanding letters of resignation from each current political appointee "in order to give the [sic] President-elect Bush maximum flexibility in the staffing of his administration." Publicly this purge sent the political message that "a new man was in charge," he explained, despite the surface continuity of one Republican administration taking over for another. Privately, however, Baker let it be known that the new president desired not just a different team but a different direction. "Don't you think you all went too far" in embracing Soviet reforms?, he caustically asked Roz Ridgway during his first visit to the State Department as the secretary of state designate. Ridgway had been lead negotiator for all five of Reagan's summits with Gorbachev. No one in the State Department was more closely associated with the embrace of *perestroika*. Baker made it clear that her services would no longer be needed. The first woman appointed assistant secretary for European Affairs, she would be retired within the year.

Gorbachev fumed in response. He cared nothing for the victims of Baker's purge, but he understood that Bush's prospective pause would consume the one commodity he could ill afford to lose: time. The longer Bush waited to endorse *perestroika,* and the longer foreign governments and investors waited to send aid and extend credit, the less likely his prospects for success. "What are they waiting for?" Gorbachev ranted to his staff. Bush "wasn't drawing the proper conclusions." Had he not read the United Nations address? Did he not understand that delay could spell the end of everything? By April the Kremlin's spokesman ominously told reporters, "Time has its limits," and so too did Soviet patience. Perhaps Bush wanted *perestroika* to fail, Gorbachev mused aloud to his inner circle, taking any hope of a Soviet resurgence with it, while he sat back and watched.

Jack Matlock wondered the same thing. One of the few Reagan appointees to remain in Bush's service, he retained his post as ambassador in Moscow yet remained far from the new president's orbit. Bush, he perceived, intended to rely on old friends and a close network of advisers before turning to anyone from "the bureaucracy." Yet it couldn't be that Bush didn't trust his own government, Matlock reasoned. He had too many years in its service for that. Those years seemed to have helped Bush learn to trust a few voices more than the rest — Baker, Scowcroft, Gates, Nixon — each of whom appeared in one way or another to believe that "perestroika was a giant hoax, designed to strengthen the Soviet Union militarily while disarming the United States and the West." Scowcroft had made that precise argument in his first post-inauguration interview. Yet from Matlock's Moscow vantage point, "the facts gave no support to this interpretation." Gorbachev was undoubtedly mercurial. But to Matlock's eyes, the Soviet leader was nothing if not sincerely committed to reform. More important, his opponents were worse.

Matlock argued as much to Bush in early March, in an Oval Office session that took weeks of lobbying merely to arrange, urging the new president to offer broad economic support immediately for struggling Soviet industries and an early superpower summit in order to reassure the communist world's reformers that Washington was on their side. Resigned to holding the meeting — he couldn't put his own ambassador off forever — Bush did not seem to want to be there at all. "Interesting," he blandly offered as the presentation concluded, more out of rote politeness than anything else.

There would be no summit. Of course he wanted to get to know Gorbachev better, said Bush, but Gorbachev would proclaim any summit proof of American approval, enabling him to declare the Cold War formally over and to make further strides in Europe's fickle court of public opinion without having enacted any meaningful change in the continent's strategic situation. Removing 500,000 troops was nice. But what about the rest? So long as Gorbachev had the momentum, the Americans would lose just by showing up, Bush reasoned, and lose in particular if the meeting failed to produce any dramatic results, with the press sure to pin blame for the absence of headlines on the new man. He would meet Gorbachev, but only when he was ready, and certainly not yet.

The new administration's initial marching orders couldn't have been clearer, Matlock informed his staff: "Don't just do something . . . stand there!" He was not alone in faulting Bush's White House for failing to capitalize on the forward momentum in Soviet-American affairs. Progress with Moscow had been won the hard way, by seizing opportunities, Ridgway later recounted. Like many Reagan acolytes purged in the transition, she focused her anger on the new president's inability to live up to their leader's example. Bush and those around him "damn near lost it" by "stopping everything and looking again," she said. "Maybe it was prudent" to pause, "but the fact of the matter is the process [of Soviet-American relations] had a rhythm to it, and had been moving right along" before Bush ordered it paused.

Yes, they had momentum, Scowcroft's staff countered, but did Reagan's strategy also have a goal? Forward motion in the wrong direction only left one lost, or worse. "A lot has happened in the relationship in an ad hoc way," Bob Gates complained when doling out bureaucratic assignments for the strategic review. For the past few years, "we've been making policy — or trying to — in response to what the Soviets are doing rather than with a sense of strategy about what we should be doing." It was therefore time to take a "conscious pause" while Bush's new team "looks over the landscape and reconsiders its position."

Hundreds of policy makers eventually took part in the exercise, drawn from the departments of State, Treasury, Defense, Agriculture, and Commerce, the Central Intelligence Agency, the Joint Chiefs of Staff, the Arms Control and Disarmament Agency, and the U.S. Information Agency. Scowcroft's National Security Council oversaw the process, tasking the bureaucracy with assessing three issues in particular. National Security Review 3 (NSR 3), the most straightforward of the three, called for a "comprehensive review of Soviet-American relations." Composed by committee, such directives were by tradition written as if from the president's own pen, though in this case Bush did in fact insert a specific question of his own. "What kind of relationship [with the Soviets] do we wish to see in the year 2000," he asked, and what steps could he take to improve Soviet-American affairs in the short and medium terms?

He wanted the answer within a month, along with their assessment of six interrelated points. First, to determine, once and for all, Gorbachev's ultimate goal. Second, to rate Gorbachev's odds of success, and in turn "the impact of both Soviet domestic and foreign policy if Gorbachev were to be removed from office." Third, policy makers were asked to predict likely Soviet moves in the international and military arenas, "for example, at what point might the Soviets feel compelled to intervene in Eastern Europe to suppress uncontrolled reform movements?" Fourth was a blunt question: Was it even in the American national interest to "'help' the Soviet leadership in the attainment of some of its internal and external objectives?" This was a crucial question indeed because it forced policy makers to consider one of Scowcroft's principal concerns: that success for Gorbachev, and in turn a more powerful Soviet state, might reignite a Cold War if changes in Soviet rhetoric outpaced any real shift in Moscow's strategic priorities. Fifth, Bush also asked policy makers to consider what sources of "leverage" Washington and its allies might have over events inside the Soviet Union. The document concluded with a plea for specific objectives and programs in the three-to-five-year and ten-year ranges, all designed to help answer a profound sixth question: "How can the U.S. get out in front of the

USSR in making proposals to affect the relationship?" Not only did Bush want to know what to do, but also he yearned to discover how he might reclaim the initiative.

The second National Security Review (NSR 5) focused on Europe, and on the challenges to continued American leadership in that vital region. Washington's military prowess had long given it political leverage, but only so long as the continent's citizens remained both divided and afraid, and thus believed themselves in need of American support. European unity and a new sense of security, however falsely derived from Moscow's apparent transformation, risked undermining what Bush's directive termed "the central importance of American leadership." He sought a list of "new initiatives that the U.S. might undertake to deal with popular perceptions in the Alliance of a significantly diminished Soviet threat." Again the question boiled down to this: What could Washington do to regain the initiative?

The official review of policy toward Eastern Europe, NSR 4, was the most profound of the lot. Noting that "the potential for real and sustained change in Eastern Europe is greater now than at any time in the post-war period," the directive further revealed how closely wedded the Bush administration was to several key ideas that had guided American policy makers for generations. These accepted truths were not only descriptive but prescriptive as well, providing the country's foreign policy establishment with a common framework for understanding the world while simultaneously offering instructions for continued success. Much like Bush's own core faith in freedom and free markets, these were there not to be interrogated but instead imbibed, consciously or not, because they allowed policy makers a consistent approach for tackling the critical issues of their day. NSR 4 noted, for example, that successive American governments wished "to see popular aspirations for liberty, prosperity, and self-determination" met behind the Iron Curtain. It simultaneously stated bluntly, "We are sure those aspirations cannot be realized as long as the Soviet occupation of East-Central Europe continues."

These statements were not as simple as they might appear at first blush. That Eastern Europeans wanted freedom the document's authors took as an obvious point of departure. Only the Soviets kept them in chains. The same could be said of peoples around the world: each wanted the freedoms Americans enjoyed. They only needed the chance to choose. When looking at Eastern Europe, therefore — though we shall see this same dynamic play out in other parts of the world as well—Bush's team began from the initial assumption that the region's people longed to be part of the West, and to rid themselves of their Soviet occupiers and the illegitimate governments the Soviets kept in power. Ultimately, it is not too much to say, they longed to be American. They would have freed themselves already, this logic ran, if only they could.

Previous generations of American policy makers had shared these beliefs, yet none had found any easy formula for promoting democracy behind the Iron Curtain without raising the likelihood of a Soviet crackdown or risking outright war. The term "Eastern Europe" is itself misleading, suggesting a false identity among peoples as distinct as Poles, Serbs, Slovaks, and Hungarians. Each bore its own history and relationship with communist rule, including a fact many wished to forget by 1989 (and after): their initial embrace of both Soviet liberation from their Nazi occupiers and of the socialism it brought. In Hungary and Poland, for example, nearly half the surviving populations favored some form of revolutionary change in 1945, hoping for elimination of Europe's archaic system of power politics that had for generations produced successive wars. As the Hungarian democracy activist and politician Gáspár Miklos Tamás put it, while "it is true that the Communist Party dictatorship was brought to the small East European countries by the victorious troops of Stalin," nevertheless "we should admit that we were ready for it." Czechoslovakia took three more years to fully join the Soviet camp (ultimately by means of a pro-communist coup). By contrast, having waged war against the Soviets, the East Germans, Romanians, and Bulgarians weren't given a choice.

Popular acceptance of communist rule differed greatly across national boundaries. Protests erupted against the ruling regime in Eastern Germany in 1953. Hungary boiled over in 1956, Czechoslovakia a decade later in 1968, while Poland offered the most visible example of anticommunist dissent by the early 1980s. Romania and Bulgaria, for their part, remained largely free of political strife throughout the same period, becoming more personal satrapies of their rulers than true socialist states. Yugoslavia, in contrast to all of its neighbors, proved to be beyond Moscow's complete control from the start.

In each case of visible protest, Moscow intervened to restore order and to reassert its rule. Soviet troops violently put down protesters in 1953, in 1956, and again in 1968, and only Warsaw's willingness to impose martial law precluded another Soviet crackdown in 1981. Western observers repeatedly, and to a large extent gleefully, interpreted such unrest as evidence of an enduring regional desire to overthrow Soviet control, leading to the widespread belief, largely impossible to refute for lack of evidence, that Eastern Europe's natural zeal for democracy required only opportunity. "In Europe," Reagan declared in 1987 in one representative example, "only one nation and those it controls refuse to join the community of freedom." Everyone knew of whom he spoke.

Yet the Soviets were too powerful for the United States to confront directly in their own strategic backyard, leaving American policy makers little to do but watch with horror whenever the Red Army crushed dissent. Sometimes they watched with a sense of guilt as well. Hungarians rose up in 1956 at least in part with the expectation of direct American aid, having heard Secretary of State John Foster Dulles call for "rolling back" communism and Voice of America broadcasts that seemed to promise aid for any who took up the cause. President Eisenhower squelched any such hopes once fighting erupted, however; he feared the potential consequences of wading into the fray behind the Iron Curtain. Hungary was "as inaccessible to us as Tibet," Ike complained, adding that any attempt to deliver aid to Hungarian insurgents was sure to lead to a broader Soviet-American conflict, perhaps even a nuclear one. "Those boys" in the Kremlin were "furious and they're scared" at the prospect of their new-found empire unraveling, Eisenhower reluctantly concluded. "And just as with Hitler, that's the most dangerous state of mind they could be in."

Left without arms or equipment, Hungary's revolutionaries stood no chance against Soviet tanks. Upward of 30,000 died in the ensuing crackdown. Another 100,000 fled. "Our troops are fighting. Our government is in place. I am making this fact known to our people and the world," Imre Nagy, leader of the short-lived revolutionary regime, pleaded even as the Red Army slowly surrounded his office. Where was their promised support?

Three hours later his broadcasts ceased, leaving little doubt who was now in charge in Budapest. Nagy was captured, tried, and executed. His fate "should be a lesson to all other leaders in socialist countries," Soviet premier Nikita Khrushchev warned while snidely cautioning would-be revolutionaries against believing American promises, which he called "rather in the nature of the support that the rope gives to a hanged man." American journalists reached a similar conclusion. Their government had not promised aid in the technical sense, but technicalities mattered little to desperate people who eagerly heard what they wanted to hear. "The real lesson," *Time* editorialized, "was that the U.S., for all its sympathy (a quality easy to ridicule when it is not backed up by

something stronger) was not prepared to go to the rescue of an armed uprising in any satellite. On the technicalities the U.S. might not be guilty of false
encouragement, but could hardly be happy to leave it at that." Hungary's tragedy illuminated a fundamental problem for Western policy makers: It was one
thing to defend democracy, another to go to war to extend it.

A similar story of protest leading to repression, and of American encouragement turning to helpless sorrow, replayed in every subsequent decade. Writers
and students proclaimed a "Prague Spring" in 1968, promising greater political
and social liberties, perhaps even self-determination in meaningful elections.
Westerners cheered. Then the tanks rolled in. More than 200,000 Warsaw Pact
troops entered Czechoslovakia during the night of August 20–21, 1968, with
the Soviet Red Army in the lead. Hundreds of civilians were killed, thousands
imprisoned. Tens of thousands fled west. Soviet leader Leonid Brezhnev declared an ominous doctrine as justification: no state could ever leave the communist family. Because no reasonable or sane people would ever reject socialism, any that did must be either deluded or unwittingly under assault from
counterrevolutionaries. In either case they would be forcefully saved for their
own good. This "Brezhnev doctrine" guided Soviet policy until the late 1980s.
Soviet tanks remained in Czechoslovakia until 1991.

The same story was repeated in Poland in 1981, with one important exception. Protesters took to the streets demanding reform, and the government appeared ready to topple. Just across the border, rows of Soviet tanks revved their
engines, prowling back and forth like great cats ready to be sprung from their
cage, lest the Poles fail to get the message. Documents subsequently secured
from Kremlin archives reveal that few within the Soviet leadership were eager
to cross the border, fearing international condemnation and further drains on
their strained forces already engaged in Afghanistan. Their bluff nonetheless
worked. Warsaw cracked down on its own. "You must not compare surgery
with psychiatry," the head of Poland's military explained. To avoid the "catastrophe" of invasion, he had chosen instead "the lesser evil."

American planners knew this pattern of history all too well and understood
that anything they did to encourage reform increased the odds of a violent
Soviet response. "The thing that bothers me," Reagan had lamented to his national security team as they sat in the White House Situation Room monitoring
reports of Poland's imposition of martial law in 1981, "the constant question is
—that we continue to deplore, but isn't there anything we can do in practice?"

No good solution seemed forthcoming but words of disapproval and condemnation.

Bush had been in that frustrating meeting as well. Like the rest of Reagan's team, he had little to offer in response, save to inject his typical note of caution. "Set the tone, say what you have done, but stop short of details," he advised Reagan. Condemn, but don't act. And don't be specific. "This is not a weak position. It is a responsible position." Just as in 1939, Poland was once more resigned to its geographic fate. Too close to the West to be ignored when threatened, it remained too far away to be helped with anything more than words.

Now, eight years after those frustrating sessions debating a response to Poland's martial law, Bush was in charge, and yet another "breeze of freedom" seemed ready to blow through Eastern Europe. But was it destined to replay the tragedies as well? Only two days before Khrushchev ordered his tanks into Budapest, Eisenhower had boasted of "the dawning of a new day" in Eastern Europe. Mere days before the crackdown in 1968, the Central Intelligence Agency told Lyndon Johnson that the Czechs "seem to have been able to preserve the essential substance of [their] democratic experiment." Now Gorbachev appeared willing to transform this same Soviet state that had repeatedly brutalized its neighbors, but it remained unclear if he could keep this promise. The region was the Kremlin's "end zone," Bush reminded Canadian prime minister Brian Mulroney in February. We must not "stir up revolution," he warned, lest the awful pattern of reform and repression repeat itself once more. The last thing he wanted to do as president was to stir up anticommunist protesters to the point where they faced violent repression, he told Germany's president only weeks into his term. He "recalled the tragedy of Hungary in the 1950s" and "did not want to exacerbate problems" behind the Iron Curtain. "The traumatic uprisings in East Germany in 1953, Hungary in 1956, and Czechoslovakia in 1968 were constantly on my mind through these tumultuous months," Bush later recalled. "I did not want to encourage a course of events which might turn violent and get out of hand and which we then couldn't — or wouldn't — support, leaving people stranded at the barricades."

This history notwithstanding, Bush's early 1989 strategic review noted, "the time may have come when creative American policies can make a more significant difference" in promoting freedom in Eastern Europe." Previous generations had been unable to liberate the region from communist control, but this time seemed different. Never before had the region's uprisings originated in Moscow. If 1953, 1956, and 1968 all proved the innate desire burning within Eastern Europe for democracy and freedom, the flame only extinguished by overt Soviet force, then 1989 might finally be the moment when freedom would blaze throughout the Soviet bloc, because it was the Soviets themselves who'd struck the first match. But would they let it burn? Thinking once more in terms of short (three-to-five-year) and long (ten-year) increments, Bush demanded a list of suggestions that would "address how to balance encouraging pluralism and greater popular participation on the one hand and the potential for dangerous unrest and instability on the other." Bush personally framed the question more directly as "figuring out exactly where that line was, and what was likely to be seen by the Soviets as provocative."

These were not easy questions to answer, and the clock was ticking. Britain's Margaret Thatcher used the occasion of her first phone conversation with Bush in the Oval Office, the type of call typically more congratulatory than substantive, to urge no delay in continuing Reagan's policies. Speed was of the essence lest "the euphoria" surrounding Gorbachev swell beyond control, she said. This was among "the most pressing problems the West had to tackle," with the allies' ability to offer a suitable alternative crucial to "maintaining a unified Western position on East-West issues."

Thatcher was in a tough spot. Increasingly unpopular at home after nearly a decade in office, she had backed Gorbachev and declared the Cold War over. Bush was not sure about either of those positions. She also was Reagan's friend, which made her further suspect in Bush's eyes, given his desire to create his own identity beyond his predecessor's orbit. Yet Thatcher knew, as would any postwar British prime minister, that her own international influence derived in large part from the presumption of a special relationship between the United States and the UK, and in particular between Downing Street and the White House. She consequently took great pains to compliment Bush during that first phone call, praising his "past statements for dealing with the Soviets," which "were right on the mark," even though they contradicted her own. She additionally authorized her foreign secretary to urge that Gorbachev be held to account for his promises, just as Scowcroft and others in the administration desired, demonstrating that at every level of the British bureaucracy there existed a similar desire to keep Anglo-American relations strong. "We must not confuse hope or even expectation with reality," Geoffrey Howe, Britain's top diplomat, consequently urged in January during a Whitehall address. "The Soviet Union has a well-stocked hat full of well-armed rabbits . . . and will go on surprising us by drawing rabbits from that hat for many years to come."

Other American allies, especially key players in NATO, feared Bush's pause would be counterproductive, making it all the more difficult to wrest the strategic initiative back from the Soviets. Presidents, not general secretaries of the Soviet Union, typically drove NATO's agenda, Germany's Wolfgang Schäuble complained during an Oval Office session. A valuable voice within Helmut Kohl's inner circle, he thought the alliance needed a "comprehensive concept of dealing with Gorbachev," which could, for all practical purposes, originate only in Washington. If Kohl felt compelled to move first, Schäuble left unstated though clear to everyone in the room, he might not be able to take the rest of the alliance with him.

Bush pleaded for patience. "Each time Gorbachev makes a forthcoming, interesting speech it was all the more important for the alliance to stay together," he implored Schäuble. "Convey to the Chancellor [Kohl] that our policy reviews were not born out of fear but instead rested on our judgment that there were new opportunities in East-West relations." American initiatives would be both comprehensive and well worth the wait, he promised, adding, "We are in a thoughtful mood here in Washington."

Bush assiduously worked the phones throughout his first weeks in office, calling world leaders at scheduled and unscheduled times, during moments allotted for his lunch or merely when a hole unexpectedly appeared in his day. Determined to build trust by listening, he heard a near constant lament that the pause was being perceived in Europe as a vacuum of American leadership, something he was at pains to dispute. "I don't want to be cast as the person stalling better East-West relations," he told Thatcher, similarly admitting to Denmark's prime minister that he was tired of "playing defense all the time" when it came to Gorbachev. Even Bush's closest international supporters pleaded for action. "They've got us figured out — we don't have them figured out," Mulroney privately complained to Bush about the Soviets in mid-February. It was Bush's first foreign trip, testament not only to the intimate Canadian-American relationship but also to Bush and Mulroney's mutual affection. Having become friendly earlier in the 1980s, the pair would only become closer in the months and years to come as the Canadian offered Bush a crucial international sounding board, especially when Atlantic and NATO issues arose.

Mulroney was therefore not shy about voicing his concerns, even as their partnership first developed. "They're outperforming us in [the] geo-political sense," he warned Bush less than a month into his friend's presidency. "They [Gorbachev and his advisers] sit in Moscow and go right to the heart of our weaknesses like you did to Dukakis with the Boston Harbor speech!" It was a reference to one of Bush's most successful campaign maneuvers, when he'd painted the Massachusetts governor as incapable of even cleaning up his own state. Washington needed new initiatives, Mulroney implored. It needed to lead, not merely to react.

Washington's key European allies did little to hide their growing impatience. "We understand the need for a transition period, but enough!" an unnamed cabinet member from a "major European ally" told the New York Times. "The Russians have pulled out of Afghanistan, they're making unilateral reductions in Eastern Europe and Gorbachev is popping up everywhere. Bush has to start taking some initiatives of his own, or the Soviets will gain the strategic advantage." The Soviets might even move to enhance their influence in new areas, such as the Middle East, where Foreign Minister Shevardnadze was soon to begin a widely publicized nearly two-week trip, the longest by a Soviet foreign minister in recent memory.

Bush had no choice but to confront these criticisms directly in mid-February when reporters put them squarely to him at a press conference. "There is widespread perception that you don't have a foreign policy," famed White House reporter Helen Thomas began, "that you have permitted the Russians to move into the vacuum in the Middle East . . . that your go-slow attitude really says: let's let the Russians grab the ball." It was less a question than an articulation of the growing consensus in Washington and beyond.

"Well, I never heard such [an] outrageous hypothesis," Bush responded, chuckling. "We are reviewing appropriately," he added after the laughter died down, though he had nothing more to say. Longtime diplomat Richard Holbrooke, a leading foreign policy voice in Democratic circles, offered an emblematic critique. "Largely because of Gorbachev," he said, "everything is in motion everywhere, except in Washington."

The White House expected Democrats to complain. More troubling was criticism from Bush's right. He'd never embraced that wing of his party, and they'd only grudgingly participated in his electoral coalition. Willing to give Reagan free rein to negotiate with the Soviets—they never doubted his conservative bona fides—they now critiqued Bush for exhibiting the same level of caution toward the communist world that true conservatives so often advised. He was already to their minds too bipartisan, especially after Baker spoke enthusiastically about the need for reaching across the aisle to ensure "continued American leadership" in the world. Reagan had famously done much similar reaching of his own, but many partisans preferred hagiography to history. Deputy chief of staff Andy Card warned that spring, for example, that the lobbying group Gun Owners of America was circulating a letter critical of Bush's policies among key conservative leaders. Their chief complaint had nothing to do with firearms but instead centered on "the president's soft attitude in dealing with the Soviet Union." They did not specify which of Bush's policies they were criticizing. Instead, his general "attitude" had raised their ire. Conservative leaders gathered in early February at the Council for National Policy meeting in Orlando offered the same complaint, a White House staffer informed chief of staff John Sununu. They were "increasingly agitated about the direction of George Bush's presidency" and were "irritated with the lack of vision and direction" for dealing with the Soviets. Bush had been in office less than a month, yet he was already being subjected to the same critique that had plagued his entire career: that he was indecisive, a weathervane, and, yes, a wimp.

The cold truth is that Bush's team sensed an opportunity but lacked a plan. "There are remarkable changes taking place," Baker advised Bush in early February, but "it's too soon to know whether Gorbachev will succeed." The Soviet leader faced tough political battles and hard choices ahead, but "we do Gorbachev no favors when we make it easier to avoid choices." In the final analysis, "his success doesn't depend on us," and "we can affect his prospects only on the margins." One thing he believed Washington could do was to quietly force Gorbachev to add substance to his rhetoric. "We must challenge him with our bold initiatives — e.g., on conventional arms reduction, on chemical weapons, on missile proliferation, and other transnational issues. Let's honestly probe, and let's challenge him to be bold in actions, not only in words."

Yet they had no bold initiatives of their own, leading to growing exasperation in the White House when the Soviets repeatedly announced new programs or reductions. They seemed to be "searching — searching for ideas," Baker complained to Bush after his first meeting with Foreign Minister Shevardnadze, "debating what's possible, and scrambling to come up with initiatives." Gorbachev's cadre didn't seem to care if any particular program worked, only if it added to the general deluge of new ideas that provided the impression of activity. There was "no question Gorbachev['s] initiatives are having [an] effect in Europe," Baker warned, which "puts a premium on new features or elements in our own proposals."

"You're right," Bush told Mulroney at nearly precisely the same moment. "We must take the offensive" and "not just be seen as reacting to yet another Gorbachev move." The Americans were losing the public relations battle and had "to do it to keep public opinion behind the alliance." But how? Results of the pause were still weeks away at the end of February, as Bush prepared for his first extended overseas trip. Unsure what to do about Europe, he flew instead to Asia. It was the first time in generations a new president had crossed the Pacific before the Atlantic, and even as Bush confronted Asian allies and adversaries on the brink of tectonic shifts of their own, the uncertainty about what was going on behind the Iron Curtain remained paramount in his mind.

6

"A SPECIAL RELATIONSHIP THERE"

THE CHINA BUSH VISITED in 1989 was little like the one he'd first known as ambassador, though still reeling from its first communist leader's influence. "Ours will no longer be a nation subject to insult and humiliation" from imperialists and those who yielded to them, Mao Zedong proclaimed in 1949 as the American-backed Nationalist regime he'd fought for decades fled across the water to Taiwan. The Chinese people, he declared, "have stood up."

Victory had brought little tranquility, however. Within a year, Chinese and American troops were slaughtering each other in Korea, and an angry mutual antagonism thereafter permeated every aspect of Chinese-American relations. Policy makers in Washington refused to recognize Mao's regime. Their counterparts in Beijing simultaneously castigated the United States as the leader of the imperialist forces who'd never rest until China was once more under their control. Only after its people "threw off the yoke" of communist rule, Dean Acheson had earlier stated, could the hundreds of millions of Chinese men and women who populated the mainland be allowed to rejoin the family of nations.

Successive American administrations followed his lead, curtailing all trade and cultural contacts, and badgering allies to do the same, hoping isolation might wreck China's economy, leading to discontent, dissent, and then popular revolt. China was too big to reconquer otherwise. At the same time, Mao's clique appeared too dangerously fanatical to merely contain, as seemed possible for the Soviets, whose quest for global revolution made them belligerent but apparently not suicidal. Bring on nuclear war, Mao once demanded instead. Socialism would triumph from its ashes. Each side might lose hundreds of millions of lives. China had hundreds of millions to spare.

Personal snubs followed as well. Meeting in Geneva in search of a diplomatic solution to East-West strife in Indochina in 1954, Secretary of State John Foster Dulles warmly welcomed his French and British counterparts, and politely if stiffly greeted those from the Soviet Union and North Vietnam, yet rejected Chinese foreign minister Zhou Enlai's outstretched hand. The Chinese remembered, as did Richard Nixon, whose historic visit to Beijing in 1972, and the ensuing reopening of relations frozen since 1949, began when the president descended the steps of Air Force One with his hand dramatically extended in friendship. That "week that changed the world" restarted trade and other relations between the two countries and helped solidify China as a tentative American ally against the Soviets, who had long since wearied of partnering with revolutionaries they could not trust. Nixon called this move to isolate Moscow and ensure the long-term fissure within the communist world "triangular diplomacy." Mao called it survival. "Didn't our ancestors counsel negotiating with faraway countries while fighting those that are near?" he responded when underlings questioned his willingness to deal with the Americans. The same applied to his critics at home, weary of his failed economic and political projects. China's Great Leap Forward of the 1950s brought more famine than industrialization. His ensuing Cultural Revolution of the late 1960s and beyond, designed to purify Chinese communism from the stain of Western and Soviet thought, instead disrupted millions of lives and nearly brought down the entire state.

Bush knew this history, and Mao's primary successor as well. Ever since he surprised his friends and family by requesting to serve in Beijing in 1974, Bush had worked hard to forge a relationship with Deng Xiaoping. He traveled multiple times to China in the late 1970s as a tourist and for business, always making time to renew acquaintances. Deng returned the favor in 1979, stopping in Houston to see Bush (and to take in a rodeo) as one of only a few destinations on his American tour that included a visit to the White House to formally establish diplomatic relations. Only Mao had a greater influence on modern China than Deng, a veteran of the legendary Long March in the midst of his country's civil war, the Chinese equivalent of having served under Washington at Valley Forge. Deng's organizational talents were recognized at a young age. So too his nationalism. Awarded a coveted scholarship to study in France, and then trained both in a trade and in communist theology in a small metalwork factory once his exchange program ran out of funds, by the age of twenty, and after further study in Moscow, the young man had become, in the words of his best biographer, "a hardened and experienced revolutionary leader" whose "personal identity had become inseparable from that of the [communist] party."

Deng rose quickly through the ranks after returning home to join the communist struggle for power, served as a Red Army political commissar during the war years, and was trusted to act as one of Beijing's two emissaries to Moscow during the Hungarian crisis of 1956. He was there to witness Soviet officials debate whether to forcefully repress Hungary's uprising, joining in condoning the Soviet crackdown while also participating in his delegation's withering appraisal that Moscow's "big-power chauvinism" was at least partly responsible for unrest throughout the Soviet bloc. A lightning rod for both change and criticism, with an innate ability to be in the thick of things when great decisions were made, he was twice banished to the countryside as political winds shifted. Few Chinese leaders survived a single exile. Deng returned each time with new resolve, ultimately winning the struggle for party dominance after Mao's death. After 1978, no one in the country wielded greater power.

Deng used that power to implement ambitious reforms once deemed dangerously heretical. Witness to the deprivations wrought by Mao's anti-materialist rhetoric, he successfully argued that the only way to achieve "socialism with Chinese characteristics" was to incentivize achievement. Why was it, he asked mockingly, when people were hungry and the state demanded greater production, that a farmer who successfully raised three ducks won praise as a good socialist, but with five he was persecuted as a capitalist? The Chinese "mustn't fear to adopt the advanced management methods applied in capitalist countries," he urged, but should instead "seek truth from facts," employing whichever policies worked best regardless of doctrine or dogma. "It doesn't matter whether it is a white cat or a black cat," he famously argued. "A cat that catches mice is a good cat."

China's economy surged in response. Growing at nearly 9 percent a year under Deng's tenure, overall gross domestic product expanded from little more than $100 billion annually to nearly four times that amount by the close of the 1980s. This growth mirrored a rise in China's overall standard of living. Per capita consumption for workers in non-agricultural fields grew nearly 6 percent annually during the same period. Annual gains for agricultural workers approached double digits. Per capita income for the country as a whole tripled. Exports increased tenfold. Overall China became wealthier with Deng in charge, with all the attendant increase in aspirations for education, health care, and ultimately political participation.

Deng had not encouraged this last development. He was no democrat. Like Gorbachev, he wanted to save socialism, not replace it. Unlike his Soviet counterpart, however, who loved nothing more than a contest of ideas, Deng had little patience for debate. While still consolidating his power in 1978, for example, he tolerated additions to Beijing's "Democracy Wall," erected as a forum for citizens to air their grievances, until contributors began assailing his own policies. "Centralized power flows from the top down," he had written when studying in Moscow. For the communist movement to succeed, he considered it "absolutely necessary to obey the directions from above." Prosperity was good. Popular rule was not.

Bush considered this mercurial man a personal friend, or at least a long-time acquaintance. They made an odd-looking pair, the lanky Texan next to the gnome-like Deng. Claims that the latter stood five feet tall were generous. "He was a short man," Bush dictated in his diary after their first conversation in 1974. It was most often the first thing that leaped to mind when one met Deng, whose feet would dangle in the air whenever he sat in the side-by-side overstuffed chairs Chinese officials employed for formal diplomatic discussions. The smoke-scarred rasp in his voice exaggerated his Sichuan accent from deep in China's hinterland, further highlighting the differences between the two men. Bush hailed from urban privilege, embraced the dogmas of his youth as reverently as Deng questioned his, and forever longed to forge the kind of frontier political identity Deng never fully shed.

Their personal paths to power could hardly have been more dissimilar, yet from their first encounter each nonetheless recognized something worth watching in the other, in particular the likelihood that each would remain a key player in their respective national politics. "Because I lived there [China]," Bush told Deng during his brief return to private life after leaving the CIA, "I understand as well as anyone in the U.S." the problems Chinese leaders faced. Therefore, he hoped that "maybe someday there will be a useful role to play" in keeping Sino-American relations strong.

Reagan inadvertently provided just such an opportunity in 1980, immediately after formally accepting the Republican Party's presidential nomination, when Beijing took offense at his unabashed support for Taiwan, prompting the campaign to quickly detail Bush to repair the breach. Political cartoonist Patrick Oliphant captured Bush's role perfectly, picturing him astride a bucking bronco with Reagan's head, having just crashed into a china shop. "Take it easy, gentlemen," he depicted Bush telling its irate owners. "I can explain everything." Similarly tasked with explaining Washington's policies toward Beijing throughout the 1980s, and in particular with negotiating with Deng directly over American military sales to Taiwan, Bush proved his partner's clear favorite in the contest to become Reagan's successor. "We had a lot of contact with him [Bush]," Deng told American reporters in 1988. "I hope he'll win the election."

Bush's aides were thus little surprised following his election when he wondered aloud about crossing the Pacific for his initial substantive overseas trip. Cold War presidents typically visited Europe first, reinforcing the region's strategic importance. Yet Bush wanted not only to recognize his personal ties to China but also to highlight that the new "breeze of freedom" he perceived in Eastern Europe was blowing through Asia as well, where democracy indeed seemed on the rise. The kleptocratic regime of Ferdinand and Imelda Marcos had fallen to wholesale calls for political reform in 1986, for example. Long-term American clients, the pair ultimately flew into exile aboard a U.S. Air Force jet, taking hundreds of millions of dollars with them. Filipinos had a new, duly elected president within a month.

Pro-democracy demonstrations had similarly rocked South Korea's authoritarian regime in 1987, with American influence once more playing a critical role in a democratic transformation. Marchers chanting for reform swelled into the millions in mid-June. Police ranks swelled in response, poised to disband the movement by force. They'd done so before. This time, however, American ambassador James Lilley, the same Lilley who had worked with Bush a decade before in Beijing, was among the foreign voices who told the ruling regime it was time for a change. Backing up Lilley's words, the ranking American military commander on the peninsula hinted publicly at plans to prevent, by force if need be, South Korean military reinforcements from entering Seoul if violence erupted. With more than fifty thousand well-armed troops at their command, the Americans were not simply going to stand aside as protesters were slaughtered in the streets. The regime blinked. By October, South Koreans had a new constitution and direct elections.

Seemingly every place in Asia the Americans touched generated in time an indigenous democratic surge. South Korea, the Philippines, and also Taiwan, where martial law lasting almost four decades was lifted in 1988, opening a new phase in the island's transition toward democracy. Perhaps China might be next? Bush claimed as much when accepting his party's nomination. "The spirit of democracy is sweeping the Pacific Rim," he boasted at the Republican convention. "China feels the winds of change . . . And one by one, the unfree places fall, not to the force of arms but to the force of an ideal: Freedom works."

Perhaps. Or perhaps just as in the Soviet Union there remained time for communist hard-liners to beat back the wave of democratic reform before it swept them fully from power. Bush's optimistic words struck fear in the hearts of China's ruling class, who had never signed on for democracy. Only for prosperity. The future of reform, in essence China's prospects for a more democratic future in line with its more market-focused economy, was therefore not yet so fully entrenched that reversal was impossible to fathom in 1989. The men who brought China this far might well determine that the country had gone far enough. Bush believed that nothing could hold back democracy forever. But he was far away, and they were in charge.

For the country that was home to nearly one fifth of the world's population, this was no small matter indeed. "The importance of China is very clear to me," Bush told former national security adviser Zbigniew Brzezinski in December after his election. "I'd love to return to China before Deng leaves office entirely. I feel I have a special relationship there." He knew Deng and believed he sincerely desired to better his people's fate, just as he trusted that China, like all places, would eventually turn fully democratic so long as its current reforms were not knocked off course. Perhaps the best way to keep China's internal transformation on track, Bush calculated, was to continue its international reintegration, in essence expanding the engagement policy initiated in 1972 by his mentor and largely supported by every president since. "We have been looking at this area almost exclusively in a bilateral relationship," but it would be better to bring China into a "Transpacific Partnership" involving all the major Pacific powers, including Japan and South Korea. Both were democracies, and allies, and thus both prime examples of the potential and practicality of the new breeze of freedom in Asia.

"Frankly," he continued in his post-election missive to Brzezinski, who had become a valued sounding board since backing Bush for the White House despite his long-term affiliation as a Democrat, "I think such a summit [with China] would help with Europe in a perverse sort of way." Moscow longed to improve its relations with Deng's regime as well as with the West, reducing the costs of the Soviet Union's other Cold War front, while, it was hoped, profiting from the country's growing economic opportunities. Gorbachev planned to visit Beijing in mid-May. As with Nixon a generation before, this first visit by a Soviet leader in decades was designed to mend fences, leaving Bush increasingly eager to stake his own claim to Beijing's allegiance before the peddler of *perestroika* could make his case for stronger Sino-Soviet ties of the sort he had already promised Western Europe.

Fate provided just the opportunity to break with precedent by visiting Asia first which Bush required. Emperor Hirohito of Japan died on January 7, 1989. He'd be buried in February, coincidentally before Bush's first anticipated trip across the Atlantic. On the throne for nearly sixty-three years, Hirohito had overseen some of the flushest times in his nation's history. Also the worst. Japanese soldiers and sailors, fighting in his name, had conquered much of the Pacific during World War II, leaving bitter memories of occupation, rape, and carnage. His navy had attacked Pearl Harbor, and his nation had fought the United States the longest of any of the Axis powers. Yet he remained the only Axis leader to retain power after the war, because his American conquerors considered him useful. "The emperor system should continue so long as the occupation does," General Douglas MacArthur's staff recommended. Hirohito, they explained, could keep his people calm. It was better to deal with a compliant possible war criminal than to risk unrest.

The decision still rankled nearly half a century later. He "should have been shot or publicly chopped up at the end of the war," New Zealand's defense minister raged upon hearing of Hirohito's passing. South Korean voices were equally critical. "No matter how much Japan denies his responsibility," editorialized *Dong-a Ilbo*, a leading Seoul newspaper, "it won't be able to deny the fact that wars were declared in his name." Japanese prime minister Noboru Takeshita's ongoing refusal to accept Japanese culpability for the war didn't help. Assigning guilt was "up to the historians," Takeshita argued. South Korean students stoned the Japanese embassy in Seoul in response. In China, crowds burned Japanese flags. Australian prime minister Bob Hawke renounced plans to lead his country's official delegation to Hirohito's funeral. New Zealand's prime minister did the same. Britain's government downgraded its own delegation in response to public outcry, replacing heir to the throne Prince Charles with his father.

Bush accepted his invitation without hesitation. "I know I'm doing the right thing, to represent the United States of America at this funeral," he told reporters. "We have a strong relationship with Japan," he explained when pressed to justify his decision. "And what I am symbolizing is not the past but the present and the future by going there." Determined to focus on the relationship he hoped to build, and aware that Japan's rising power made it an increasingly indispensable ally as well as the object of American anxiety, he could not help but feel the weight of the past once seated for the lengthy funeral rites. "I can't say that in the quiet of the ceremony," he told reporters, "my mind didn't go back to the wonder of it all, because I vividly remember my wartime experience. And I vividly remember the personal friends that were in our squadron that are no longer alive . . . But my mind didn't dwell on that at all. And what I really thought, if there was any connection to that, isn't it miraculous what's happened since the war. [In Japan] we're talking about a friend, and we're talking about an ally."

Not everyone was so willing or eager to move on. "I expect some families who lost loved ones in World War II might not share my view on the importance of reconciliation," he wrote a trusted friend years later, "about forgetting the brutal past; but given the importance of the US-Japan relationship and Japan's commitment to democracy and freedom I am sure I am right . . . Besides, isn't it good to heal old wounds?" Sixty-three percent of Americans polled approved of his attending Hirohito's funeral, though the rate was a mere 41 percent of those over the age of sixty-four. His interest in forgiveness thus outpaced that of many in his generation, of whom more than a third admitted to harboring ill feelings toward the country that bombed Pearl Harbor, and toward its emperor. (Nearly half of Japanese polled reported similar hard feelings toward the nation that atomized Hiroshima and Nagasaki.) "He knew if anybody was going to bury the hatchet," Bush's communications director later recalled, "he had the credibility to do it."

The trip also offered precisely the opportunity of a quick visit to see Deng that Bush desired. Too quick for some. Newly hired White House staffers unused to travel procedures left a CNN camera crew behind when Air Force One made a brief refueling layover in Alaska. Such mishaps notwithstanding, the trip put Bush in Beijing nearly two full months before Gorbachev, whose presence dominated the president's agenda. The "key purpose" of your visit to Beijing, Baker advised the president, is "to consolidate your personal ties with

China's leaders" in order to "offset [the] impression that Sino-Soviet rapprochement is at our expense." Keeping the Chinese leaning to Washington's side despite *perestroika*'s appeal was what mattered most. The American ambassador
in Beijing was more direct. Soviet reforms put each of Moscow's relationships
in a new light, Winston Lord and his embassy staff believed, and therefore Bush
needed to "obtain Chinese assurances that Sino-Soviet relations will not upset
the strategic balance."

Lord had even more to say privately. A decade younger than Bush, Lord had
taken part in Kissinger's first secret trip to China in 1971 and had accompanied
Nixon the following year. Wealthy and well connected by birth, like Bush a
Skull and Bones man from Yale, he also like Bush never learned Chinese or
studied much of the country's history before his 1985 appointment. As one of
Kissinger's protégés, he had instead learned to be suspicious of those who spent
too much time studying the trees of one country to the exclusion of the global
forest. "A China expert is an oxymoron," Lord frequently quipped. More than
just a man with a similar outlook and background, he was also a friend. The
pair frequently played tennis whenever they were in the same city, and he thus
felt comfortable enough with the new president to send a far more personal
letter of advice beyond his embassy's official report.

Bush needed to appreciate just how far the Chinese had come in such a
remarkably short period of time since Deng began his reforms, Lord stressed,
but he also needed to appreciate the depths of their current problems. "Since
you were here last three years ago," Lord began, "China has prospered in the
world and floundered at home." Economic productivity and affluence were on
the rise, and "any Chinese will tell you they are better off than when you were
first here" in 1974. Yet "many Chinese will tell you they are worse off than when
you were last here" in 1986. Signs of both abundance and discontent were everywhere. "In both the cities and countryside there is growing cynicism over
shifting political winds, stop-and-go economic policies, 'back door' influence,
nepotism at the highest levels, and outright graft." Yet "within a stone's throw of
Chairman Mao's mausoleum sits the world's largest Kentucky Fried Chicken,"
proof that China now engaged the global economy as at no other time since
the 1940s.

That chicken outlet, however, was symbolically fraught. Tens of thousands
of ordinary citizens filed past the restaurant after it opened in 1987, wondering
at the mysteries within. Strange and enticing smells wafted out into the street.

Yet smells and glimpses were all most could afford. The price of a meal, a bargain by American standards, surpassed what a typical Beijing worker could pay. China's new indulgences existed only for a fortunate few. Indeed, too few. Poor attendance forced city authorities to issue special "foreign exchange certificates" for party members to dine at KFC lest the outlet fail to meet its guaranteed sales. The masses had no such access, but instead the growing sense that for the first time in their lives, some in their midst enjoyed luxuries most could never hope to attain.

Deng's government recognized the problems that wealth disparity posed, Lord wrote, and the ensuing potential for social unrest within a society long tutored to believe in economic equality. Yet the state lacked remedies. The cadre that marched with Mao and governed with Deng were out of ideas and exhausted. Retirement beckoned, and thus Bush's forthcoming trip would be, he urged, "a very important time for you to sketch the agenda of our bilateral relationship . . . with the next generation of leaders who will dominate the coming four years." He suggested twenty-two points of contention in the Sino-American relationship that Bush could raise, ranging from Taiwan's future to intellectual property rights, noting sardonically that the president might wish to pay particular attention to the last point. His campaign book was now in its eighth Chinese edition, Lord reported. "The bad news is that the author is receiving no royalties — a prime example of why we must press the Chinese to protect intellectual property."

None of these issues would be resolved over the course of such a short visit, Lord counseled, but raising them would "lay down a few brush strokes, to be filled in during the coming months and years." The Chinese were not suddenly going to free their political prisoners or allow freedom of speech and conscience, or even rethink their intellectual property laws merely because Bush asked. He had to raise these issues nonetheless, in part because domestic politics back home demanded as much, and to an even greater extent because only through steady but respectful discussion would China's incoming generation of leaders appreciate just how much Americans cared not only for Chinese prosperity but for the country's political growth as well. "We will — and should — continue to raise human rights issues," Lord consequently advised, but judiciously. "The Chinese do not seek to impose their concepts on us and thus will continue to resent what they perceive as our attempt to do so on them."

In this one line Lord crystallized the central difference between Chinese and

American culture, and the source of so much friction. Both considered themselves exceptional nations, but for different reasons. Chinese leaders thought their long history made their culture impossible to replicate, or even for any foreigner to truly understand. Foreign advice was thus useless or malicious, designed either out of ignorance or from a desire to control. What worked in China would work only for China, whose leaders, unlike their American counterparts, had little time or inclination to tutor others. "Let us love our country and restore our Great Wall," Deng proclaimed.

American leaders traditionally considered their nation exceptional as well, by virtue of ideal values suited for any circumstance. Theirs was a "shining city on a hill," standing like a beacon for all others, a Pilgrim leader declared in 1611. "And she's still a beacon," said Ronald Reagan in his farewell address, "still a magnet for all who must have freedom, for all the pilgrims from all the lost places who are hurtling through the darkness, toward home."

One country's symbol was a wall, the other's a beacon. Americans believed the world improved when people learned to be like them and considered it their highest duty to promote their own values. Bush certainly did. Their Chinese counterparts longed for a world that finally recognized the futility of trying to change them. Deng had spent his entire life wishing for Chinese independence, not just from outside control but from foreigners who thought they knew best what his country required.

Lord understood both sides. He thought Bush had little prospect of changing Beijing's values, and that China's notoriously thin-skinned leadership might take offense if he tried. But neither should Bush back away from promoting American values. Before pushing too hard on the Chinese to liberalize faster, Lord counseled, Bush should recognize the great strides they'd already made, many quite recently. He was scheduled during his whirlwind visit to Beijing to visit his old church, for example, site of his daughter's baptism; he would be given the opportunity to deliver an address directly over Chinese airwaves; and finally, after much negotiation, the White House had won the opportunity for Bush's motorcade to stop "spontaneously" so he might press the flesh with ordinary Chinese. Reagan had been allowed none of these opportunities when visiting in 1985. He could therefore explicitly tout the merits of free speech, free exercise of religion, and democratic freedom directly to the Chinese people as no president had ever done before, noting as well the country's progress on each of these fronts. Before demanding even more from a leadership tired from the exertions of transforming their country, a leadership that believed they had already bent over backwards to accommodate the new American president's agenda, Bush therefore needed first to ask what issues mattered most to his own long-range agenda for Sino-American relations. He would not secure everything he wanted in a single trip.

Gorbachev topped the list. The Gorbachev phenomenon was Lord's second and more significant reason for writing Bush on the eve of the president's visit. *Perestroika* was a quandary the Chinese and Americans shared, as neither could say for certain if it would succeed, or what its aftermath might bring. But whereas changes in the Soviet bloc disquieted yet also thrilled the Americans, those same reforms terrified the Chinese, who feared that their workers and students might find Gorbachev's words as catalyzing as their counterparts in Eastern Europe did. Bush could therefore put their unease over *perestroika* to good use as a means of shoring up Sino-American relations. He needed to talk about human rights and democracy, quietly and cautiously, but if he could hope to accomplish only one thing during his trip, it had to be maintaining the continued validity of the triangular diplomacy Nixon had initiated nearly twenty years before. Indeed, the three top items on Lord's long list of potential points for discussion were in fact different articulations of this point: "our intentions toward the Soviet Union; Chinese intentions toward the Soviet Union; comparison of assessments on Gorbachev's policies and prospects."

Bush took Lord's advice to heart, in large measure because it reflected his own view that Washington's best hopes for China lay in simply keeping Beijing on its current path, though Japan came first. Leaders from 158 nations attended Hirohito's funeral, providing ample opportunity for the new president to move his personal diplomacy beyond mere phone calls. Ensconced in the American ambassador's residence in Tokyo like some satrap of old, ready to receive homage, Bush met in rapid succession Japan's prime minister; the kings of Spain, Jordan, and Belgium; and political leaders from Thailand, Israel, Egypt, Portugal, India, the Philippines, Italy, Turkey, Zaire, Germany, Singapore, and Nigeria. Save for attending funeral ceremonies, he left the compound only for an audience with the newly installed emperor Akihito at the Imperial Palace. Baker sat in on some of the meetings, Scowcroft on others. Most of the time, however, in direct contrast to Reagan's diplomatic routine in his last years in office, Bush conducted these meetings alone, aided only by a translator or note taker. He and France's François Mitterrand shared a private lunch, for example, joined only by their interpreter. Not surprisingly, Gorbachev dominated their conversation.

Several themes emerged from this diplomatic speed-dating. The first was the way Bush listened. Remembering the old adage that people typically enjoy conversations most when talking, Bush warmed each guest up by asking how the world looked through his eyes. "We always appreciate your sage view of the world scene," he told Israel's president Chaim Herzog. Tell me your "opinion of events in the region," he began with King Hussein of Jordan, similarly beginning his discussion with Portugal's Mario Soares by noting that he "looked forward to learning his [Soares's] views on many issues." These were pleasantries and expressions of good manners. But they were also revealing. Not every world leader listened before speaking. Reagan had hewed closely to his notecards, no matter where the conversation ran. Gorbachev could drone on for hours on almost any topic; Deng too preferred lengthy monologues to give-and-take discussion; Margaret Thatcher frequently let foreign visitors know the moment they entered the room if the conversation would be difficult, silently staring down foreign diplomats till they nervously began to speak. While Bush was rarely shy about expressing his own opinions, his broader goal in these sessions was to demonstrate that he would operate as president as he had as United Nations ambassador, building relationships in times of calm that might prove useful in moments of crisis, recalling all the while that each nation cared as deeply about its own sovereignty as Americans did for theirs.

Nearly all of these conversations turned to talk of Gorbachev's momentum, and ultimately to pleas from Bush for patience when his allies urged him to do more in response. "When the U.S. finishes its policy review," he told Japan's prime minister, "there will be nothing that adversely affects Asia, especially China or Japan." Speaking to Italy's president, Bush promised that he "intended to see that Gorbachev did not achieve dominance of public opinion in Europe," a point he reiterated to Germany's president with the comment that "we don't want Gorbachev to win a propaganda offensive." Bush admitted he "was not sure what was in fact happening in the Soviet Union," but for NATO to survive, "we must stay together and not be naive." The Americans had best not wait too long, Turkey's prime minister warned in response. "If Gorbachev were successful, the change would be bigger than that of 1917."

Chinese leaders offered similar warnings once Bush landed in Beijing. He had known many of them for years and was welcomed back as their *lao pengyou*. Most easily translated as "old friend," the term has particular meaning, afforded only to influential foreigners with long experience working with the Chinese. Respected enough that they might speak plainly, such friends could be trusted as interlocutors between China and the broader world. Nixon, Bush, and Kissinger bore this distinction. Carter and Reagan did not.

Such trust, however, carried a price. Chinese officials typically turned to their "old friends" whenever they had an important message — or threat — to deliver. Frequently given favorable treatment, old friends were expected to perform favors as well. China's leaders made requests of Nixon and Bush they would never have asked of Reagan. "The Chinese seem to feel comfortable only in dealing with those who share a basic inclination to establish positive *guanxi* [interpersonal connection] at the human level," Richard Solomon wrote at the end of a diplomatic career that included many negotiations with the Chinese. "Yet these same friends are the ones who receive the brunt of the pressure when there are problems to be resolved, for the Chinese assume — not without reason — that those who see value in the US-PRC relationship are the ones who will work to resolve the problems and thereby sustain their status as friends."

The term *lao pengyou* echoed throughout Bush's late-February meetings. "Every time you come" to China, President Yang Shangkun told him, "you discuss major issues with Chinese leaders. So you've made great contributions to the development of Sino-U.S. relations and to cooperation between our two countries." They felt comfortable with him as with few other world leaders, said Yang, and even fewer American presidents. After small talk at the start of a separate meeting, Zhao Ziyang, general secretary of the Communist Party's Central Committee and a man American analysts considered Deng's most likely successor, abruptly launched into nearly identical remarks, his brusque turn making plain that his words were both scheduled and well rehearsed. Bush was, he said, "the best witness to the development of Sino-U.S. relations in the last decade."

"You shouldn't take such talk seriously," Bush told this author in 2005, during yet another trip to Beijing. Hurtling through closed streets in an official limousine, more than a decade removed from office, Bush continued to receive the kind of treatment typically afforded visiting heads of state. No doubt having a president for a son helped. There to attend a business and academic conference, he spent several days beforehand making the rounds of local leaders, appearing at charity functions, and even privately meeting with China's high command.

Chinese officials fell over one another with praise at every stop, calling Bush their partner and friend. He reciprocated in kind from the podium but was clearly unimpressed once back in the confines of his car. "I do have a relationship there; there is warmth," he acknowledged. "But those kinds of welcomes, those over-the-top words, they're for show; it's what they do." Barbara Bush was blunter. She typically was. "They say all kinds of things," she added, "especially when they want something."

Following the ritualistic niceties, China's leadership collectively made two points during Bush's series of meetings in 1989. The first was to put to rest any American anxieties over a renewed Sino-Soviet collaboration. "Normalization of Sino-Soviet relations will not be like in the 1950s," Zhao said. "Any military alliance or military relationship is out of the question." This would be a refrain heard often during Bush's visit: that Gorbachev's impending trip augured merely an end to an era of direct antagonism, nothing more. It was good to have a safer border and better relations with the Soviets, Zhao continued, but the Chinese were as concerned as the Americans over where *perestroika* might lead. "We must watch the deeds of Mr. Gorbachev and not just listen to his words," he cautioned. True change took years, even generations, President Yang similarly told Bush. "It would be at least ten years before they will see any results, because it has taken China that long [since the 1978 third plenum] to see results." Anything faster was in their opinion too fast, even reckless, especially as "the process of democratization" the Soviets had initiated without first putting their economic house in order might "provoke ethnic problems." In his opinion, "the Soviet Union should mainly concentrate on the economic problems of the country," as Chinese leaders had done, though he could not resist noting that Gorbachev's push for democratization "may suit the taste of the U.S."

The message behind these critiques came from sad experience, which China's current crop of leaders simply refused to repeat. Little more than a decade removed from the Cultural Revolution, they knew what could happen when the impulse to reform got out of hand. Roving bands of youthful Red Guards inspired by Mao claimed full license to deploy any means necessary in their pursuit of ideological purification. Justice was swift, appeals nonexistent, punishment dispensed without delay. Tens of thousands died. Millions more deemed unreliable or prone to bourgeois tendencies had their lives permanently disrupted, many ruined, when plucked from their homes and families for "reeducation." Business and industry ground to a halt. Museums were shuttered or destroyed. College campuses were transformed into ideological boot camps where facts and figures, even in science and engineering, mattered less than dogma. No one was immune. Between 60 and 70 percent of public officials were removed from office during the first two years of the Cultural Revolution, a rate higher than in the Soviet Union during Joseph Stalin's purges in the 1930s. Of thirteen members of the Central Committee in 1966, only four remained three years later. Former president Liu Shaoqi was tortured and left to die in prison. The Korean War's chief military commander was similarly beaten by an angry mob and left to suffer in the street with a broken spine. Even authoritarians found the intensity of the violence appalling. "The Hitlerites could have learned something" about cruelty from the Red Guards, one Cuban diplomat in Beijing observed.

Lao She's fate was sadly all too typical. Widely considered China's greatest living writer, in late August 1966 he was paraded through the streets leading to Beijing's Confucius Temple by an angry band of Red Guards. Forced to kneel in the center courtyard, he and other writers were ridiculed and whipped with leather belts tipped with heavy brass buckles. It took hours for the revolutionaries, mostly fifteen- and sixteen-year-old girls, to expend their fury. Finally released, Lao rose early the next day, walked to a nearby pond, and read poetry and wrote notes till the sun set. Unable to live in a world where children could be cheered for beating their elders, he loaded his pockets with heavy stones and walked straight into the water.

Deng suffered exile as well, sent to the hinterland to relearn the virtues of manual labor. His eldest son's fate was worse. Tortured into confessing to capitalist leanings while his family watched, he subsequently leaped from a high dormitory window in order to escape his tormenters. Some contend he was pushed. Denied admission to a local hospital, he lay on the pavement for hours, destined to spend the remainder of his days confined to a wheelchair, an inescapable reminder for his father of the costs of chaos.

No one who endured this period would ever again lightly embrace calls for unfettered reform. Deng's cohort cracked down in 1985, for example, when patriotic commemorations of the anniversary of Japan's 1931 invasion of China spontaneously morphed into popular calls for an end to state-sanctioned nepotism. The best students should get the best posts, protesters demanded, not just the best connected. By mid-September, the movement that had begun on outlying campuses converged on Beijing, as students prepared to plead their case in Tiananmen Square, the city's epicenter and the nation's symbolic heart.

Deng's government ordered the protest halted. Always averse to unauthorized public demonstrations, the marches displayed two elements in particular that sent shudders through his ruling band. First, for the first time since the Cultural Revolution, student protests had elicited sympathetic responses from workers, with several prominent working-class leaders joining in the march. Students were one thing. When budding intellectuals began cooperating with common laborers, however, as had occurred in Poland with Solidarity's rise, authorities feared a broader assault on their rule. Second, and more ominously, the marches appeared coordinated, revealing an underground communications network largely unknown to the authorities. As observers at the American embassy in Beijing wrote home, "The communist party has lost the initiative on campus."

Protests were resurrected the following year, once more seemingly spontaneously throughout the country, though with greater emphasis on democratic reforms. An astrophysicist named Fang Lizhi argued in front of an enthusiastic crowd that people possessed rights not granted by the government. The party promised "relaxation of control," but any freedom granted by a regime could just as easily be revoked.

A world-class physicist before turning to political activism, Fang would ultimately play an outsized role in the broader story of Bush's relationship with China. Once China's youngest full professor, he'd won coveted opportunities to study abroad, and earned the ire of party leaders. His calls for separating ideology from scientific research prompted his expulsion from the Communist Party in 1957. That experience, coupled with time spent at Princeton University in the early 1980s, where he witnessed scientists discussing politics without reproach, led Fang to adopt Albert Einstein and later Andrei Sakharov as role models. By 1985 he was openly traveling to campuses around China, calling for free inquiry and freedom of speech, arguing that only pluralism — and not the imposition of order from above — could prevent another Cultural Revolution. That December he openly rejected Marxism. "It is an undeniable fact that not a single socialist country has succeeded since the end of the Second World War," he told a crowd at Shanghai's Tongji University. "Socialism has failed, from Marx, Lenin, Stalin, and all the way to Mao."

Crowds cheered. Deng's clique fumed. A thousand students protested at

Fang's university in Hefei on December 5, 1986. Similar rallies sprang up the following week in Xi'an, Tianjin, Nanjing, and Shanghai. By month's end the protests had spread to more than 150 campuses in seventeen cities. Thousands marched down Shanghai's famous Bund carrying banners demanding "Give Me Liberty or Give Me Death." Jiang Zemin, the city's mayor (and the country's future leader), ordered the protesters to disperse. A barrage of catcalls and jeers forced his retreat. When local police seemed powerless to control the marchers, the central government stepped in with troops. Implored by their professors, the students grudgingly but obediently returned to their campuses, giving authorities in Beijing the impression that the threat of violence could trump any calls for greater freedom.

The ensuing political backlash reached all the way to the top, claiming Hu Yaobang, general secretary of the party and Deng's nominal second in command, who was removed from office for the "mistake" of being too soft toward the protesters and their ilk. A well-known proponent of political reform, Hu also led an increasingly vociferous faction of youthful bureaucrats eager for their senior colleagues to share the rewards of patronage and nepotism. His expulsion afforded Deng's old guard the opportunity to consolidate their power even further while warning proponents of change that dissent would not be tolerated, no matter if protests came from the streets or from the bureaucracy, and no matter how good the intentions reformers might claim. Deng argued that just because the Chinese people increasingly had access to Western prosperity, this did not mean they were entitled to the West's anarchic freedoms. "Bourgeois liberalization," his term for political changes seen in decadent capitalist countries, "would plunge the country into turmoil once more," he told the party elite.

Their message for Bush in early 1989, repeated in each of his conversations with China's top leadership, was plain: they would never allow the kinds of haphazard reforms Gorbachev so blithely offered without thought to the potential consequences. "China would not welcome the kind of labor problems that Poland is experiencing with Solidarity," Li Peng declared during his session with the president. General Secretary Zhao made the same point. "In view of the tremendous beneficial changes brought by reform," he noted, "and the improvement in living standards, there is no reason or basis for changing current policies." A small number of agitators might call for more, and for faster change, perhaps even for "the introduction of a Western political system," a proposition that "does not tally with the realities of China." Such pleas would be stifled before they got out of control. If their ideas were "carried out," Yang told Bush, "chaos will result, and reform will be disrupted."

China's leaders also delivered a warning that foreigners, and the Americans in particular, should do nothing to support such agitators, no matter how sympathetic their goals. "Some press people in the West and the U.S. feel warmly toward those in China who advocate a Western political system and have great interest in them," Zhao said. "In our view," however, "if there are Americans who support those Chinese people who are opposed to the current policies of the Chinese government, they will hurt reform as well as Sino-U.S. friendship."

Zhao's threat was clear: if you meddle, we will both crack down and align with the Soviets. "Mr. President, you know that it is the Chinese government and the people who are promoting reform in the light of the actual conditions in China," Zhao told Bush. "It is not these others," he continued. "You know well China's history and its realities . . . I hope the United States government will pay attention to this question for the sake of Sino-U.S. friendship, the stability of China, and the success of reform."

It was left to Deng, last on Bush's formal itinerary, to drive the point home lest the Americans depart with any illusions about how much patience the Chinese would have with any outside interference in their internal affairs. "With regard to the problems confronting China, let me say that the overwhelming need is to maintain stability," Deng told him. "Without stability, everything will be gone, even accomplishments will be ruined." With all that was changing throughout the world, he concluded, "we hope our friends abroad can understand this point."

"We do," Bush said. "All right, then," Deng offered. "Let's have lunch." Their meeting thus came to an abrupt halt, their formal discussion time expired. But Deng had said all he had desired. The Chinese would change, if at all, at the government's chosen pace. Unauthorized reformers would be rooted out, and any foreign encouragement of democracy would be considered the same as fomenting revolt.

They were not yet done sending messages, however. A banquet remained, with Bush the official host. Organized by the American embassy, it would be an old-fashioned Texas barbecue, or at least as near as the Foreign Service and the Beijing Hilton could produce. Back in 1975, then-ambassador Bush had worked for more than a month to secure enough Coca-Cola and hot dog buns to throw a Fourth of July picnic. Authentic supplies were at least easier to find in 1989. Checkered tablecloths were laid out. Waiters circulated with colorful bandanas around their necks, and a massive Texas flag stood watch over the festivities. Steaks and Lone Star beer were imported for the occasion, as was a country-western band, though history has left unrecorded how Chinese officials reacted to their extended rendition of "Your Cheating Heart."

What history will record of this dinner was that it nearly took place without any official Chinese representatives at all. Tasked to invite leading artistic and educational figures, and eager to find some quiet way of demonstrating American support for the country's budding democracy movement, Winston Lord's staff ran the potential guest list by the White House and State Department, highlighting Fang Lizhi's name along with those of other potentially controversial invitees. No one back in Washington objected, so his invitation went out with the rest. "We said the Chinese won't like this," Lord said a decade later, "but frankly we did not expect an explosion in the reaction."

"Who IS Fang Lizhi?" Bush yelled at Lord's staff when news arrived that Beijing's top leadership refused to attend the banquet if Fang's invitation was not rescinded. A real China expert would not have needed to ask. He learned only en route from Tokyo that he was about to throw a dinner party with empty seats at the head table. His anger boiling over by the time he landed in Beijing, the normally mild-mannered president laid into his ambassador as Air Force One idled on the tarmac. Employing a lifetime of training in the delicate language of diplomacy, Lord would later say only that Bush "was distinctly unfriendly" during their exchange.

He ordered Lord to solve the problem. Behind the scenes throughout Bush's entire time in Beijing, negotiators from both sides exchanged heated recriminations, proposals, and plans for a way out of the potential spectacle. Neither side could easily back down. Chinese officials claimed their stance was a matter of national honor, calling it outrageous that a visiting foreign dignitary would so publicly court a well-known critic of the regime. Conversely, Bush's team knew they would be pilloried back home if a guest of the Americans was barred from attending because of his pro-democracy views. "If I had to list four or five of the worst moments of my career," Lord later conceded, "one would be the Bush banquet and Fang Lizhi in February 1989." The dispute colored Bush's entire stay, for he knew that no matter how successfully he renewed his personal ties with China's leaders or won guarantees that no great Sino-Soviet alliance was in the offing, if the Americans failed to gain Fang's admission to dinner, the entire trip would likely be remembered only as an embarrassment.

A deal was eventually struck mere hours before the first guests were to arrive. President Yang and other ranking Chinese officials would attend, as long as their estranged physicist was seated as far away from the head table as possible, out of sight, and especially outside any possible camera angle that could capture his image along with those of his antagonists. The crisis seemed resolved. "We are throwing our hats in the air and figure everything is fine," Lord recalled. Their president would not suffer embarrassment. The press wouldn't be any the wiser about how close the entire trip came to disaster. Their careers would continue.

The Americans should have been more suspicious. "They just said they were coming," Lord continued, referring to China's leaders. "They didn't say anything about Fang not coming." They had other plans for him. Chinese police thwarted his every effort to make it to the banquet hall. His driver was stopped, his car impounded for a series of obscure and unexplained moving violations. Ejected onto the sidewalk, Fang and his companions next tried a bus, but officials had ordered every driver on their route to bypass them. They tried to hail a cab, but found those drivers had been threatened as well. He never made it to the Hilton.

Lord thought the banquet a roaring success until he heard the news. "My heart stopped," he later remembered. "I knew the press was going to get ahold of this. It turned out to be a disaster beyond my wildest dreams . . . Fang holds a press conference at a hotel. That's all the press cares about. Nothing else about the trip. It is all down the drain." In the world of public perception, Lord was right. "President Bush's whirlwind tour through Asia, originally conceived as a simple reiteration of old friendships and alliances, has run aground with a diplomatic furor over human rights in China," the *Australian Financial Review* concluded. The *Financial Times* headlined, "Chinese Human Rights Row Flares During Bush Visit," while the *Washington Post* blared, "China Rebukes U.S. Over Dissident." The *Wall Street Journal* was more pointed: "Bush's Visit Is Marred by Flap Over Chinese Dissidents."

"We end up with the worst of worlds out of this," Bush lamented to his diary on the flight home. "Newspapers are all over the story, 'human rights abuse.' They won't point out that two of the [other] dissidents were there, and that

China has come a long way . . . and now we're scurrying around trying to figure out how to handle it with our press."

The banquet affair highlighted Bush's failure to mention, even once, any human rights issue during his formal talks with China's top leadership. He had promised to raise matters of religious freedom and democratic reform, but as press secretary Marlin Fitzwater later conceded to the White House press corps, "human rights was not discussed." The documentary record, released two decades after the event, makes it plain: when it came to this entire range of issues, Bush said nothing.

China's threats effectively muzzled him. Determined to demonstrate the continuity of Sino-American friendship despite Gorbachev's overtures to Beijing, Bush kept human rights off the agenda lest he scuttle Lord's sensitive behind-the-scenes negotiations, whose failure would ensure the kind of international incident that would call into question his entire approach to China's rise and reforms. If Deng's government could not even resolve a seating arrangement with their old friend in the White House, what hopes might there be of countering *perestroika*'s reach? Ultimately Bush gained little for holding his tongue, given the public embarrassment Fang's failure to appear generated in any event. Baker was tasked with discussing China's human rights record during his meetings, so the topic was at least raised on some level, but not, Fitzwater admitted to the press, with "a big, public fanfare and a lot of noise" of the kind generated by a president's personal commitment. Bush "believes quiet diplomacy is the appropriate course for raising human rights issues in China," Fitzwater added.

Bush's silence reveals much about his overall strategy toward Asia. Believing China on the right path, and profoundly impressed by its gains, he feared catalyzing repression more than he desired to speed China's reforms. It had, he wrote in his diary, "come a long way" since 1974. It would get to democracy eventually, he believed, not only because the breeze of freedom sweeping Europe had reached the Pacific as well, but also because he simply could not fathom how a regime might promote economic liberty without commensurate political change. Democracy was the world's future. He was certain of it. Some states would discover "what worked" faster than others, but even once-belligerent regimes like Beijing's had no choice but to eventually accept the inevitable.

So long as they did not sidle up to the Soviets, therefore, Bush was perfectly content to keep out of China's internal affairs, just as its leaders desired. So long as the country continued its peaceful march toward free markets and what he considered intertwined political freedoms, he believed there was little an American president could or should do to influence what happened within its borders. Time was on his side. "At the start of my administration, we are engaging in a total review of many policies, including arms control negotiations with the Soviet Union and what to do in the Middle East," Bush told Li during one of their formal sessions together. "China is an exception: we already know where we are, where we want to go, and what we want to do . . . [T]his relationship stands on its own, and will not be affected by rapid changes on the international stage."

Little did he know how wrong those words would soon prove. China's protest movement had been silenced but not eliminated. It would rise again. Deng's government reaffirmed weeks after Bush's visit its determination to crush even well-intentioned dissent. Given how troubling the "unrest in Eastern Europe" appeared within Beijing's halls of power, party officials proclaimed in March, "every effort should be made to prevent changes in Eastern Europe from influencing China's internal development." Fang and those like him had other plans. "A tide was running which could not be stopped," he told reporters after Bush's departure. All that was required to unleash the deluge was a crack in the dam.

7

CHENEY RISES AND THE PAUSE ENDS

BUSH PAUSED, but the world did not. Eastern Europe's democratic revolutions picked up speed throughout the spring of 1989 as American policy makers dithered and debated. Signs of progress were everywhere in Western eyes. So were points of fissure and disorder. In Poland, former political enemies, literally jailers and those they'd jailed, hammered out the details of a new government together. Beset by financial woes and the nonstop threat of protest strikes, Communist Party officials agreed, for the first time, to hold open elections, and to relinquish their monopoly on power. Hungary's government similarly removed the doctrine of party supremacy from its constitution, stripping the words "Marxism" and "Leninism" from the document while instituting a new national holiday commemorating its 1848 independence from Austria in place of the requisite annual commemoration of the 1917 Bolshevik Revolution. Thirty thousand people marched through Budapest in the government's official celebration on March 15. Three times that number joined an alternate parade sponsored by opposition groups, their path through the city marking six key locations from the 1956 uprising, the stations of the cross of Hungarian suffering and sovereignty. Police and troops warily lined the route at first. By day's end, most of them had joined the march.

Soviet citizens went to the polls as well, not in yet another sham election between handpicked party chiefs but, for the first time in their nation's history, with real power at stake. A fortnight after the dueling Hungarian marches, in the Soviet Union more than 170 million people voted to select the 2,350 members of the country's new super-legislature, whose prime responsibility would be to forge a new 542-member Supreme Soviet. Twenty percent of Communist

Party candidates lost, an unprecedented number in Soviet elections, including many who had run unopposed: Leningrad's party chief received 110,000 votes, but nearly 130,000 of his constituents crossed his name off the ballot. Moscow's party chief suffered the same fate, much to the delight of Boris Yeltsin, the election's biggest winner. A former protégé turned political enemy of Gorbachev's, Yeltsin despised *perestroika*'s pace. He thought reforms should come faster. The 89 percent of Muscovites who voted for him to represent them seemed to agree.

Intended by Gorbachev to provide popular endorsement for his reforms, the election results were instead a clear repudiation. He was the party's head; its reforms were his, and so too the frustrations they had generated. He nonetheless claimed victory even in defeat, interpreting the results as proof that what voters truly wanted was not to retreat from *perestroika* but to surge forward with renewed speed. "The people are once again ahead of us," he told his inner circle. "All problems must be resolved through *perestroika*. Not instead of it, not by digressing from it, and not by twisting the line of *perestroika*."

The referendum laid bare the Kremlin's waning hold on power. The expected hardships of economic transformation had arrived, but not yet any signs of payoff or progress. Three fourths of Russians polled believed that *perestroika* needed to move faster. Fewer than a third, however, were willing to accept any more sacrifices. "The planners of *perestroika* are baffled," a noted professor of Russian studies at Columbia University concluded. "They don't know how to proceed because they found the economic situation far worse than their worst expectations." Goods seemed scarcer with every passing month. Food proved increasingly hard to procure, especially in cities. Just one in three Russian consumers was issued a ration card for meat; the rest could only dream of it. "Shortages attack us literally from all sides," complained the daily *Vechernyaya Moskva*. Independent economists calculated that the current Soviet diet was poorer than in the year before the Russian Revolution. Communists had been in power for seventy-plus years, had beaten back the Nazis, put men and women into orbit, and boasted of the ability to churn out missiles like sausages. Yet there were no actual sausages.

Rumbling stomachs in the republics fueled calls for independence. Nationalist fervor bubbled over in Georgia in particular, where secessionist demands for separation from the Soviet Union mixed with pleas from minority groups for a new Abkhaz Republic. Protests reached their peak on April 4. Soviet troops joined local police to disband crowds that had grown to nearly twenty thousand. Enraged protesters swung shovels in anger. The army fired in response, first with rifles, then with machine guns. Hundreds fell wounded. Twenty died.

Gorbachev recoiled at the news. He abhorred violence as a matter of principle, and he had not ordered any crackdown. Indeed, he'd been out of the country. Yet blood had been spilled by troops acting in his name. He thereafter explicitly ordered commanders to refrain from violently countering crowds, telling the Politburo: "We have accepted that even in foreign policy force is to no avail. So especially internally—we cannot resort and will not resort to force." His trusted foreign minister, himself a Georgian, complained that "the Army should not be drawn into resolving internal political conflicts" in the first place, and that deploying combat troops to monitor political demonstrations made conflict inevitable. They were trained to use firepower, not to respect newfound rights.

For Prime Minister Nikolai Ryzhkov, however, the real issue was what the incident demonstrated about the Kremlin's fleeting control. "Armed forces were used, and the General Secretary learns about it only the next morning," he complained to the Politburo's Central Committee a week after the incident. "How then do we appear before Soviet society, before the whole world? In general, wherever you turn, things go on without the Politburo knowing about them. That is even worse than if the Politburo had made bad decisions."

Gorbachev's cadre left the meeting determined to reassert their command over a reform movement they knew could easily take on a life of its own. They also knew that despite his prohibition on further violence, the army insisted on bolstering its forces in Tbilisi. One never knew when armed troops might be needed again, military commanders argued. It was an implied threat that a different general secretary, a tougher one, perhaps, might yet warrant their use. Watching quietly from the wings, Anatoly Chernyaev wrote in his diary that Gorbachev felt "that he is losing the levers of power irreversibly . . . he has no concept of where we are going . . . his declarations about socialist values and the ideals of October, as he begins to tick them off, sound ironic to the cognoscenti. Beyond them is emptiness." Chernyaev was one of Gorbachev's closest allies and friends. His enemies said far worse.

Bush received daily reports detailing the growing unrest behind the Iron Curtain, yet as far as the public could discern, he did nothing in response. Embarrassed and largely cowed into silence by the Chinese, upstaged at every turn by Gorbachev, and increasingly accused of lethargy by his closest allies, he instead stuck to his script. "Things are moving," Bush told reporters after hearing the good news from Poland. "And I think it's a sign of the change that democracy and democratization, if you will, and elections and parliaments and congress — is on the move." Yet "what it means to the other Eastern European countries, I simply can't tell you. In terms of my own plans, we have not formulated any plans yet."

Meanwhile he was stung by an unprecedented political defeat. The Senate rejected John Tower's nomination for secretary of defense. No previous cabinet nominee had been rebuffed since 1959, and never before in American history had the Senate denied a newly elected president's choice. Tower had spent three terms as their colleague and understood Pentagon budgeting as well as anyone in Washington. Yet he had also made enemies, and found few defenders when charges of drinking and philandering surfaced. Reports of financial improprieties soon followed. Bush had known Tower from back in the days when, as one commentator quipped, their Texas Republican convention could be held in a phone booth. He considered him a friend. He also did not want to cave in his first major showdown with the Democratic-controlled Congress. "The hell with it," he replied when Tower offered to withdraw. "Let's fight it out." Democrats controlled both houses with solid majorities, however. "They say it's the second time you've been shot down over the Pacific," a reporter joked with Bush when news reached Air Force One en route to Japan that Tower's nomination had failed to win committee approval. He didn't laugh. Loyalty wasn't a laughing matter. "He is my choice, my only choice, and I am standing with him," Bush declared.

No amount of lobbying could save Tower when the full Senate gathered to vote, even as the nomination fight exposed ancient rivalries and ribald tales of the upper chamber's old boy culture. There was a "fishy odor of hypocrisy" wafting over a "cantankerous Senate," one political columnist noted. "There were pious speeches on the Senate floor as member after member harrumphed about the nominee for secretary of defense drinking too much," even though "among those listening to such speeches (and voting against Tower) were more than one senator who has had trouble negotiating the way to his seat in the chamber." Bush faced more lopsided congressional opposition than any other twentieth-century president, and the final tally played out along party lines. Forty-seven Republicans voted in favor of Tower's nomination. Fifty-three Democrats voted against.

Tower's defeat altered the fates of two men who would each play a crucial role in Bush's presidency and beyond. The first was Dick Cheney. A five-term congressman from Wyoming, Cheney was well known in Washington since his time as Gerald Ford's chief of staff, and particularly well known to Bush's inner circle. Scowcroft had worked for him at the White House, while he and Baker had become close enough to share fishing expeditions in Wyoming, a place Cheney called home and Baker came to love. "On a couple of these pack trips, we've even bunked in the same tent," Baker later recalled, quipping, "Dick washes dishes, I dry."

Cheney aspired to be House speaker, and during his time in Congress had developed a reputation as a man one could trust and negotiate with. Conservative to his core, he'd once lambasted a reporter in a late-night phone call for having dared refer to him in print as "moderate." Yet he also routinely put aside his ideological blinders when it was time to make a deal. "Whenever there was a closed-door meeting on Capitol Hill where congressional Republicans were working out their policies or strategy," journalist James Mann noted, "Cheney was probably inside," counseling his colleagues to find the bargain they could accept rather than continuing some quixotic quest for perfection. "Principle is okay up to a certain point," he once advised Gerald Ford. "Principle doesn't do any good if you lose."

Whereas Tower sought to bring budgeting expertise into Bush's inner circle, Cheney's arrival brought something different: a sense of certainty. It was his core characteristic. Scowcroft offered skepticism; Bush, loyalty; Baker, the art of the deal. Unlike the others largely averse to ideological answers to complex questions, Cheney always knew he was right, even if he was inconsistent. A self-proclaimed "values" Republican, he toted to Washington a rap sheet of drunk driving convictions. Intellectual and well read, he'd twice flunked out of Yale before ultimately abandoning his studies altogether when Washington beckoned. Hawkish in the extreme, he had nonetheless secured five educational and family draft deferments in order to avoid putting on a uniform himself. "I had other priorities in the '60s than military service," he later explained. After 2001, and in particular following the terrorist attacks on the American homeland that took place while he was vice president, and indeed in the bunker buried beneath the White House, Cheney discarded much of his penchant for compromise. "If you are a man of principle," he noted after leaving office, "compromise is a bit of a dirty word."

This was not the man Bush nominated for secretary of defense in 1989, who instead had won bipartisan respect both for his competence and for his willingness to negotiate, though not on everything: he believed that Soviet reforms were still too tenuous for the United States to reduce its military posture. Scowcroft lauded each of these attributes when suggesting Bush take a long look at Cheney as Tower's replacement. They wanted someone quick, someone they knew and could trust, and someone likely to be confirmed. Senators voted Tower down on the morning of March 9. Cheney was in chief of staff John Sununu's office that afternoon. A private interview with Bush followed. They discussed Cheney's health (he had suffered a third heart attack the previous year) and drunk driving arrests, no small matter given the issues that had just sunk Tower, which Cheney ultimately explained away as youthful indiscretions. Asked by senators why he'd changed, Cheney answered simply, "I got married." That seemed a reasonable enough answer to anyone who knew his wife, a powerful Washington insider in her own right. By four o'clock that afternoon the nomination was his.

Bush had only one reservation: Cheney's departure opened a spot in the House leadership for Newt Gingrich. Cheney's equal in conservative certitude, Gingrich exhibited none of Cheney's penchant for compromise, or any of the party loyalty Bush espoused. "The question is — will he [Gingrich] be confrontational; will he raise hell with the establishment; and will he be difficult for me to work with?" Bush wondered in his diary. Gingrich was "a bright guy, an idea a minute," but he'd frequently need reminding that he "hasn't been elected president, and I have." Bush had little choice, however, but to accept Gingrich's rise as the price for bringing Cheney into his cabinet. He would come to rue that cost.

Cheney played a large role in forging and implementing Bush's foreign policy. Quickly confirmed by the Senate, he fit in as seamlessly as expected, joining his fellow Ford veterans Scowcroft and Baker for a weekly breakfast meeting to review the key issues of the day and hash out any potential differences. There were few issues this troika of old friends couldn't resolve. Cheney believed like the rest in the innate superiority of free societies and free markets over all potential competitors. His reputation for negotiation notwithstanding, here was an area in which Cheney refused to compromise. "You can't expect them [the Soviets] to accept all our terms," a fellow congressman had lectured him during a trip to Moscow in 1983. "You can't expect them to surrender." Actually, "yes, yes I can," he responded quietly.

Cheney's nomination troubled Moscow. Gorbachev and his aides had hoped Bush would select a moderate for this key post, or at least someone more like Tower, with a technocratic bent, and thus perceived his selection as further evidence that the strategic review merely masked Bush's intent to return the Cold War to its pre-Reagan days. "If you buy the argument" that *perestroika* would lead to good, Cheney told the Senate Armed Services Committee during his confirmation hearings, then of course Gorbachev deserved Washington's unbridled support. This was not an argument Cheney bought. He believed it wiser to continue to hold the same hard line against the Soviets that he believed had brought them to their knees. "Obviously," he noted, "if you had a Soviet Union that was revitalized, one with a strong, healthy economy and the same kinds of policies towards the rest of the world that have dominated during the postwar period, that would not be good from the standpoint of the United States. It would mean we were faced with an adversary far more capable than the one we're faced with today." It was not yet time to ramp down American military spending, he proclaimed at his swearing-in ceremony, or merely to take Gorbachev at his word: "There are those who want to declare the Cold War ended. They perceive a significantly lessened threat and want to believe that we can reduce our level of vigilance accordingly. But I believe caution is in order."

Those words stung in Moscow, as did the image of Bush, smiling and nodding as Cheney spoke. The new president wasn't Reagan, Gorbachev complained bitterly to Margaret Thatcher. He wasn't even someone interested in cooperation. Gorbachev had worried about this ever since Governors Island. Now he was sure. "There is a point of view emerging in the White House that the success of our perestroika, the development of a new image of the Soviet Union, is not beneficial for the West," he told her. "I tell you frankly, we are concerned about it."

Thatcher, who'd once pleaded with a bellicose American president to give Soviet reforms a chance, found herself four years later pleading with Gorbachev for similar patience as the new president found his bearings. After he left, however, she fired off yet another plea for action to the White House. She was not the only one. There is "a lot of criticism that your administration is in drift, there is malaise," a reporter challenged Bush at a press conference in early March. "Nobody knows where your administration is going," another blurted. Bush countered that in the few weeks he'd been in office, he'd already addressed major domestic issues and sent Congress a new budget. "I've taken a substantive foreign policy trip that took me not only to three countries, where I met with, I think, some 19 representatives of 19 countries and talked about their objectives and mine," he continued in his own defense. "Our Secretary of State has not only touched base with all the NATO leaders but has had a productive meeting with Mr. Shevardnadze. The defense reviews and the other foreign policy reviews are underway." His administration was working hard, he insisted, but he refused to move until ready. "I will not be stampeded by some talk that we have not come up with bold new foreign policy proposals in 45 days — not going to be."

Behind the scenes, however, despite Gorbachev's fears, and despite the apparent warning signal of Cheney's elevation, Bush, along with his inner circle, was coming to believe that the Soviet leader represented their best hope for a meaningful recalibration of Cold War tensions. They didn't have to like him, or trust him, in order to understand that whoever took his place would likely be far worse. More important, Bush's team reasoned, Gorbachev offered an opportunity not only for a potential end to the Cold War but also, more fundamentally, for the United States to fully solidify its leading role in Europe and beyond, irrespective of the forecasts of imperial overstretch, European unity, or communism's decline.

This shift, more cognitive than practical, is best seen in the way the NSC's senior specialists for the Soviet Union and Europe, Condoleezza Rice and Robert Blackwill, framed and then reframed their most pressing strategic needs during Bush's first months in office. Tasked upon joining Scowcroft's staff with forging a road map for "getting ahead of Gorbachev," they initially cautioned against trying to match the Soviet leader move for move, which merely played to Gorbachev's strengths as a popular figure (globally, anyway) with a flair for the dramatic. "The worst thing that we can do at this critical juncture is [to] make policy precipitously or in piecemeal fashion," they advised a month into Bush's term. "He is very good. We have to be better." Bush endorsed this point, underlining and adding a checkmark when reading the memo. Taking his president's cue, Scowcroft in turn admonished his team to think more grandly. If they were not going to match Gorbachev, but instead hoped to best him, they needed an ambitious plan of their own. "You are being too conservative," he teased. "You've become a bunch of old curmudgeons."

Another month of debate helped Scowcroft's staff realize that they had to do more than merely offer a reasonable alternative to *perestroika*. That wasn't thinking big enough. They needed to see the present instant for what it really was: a flexible hinge moment in world affairs unseen since the closing days of World War II, when Europe and Asia lay largely decimated following fascism's defeat. The world in 1945 beyond the Western Hemisphere appeared broken, leaving for American policy makers a task that, according to Secretary of State Acheson, was no less formidable "than that described in the first chapter of Genesis," as Scowcroft's staff reminded Bush in March, when God had "create[d] a world out of chaos." Truman's crowd put half the world in order. The other half followed Moscow into darkness.

Nearly a half century later, standing at the end of a long line of Cold War administrations that had each worked toward this day, Bush's team perceived the opportunity to finish the job. "When those creators of the 1940s and 1950s rested, they had done much," Scowcroft's staff, and in particular the NSC's chief arms negotiator, Arnold Kanter, wrote. The financial, political, and military institutions they forged rebuilt the free world and kept it safe for generations. "We now have unprecedented opportunities to do more, to pick up the task where they left off, while doing what must be done to protect a handsome inheritance."

The audacious argument, as bold as any issued in the West Wing in decades, struck a nerve within perhaps the most cautious man to occupy the Oval Office during that same period. "Brent, I read this with interest!" Bush wrote, having underlined, asterisked, and marked his way through his copy of the document. Though he was prudent to the core, this seemed no time for prudence. He finally saw the need to think, and ultimately to act, just as boldly as the moment demanded. But how? What he might do remained unclear. Yet a key cognitive shift had taken place: Bush, months into his term, was finally willing to act — even if only his closest advisers knew it.

Similar thinking was under way at the State Department, especially for Baker, whose attitude toward *perestroika* was transforming in line with his growing personal relationship with Eduard Shevardnadze. The Soviet foreign minister was not a typical diplomat. Indeed, he was not a diplomat at all. Friends since the 1960s, he and Gorbachev had bonded over their shared belief in the necessity of reform at a time when the notion was considered heretical — and dangerous. "Gradually," and no doubt cautiously, "we opened up to each other," Shevardnadze recalled, "beginning to confide our secret thoughts." For Gorbachev, reform meant streamlining an ineffective and corrupt economy. For Shevardnadze, it meant eradicating corruption in his home republic of Georgia, his signature issue within the regional government until Gorbachev gave him, as he put it, "the shock of my life," asking him to replace Andrei Gromyko, their country's iconic foreign minister. Gromyko had served since the 1950s, meeting with every American president since Eisenhower, and surviving every Soviet regime in turn. No one more represented Soviet respectability and the status quo.

"Who could have expected a provincial party leader from Soviet Georgia to be chosen to replace Gromyko?" Gorbachev's favorite interpreter later asked. The stoic and unflappable Gromyko knew every global figure worth knowing. His replacement knew none of them, and bore both a reputation for candor and a face that betrayed his every emotion. He laughed, for goodness' sake, something no one could recall seeing Gromyko do in public. He even genuinely liked other people. "Civilized person-to-person relations are above ideology or class or particular interests," he told a reporter in 1990. "This is what guides me when I talk to [foreign diplomats such as] James Baker, or to Douglas Hurd, or Hans-Dietrich Genscher. They are partners." Appointing Shevardnadze, Gromyko's opposite in every way, allowed Gorbachev to demonstrate as words never could that the old days were over. Shevardnadze even looked and sounded different from his predecessor. With a shock of white hair, flashing eyes, and a heavy accent, "he reminded me a bit of Albert Einstein," Baker admitted.

The two diplomats met for the first time in Vienna in early March, in the same room in which John Kennedy and Nikita Khrushchev once verbally sparred. That summit had ended in shambles, directly setting the stage for the Cuban missile crisis a year later. This one went better, with both men not merely talking but truly striving to listen. One "can't help but get [an] impression of the pressure Shevardnadze and Gorbachev feel to make perestroika succeed quickly," Baker reported to Bush that night. "They're leaders in a great hurry, possessing a sense of urgency, but lacking a plan. As a result, they are searching — searching for ideas, debating what's possible, and scrambling to come up with initiatives." There was "no alternative to success," Shevardnadze told Baker. It was precisely the same phrase the Polish and Hungarian ministers had employed earlier in the week.

Something about Shevardnadze's message resonated with Baker, some indefinable quality of sincerity that encouraged an unusual degree of trust, indeed a spark of "the" trust, within Bush's most hard-nosed negotiator. What was happening in Eastern Europe was real, Baker told Bush on his return to Washington. And it was happening fast. "The pace may surprise us and create a new reality in Eastern Europe," he warned as the two sat alone in the Oval Office. "You must be able to present a bold, sweeping proposal" lest Washington get left behind. "In international politics, as in domestic politics, a sitting target is usually the easiest target," he later wrote. "The more we were moving, the harder it would be for Gorbachev to gain advantage over us."

The question still remained: How to act? The much-ballyhooed strategic review finally reached the White House on March 15 but contained no answers. Derisively dubbed "status quo plus" by White House staffers, it essentially advised staying the course, without broad initiatives, or really any initiative at all. It was "mush," Baker concluded, produced by a sclerotic national security apparatus "incapable of thinking things anew." Gates conceded they should have expected as much. "It was a triumph of hope over experience," he complained, to think "the bureaucracy might come up with some new initiatives or approaches." They had hoped for new ideas, but "mostly we got back studies that said do more of the same," Scowcroft later agreed. "It's not surprising that the bureaucracy thought everything was going well, because they had designed the policy."

Despite countless hours of analytical work, the strategic review also failed to tell Bush the one thing he most wanted to know: How likely was Gorbachev's success? "Some analysts see current [Soviet] policy changes as largely tactical, driven by the need for breathing space from the competition," read one representative document from the spring, a National Intelligence Estimate of Soviet capabilities and intentions. In their eyes, no matter what Gorbachev said or promised, "the ideological imperatives of Marxism-Leninism and its hostility toward capitalist countries are enduring." At the same time, "other analysts believe Gorbachev's policies reflect a fundamental rethinking of national interests and ideology."

Translated into plain language, Gorbachev might truly be trying to change the world, or he might not. Another intelligence report gave *perestroika* a "50–

50" chance of success, suggesting in turn that, given the Soviet leader's precarious position, caught between radical reformers and increasingly recalcitrant conservatives, American policy should not strive to "either help or hurt."

This was not helpful. The best minds within the U.S. government had spent weeks in an all-encompassing study which advised Bush that perhaps he should embrace Gorbachev, or perhaps not; he should push for further reforms, though there was virtue in holding fast; he might encourage a conservative backlash against Gorbachev by doing too much, or he might embolden Gorbachev's critics by doing too little. "Do you have any reply to Mr. Gorbachev's contention that the foreign policy review is taking too long?" a reporter asked at a press conference soon after. "The answer is no," Bush fumed in response. "Let me simply say that we're the United States of America. And I will be ready to discuss with the Soviets when we are ready." But this was April 7, three weeks after the "mush" of the review had been presented. Its contents remained largely secret, but news of its delivery to the president was not.

"This is an environmental briefing," Bush pleaded when the reporters persisted. A supertanker had just run aground in Alaska, soaking the coastline with a thick blanket of oil. The *Exxon Valdez* incident had all the makings of a major disaster, and Bush wanted to discuss his government's response. Yet reporters kept bringing up Gorbachev. "Mr. President, Mr. Gorbachev today made another arms control gesture . . . what's your response or reaction?" Bush answered in the deliberate tones of a man struggling desperately not to blow his top. "I've given you my response," he said. "We'll be ready to react when we feel like reacting and when we have prudently made our reviews upon which to act."

Most histories of this period are derisive of "the pause," noting its failure to produce any decisive new policies, and the brake it placed on Soviet-American forward progress after Reagan. Anatoly Chernyaev termed it part of "the wasted year." Ambassador Matlock, for another example, headed this section of his memoirs "Washington Fumbles," offering the image of one administration dropping the baton handed to it by another. On the surface, these critiques are correct: the strategic review itself was a failure, producing nothing new and little of analytical value. "It was pretty much like asking an architect to review his own work," as Baker later put it. "He might change a door here or a window there, but it would be unlikely for him to question the basic foundations on which the structure stood."

Beneath the surface, however, Bush's inner circle employed the pause to attain that most precious of quantities for harassed policy makers: time. Despite combined decades of Washington experience, they needed time in the wake of a contentious election and transition to think anew about the unusually tumultuous world they'd inherited upon taking office, and time in particular to discern what amidst the cacophony of events truly mattered and what was merely noise. The pause provided that space. Bush's team came to recognize during these months the opportunity before them, and homed in on the necessity of sustaining American power overseas, and in Europe especially, as their fundamental goal. Europe had been peaceful with American oversight, they reasoned, and nothing Gorbachev offered would prove worthwhile if the price ultimately turned out to be an American withdrawal.

Rather than ask if *perestroika* would succeed, or if Gorbachev would even survive, during the pause Bush's inner circle realized that their prime goal had to be the continuation of what had made the Cold War peace endure: their own leadership, and commitment, across the Atlantic. The outbreak of democracy in Eastern Europe mattered less in the end than the security that derived from the American military presence overseas. Democracy was largely inevitable, after all. The lesson of 1945, and the inheritance left by Truman, Acheson, and the rest of the American team "present at the creation" of the postwar world, was that peace could be assured only through American strength. That cadre had saved half the world. After months of thought, Bush's team perceived at last an opportunity to bring the whole world into their fold, offering a new hand of openness to the men and women of the communist world who appeared, at long last, headed in the right direction, so long as Americans maintained a hand on the helm. "The postwar strategy of containment has succeeded beyond our best expectations," Scowcroft told Bush by spring's end. "It is at this time, with all its promise and uncertainty, that West European leaders will be looking to you to set the tone and offer guidance for the Alliance, and the free world." Gazing across the Atlantic, and across the globe, he found it "difficult to recall when so much was at stake," he argued, "and, at the same time, when there were so many opportunities."

Publicly fruitless, the pause was not pointless. Without it, Bush quite likely would have continued to offer piecemeal solutions designed to assure restive allies and to slow Gorbachev's momentum. With the pause, Bush's inner circle had time to think in uncomfortable terms. They also learned to distrust their own bureaucracy, or at least to put little faith in its ability to think creatively. The list of people who would come to influence policy grew smaller as a consequence. The National Security Council and Baker's personal staff in particular would wield more power and influence as a result, having convinced the man whose opinion mattered most that it was finally time for a change.

COGNITIVE SHIFTS in the White House meant little if not followed up by visible change, though Bush's first attempt to outline a strategic response to Gorbachev was downright milquetoast in its own right. Speaking at a series of college commencements in May, Bush outlined a new way of dealing with the Soviets and the Cold War "beyond containment," whereby Soviet deeds would be met with equivalent American responses. "The Soviet Union says that it seeks to make peace with the world, and criticizes its own postwar policies," Bush told the graduating class at Texas A&M University on May 12. "These are words that we can only applaud. But a new relationship cannot simply be declared by Moscow, or bestowed by others. It must be earned." The Soviets could win the West's trust by further reducing their military footprint in Central and Eastern Europe, and by continuing to work for diplomatic solutions to destabilizing struggles in Afghanistan and Central America. They could show good faith by further improving their human rights record, strengthening their environmental protections, and helping fight the international drug trade. As a show of Washington's good faith, Bush proposed a new "Open Skies" regime, whereby unarmed surveillance aircraft from each nation would be allowed to overfly the other, opening up each other's lands to scrutiny in order to prove beyond doubt that neither had anything to hide, or ultimately to fear.

There was little here that was new, and certainly nothing innovative, and this final suggestion in particular epitomized everything that appeared wrong with the administration's thinking as the spring of 1989 turned to summer. "Open Skies" was old news. It had been proposed by the Eisenhower administration more than thirty years before. Having urged his staff to think big, to throw off their curmudgeonly old ways in order to find new solutions to unprecedented problems, the best Bush could offer the world in response was a dusted-off old idea from a previous era, better suited to the technology of the 1950s than the 1990s. To say he'd move "beyond containment" was like telling the world: The policies of the last forty years? Yes, that, and just a touch more.

"It was cautious and prudent," Scowcroft subsequently argued, defensively, of their initiatives. "We were shifting policy from the old and narrow focus on strategic arms to a wider dialogue designed to reduce the threat of war and bring real peace . . . All this was aimed at encouraging a 'reformed' Soviet Union, ready to play a trustworthy role in the community of nations — one far less threatening to the United States and its allies."

The key to Scowcroft's assessment, and the key takeaway from the entire period of the pause, boils down to his use of the words "shifting policy." The shift that truly mattered was happening within the minds of Bush's inner circle, even though Bush's public rhetoric and rationale sounded little different from what he, or generations of earlier presidents, had always uttered. "A single idea — democracy . . . is why the communist world, from Budapest to Beijing, is in ferment," Bush announced in late May. It was still not clear in his mind why communist leaders had unleashed this dangerous idea of democracy in their midst in the first place. "But whatever their motivations," he said, "they are unleashing a force they will find difficult to channel or control." The United States, he added, would continue to watch carefully what happened behind the Iron Curtain, but rather than just contain the communist world, as had been the case before, his administration would not rest until the European continent was once more wholly free. Rather than meet the Soviets solely through strength, however, seeking their collapse, he proposed to meet them in a spirit of friendship, even if their fundamental goal remained unchanged. "The Cold War began with the division of Europe," he said. "It can only end when Europe is whole."

Not everyone within his administration was fully on board with the consequences of this cognitive shift. Dick Cheney was not. Buried deep in the archives of the Bush administration lies a speech he planned to deliver the day before Bush declared his intent to move "beyond containment," and his hope that *perestroika* would succeed.

Cheney thought differently, not so much about the goal of American policy but about the means of achieving it. "We hear that world politics is undergoing a fundamental change, that the Soviet threat had diminished, if not totally evaporated, and that in response to all this we must reshape our strategy for keeping the peace and securing our national interest," he planned to say. "We have every reason to hope that Soviet military power takes on a less threatening posture. And we have every reason to hope that people now subjected to dictatorial rule may be given a breath of freedom."

"Hopes, however," in Cheney's words, "cannot rule defense policy." In his view, "we cannot be certain what the outcome of *glasnost* and *perestroika* will be, nor can we be certain that a more modern Soviet economy means a less threatening Soviet military. We hope the Soviet changes are sincere and permanent." But, he argued, "it would be dangerous — extremely dangerous — to believe we should abandon a policy that works, just because we have some reason to hope." Containment still had value, as did vigilance. The Soviets remained as dangerous as ever, and despite its friendlier tone, communism remained just as evil as Reagan had once preached. In the final analysis, "let us not forget that the Cold War may not be over."

He never got to utter those words. Bush wanted to move beyond that doctrine, beyond the idea he'd clung to throughout Reagan's second term and throughout his campaign to be Reagan's successor, the idea that the "jury was still out" on *perestroika*. He had decided the verdict was in, and that was the most important cognitive shift of all. "This guy *is* perestroika," Bush told his staff. Gorbachev's success or failure meant the same for his policies. They were intertwined. Bush was not yet ready to say he wanted to help his Soviet counterpart; indeed, he did not yet know what form help might take. But he knew, at the least, that Gorbachev was worth the try, and worth a new approach in turn. Cheney could not have disagreed more. "We must have the self-confidence to admit that what is happening in the Soviet Union is at least in part a result of our own post-war policies," his planned speech concluded. "It would be supreme folly to quit the struggle on what may well be the eve of a less threatening world."

The White House barred him from saying it. "It serves too well its intended purpose," Scowcroft's staff warned once they read an advance copy, "strongly suggesting that no major changes are needed" in Washington's global posture. This was precisely the opposite of what Bush was about to argue, prompting Scowcroft to tell his old friend, delicately but firmly, to get himself a different speech. Cheney complied, delivering a remarkably bland set of boilerplate remarks instead of the rousing call to arms he'd intended.

By May, Bush was finally ready to move, not just beyond containment, but directly against the Soviets in the competition for the good opinion of the European public, whose continued consent he would need if he were to maintain an American presence in their midst. "We've got to get ahead of the problem by the time of the NATO meeting" set for late in the month, Baker urged. "We've got to be in the position of making a proposal that is really serious and that is taken really seriously." Baker hoped for a 25 percent cut in NATO's overall troop levels. The Pentagon offered at first a 5 percent reduction, reluctantly consenting to 10 percent when prodded. Bush wanted more. "I want something that the U.S. can do alone, by itself, on its own," he told his staff, instructing them to take the cuts from American forces alone rather than from NATO's overall structure. Fiscal hawks and military doves in Congress would both appreciate the savings. "Don't keep telling me why it can't be done," he implored when Cheney and the Joint Chiefs objected, finally ordering the troop cuts despite their opposition. "I want to show that the United States is leading the alliance," he insisted, berating his defense secretary and the Pentagon brass. "Tell me how it can be done."

Even the French applauded when Bush finished outlining his proposal. The new president displayed "imagination," François Mitterrand told his colleagues in Brussels, "indeed, intellectual audacity." Though not formally part of NATO's operational structure since the late 1960s, the French continued to coordinate with the organization on a strategic level. Perhaps he'd not believed Americans capable of thinking in terms of cuts at all, but Mitterrand no doubt appreciated that Bush had tipped off Europe's most influential leaders — himself, Thatcher, and Kohl in particular — about his plans before the summit began. Britain's prime minister reacted poorly at first but made peace with Bush's proposal lest she strain the special relationship she counted on for influence. "She was tense," Bush dictated to his diary, and lectured the group at length on the necessity of supporting freedom. "Why does she have any doubt that we feel this way on this issue?" he wondered, before noting that Thatcher was "a good friend of the States, but she talks all the time when you're in a conversation."

The NATO summit in Brussels in late May, Bush's first as president, coming four months after he'd taken office, proved a great success, particularly in the realm of public relations. By the time the sessions ended, he had shown more flexibility and initiative than most expected from a status quo politician long dubbed a "wimp," and had given at long last some substance to the idea of what "beyond containment" could mean practically, even if the total potential reduction of American forces had been whittled down by the end to a mere 7,500 troops. He had refocused debate away from Gorbachev's proposals to an idea from the other side of the Atlantic. The pause was over for good, as were fears that the Americans, long leaders of the alliance, would be merely passive observers under his charge. Bush had ridden to the organization's "rescue like the proverbial U.S. cavalry," one left-leaning London newspaper noted, albeit "at the last possible moment." The *New York Times* was enthusiastic, praising "the willingness of this deeply cautious man to aggressively seize the moment."

Bush was elated by the response. "Here we go now, on the offensive," he crowed to his staff as they departed the Brussels summit.

By week's end, however, few remembered.

8

FROM A FUNERAL TO A RIOT

"IT WAS A TIRED but happy team who climbed aboard Air Force One" after the NATO summit, Scowcroft recalled. Some slept. Others stared out the window at the dark Atlantic below. Bush characteristically took out his pen. "There is something very reassuring to me (comforting — if that's not too wimpish) to have you running our foreign policy," he wrote Baker. He scribbled Scowcroft something similar: "After the euphoria of the trip wears off — and it will as the Monday morning quarterbacks start second guessing the plays, I will remember the sound advice you have given me." For the man who spent more hours at work in the White House than any other, he had some simple advice: "Get some rest now — a lot of battles lie ahead."

On the other side of the planet, China roiled with uncertainty. Deng's reforms had brought prosperity but also unrest. Ten years into his economic program, living standards were improving at an unprecedented rate across China, but not uniformly. There had been one television set for every 327 Chinese citizens in 1980. By 1985 there was one for every twenty-five. (By comparison there were close to two television sets per American *household* during the same period.) That statistic reveals the great strides taken under Deng's leadership but also explains an unintended consequence. Those television sets enabled ordinary Chinese citizens to see their country as never before, in particular the details of the vastly different lives experienced by farmers in the south, miners up north, and bankers on the coast. What they saw most of all were displays of material wealth unfathomable during the lean times of the 1950s, and downright treasonous during the Cultural Revolution. Many of those watching eyes belonged to people still living a lifestyle closer to that of the nineteenth century than the late twentieth. Wealth was good, Beijing's leaders proclaimed; but it was clearly better for some than for others.

Students perceived these disparities most keenly. Raised on traditional notions of communal equality, alongside newly sanctioned ideas that anyone could better his or her life in the new China, many balked at the realization that connections remained a better guarantor of success than grades or talent. Protests against nepotism and state-sanctioned corruption erupted in 1985 and 1987, each time snuffed out by a government scarred by memories of similar calls for reform in the past.

J. Stapleton Roy knew this history well. Raised in Nanjing, he was a true "China hand": bilingual, comfortable across the Pacific, and born into the family business. His parents had been missionaries. Expelled during the Korean War, they continued their work from Hong Kong but remained as close to their prospective flock as possible. His older brother took his upbringing into the classroom, teaching Chinese literature at the University of Chicago while devoting his life to completing the definitive translation of the sixteenth-century Ming era's most influential novel. "Stape" found his calling in the Foreign Service, where he rapidly climbed the ranks and served as the State Department's director of Asian affairs until his 1986 posting as ambassador to Singapore.

Baker called him back to Washington in 1989 to serve as the department's executive secretary, responsible for the organization's day-to-day operations, but also to lend a hand in Sino-American relations. Roy accompanied Bush to Beijing in February, for example, and largely negotiated the short-lived truce that led the Americans to believe that Fang Lizhi would attend Bush's final banquet. It was Roy who sat down with a Chinese embassy official back in Washington two weeks later, hoping to unravel precisely what had gone wrong and why China's leadership had been so willing to risk trade and strategic ties, and potentially to offend their "old friend" Bush, rather than dine in the same ballroom with an agitating physics professor.

"The reason the Fang issue had been so sensitive was precisely because of the volatility of the situation in China," Zhao Xixin, the principal minister for China's embassy, explained. Like many Chinese officials, he had no qualms about openly discussing his country's problems, so long as meddling foreigners refrained from offering solutions. There was rampant "dissatisfaction with inflation, special favors for the children of high level officials," he continued, "and other issues had reached the point where a person like Fang could, if of-

fered encouragement, generate an outpouring of mass sentiment." His generation viewed any outpouring of mass sentiment warily, Zhao reminded Roy, and
the current protest movement Fang represented, even if largely dormant, "was
of particular concern to the leadership" of China, "and why Deng Xiaoping
had stressed to the president [Bush] the need for stability in China." If stability
could not be encouraged or purchased, he stated bluntly, it would have to be
enforced.

Roy reassured his guest. The American people, and this president in particular, had no interest in destabilizing China. Bush had made this point repeatedly while in Beijing, and it was a point the new U.S. ambassador, James
Lilley, would surely stress when he took Winston Lord's place. Wishing to put
the embarrassment of the Fang banquet behind them in order to focus on the
larger issues in their relationship — such as trade, technology controls, and,
most important of all, how to deal with changes in the Soviet Union — Roy
and Zhao simply agreed to move on, each hoping that whatever happened next
in China's long economic transformation, it would at least be calm and quiet.
Privately Roy hoped as well that whenever protests erupted anew, as they were
sure to do, Washington could somehow avoid the blame.

What happened next was anything but calm. Hu Yaobang died on April 15.
Long considered a champion of student causes within the party leadership,
he was also their martyr. "When a disturbance breaks out in a place," Deng
charged in a speech that effectively ended Hu's career, "it is because the leaders" like Hu "didn't take a firm clear-cut stand." Force, Deng insisted, was their
best defense against "bourgeois liberalization," a catchall phrase employed to
castigate any who took the call to prosperity as a license to criticize the government. "When necessary we must deal severely with those who defy orders," he
had earlier preached. "We can afford to shed some blood. Just try as much as
possible not to kill anyone."

They'd been friends for over four decades, but Deng had quietly been planning Hu's ouster for months. The latter's anticorruption campaign frightened
powerful cliques within the Politburo's senior leadership, for whom accusa-
tions of nepotism hit too close to home. Ejecting Hu from the party's highest
ranks consequently ensured that Deng would retain their continued support.
Allowed to keep his seat in the Politburo, albeit without power or portfolio,
Hu soon began to suffer from depression. His weight plummeted. He ceased
exercising, reading, or following the news. Ignoring Deng's overtures to renew

their friendship, perhaps over a relaxing game of bridge as in happier times, he collapsed at a Politburo meeting on April 8, 1989, from a heart attack. At the memorial service later that month, Deng extended his hand, and his sympathies, to Hu's widow. She refused both. "It's all because of you people," she spat.

Mourners almost immediately began converging on Tiananmen Square, as student leaders seized an opportunity to voice their dissent. Party officials traditionally relaxed the normal prohibitions on popular expression following an influential leader's death. Tiananmen Square had filled in 1976 upon news of Zhou Enlai's death. The protest was grudgingly tolerated at first, but more than four thousand were ultimately arrested before the square was once again under government control.

Eight hundred mourners gathered the day after Hu's death was announced, ceremoniously laying wreaths and reading poems composed in his honor. "Why are we here?" the leader of one student activist group asked another when they stumbled upon each other in the crowd. I thought it was you guys who initiated this, he heard in bewildered response. Neither had instigated the rally. It was a spontaneous combination of grief and pent-up frustration, with more the air of a carnival than a revolution. "I saw many students marching," one participant recalled. "I asked a female student what they were doing. She told me that they were demonstrating . . . I followed them by bicycle." The parading crowd was a spectacle, its energy intoxicating. "When students chanted 'down with official corruption,'" she continued, "a few people on the street shouted back, 'we support you!' Then we laughed together. The whole situation made me very excited."

Formal demands slowly emerged. Posters erected around Beijing University called for an end to Deng's campaign against bourgeois liberalization. Calls for abolition of nepotism soon followed, as well as for an end to the stationing of "political guides" in student dormitories, typically party loyalists whose reports on a student's fealty oftentimes proved more critical than grades for determining initial job placements. Banners demanding freedom eventually flew, though as Deng's biographer Ezra Vogel has sagely noted, one should be wary of applying too universal a definition to the word. "Freedom" in the West typically meant the ability to vote, pray, work, and speak as one desired. Anyone unversed in the nuances of Chinese affairs might therefore naturally see in these banners, many written in English in order to maximize international appeal, proof of the same wave of political liberalization then sweeping Europe, or confirmation that students in Beijing, like leaders in Washington, believed that democracy's final triumph and the end of history were nigh.

Bush and those closest to him were particularly likely to draw this conclusion. "*Glasnost* and the Beijing demonstrations proved that the democratic way is on the march," Bush told reporters in May, for example, "and is not going to be stopped." Baker made the connection between protesters in Eastern Europe and China even more explicit, seeing in each nothing less than a global endorsement of the system Americans led. "They may have that name [Gorbachev] on their lips," he told Bush, "but they have the policies of the West in mind. It is the philosophy of the West that they are advancing, and it is the values of the West that they are seeking."

This conflation of *perestroika,* the Chinese protest movements, and American idealism was simply wrong. Some words meant different things when used in Washington, Warsaw, Budapest, or Beijing. The "freedom" the students rapidly gathering at Tiananmen desired was not Bush's or Baker's, or even Gorbachev's. It was instead the ability to choose their careers more fully and freely, without party corruption and connections beyond their control. Their version of "democracy" is best understood less as the right to choose their own leaders than as a yearning for equal opportunity. What they largely wanted, at least initially, was not the government's downfall or a new revolution. They did not even wish to discard communism. They wanted instead the freedom promised to them in the Chinese constitution, and by Deng's own rhetoric, and hoped to make those more than just empty words but instead the basis of a new, cleaner, less corrupt, and more transparent China. "Students actually spent little time discussing election systems," Vogel has concluded. What they wanted was instead best described by a large protest poster hung at Beijing University: "Our demand: dismiss the incompetent government, overthrow autocrats, establish a democratic state, and found our society on education."

Many of the marchers who gathered to mourn Hu and simultaneously call for reform felt this was the patriotic test they'd been preparing for their entire lives. "We learned our political lessons too well," Ling Shiao later remarked. A participant in the early days of the protest, she operated one of the movement's few fax machines. "We'd learned in school and from our political education how to be revolutionaries," she recalled, "how to rally support for a popular cause, and how to spread messages throughout the people, learning at the same time that China's growth throughout the twentieth century had been fueled by one revolution after another." Not, she stressed, a revolution to overthrow the regime — not at first — but rather to save it.

The entire tone of the protests changed on April 18, three days after Hu's death, when the carnival atmosphere gave way to a more direct confrontation with government forces. Several thousand students marched two miles from Tiananmen Square to the Xinhua Gates at Zhongnanhai, the seat of the government, adjacent to the old imperial Forbidden City, demanding a hearing with the party's highest leadership. It was akin to a mob circling the White House or the Kremlin. Those on the outside cheered. Those within trembled. Premier Li Peng, recalling Deng's earlier admonitions against letting protests fester, argued for an unyielding response. Trained as an engineer in the Soviet Union, he believed in rules. The protesters were not authorized to be in the streets, Li argued. They should be disbanded at once.

General Secretary Zhao Ziyang, at the moment perceived by Li as a principal political rival, advised a calmer approach. It was in his nature to be flexible. Charged during the late 1950s with implementing ill-conceived agricultural policies that caused millions to starve, he had been beaten years later by Red Guards. They placed a dunce cap on his head before sentencing him to four years as a pipe fitter in the hinterland. The experiences of both decades haunted him, reinforcing his belief that dogmatic obedience to any cause led not to order but to suffering and chaos. Now he argued for treating the students with moderation. The crowds just want to feel they've been heard, he reasoned. Condemning their well-intentioned desires would only fuel their rage. Let's listen, he said — or at least pretend to — and let their zeal simply peter out.

Zhao's judicious position initially held sway, to a degree. Troops were dispatched to Zhongnanhai as a precaution but remained behind the gates. Party leaders largely refused to meet with the protesters, however, recalling that Poland's rulers had begun to lose power the moment they negotiated with opponents as equals. They demonstrated their general tolerance in less visible ways, ordering Tiananmen Square routinely cleared of the crowd's trash, for example, and having sanitation facilities installed for the protesters' convenience, all in the hope that without further provocation the students would vent their collective anger and then return to class.

The *People's Daily* warned "not to mistake the regime's forbearance for weakness," but the crowds instead only grew larger. They sang patriotic tunes and also the "Internationale," anthem of revolutionaries since the late nineteenth century, overtly signaling that they were now China's true patriots and heirs to communism's original values. Meant to terrorize kings, the song's words would strike fear in the heart of any regime: "No one will give us deliverance; no god, no czar, no hero; we'll arrive at our freedom only by our hand."

Party officials hoped the agitated students could be mollified if Hu received a hero's funeral, but the gesture only seemed to embolden them. More than a million citizens lined the route of the procession, marching with a massive portrait of their fallen leader, which they placed directly opposite Mao's in Tiananmen Square. Deng and his cadre left the formal ceremony at the Great Hall to the sight of more than 100,000 students silently standing in protest. A palpable sense of fear permeated China's leadership by the following day, as reports of vandalism and scuffles between protesters and police steadily poured into the government's headquarters. Rumors surfaced of Fang Lizhi's involvement, though none could be verified. "The turmoil on the campuses has already started to spread to the rest of society," Beijing mayor Chen Xitong warned. "The outlook is grim," and the protesters showed no sign of departing. On the contrary, their numbers kept swelling.

By April 25, a week after ceding control of Tiananmen to the crowds, Deng had seen enough. In Hungary and Poland, lingering protests had led to each government's collapse, he warned his colleagues. It was time for the party to reassert its unbridled authority. Mayor Chen joined the chorus. "At Peking University some students have imitated Poland's Solidarity to form their own Solidarity Student Union," he said. "These actions seriously harm social stability and unity, and they disrupt social order," Li Peng similarly warned. "Those of us on the Standing Committee all believe that this is turmoil and that we must rely on law to bring a halt to it as soon as possible." Vice Premier Yao Yilin agreed. "The nature of this student movement has changed. It began as a natural expression of grief, and has turned into social turmoil." For President Yang Shangkun the message was clear: "It's crucial that we maintain social order throughout the country, especially in the capital. We certainly can't allow a few people with ulterior motives to make use of this movement to manufacture turmoil."

Turmoil. More than just a word, it was an ominous code, recalling at once painful memories of the Cultural Revolution and the collective vow to prevent its recurrence. "This is no ordinary student movement," Deng told his fellow party leaders. "We've been tolerant and restrained," but because of this restraint, "a tiny minority is exploiting the students; they want to confuse the people and throw the country into chaos." The students claimed to desire only a better China, but in truth, "this is a well-planned plot whose real aim is to reject the Chinese Communist Party and the socialist system at its most fundamental level . . . We've got to be explicit and clear in opposition to this turmoil."

His words appeared the next day in a scathing editorial run by the *People's Daily* and were read aloud on television and radio. Its message was impossible to miss, summed up by its headline, "The Necessity for a Clear Stand Against Turmoil." Enough was enough. The students had to disband. The government would respond to their complaints and implement reforms accordingly, but it was time for the country's youth to return to their campuses. Any who subsequently remained in the streets clearly wanted more than an end to corruption; they wanted an end to China's communist revolution altogether. They were not heroes but traitors.

Deng's editorial only stiffened the students' resolve, as they rejected the idea that their efforts to improve their country could somehow be treasonous. Demonstrations arose in response in Shanghai, Tianjin, Hangzhou, Nanjing, Xi'an, Changsha, and Hefei. More than three thousand students from Jilin University occupied the offices of the provincial party committee. Railway officials reported a surge in passengers traveling to the capital. Many ominously demanded free passage as heroes of the people, a practice unseen since the days of the Red Guards.

The movement was growing, and becoming increasingly vitriolic. Deng found himself targeted for attack, mocked by students who carried banners displaying his previous statements on behalf of free speech and democracy. They called him a hypocrite and worse. "All innocent and well-meaning young students should understand that in any large mass event dragons mingle with the fish," the *People's Daily* editorialized the next day. "If you believe the insult, slander, and attack on party and state leaders in all the protest posters that appear everywhere, if you buy all the rumors about 'seizing power' and 'taking over,' if you skip classes and go networking, our country could well fall into chaos again." Having issued its final warning to little effect, the government effectively issued it again. The protests won't dissipate through mere words alone, Li Peng warned the Central Committee. "We must prepare for a long battle."

* * *

SEATED AT LUNCH in a posh restaurant near the Capitol building in Washington, James Lilley could pose only the most basic of questions to the man seated across from him: "Is this for real?" Winston Lord nodded in reply. He'd just returned from China. Lilley was about to take his place. "Are these demonstrations really against the government?" he probed further. From what could be gleaned from afar, said Lilley, "this isn't the China I know from the 1970s."

Lord nodded yet again. The calls for transparency that had begun in Beijing had spread to the hinterland. Whatever the protest movement was, it was no longer merely an isolated response to the passing of a beloved political figure but instead had all the makings of a mass movement for change, and all the ominous markings of a revolution. As he saw it, the dissatisfaction with socialism and desire for greater democracy that permeated Eastern Europe appeared to have finally infected China.

The size and scope of the protests took the Americans by surprise, even those who had been tracking China's budding democracy movement for years. "Frankly, we did not predict Tiananmen Square," Lord later conceded. "I don't think anybody did in terms of the massive demonstrations that did take place not only in Beijing but in over 200 cities." After four years as Ronald Reagan's envoy, he had come to believe that China was on the right path, and Sino-American relations as well, with bumps in the road like differences on trade or Tibet less important than the economic and strategic opportunities the two countries shared. "For the sake of world peace," China's foreign minister had toasted Lord during a farewell dinner, "good Sino-U.S. relations as well as friendship between our two peoples are indispensable."

That was a week ago. Lord now realized that their two nations approached much more than a mere bump, as the American zeal for democratic reform seemed about to run headlong into a band of Chinese autocrats determined above all else to retain power. He thought Bush had to back the protesters. That is what American presidents did when people called for democracy. Just as surely, Deng would have to react. Whatever happened next, he told Lilley, might well shape Sino-American relations for decades.

Bush couldn't have cared less what Lord thought at this point. He refused to take his longtime friend's calls after the embarrassing Fang Lizhi affair. Scowcroft wouldn't even read Lord's cables. Barbara Bush, who had known and trusted Lilley since 1974, listened intently as he explained why Deng couldn't simply crack down on such large crowds but dismissed his arguments once she learned he had been talking to Lord. In their long marriage, George Bush largely made and maintained their friendships. It was Barbara who ended them, and who kept track.

Lord was their fall guy for all that had gone wrong when Bush was last in Beijing. "They didn't do their homework," he later lamented, "and then they stuck it to me" when the February trip nearly turned sour. There was no other scapegoat readily available. Baker had wanted nothing to do with China in the first place. Preoccupied with countering Gorbachev in Europe, he was perfectly happy to leave Chinese affairs in Bush's hands, reasoning that the president knew the political landscape of his old stomping ground as well as any analyst. "In the case of China," Baker later recalled, "it's fair to say that very few policy initiatives were generated either by State or the National Security Council Staff during my tenure. There was no real need. George Bush was so knowledgeable about China, and so hands-on in managing most aspects of our policy, that even some of our leading sinologists began referring to him as the government's desk officer for China." The arrangement clearly suited the pair. "Within the administration," observed one NSC staffer, "there wasn't really much debate. I mean [the president] was the China desk officer . . . and there was always a sense that you couldn't really influence policy because he was really invested in these issues."

Bush, however, was no China hand. He had never learned the language, or delved deeply into Chinese history. He knew Deng, and Beijing's diplomatic circuit in 1974, but he had never been alone in the room with him without someone else translating their words. Neither had he the time for more than a passing interest in what had taken place in China since. But Lilley knew he had little to gain by forcing Bush to confront his shallow understanding of events in Beijing, and thus thought his best service would be as the president's eyes and ears on the ground in the Chinese capital, especially given his faith that he could send Bush private messages outside the State Department's cable system if need be. It would be just like old times in Beijing, when he'd sent secret messages from Bush back to Ford, beyond Henry Kissinger's ever-prying eyes. As he continued his rounds in Washington before departing, he found little consensus within the administration over what the protests meant, or even if they mattered at all. He met with Undersecretary of State Robert Kimmitt on April 26, for example, only to learn that the issue of student unrest wasn't even on their agenda. It was the same day Deng ordered China's students to disband

and return to class, enraging the protesters and sealing their fate. In the calm of Kimmitt's Foggy Bottom office, however, the focal points of discussion were matters of trade and military relations, regional issues such as Cambodia and Tibet, and the upcoming visit to Washington of Wan Li, a ranking Chinese official, scheduled for the following month.

Lilley landed in Beijing a week later only to feel that he had "stepped on a volcano." Packed streets ended abruptly in makeshift barricades as students and their supporters prepared for government assaults. Whatever discipline and restraint had governed the protest's first weeks were by now long gone. Newly arriving protesters felt little allegiance to student leaders they had neither met nor endorsed. Many had traveled for days, building energy and emotion along the way. Veterans of earlier marches preached organized calm. Newcomers longed to scream. Mob rule seemed increasingly likely. There were "several coups an hour," Ling Shiao reported, as would-be leaders grabbed the microphone from one another. Aspiring speakers took to surrounding themselves with increasingly large phalanxes of burly bodyguards, capable of clearing a path through the crowd or, when necessary, cowing opponents into submission. The rapidly deteriorating physical environment did not help. Government workers no longer patrolled the square picking up trash. Parts of it looked like a tent city, others a religious revival, still others a giant open-air classroom. The sanitation system failed. When the wind blew the wrong way, the place smelled like a sewer.

Wanting to see it all with his own eyes, Lilley hopped on a bicycle and headed for the square even before unpacking. He went alone, much to his staff's surprise, without a guard, an interpreter, or even a map. Their new boss was no staid Winston Lord, most comfortable in a Brooks Brothers suit and wholly reliant on aides to give him directions and to translate even basic Chinese. The son of a Standard Oil executive stationed in China in the 1930s, Lilley was instead fluent in Chinese, a bear of a man who grumbled when he talked, trimmed his fingernails with heavy scissors during meetings, and preferred a city's back alleys to its cocktail parties.

Embassy staffers wondered how two Yale men could be so different. While Lord had been schooled by Kissinger in diplomatic nuance, Lilley's formative education came in the jungles and slums of Southeast Asia, where he'd trained anticommunist rebels for the CIA throughout the 1960s before helping establish the U.S. liaison office in Beijing. He was the office's undercover spy, something Bush learned upon his own arrival in 1974. The two men subsequently bonded over whiskey sours, tales of their alma mater, and Ping-Pong, though Bush never managed to beat Lilley's son. While he worked through the day as a foreign service officer, Lilley's real mission, most often conducted at night, was clandestine: to establish his country's first on-the-ground intelligence network in China since the time of the Korean War. He searched out people willing to pass along information that might very well get them jailed, internally exiled, or worse, meeting in dark places where a knock on the door or the illumination of a flashlight might be the first sign of a police raid and a life suddenly extinguished.

Their time together in Beijing proved short-lived. A Washington reporter blew Lilley's official cover in the fall of 1974, forcing his expulsion from the country, even though the Chinese already knew his real identity. Kissinger and Zhou Enlai had secretly agreed that each nation could include one intelligence officer on its embassy staff, in the kind of informal arrangement frequently employed during the Cold War to limit dangerous surprises. "Henry actually told them [his State Department aides] that this deal had been made personally with Chairman Mao," Lilley later joked. No matter that the Chinese government knew of his true employer, he could not stay once outed. Unable to see his subordinate off at the train station, Bush instead characteristically sent a handwritten note. "As you leave Peking tuck away into your heart of hearts the fact that you three have done well," Bush wrote of Lilley and his wife and son. "You've made friends. You've given to your country in a tough assignment. Bar and I will miss you much."

The pair would work together again when Bush headed the CIA, remaining friends during the 1980s, when Lilley served in the White House before his appointment as de facto American ambassador to Taiwan and official ambassador to South Korea. He was an obvious choice to be Bush's new man in Beijing. By 1989 Lilley was at the pinnacle of his career and, when wading into the mass of protesters, in his element as well. Having seen Taipei and Seoul rocked by demonstrations, he quipped, "Demos seem to follow me around."

He could walk Beijing's streets in a way China's current leadership could not. More than 150,000 demonstrators pushed past police barriers the day after Deng's editorial was published in order to march through the city's streets. "You know you can't stop us," one student leader challenged the police commander who stood in their way. "You're only three deep, and there are thousands of us. Why don't you let us pass?" Then he warned, "If you don't let us pass, we're going to rush your line, and none of us will be able to control the students."

The logical appeal worked. "He didn't say a word to us," the lead protester later recalled, "just ordered his men to move aside. Wang Dan [another prominent student democracy advocate] and I looked at each other and smiled, surprised it had been so easy." It was in fact deceptively easy. The police were largely unarmed, and by some accounts intentionally undermanned, lest they act with the overconfidence that often accompanies a gun or masses of comrades. Many sympathized with the students' demands in any case. "We absolutely have to stop the student movement from getting bigger!" Bo Yibo told the Politburo's Standing Committee in response to this news on May 1. "Students are networking all over the country. This is serious!"

The growing protest movement laid bare the struggle for succession at the heart of China's ruling circle. Zhao Ziyang once more called for restraint. The protesters had a point, he argued. "The vast numbers of youth . . . hope to promote democracy and call for punishing people who are . . . guilty of corruption. This is also the exact intention of our party." The Politburo must lead by example. "My children should be the first to be investigated," he volunteered. "If they're guilty, they should be punished, and the same goes for me if I am found to have any involvement." The offer made many of his senior colleagues cringe. They had long taken advantage of their privilege. Zhao thus made himself as dangerous in the eyes of some Politburo members as the crowd outside.

Corruption was less the issue than counterrevolution, Deng maintained in response. Just look at Eastern Europe. "These people . . . have been influenced by the liberalized elements in Yugoslavia, Poland, Hungary, and the Soviet Union," Li Peng argued. "Their motive is to overthrow the leadership of the Communist Party and to forfeit the future of the country and the nation." He wanted to act. So too did Deng, who rarely made a move without first ensuring widespread support. "We are not afraid to shed a little blood, or to lose face," he told party elders, slowly building consensus on action, "since this will not seriously harm China's image in the world."

Zhao pleaded with them to see events in Europe as instructive rather than frightening. "Times have changed," he said. "Democracy is a worldwide trend, and there is an international countercurrent against communism and socialism that flies under the banner of democracy and human rights." They could never hope to sail against the stream of history, so they had better try to ride its current as best they could, because ultimately, "if the Party doesn't hold up the banner of democracy in our country, then someone else will, and we will lose out."

While the Politburo debated, the protesters' radicalization intensified, their optimistic chants increasingly turning to vitriolic barbs, then to flat-out insults. Reluctantly agreeing to discuss their demands on May 19, for example, Li Peng found himself confronted by a youthful counterpart still dressed in a hospital gown after treatment for the effects of his hunger strike. For all practical purposes, Wuer Kaixi was wearing pajamas to a meeting with China's premier. Li's was a "haughty and high-handed attitude," the youth charged at the meeting. He, conversely, hoped to "create an equal status between the people's movement and the government" by eschewing formalities like proper clothing, because he considered the government rife with "reactionary warlords, reactionary government, and fascist military" and thus unworthy of the people's respect. Each of these was a term used by previous generations of communists to debunk the prior regime's legitimacy. Student leaders similarly mocked party officials by kowtowing, as though to China's emperors of old, when delivering their manifestos. It was an infuriating gesture to men who had told themselves over the course of a lifetime that they were the solution to the decadent imperial system.

Li listened passively to the insults hurled his way by a man barely one third his age, seething inside as his public shaming was broadcast live on national TV. Who was this mere child to question one of the regime's most accomplished leaders? What had any of this "hothouse generation," raised in the fertile soil provided by Li's comrades, done besides sit in comfortable classrooms? They knew nothing of long marches or sacrifice. Apparently they knew nothing of respect as well. "We were appalled at the rantings of the student leaders," one senior Chinese diplomat later recounted, and offended by the very notion of being lectured by anyone who "had been a month-old infant when the Cultural Revolution began."

Hope for a peaceful resolution faded in Washington, which by mid-May could no longer ignore the growing turmoil on the far side of the world. After reading intelligence reports from Beijing, Gates told his staffers a clash seemed in the offing because the Politburo appeared fearful of losing not just power but their very lives. "They were afraid that these demonstrators would move on the leadership compound," he recalled. "There began to be an actual physical fear on the part of the Chinese leaders for their own survival." Lilley echoed his concern, warning Washington that no previous group of party elders had ever countenanced such sustained critique without striking back. A true reformer, Deng was also an "Old Testament" leader, Lilley reminded them, perfectly willing to punish sinners, especially those who violated the commandment to respect one's elders. Deng knew whom the students meant by their banners that read "Down with the Dictatorship of the Individual" and "Government by Old Men Must End!" He also knew how to respond. "Revenge was in his nature," Lilley warned.

He and other Western diplomats did what they could to ease tensions, reminding their Chinese counterparts that the world was watching. Opportunities for face-to-face discussion dwindled, however, as the government hunkered down, and by mid-May few Chinese officials would accept his call. Finally arranging a meeting with Li Peng, Lilley dared not raise the issue of the protests directly (even though echoes of the chanting masses outside could be heard throughout the meeting), lest he be accused of meddling in China's internal affairs. He instead tried an indirect approach, recalling the student protests that roiled his alma mater in the 1960s. Lilley noted in particular the conclusion reached by Yale's president at the time: "If you lose your youth, no amount of crisis management will make much difference in the long run."

"No government in the world would tolerate this kind of disorder in the middle of its capital city," Li angrily shot back, calling their meeting to an abrupt halt. Lilley tried anew with Foreign Minister Qian Qichen. "If things go wrong," he warned, "the Western media will go after you like a mad dog." In Lilley's retelling, Qian just smirked in response.

Bush pleaded for restraint as well. He had been mildly encouraging of the protest movement at first but now sensed the crisis would not end well. "I have words of encouragement for freedom and democracy wherever," he told reporters on May 4. While "I wouldn't suggest to any leadership of any country that they accept every demand by every group . . . I will say that as I reviewed what the demands are today, we can certainly, as the United States, identify with them." Responding to another question, he went out of his way to couple events in China with those in Europe and beyond. "Democracy is on the move," he said, and "we're going to be on the side that is winning and the side that is right, fundamentally right. Freedom, democracy, human rights, these are the things we stand for. So, I would encourage every government to move as quickly as they can to achieve human rights."

It was Bush's final comment of the May 4 news conference, though also the first time a reporter had raised China in the session, after thirty-seven questions about NATO and other topics. The American press and public were only vaguely aware of the furor brewing in China at the start of the month. When talk turned to foreign affairs, it was Europe, and Gorbachev, that remained foremost in their minds.

That would change by month's end, as Tiananmen increasingly seized global attention. Bush's words grew more cautious as Beijing's government and its critics appeared at an impasse. "I think this perhaps is a time for caution," he said on May 22, three days after Li Peng's quickly infamous dressing-down by student leaders. France's François Mitterrand stood at Bush's side, having journeyed to Kennebunkport for a day of meetings and boating. Yet despite the presence of a major ally and the looming NATO summit the following week, China dominated the discussion. "Do you have any message for the students, other than that the United States supports freedom of speech and freedom of assembly?" the first reporter asked. "I don't want to be gratuitous in giving advice," Bush responded, "but I would encourage restraint."

"We aspire to see the Chinese people have democracy," he continued, but "we do not exhort in a way that is going to stir up a military confrontation . . . And so, as we counsel restraint and as we counsel peaceful means of effecting change — that is sound advice. And I think to go beyond that and encourage steps that could lead to bloodshed would be inappropriate." Ruminating further in response to a follow-up question, Bush grew more philosophical: "I lived there; I saw a society totally different than the one that exists in China today. China has moved, in some areas, towards democracy. Now, the quest is — and the appeal from these kids is — to move further. And so, I am one who feels that the quest for democracy is very powerful, but I am not going to dictate or try to say from the United States how this matter should be resolved by these students. I'm not going to do it."

Bush fell back on first principles: first do no harm, prioritize stability, and let history run its course. "I am old enough to remember Hungary in 1956," he told reporters. "I do not want to be a catalyst for encouraging a course of action that would inevitably lead to violence and bloodshed." Indeed, asked if there was "anything concretely the West can do to help the process without interfering," Bush conceded, "Not that I can think of . . . sometimes, well-intentioned statements can be counterproductive." The Chinese were famously sensitive about perceived slights to their national sovereignty, he added, noting, "I do not think any outside country should dictate what happens inside another country."

Baker delivered a far different message in private when Wan Li came to Washington on May 26. The highest-ranking Chinese official yet to visit the Bush White House, Wan chaired the People's Congress Standing Committee. He was also one of Bush's old tennis partners from his Beijing days, and thus a man the Americans believed they could speak with candidly. "The United States is watching with extreme interest what is happening in China today," Baker said, and "supports democratization and the freedoms of speech, assembly, and association everywhere in the world." There could be no question of where the administration's sympathies lay, nor should the Beijing regime have any doubt as to the attitude of the American public, who in Baker's diplomatic words would have an "adverse reaction" to any "setback" for those causes in China. At the same time, Baker continued, the administration was not unsympathetic to the regime's plight. "Instability in China was not in the best interests of the United States," he told Wan, and expressed the hope "that both sides would exercise restraint and avoid bloodshed."

Wan made it clear that such advice was not just unwarranted but unwanted. "The present unrest in China was the biggest since the founding of the People's Republic," he said. "The Chinese government has adopted measures to stop the current unfortunate unrest in order to restore stability and unity." While it was true that "the Chinese constitution stipulated free speech, association, and assembly as well as democracy," it was equally plain that "these required order." Thus, while he of course agreed with Baker that each side should "exercise restraint" and avoid "violence or bloodshed," he was not there to make promises. "There might be a possibility that bloodshed could not be avoided," Wan flatly stated. "Some things were unavoidable, independent of man's will." It was to be hoped that in the next day or two the crisis would subside. "However, one should not exclude the possibility of unfortunate incidents." Conceding the point that it was not up to outsiders to decide how the Chinese handled their internal affairs, Baker wrapped up the meeting with a final reminder that "it would be significant for Sino-US relations were restraint to give way to violent measures."

When Wan met Bush in the seclusion of the president's personal dining room, he confessed to being "deeply worried." The two men had hoped to renew their tennis rivalry on the White House court. Circumstances prevented a rematch, as no one in the West Wing judged that image worth risking as students and police stared each other down back in Beijing.

They could, however, still meet for tea. Bloodshed had not yet occurred, Wan said, his voice tinged with fear, and "under no circumstances should there be bloodshed." But with protesters calling for Deng and Li to resign, there was no telling what might happen if the students failed to disperse. "It was Deng Xiaoping who first instigated the talk of democracy," Wan noted, "but in recent years, the Chinese have slowed the pace of democracy." He knew that calls for reform had merit, and hoped to return home and renew that push for "democracy and free expression by the people." Watching his old friend clearly struggling under the weight of events, Bush offered sympathy rather than the warning he'd planned. He could never back away from endorsing democracy, but he promised Wan he'd keep his public comments in check lest he make matters worse.

Bush never raised the specter of damage to Sino-American relations if blood were spilled. He never delivered any kind of threat of sanction or retribution should Deng's government crack down with force. He felt that message had already been conveyed by Baker in Washington and Lilley in Beijing. His goal during these sessions, as China's old friend, was instead to provide wise counsel above the fray by reassuring the Chinese policy makers he knew that he understood their plight, and that he would do everything he could to give Deng and those around him the room to make the right decisions. He also knew from his intelligence reports that not everyone within the regime backed a hard line, and he did not want to do or say anything that might in any way make it harder for proponents of a peaceful resolution, including Wan, to marshal their own forces. Given the chance to speak directly to a man whose voice carried weight in Beijing, Bush effectively offered more comfort than warning.

He nonetheless was more direct with Deng, or at least as direct as possible, composing a long personal letter while on Air Force One. It was precisely the type of move Bush's sense of personal diplomacy prescribed, being at once frank, friendly, and most of all private — at any rate as private as two world leaders could expect. Anything he wrote Deng would of course most likely be read by the latter's closest advisers, and maybe even made public. But he hoped nonetheless to appeal to Deng's sense of friendship.

"I have no desire whatever to be seeming to interfere in the internal affairs of the People's Republic," Bush began, acknowledging the sensitivity of China's leaders to perceived impositions on their sovereignty. "And I want you to know that I am determined to achieve the closest possible relations between our two countries." Yet he also felt compelled, he added, to "express to you my hope that there would be no outcome with respect to the student demonstrations which would interfere with my ability to pursue the kinds of policies which would promote the goals we seek in our relationship. Specifically, it would be my hope that any solution you decide upon would avoid violence, repression, and bloodshed." There would be consequences to violence, he wrote, prompting a public backlash in the United States sure to reduce if not wholly eliminate his ability to sustain Sino-American relations. Leaders like Bush and Deng, old friends charged with safeguarding their nations, should do nothing to tie their own hands. Most of all, Deng must have no doubt of one thing: the world was watching.

9

CRACKDOWN

THE LAST THING THE Chinese leadership needed at that moment was a visit from the world's most famous reformer. Yet Gorbachev was due to arrive in mid-May, amid a flurry of international media eager to document the long-anticipated end to Sino-Soviet antagonism. Each of the major American network anchors (Dan Rather, Peter Jennings, and Tom Brokaw) would be on hand, for example. Having come to Beijing to document a diplomatic summit, they were not about to ignore the more enticing story of a democratic uprising playing out before their cameras. Whatever happened next would be filmed. "When Gorbachev's here, we have to have order at Tiananmen," Deng demanded four days before his arrival, saying, "Our international image depends on it." His deputies agreed. "Tiananmen is our national face," Yang Shangkun responded. "Especially when Gorbachev's here, we just can't let it turn into a stinking mess." Soviet officials weren't happy with the situation either, fearing both that they would confront angry mobs and that Chinese leaders might blame *perestroika* for their troubles. "We don't want our own Fang Lizhi incident," one Soviet diplomat fretted to his American counterpart.

Deng ordered the streets cleared. But do it peacefully, he instructed, so foreigners would have nothing terrible to report. The crowds grew instead, as did tensions within Zhongnanhai's high walls. "The students have moved steadily from sticking up posters to demonstrating to occupying radio stations to forming illegal organizations—and now they defy us with a hunger strike," Chen Yun told Vice Premier Bo Yibo. "If we don't call this turmoil, how can we face the memory of the tens of thousands of martyrs who shed their blood for China's revolution?"

Unable to keep the protesters from Tiananmen, Deng's government tried instead to keep Gorbachev from the protesters, altering his meticulously choreographed itinerary on the fly. Scrapped was the elaborate formal reception planned for the Great Hall of the People. Yang Shangkun officially welcomed Gorbachev at the Beijing airport instead, though no one thought to tell the Soviets of the change until their plane was well into its final approach. Dignitaries and reporters were hastily bused from the capital to witness the event, where they found neither seats nor bleachers. There wasn't even time to roll out a ceremonial red carpet long enough to meet the plane's staircase. The longer official one remained back in the city.

It was not an auspicious start. Deng insisted on holding his personal sessions with Gorbachev inside the Great Hall despite the throngs of students surrounding the building. He had worked too long to repair Sino-Soviet relations to surrender his triumph to a mob. Outside, more than three thousand hunger strikers stood watch. Two hundred had thus far been rushed to the hospital. The crowds' chants echoed throughout the building. Demanding an audience with the visiting Soviet delegation, and with Gorbachev himself, students hurled themselves at the doors of the Great Hall. Bones were broken. Gashes were opened. Windows were smashed. Outside, cries of pain mixed with chilling calls for action. Inside, tensions grew. Chinese security officials screamed for order while their Soviet counterparts huddled to develop an evacuation plan should the mob burst through. Deng continued all the while as though nothing were out of the ordinary. Yet cameras from the world's major news outlets filmed everything.

"Where is our Gorbachev?" students chanted, the image beamed to the world. "Where Are You, China's Walesa?" a large banner read, referring to the inspirational leader of Poland's Solidarity movement. The more they chanted, the longer Deng seemed determined to hold Gorbachev's attention. "We do not believe that our views were always correct," Deng acknowledged, before launching into a lengthy lecture on why the Soviets were largely to blame for the great rift in the communist world. His ensuing monologue stretched well beyond an hour. Though on the outside he seemed calm to the fullest, Deng's tension ultimately betrayed him. His hands shook, food dropping from his chopsticks throughout their lunch banquet. He seemed tired and old.

Just as they had with Bush earlier in the year, Chinese officials scheduled individual sessions with Gorbachev, whose conversation with Zhao Ziyang proved particularly significant. The last major voice within China's ruling circle for restraint in dealing with the protesters, Zhao twice revealed too much. "We still need Comrade Deng at the helm when it comes to most important questions," he told Gorbachev, directly violating his senior comrade's desire to appear, officially at least, as but one voice among many. The gaffe further eroded Deng's confidence in Zhao's abilities, as did word that he and Gorbachev had discussed the sea of dissenters mere yards away. Deng had assiduously avoided the topic during his own talks and expected underlings to do the same. Even worse, Zhao had apparently raised the subject on his own. "There is the sense of a lack of mutual understanding between Party and state institutions on one side and young people and students on the other," he told Gorbachev. There were four generations of Chinese in the country, the oldest of which had joined the revolutionary triumph of 1949, and "mutual understanding among them is very important. I belong to the second generation, the students to the last, and Deng Xiaoping to the first." None seemed to understand the others clearly, he added. "We are running into the same general problems," Gorbachev said, nodding in agreement. "We have our hotheads, too."

Gorbachev's visit ended without major incident but was nonetheless profoundly embarrassing to China's rulers. His departure removed their last barrier of restraint. The Politburo's Standing Committee held yet another emergency session late that night. "It's what you could call anarchy," Yang Shangkun exclaimed. "We are having a historic Sino-Soviet summit and should have had a welcoming ceremony in Tiananmen Square, but instead we had to make do at the airport . . . Pretty soon we won't be able to call this capital of ours a capital anymore!" The next day Deng declared, yet again, that he had seen enough. "Of course we want to build socialist democracy, but we can't possibly do it in a hurry, and still less do we want that Western-style stuff," he told underlings. "If our one billion people jumped into multiparty elections, we'd get chaos like the 'all-out civil war' we saw during the Cultural Revolution. You don't have to have guns and cannon to have a civil war; fists and clubs will do just fine.

"After thinking long and hard about this," Deng continued, "I've concluded that we should bring in the People's Liberation Army and declare martial law in Beijing." Zhao tried one more time to change Deng's mind. "Comrade Xiaoping," he demurred, "it will be hard for me to carry out this plan. I have difficulties with it."

"The minority yields to the majority!" Deng erupted in response. Zhao knew then he had lost. Martial law would go into effect on May 19, the second day after Gorbachev's departure. Zhao left the meeting and wandered, dazed and weeping, through the sea of protesters gathered in the square. "We have come too late," he told them, imploring them to leave while they still had the chance. "I have to ask you to think carefully about the future . . . I beg you."

Zhao's plea sealed his fate, demonstrating in Deng's eyes intolerable weakness. "Did you see that Zhao went to Tiananmen and spoke?" he asked Yang Shangkun. "Did you hear what he said? Tears were streaming down his face, and he really tried to look mistreated. He's flouted Party principles here — very undisciplined." Zhao was subsequently confined to house arrest. He never emerged. Deng lived another eight years. Comrades for decades, the two men never spoke again.

American officials monitored it all, paying particularly close attention to reports of troop movements closing in on the capital. Before the regime could do anything in response, they reasoned, it would need to cock its fist. "The events of May 19 constituted a coup engineered by Deng Xiaoping," Lilley reported from Beijing as word leaked out of Deng's decision and Zhao's defeat. "We are in the midst of a major power struggle in China." Rumors abounded. Zhao was under house arrest (true). Hard-liners had diverted Wan Li's return flight from the United States to Shanghai, barring him from Beijing until they could be confident he wouldn't disrupt their plans (true). Deng had left for Hunan, in Lilley's words, to "rally a reluctant military behind martial law and military intervention." That piece of information proved only partly correct. Deng had not left. Neither, however, was he convinced that every officer of the People's Liberation Army (PLA) would follow his orders.

Bush's advisers increasingly feared they were witnessing the first stages of a full-scale civil war. Conservatives and reformers seemed ready to square off within the government, each bringing his own loyal military and police units to bear. There was no guarantee that soldiers would willingly strike their own people, Lilley advised Bush, no way to feel confident that some units might not fire on those who did, and no guarantee that Li Peng and Deng would remain in power even if they succeeded in clearing the square. In Lilley's dire warning, the conflagration on the horizon might make the simple trampling of democratic protesters seem whimsically tame. The world had never witnessed a nuclear power come apart at the seams. Not, perhaps, until now.

Ranking Chinese officials feared the same, some even admitting as much with remarkable candor. There were three possible outcomes to the crisis, Ji Chaozhu lamented to the American consul in Hong Kong. None were good. Raised in New York, Ji had attended Harvard before joining the revolution, quickly climbing the ranks as a translator for both Mao and Deng, owing to his ability to mix linguistic proficiency with far more personal sensitivities. Mao was nearly deaf by the end. Ji spoke directly into his ear. Deng appreciated how Ji would bend at the knees rather than the waist when whispering into his own ear, making his short stature less obvious to the cameras. Trusted and well liked, Ji had recently been named ambassador to Great Britain.

Ji took a distinct risk discussing his party's inner workings so openly with his American counterparts while on a Hong Kong layover, especially as he predicted the crisis would lead to "a genuine revolution" and "major changes within China, including some leadership changes." Repression might not work, "given the magnitude of the demonstrations," he said, and China would then likely "enter a period of turmoil and indecision, with no one clearly in charge." According to his American contact, Ji "clearly supported the demonstrators and their cause, and was hoping for an outcome that would bring greater liberalization and democracy to China." Not everyone within Beijing's high command agreed with what was about to take place, he warned, but all knew what was coming. Now, so did the Americans.

On May 19, government loudspeakers on the square ordered the crowds to disperse by midnight. "We have no alternative but to call some troops to Beijing," Yang Shangkun declared in a broadcast message. Western broadcasters were summarily shuttered. CNN's directors protested that they had "official" license to stay on the air. An officious Chinese bureaucrat simply wrote out a new "official" cessation order. The network's transmission from Beijing turned to static, but not before the entire scene had been captured live, its replay boosting ratings back home and around the world.

Dan Rather of CBS found himself in similar straits. "Midnight, the deadline passes. More reports and rumors of troop movements, many close by," he reported live from the square. Dramatic even in tranquil times, his gripping broadcasts, now coming from the midst of the protests, repeatedly interrupted regular programming back home. Midnight in Beijing was noon in Washington. Whatever happened by daybreak would appear on the prime-time news on the American East Coast.

12:30 AM: Premier Li Peng appears on radio and television, speaking to an audience of top officials and over loudspeakers to the crowds. He says anarchy and riots are threatening the state, threatening to wreck the economy. He blames it on a small group of conspirators trying to overthrow communist rule. He declares a kind of martial law to put down the protest. Those listening on loudspeakers jeer and chant, "Li Peng come down! Li Peng come down! . . ."

2:00 AM: The army still hasn't arrived. Truckloads of strike supporters begin coming in again. The students dare to hope . . .

4:00 AM: Reports increase that the army has tried to move closer but has run into passive resistance from the students . . .

By dawn's early light, the students were still here. They had won the night.

Rather's crew then lost contact with their producers back at broadcast headquarters, a maze of offices on the fifth floor of the Shangri-La Hotel. Fearing the worst, they hustled back to find government officials dismantling their satellite feed. Urged to buy time so additional footage could be uploaded to the network, Rather walked the increasingly agitated bureaucrats around and around their ramshackle studio, a camera crew beaming home their half-hour argument. Eventually the jig was up. "This is Dan Rather, CBS News, Beijing," he began, speaking directly to the camera. He never finished. The feed went to black. "When they shut off live television," he knew "it was only a matter of time before they would move in," he later told an interviewer.

What happened that night and over the next week was something wholly unexpected. The troops moved in, but the protesters refused to budge. More important, Beijing itself rebelled. Over fifty thousand soldiers attempted to breach the city limits on the night of May 19–20, but citizens from all walks of life rallied, surging into the streets, shutting down each of the major and lesser routes in and out of the central area. A full moon helped their cause, offering just enough light for organizers to quickly reposition their meager resources against the expected onslaught. Ordinary civilians erected barricades and turned back the troop-laden trucks and buses. They lay on railway tracks and against shuttered subway entrances, preventing hundreds of soldiers from emerging in the middle of Tiananmen, effectively trapping whole units underground with no option save retreat.

The masses won the first round, but they weren't facing the army's best troops, primed for battle. Most soldiers sent into Beijing that night arrived unarmed, a testament to the government's hope that violence could be avoided. Many appeared disoriented, lacking adequate maps of alternate routes beyond their designated roads. Once stopped by unyielding crowds, many of the youthful soldiers—largely less educated than their university-trained opponents—simply wandered in a daze as crowds swarmed their vehicles. Rarely in their lives had China's citizens stood up to anyone wearing a uniform, yet city residents badgered soldiers to fall back, to join their side, or simply to refuse to play a part in the repression. Many did join the cheering crowds. Others just stood and stared, bewildered, without orders and without direction. Some just walked off into the night. This was not how their world was supposed to work. "Even in the restroom, there was no reprieve," one soldier later recalled. "If one student would go hoarse yelling, another would take his place."

"We had not expected great resistance," Li Peng soberly recorded in his diary. The leaders had anticipated that the crowds would melt in the face of the widely revered PLA. By 4:30 on the morning of May 20, student-controlled loudspeakers instead overwhelmed the commands of their government antagonists, the square still in the people's hands. The crowd cheered.

Deng and Li schemed anew, the former fearing that the "soldiers' hearts may not be steady" enough for what would come next. They had erred in sending in soft units when hard military might was necessary, and yet the scene repeated itself the next two nights. By the morning of May 22, the last of the PLA's stymied troops were ordered out of the city. Beijing was officially in opposition hands. "Today is the people's day," the loudspeakers blared over and over. "Tomorrow is the people's day. Forever is the people's day!"

"A confrontation resulting in bloodshed is probable at this point," Lilley warned officials back in Washington. There was nothing more to do to affect matters on the ground, so American officials should begin planning for the aftermath. "As a matter of policy," he advised, "the United States should consider now taking measures to distance ourselves from the Chinese authorities who appear to be getting ready to crack down on their own people." In his view, betting on the hard-liners was a poor choice. "This current PRC government is not a strong one," Lilley concluded. "It may not last long, despite its ability to suppress an uprising. What is happening here in opposition to the authorities has a permanence about it. It is not going to go away." They could retake their city, he reasoned, but in doing so lose their country, and Bush and his advisers back in Washington would be wise to plot not just their immediate reaction but instead their long-range response to the new Chinese realities likely to emerge from the chaos, and to the new leadership likely to emerge as well. In hindsight, of course, he was wrong to predict the regime's demise. But from his street-level perspective it was hard to believe the protesters would ever go away peacefully at this point.

With violence likely, there was little left to do, Lilley argued, other than to "come down on the side of our principles and our own interests." He'd endorsed Bush's decision to remain quiet thus far but now urged him to make a forceful show of support for the protesters, and to work equally hard to preserve Sino-American trade ties no matter what pressure the public at home might bring to bear. "You might ask," he continued, "doesn't this play into Chinese hands, separating economics from politics?" It was only economic progress that had made the country's political change possible in the first place, he answered. Thus the only way to keep any hope for democracy in China alive following the impending political crackdown was to keep its circle of prosperity expanding even amidst the violence and repression. "Economic forces and growing political awareness in Shanghai, Guangzhou, Hong Kong and even Taiwan can influence the long-term evolution in China towards a freer economy — and a more open system," he wrote.

No matter what happened next, the United States and its allies needed to remain involved in China as much as possible, he advised Bush. Reflexively cutting ties in response to violence was the last thing the administration should do. The pressure to sever ties would be nearly overwhelming, but "if we are major players in this economic game and we support economic strength in China and support our natural allies who are also in Beijing" while quietly opposing Deng's hard line, "we could end up on the right side of the winners in the long term in China."

BACK IN WASHINGTON, Bush's team watched and waited. James Baker planned to spend the first Saturday of June playing golf. The weather was perfect, and the president had decamped to the Maine coast, which made it the ideal moment for a tired secretary of state to get in a quick eighteen holes. He phoned his eldest son and told him, "Grab your sticks and come over right now." There was a pause at the other end of the line. "I don't think you are going to be playing any golf today," Jamie Baker replied. "I'm sitting here watching tanks roll through Tiananmen Square on CNN."

Then followed another pause, ended only by a quiet "You're kidding me . . ." Baker's other phone began ringing, the one connected to the State Department's Operations Center. China had erupted, the duty officer calmly informed him. Armored units appeared to be firing directly into the crowds massed in Tiananmen Square. Embassy officials in Beijing reported heavy casualties in the streets and a city under siege.

The long-expected crackdown had begun. Before it was over, hundreds would be dead, thousands imprisoned, and Sino-American relations would plummet to depths unseen since the height of the Cold War. Thanks to satellite transmissions, people throughout the world learned of the violence even before the chief diplomat of the world's most powerful country. "Not since the Vietnam War had Americans witnessed such dramatic images in their living rooms," Baker later wrote, but whereas in the 1960s war footage typically took hours or days to appear on people's TV screens, "the carnage in Beijing was captured live and relayed instantly."

The bloodletting began the night of June 3, the day after another stinging embarrassment for the increasingly beleaguered regime. Thousands of largely unarmed soldiers streamed into the city in the late evening hours of Friday, June 2, in one final attempt to retake Tiananmen without resorting to lethal force. They never made it. Barricades halted their progress as crowds reminded the soldiers that the People's Liberation Army was there to protect the people, not the authorities. On one street an elderly woman openly berated a group of shamefaced soldiers staring at the ground. Another avenue saw a ring of humanity surround packed PLA buses. Taunts and jeers flew through the windows, while city residents climbed on the roof to have their picture taken with helmets and rifles swiped from within.

The rifles themselves were significant. Most of the soldiers arrived unarmed, but not all. News quickly spread through the city that the government had this time brought the means to kill, while a similar chill ran through police and army circles when they realized that their next trip into the city might be met with bullets rather than rocks. Communist China had no enshrined right to bear arms and no history of widespread gun ownership. The sight of citizens wielding weapons — AK-47s capable of firing more than one hundred rounds a minute — consequently startled officials and demonstrators alike, the latter placing their newfound weaponry on display as stark validation of their warning that the regime could no longer be trusted.

Several hours later some two hundred police rushed the crowd in the hope of seizing the lost guns and freeing the captured soldiers, wildly swinging clubs and heavy belts while firing tear gas indiscriminately. The crowd fought back. Everywhere, glass and bones were being broken. Within hours the first real battle was over, and the people had won. "History will show that this day will be a symbol of shame, a day that the people will always remember," protest leaders broadcast in triumph. "On this day the government has ripped off the last shred of the veil covering its hideous visage . . . they have used every type of weapon, from tear gas to electric truncheons. We no longer hold out any hope whatsoever for this government. We now solemnly declare: if Li Peng's government is not brought down, China will perish . . . [T]hus we solemnly state that our rallying cry is: Down with Li Peng's government! . . . Victory is near!"

"Whatever the plan was," Lilley reported home, "the result was a complete disaster." The "vaunted People's Liberation Army seemed to wilt in the face of people power." His defense attaché called it "a Chinese version of Napoleon's retreat from Moscow," to which the ambassador sarcastically added in his dispatch back to Washington, "At least Napoleon's troops did not have the humiliation of being lectured in their own language by grandmotherly women calling them 'bad boys' or 'a disgrace to the PLA.'"

Scheduled to appear on CNN that afternoon in Washington, Baker cabled Lilley for advice. Never shy about offering his opinion, the ambassador suggested making a "passionate statement in favor of democracy and against the use of military force." Americans traditionally supported citizens agitating for democracy, Lilley reminded his boss. Deng's cadre would explode at the first sign of any direct criticism from Washington, but it was a regime whose days increasingly appeared numbered. It was thus in his opinion "worth ruffling a few feathers in the short term to maintain our long-term relationship with China's future."

These were among Lilley's last calm words. Saturday was Beijing's last good day. Tension rippled through the streets as evening approached. Lilley upgraded the likelihood of violence from "probable" to "imminent" as people poured out of the city. For the first time in weeks, railroad officials reported more students departing Beijing than arriving. Those who remained in the square, whether newly arrived or the most passionate for reform, pledged, "I swear to devote my life and my loyalty to protect to the death Tiananmen Square, the capital Beijing, and the republic." Student leaders hoped the magnitude of those words would unify their cause, which showed signs of fraying. Pockets of demonstrators scattered every few yards heatedly debated the wisdom of staying, in turn questioning one another's dedication and courage. Votes were held, resolutions debated. Followed by more debates, more resolutions, and yet more votes. Having called so long for democracy, yet having little practice in its execution, the demonstrators, with no one in charge, quixotically strove for unanimous consent. Anxiety grew inside the rapidly constricting circle of protesters. Outside, a carnival atmosphere of sorts prevailed, with some bystanders even bringing their children to witness the scene. It was a warm day, and pushcart vendors did a land-office business in ice cream and popsicles.

In the square itself, students from Beijing's Art Academy had only days before erected a towering statue of a robed goddess holding a lamp aloft. Placed a stone's throw from the Monument to the Martyrs of the Revolution, their new Goddess of Democracy, which bore a striking resemblance to New York's Statue of Liberty, stared directly into the eyes of Mao's massive portrait overlooking the entire square. The similarity was no accident. Neither were the banners written in English and held up by the chanting students who circled its base. "Do you want me to take you to see it?" an American Foreign Service officer's cab driver asked her while driving her home from the airport on June 2. She had been in Hong Kong for a few days and replied that she wanted instead to be taken straight home. "Perhaps I will go see it tomorrow," she told him. "Tomorrow it won't be there," he responded.

Tensions were running high in Zhongnanhai, as China's most powerful leaders reassured themselves that what they were about to do was their duty but not their fault. Outside agitators were to blame. Especially the Americans. "Each American administration has done a great deal of mischief aimed at overthrowing the Communist Party and sabotaging the socialist system," the State Security Ministry reported to the Politburo. "The phraseology may vary, but the essence remains the same: to cultivate so-called democratic forces within socialist countries and to stimulate and organize political oppositions using catchwords like 'democracy,' 'liberty,' or 'human rights.'" Their goal was nothing less than counterrevolution. Western democracy advocates had funneled money and supplies to the Tiananmen protesters. Voice of America had helped spread their message. Hadn't Bush repeatedly endorsed the demonstrators' goals of a more democratic China, calling freedom the wave of the future? "Western capitalism really does want to see turmoil in China," Li Xiannian told the Politburo's Standing Committee. "And not only that; they'd like to see turmoil in the Soviet Union and all the socialist countries of Eastern Europe."

Foreign imperialists were the real problem, Deng told his tired colleagues, just as in the time before their revolution. "The Western world, especially the United States, has thrown its entire propaganda machine into agitation work and has given a lot of encouragement and assistance to the so-called democrats or opposition in China—people who in fact are the scum of the Chinese nation," he declared. To his way of thinking, "Western countries use things like 'human rights,' or like saying the socialist system is irrational or illegal, to criticize us, but what they're really after is our sovereignty." We have to act, he said, to save the country, and to save the world from the kind of turmoil that might ensue from open civil war and the resulting flood of refugees. "This would be disaster on a global scale," he argued. "So China mustn't make a mess of itself. And this is not just to be responsible to ourselves, but to consider the whole world and all of humanity as well."

A few minutes after 4 PM the order was given. "We have to be absolutely firm in putting down this counterrevolutionary riot in the capital," Li Peng told the Politburo's Standing Committee, its most powerful body. "We must be merciless." Armed troops were to disband the protest no matter the cost, using whatever force proved necessary everywhere but in Tiananmen Square itself. It was too symbolically sacred, and filled with too many Western camera crews. "Let me repeat: no bloodshed within Tiananmen Square—period," ordered Yang Shangkun, reiterating Deng's wishes. And if the students refused to leave? "Then the troops carry away thousands of students on their backs. No one must die in the square."

Just as Deng's regime wanted to demonstrate its willingness to kill to maintain order, student leaders likewise reasoned they had to prove their willingness to die if their movement was to survive. Real bloodshed in Beijing might rally the rest of the country to their cause, revealing once and for all the regime's true despotic nature. They could be martyrs just like the venerated heroes of 1919 and the Long March, celebrated for generations. One prominent student leader, Chai Ling, had only a week before offered a chilling interview to an Australian journalist (though it was not broadcast until months later). "The situation has become so dangerous," she said, staring directly into the camera. "The students asked me what we were going to do next. I wanted to tell them that we were expecting bloodshed, that it would take a massacre, which would spill blood like a river through Tiananmen Square, to awaken the people. But how could I tell them this? How could I tell them that their lives would have to be sacrificed in order to win?" Wiping away tears, she haltingly admitted, "What we are actually hoping for is bloodshed. Only when the square is awash with blood will the people of China open their eyes."

The news went out at 6:30 that night. Loudspeakers blared reports of renewed martial law. State-controlled radio and television broadcasts explained the details: "Beginning immediately, Beijing citizens must be on high alert. Please stay off the streets and away from Tiananmen Square. All workers should remain at their posts and all citizens should stay at home to safeguard their lives." The message was repeated for hours. Students with loudspeakers took up the challenge, calling upon the masses to assemble at key intersections. The army was coming, again. Once more the city was ready.

This time was different from the start. Helicopters rattled overhead, and the rumble of heavy engines could be heard throughout the city. Armored units converged from all points of the compass. The soldiers were primed for battle, having been shown pictures of comrades bloodied by the crowds, warned of the agony of being burned alive in tanks set ablaze by Molotov cocktails, and reminded that their own officers would shoot them down if they failed to advance. Drawn from China's hinterland, unlike previous units deployed to the city, these soldiers were expected to have fewer qualms about combating Beijingers.

The showdown first came to a head to the west of the square, where city buses blocked the Muxidi Bridge and access to the city center. The PLA arrived around 9:30. Buoyed by their previous success, citizens swarmed in front of the troops, shouting down their orders to disperse with patriotic slogans and insults. A few tried to shove iron bars into the treads of troop carriers. Tear gas canisters flew, answered by volleys of rocks and Molotov cocktails. By 10:30 PM, units of the Thirty-eighth Army Group began throwing stun grenades and firing directly over the heads of the crowd. One American student who witnessed the confrontation reported people warming their courage by telling one another, "I didn't think they would use live ammunition."

The soldiers lowered their aim. Machine guns joined the chorus. Civilians dropped. Stunned, most froze. Then they ran. The PLA's armored personnel carriers leaped forward, crushing the barricades while trampling civilians and police alike, firing indiscriminately into any large formation of people while raking the windows above street level, no doubt striking even those who had heeded the government's warnings to stay inside. "No one said to shoot," a soldier later told PLA investigators. "But it was, like, 'We're going to teach them a lesson' . . . You pulled the trigger and bang, bang, bang, it was like rain, the noise shaking the heavens."

The crowd claimed small victories. An American embassy official saw demonstrators disable an armored personnel carrier while the rest of its squad continued forward, either unaware of or unconcerned for their stranded comrades. Protesters torched the vehicle. Two soldiers were burned alive inside. The cable back to Washington describing the scene used the word "roasted." A third soldier escaped the inferno, only to be beaten to death by enraged citizens. The entire scene was filmed and later broadcast by CNN. A Chinese American journalist tried to intervene, receiving an angry blow to the head and a hospital stay as a reward. The American diplomat on site recalled wondering why so many people had suddenly become so clumsy, tripping and falling over rocks in the road. Only then did he realize they were being felled by bullets. Retreating down an alley, he turned to ask the young man beside him a question. A large red bullet hole in the man's forehead ended their conversation.

Similar scenes were repeated with tragic frequency. Xinsheng Liu, a graduate student in international affairs, ran for cover once the bloodshed erupted. He stopped when he was seemingly beyond the fray, shaking too hard to light a calming cigarette; a friend obliged, until his neck exploded in a surge of blood. The bullet missed Xinsheng's own head only because he'd stooped to catch the flame. Twenty years later he remained unable to quit smoking yet couldn't keep his hands from shaking every time he lit a new one.

Discipline broke down on both sides as individual acts of kindness and cruelty bled into one another. For every soldier left to burn, another was bravely pulled from his blazing vehicle by the crowd and sent on his way. For every city resident shot in the midst of attack, another was felled trying to rescue a comrade lying in a pool of blood in the street. Bicycles rolled haphazardly through the crowd, piloted by some trying to flee and others hoping to aid the wounded, who were often piled into the backs of cars to be driven at breakneck speed in search of medical help, their drivers inadvertently running down by-standers in their haste to depart. It was chaos. And loud. "Chinese don't kill Chinese," a band of demonstrators chanted in unison, moving slowly toward a hunkered-down squad. Machine gun fire mowed them down. "During the movement we had talked about dying for democracy," one graduate student from Wuhan University later reported, "but that was purely hypothetical." Apparently, he was unaware of the sacrifices other student leaders readily contemplated. "We never dreamed we would actually have to lay down our lives."

The first PLA units arrived at the square shortly after midnight. Fanning out along its perimeter, they once more ordered the crowd to disperse, firing into the air as incentive. Most heeded their warnings and fled. By 2 AM only a few thousand demonstrators remained huddled around the Martyrs Monument. Literary critic, professor, and future Nobel Peace Prize laureate Liu Xiaobo and student leaders tried to negotiate a peaceful retreat. They had proved their willingness to die. It was time to survive, they pleaded into bullhorns.

By 3 AM the square was entirely surrounded by amped-up PLA units, frenzied from the bloody fight of the previous hours. Self-appointed student representatives strode toward the army's lines bearing a flag of truce, hoping to negotiate safe passage for all who would take it. Just then the lights went out in the square. The time for the final assault had arrived. "We come to talk! Don't shoot!" they shouted. Within minutes the army's spokesman offered to open a hole in their lines to the southwest of the square. Any who wanted safe passage would have it. Perhaps three thousand took this opportunity. By 4:30 the troops were on the march again, finding an hour later only two hundred stragglers huddled defiantly within the carnage that had once been their protest camp. They too were forcibly ejected amidst cries of "Fascists!" and scattered renditions of the "Internationale." Soldiers overturned tents and cots in search of remaining holdouts, tearing the Goddess of Democracy to the ground while building a massive bonfire from the remnants of the tent city. Just before dawn, Tiananmen Square was, for the first time in weeks, in the government's hands.

10

UNTYING THE KNOT

THE WEEKEND THEREAFTER remembered for the Tiananmen crackdown witnessed remarkable events around the world. Poland, for example, no longer had a communist government. Party officials had expected some losses in the Soviet bloc's first legitimate elections in more than four decades. They never thought that every one of their candidates would lose, even those who ran unopposed. Yet on June 4, 1989, that is precisely what happened.

Four thousand miles to Warsaw's southeast, Ayatollah Ruhollah Khomeini, spiritual leader of the 1979 Iranian revolution that remade the Middle East, lay dead. Three million mourners poured into the streets to view his funeral procession, crushing ten and injuring hundreds. The weekend's news was more tragic to Tehran's north. A natural gas line explosion high in the Ural Mountains engulfed two nearby passenger trains. At least eight hundred were dead, many incinerated beyond recognition. Most were children bound for holiday camps on the Black Sea. The death toll from the Soviet Union's long economic slide kept rising. "What the hell is going on?" Bob Gates, in Kennebunkport with the president, facetiously asked Brent Scowcroft back at the White House. "Can't you keep control down there?"

Despite all this drama, China held Bush's attention as he quickly condemned the violence that was unfolding on television sets everywhere. "I deeply deplore the decision to use force against peaceful demonstrators and the consequent loss of life," a written statement released by his office declared. That was all for the moment, however. He wanted time to consult with his staff and allies before saying or doing anything more, and time to sort through the flood of contradictory rumors pouring out of Beijing. The crackdown continued through the weekend, Lilley's staff reported. Military forces appeared to be battling one another as well, pitting those who were following the government's orders against those who backed the crowds. Additional troops marched on Beijing, but whether to bolster the regime or to fight it no one could accurately say. Deng had not been seen in public in days. He was in hiding, some American diplomats heard, or rallying forces in the West. One source placed him in the hospital, felled by a stroke. Another said he was dead, a third that he'd fallen victim to a coup. Nearly seven thousand miles away, there was little Bush could do to pierce the thick fog of misinformation. "It's going to be difficult to manage this problem," he lamented to Baker.

Critics cried out for more. "Diplomatic messages of disapproval are a pretty puny reaction to the murdering of innocent civilians whose only crime is to want the same freedoms as we in the West take for granted," Republican representative Mickey Edwards of Oklahoma complained. "We need to do something besides talk." Suggestions swiftly arrived from across the political spectrum. Democrat Stephen Solarz of New York, chair of the House Foreign Affairs Subcommittee on Asian and Pacific Affairs, called for Ambassador Lilley's immediate recall and a halt to arms and technology sales to China. One of Congress's most liberal members, Solarz was soon joined by its most conservative, Jesse Helms of North Carolina, the ranking Republican on the Senate's Foreign Affairs Committee, who vowed to submit legislation backing his colleague's suggestions. Weak written responses were mere "folly," Helms said. "We should stand with these young people" in Beijing. Solarz reinforced the point. "If the President doesn't take the initiative in changing American policy in this regard," he vowed, "the Congress will do it for him."

Senate majority leader George Mitchell called China's action "murder" and Bush's milquetoast statements "outrageous." Other senators and representatives quickly followed suit, competing to see who could sputter the angriest critique or suggest the most strident response. Calls for cessation of military or technology sales and Lilley's recall soon led to demands for expulsion of China's ambassador, a full break in diplomatic relations, seizure of Chinese overseas assets, and a halt to all commerce and trade. Republican senator Richard Lugar even suggested that Bush ship tens of thousands of replicas of the Statue of Liberty to China to replace the lost Goddess of Democracy statue and to "encourage the symbolism" of the American icon.

"Will you, Mr. President, be able to accommodate the calls from Congress for tougher sanctions?" a reporter asked Bush at his Monday morning press conference, mere moments after he'd detailed his initial round, including an end to military contacts; cessation of arms and technology sales, including navigation equipment for China's Boeing-built jetliners; and an automatic extension of educational visas for any Chinese exchange student unwilling to return home. Critics labeled them inadequate before he'd even finished speaking. "Many lawmakers felt you were slow to condemn or criticize the violence in China before now, and many are pushing for much tougher action on the part of this country," another journalist noted.

"I've told you what I'm going to do," Bush answered in a huff. "I'm the president; I set the foreign policy objectives and actions taken by the executive branch. I think they know, most of them in Congress, that I have not only a keen personal interest in China, but that I understand it reasonably well. I will just reiterate to the [congressional] leaders this afternoon my conviction that this is not a time for anything other than a prudent, reasoned response."

Tough talk was easy for legislators, he wrote that night in his diary. He was the one who would have to make the sanctions stick and then live with the consequences. "Solarz on the left and Helms on the right want us to move much more radically," he said, fuming. "Helms has always detested this relationship [with China]," and "Solarz is the kind of guy who wants to overthrow no matter who's involved. He's the kind of guy that was delighted about the overthrow of the Shah [in Iran in 1979], not worrying about what follows on." Revolution in China, Bush feared, would be far worse.

Bush's initial sanctions represented the most severe downgrade of Sino-American relations since the Korean War, with a total cost expected to exceed $700 million. Yet they did nothing to quench the public's bloodlust or salve its collective frustration, in large measure because, as Thomas Friedman of the *New York Times* noted, the president's response was technically correct in every respect but devoid of anything more. "One person's inflammatory rhetoric is another person's poetry, and because the Administration's declarations on Tiananmen Square have contained no poetry, no memorable language, nothing seems to have stuck in the public's mind."

Other world leaders seemed better able to tap into the day's emotions. Prime Minister Bob Hawke of Australia openly wept during a memorial service for the weekend's casualties. Britain's Margaret Thatcher proclaimed in blistering words how "appalled" she was "by the indiscriminate shooting of unarmed people." Members of her cabinet openly hinted it might be time to rethink London's recent promise to cede control over its remaining colonial possessions along the Chinese coast. "Britain will continue to stand by its commitment to a secure future for Hong Kong," she ominously warned.

Most galling of all from the White House's perspective, Ronald Reagan gave the speech many Americans wished Bush had delivered. "You cannot massacre an idea," he said during a mid-June visit to London. "You cannot run tanks over hope. You cannot riddle a people's yearnings with bullets." Bush never once produced such eloquence throughout his presidency. Frequently emotional in private, he stiffened when the stage lights were turned on. He was no comforter in chief and channeler of the national mood. "Now is the time to look beyond the moment," he noted in dry abstraction, "to important and enduring aspects of this vital relationship for the United States."

Popular outrage over events in Beijing only grew. Rallies in support of the crushed student movement sprang up in several of the major North American cities boasting significant Chinese populations — Los Angeles, Seattle, Vancouver, Philadelphia, and New York. Students temporarily shut down the Golden Gate Bridge in San Francisco. Manhattan police were forced to cordon off the Chinese consulate from an angry crowd wielding baseball bats and waving Chinese flags spray-painted with swastikas. Ten thousand Chinese students converged on Washington, their march down Pennsylvania Avenue joined by twelve members of Congress and activists from across the ideological spectrum, temporarily bridging onetime segregationists like Jesse Helms and civil rights icons like Jesse Jackson. They marched past the White House to the Chinese embassy, where Deng hung in effigy across the street, the entire complex shuttered with the exception of a lone photographer taking pictures of the crowd from a top-floor window. For students studying far from home, it was an ominous reminder that the regime was watching, and would have a long memory.

Bush's comparatively weak response engendered criticism in part because he was so personally linked to the long-range policy of engagement with China, which seemed now to have borne terrible fruit. "The love affair [with China] is over," conservative columnist Patrick Buchanan opined, and "the time for realpolitik is past." Successive administrations since Nixon's had been seduced by visions of strategic counterweights and blossoming markets, he argued, and lulled by the pipe dream of China's steady progress from authoritarian rule to freedom. The weekend's events had finally revealed the regime's true face and the flaws in the approach taken by old friends of China like Nixon or Kissinger or Bush, Buchanan argued. Beijing's decade-long experiment in reform either had been a façade or was now in full retreat, ABC News anchor Peter Jennings observed, as China's conservatives seemed intent on sending the country "back, back, not forward." Senator Claiborne Pell, chair of the Foreign Relations Committee, declared that "China in the past 15 days has pushed back her position in the international community to where it was 15 years ago."

Nixon had been wrong, they all agreed. As were the supposed "China hands," including Bush. Trusting Deng's reforms to bring freedom had been wrong too. It was time once more to quarantine the worst elements of the communist scourge, in effect by locking them away until they learned their lesson. "Mr. Deng and his comrades have declared war on the Chinese people," Buchanan concluded, "and America must stand with the people as allies against Mr. Deng. President George Bush should not hesitate to confront" this horrific regime "openly and forcibly."

One image from Beijing in particular captured both the hopes and harshness of Tiananmen. Television cameras had caught it live, then replayed it repeatedly. Smoke funneled up from scattered fires throughout the city a full day after tanks had crushed the protesters' camp and knocked down their Goddess of Liberty. Soldiers jogged along Beijing's main boulevards. The sound of rifle fire echoed, keeping residents largely huddled inside, including a multinational group of journalists on the sixth floor of the Beijing Hotel, where a CNN camera continuously broadcast a clear view of Chang'an Avenue along Tiananmen's north side.

Reporters narrated the live feed as armored vehicles massed on the road's east end, their voices rising as the tanks began advancing on civilians gathered behind a police checkpoint. Renewed bloodshed, this time in broad daylight, seemed imminent. Suddenly a lone figure loped out of the crowd. "What is that guy doing?" a cameraman yelled.

Armed only with shopping bags, the man walked directly in front of the lead tank, waving it away as its massive treads ground to a halt mere yards from his flailing arms. The giant machine shuddered, pivoting first right and then left, its driver unable to get past the man, who mirrored its every move, yet unwilling merely to run him down.

"Tank man" then climbed aboard, appearing to speak directly to the soldiers inside. The sound of rifle fire grew louder, as did cheers from onlookers, both cutting through the anxious narration of CNN's reporters. Jumping down from the tank, he once more blocked its path, the stalemate continuing until two additional men wearing civilian clothing finally rushed him away. His identity remains a mystery more than a quarter century later. Were those two men helping him? Or arresting him?

His singular act of defiance held the world entranced for nearly twenty minutes. "I waited for the moment he would get shot," a photographer later recalled. "I waited, and I waited . . . and he wasn't." Tank man became an inspiration around the world. "Mr. President, you have said the genie of democracy cannot be put back in the bottle of China," a reporter challenged Bush the following day. "Do you still believe that?" He was direct in response. "Yes, I still believe that. I believe the forces of democracy are so powerful, and when you see them as recently as this morning—a single student standing in front of a tank, and then, I might add, seeing the tank driver exercise restraint—I am convinced that the forces of democracy are going to overcome these unfortunate events in Tiananmen Square."

What had happened was an atrocious crime, demanding consequences. Bush never argued otherwise. But in the face of such atrocious behavior, he believed the outside world had an obligation to sustain the growth and exposure that had spawned China's democracy movement in the first place. "I don't want to see a total break in this relationship, and I will not encourage a total break in this relationship," he insisted. "When you see these kids struggling for democracy and freedom, this would be a bad time for the United States to withdraw and pull back and leave them to the devices of a leadership that might decide to crack down further."

Engagement worked, he believed, because American ideas worked, and it was China's exposure to the American system even after decades of angry isolation that underlay its recent changes and constituted its best hope for a more democratic future. "As people have commercial incentive, whether it's in China or in other totalitarian systems, the move to democracy becomes more inexorable," Bush said. One only needed to consider the long history of Sino-American relations against the chronology of China's post-1949 development to understand. "The budding of democracy we have seen in recent weeks owes much to the relationship we have developed [with China] since 1972," Bush argued, the sentiment echoed to reporters by one of his senior advisers — most likely Scowcroft — on background. "What you have to remember is that China has come a long way," that senior official noted, explaining the administration's position. "There's a long way to go, but I think the U.S.-China relationship has been enormously helpful in making this kind of evolution."

What happened in Beijing in the first weekend of June 1989 forever changed Tiananmen Square, in American minds at least, from a place to an event but did little to change the long-term hopes of those who had long touted engagement. Their reputations were at stake in its survival; yet more important, its progenitors believed in it to their core. "Don't disrupt the relationship," Richard Nixon advised Bush during an 8 AM phone call the day after the crackdown. They spoke twice that day, with Bush calling his former boss again little more than an hour before announcing his initial round of sanctions. "What's happened has been handled badly [by the Chinese] and is deplorable," Nixon conceded. "But take a look at the long haul." In a series of newspaper columns and magazine essays the following month, he argued that "lashing back with punitive policies would be politically popular and emotionally satisfying for many people." Yet it would also "dash the Chinese people's chances for further economic progress and eventual political reform." Their only hope for prosperity and democracy, Nixon said, lay in keeping contacts with the wider world alive.

The crisis demanded "political maturity," Henry Kissinger similarly argued in print, defending in one breath both Bush and his own legacy. Without continued international engagement, China's modernization would surely end, and "unless China returns to the path of modernization, it will descend into chaos or be driven back to the practices of Maoism." He put the point more colorfully when privately advising Bush. "You can't let the reaction you're getting from fuzzy-headed liberals/bleeding hearts about the suppression of demonstrators destroy your/our China policy," Kissinger told him tartly.

Meanwhile, American officials continued to fear that China's unrest might lead to outright civil war, a worry exacerbated by their inability to determine for days either where Deng was or who, if anyone, even held the regime's reins of power. "We believe the reports of Deng Xiaoping's incapacitation (or death) are credible," Lilley's staff wrote Washington on June 6, "leaving President Yang

Shangkun the senior party and military leader." Newspapers around the world circulated the same rumor, though in the critical hours after the crackdown, American intelligence analysts went even further. Agents "learned via a telephone call (placed at 1300 EST, 5 June) with Chinese officials that (Deng) Xiaoping, China's supreme leader and head of the Central Military Commission, has died of heart failure," the Pentagon's Defense Intelligence Agency reported. "Deng's final words reportedly were, 'originally I did not want to use force.' There is widespread speculation that President (Yang) Shangkun initiated the military crackdown on the students in Tiananmen Square prior to Deng's death, supposedly intending to place the blame for the military action on the dying ruler." As far as the Pentagon's chief spies were concerned, China was rudderless and its current leadership operating with a mandate from the dead.

"The reports from China are crazy," Bush vented to his diary. "There are rumors that 'Li Peng has been shot,' and rumors that 'Deng was dead.' All of this tells me to be cautious, and be calm." New army units were reportedly converging on Beijing, though with unknown intent. "International telephone calls to the PLAAF [Chinese air force] Headquarters high leadership private switchboard are not being answered," Pentagon analysts noted, "another indication that there is great chaos among the high-level military leadership." Baker's briefers warned that "reports of clashes between army units, or between army and police, are widespread," while those few Americans capable of eluding the army's cordon around the embassy warned of units hunkering down behind fortifications circling the Zhongnanhai compound. "Leaders and army commanders who have ordered or conducted atrocities now feel they are fighting for their lives," Lilley's staff observed, noting that government spokespeople had begun to focus even more intently on Westerners as the "evil" at fault for riling up Chinese passions beyond the boiling point.

It seemed increasingly likely that whoever remained in charge would "blame the foreigners," Lilley reported, "especially Americans, for China's troubles." From his perspective, "we are in fact way up the flagpole" already, having inadvertently provided the regime with fodder for arguing that the entire protest movement was part of an overall American scheme to discredit and ultimately destroy Chinese communism. "Congressional resolutions, media and VOA [Voice of America] coverage, individual critical statements by influential Americans and our occasional private intervention all make us vulnerable to counterattack" by a regime eager to shift blame for the tumult onto a familiar foe, he argued, adding ominously that "the Chinese have at all levels signaled us to watch our step or else."

Their warnings soon went beyond mere words. Major Larry Wortzel, a military attaché in the American embassy in Beijing, was awakened in the middle of the night on June 6 by his office phone's incessant ringing. On his feet for days tracking troop movements, he'd hoped for a few moments of sleep under his desk. "Do not go to your apartment between ten in the morning and two in the afternoon," the voice at the other end of the line said. "Do not go above the second floor of your apartment building."

Still shaking the cobwebs from his head, Wortzel replied that he had no plans to go home at all. There remained plenty to be done at work. Sensing he did not fully understand, the voice repeated before hanging up, "This is very important. Do not be in that building above the second floor."

The meaning slowly sank in. "They have warned us," Wortzel immediately told Lilley. "Something is going to happen." The ambassador ordered the apartments cleared, scheduling a meeting of all embassy staff and dependents to discuss evacuation plans so that the building would be as empty as possible without raising too much alarm. He should have raised more. Seven small children remained inside, chaperoned by two Chinese nannies, their parents unaware of the meeting's true purpose.

Gunfire ripped through the windows. Troops on the street below and atop a neighboring building had opened up on the American residences. They later claimed to be returning a sniper's fire. The nannies threw themselves on the children, undoubtedly saving their lives, as shards of glass careened around the room. Thinking themselves under attack, embassy officials scrambled in response, ordering marine guards to their defensive positions and dependents into the compound's lead-lined safe rooms, and calling for the shredding of sensitive documents. Recent history offered the painful reminder of what mobs could do to foreign outposts. Britain's embassy in Beijing had been ransacked by angry crowds at the height of the Cultural Revolution, and this generation of American officials had strong memories of 1979 in Tehran.

A tense calm eventually prevailed as the assault ended, leaving only American fury. Lilley demanded an immediate session with government officials, but his complaint received only what he called "a knowing smirk," further fueling his anger. "I was only a private in the infantry," he told China's vice foreign minister. "But I know that you don't machine-gun a building from one end to the other and up nine stories to get one guy on the roof."

His Chinese interlocutor was livid. "You have insulted the Chinese government and military by saying we directly fired at you," he retorted. Lilley quickly cut him off, convinced the attack was a deliberate attempt to remind Americans of the dangers of interference, or better yet to drive them from Beijing entirely. "If you had hit one of those kids," Lilley told China's vice foreign minister in a rage, "I wouldn't be sitting here today." Sino-American relations would have been severed, exactly as China's hard-liners desired, calculating that whatever indignation foreigners might muster, they'd eventually return. In the meantime, their absence would provide all the opportunity the government required to impose the order it desired. Lilley recalled an old Chinese maxim, *guan men da gou,* or "close the door to beat the dog." Desiring to round up and punish everyone involved in the protest, authorities did not want foreign witnesses complicating matters.

Matters were already becoming complicated on their own. An equally dramatic story was meanwhile unfolding deeper within the embassy compound: Fang Lizhi was there. He'd arrived with his family on June 5, pleading for refuge. State Department instructions strongly discouraged the granting of asylum to any Chinese refugees, however, so Lilley's staff offered little more than advice, a ride back to their hotel, and a promise to check on them the next day.

"What the fuck are you doing?" Jeff Bader, deputy director of the State Department's China desk, screamed into the phone when he heard the news. He didn't care that it wasn't a secure line. Those instructions against harboring refugees had been meant for rank-and-file Chinese, not the country's leading democracy advocate! If Fang were arrested, or worse, and word got out that the Americans had denied him sanctuary, they would never hear the end of it. "Send somebody out to tell him that if he'd like to come to the embassy, he can." Acting Assistant Secretary of State William Clark personally instructed Lilley to invite Fang, along with his entire family, to be "a guest of President Bush." They quickly accepted the offer, spending the night in a blacked-out back room of the ambassador's residence, walled in by boxes of books stacked to the ceiling in hope of deceiving any would-be investigator. The next day they settled into a small apartment toward the rear of the compound, where Fang and his wife would remain for more than a year.

"The Chinese government went ballistic," as Lilley later described it, especially after White House spokesman Marlin Fitzwater inadvertently confirmed Fang's whereabouts. The telephone rang again with threats, this time at the State Department in Washington. "They would storm the compound," Lilley's staff was warned, and would not stop till Fang was turned out. "We had no choice but to take him in," a worried Bush told his diary, "but it's going to be a real stick in the eye" for the regime.

Beijing demanded Fang's immediate release into Chinese custody, something Bush could not possibly have condoned politically even if he were so inclined. Unable to keep the tanks from crushing students, he could at least save one of their professors. The embassy was technically American soil. So long as Fang stayed inside its walls, Bush had the final say. Of course, the same had been said a decade earlier as angry crowds gathered around the American embassy in Tehran. "It is awful hard for the United States, when a man presents himself — a person who is a dissident — and says that his life is threatened, to turn him back," Bush explained to reporters. "That isn't one of the premises upon which the United States was founded. So, we have a difference with them [China's leadership] on that." Asked if he had spoken directly to China's leaders about the matter, or spoken to Deng in particular, Bush then offered a wry reply. "The line was busy," he said. "I couldn't get through."

That was not exactly accurate. "I was a little pissed off," Bush confessed to his diary. He longed to talk to Deng, hoping at the least a personal phone call might settle the question of his whereabouts — or determine if he was still alive. Whereas Washington and Moscow had a hotline so leaders could speak quickly in a crisis, however, nothing of the kind connected the White House with Zhongnanhai. No American president had ever even spoken by phone to his counterpart in China, a streak that now continued despite herculean efforts by the White House switchboard operators, who prided themselves on being able to locate anyone at any time their commander in chief desired to talk.

Deng could not be found, the Foreign Affairs Ministry reported, further fueling fears of his demise. No one else in Beijing would take Bush's call in Deng's stead. "Nobody knows" who is in charge in Beijing, Bush told reporters on June 8, his frustration spawning remarkable candor. "But what I do know is that there's events over there that — it doesn't matter who's in charge — we condemn. And there's a relationship over there that is fundamentally important to the United States that I want to see preserved. And so, I'm trying to find a proper, prudent balance, not listening to the extremes that say, take your Ambassador out; cut off all food to the Chinese people so you show your concern. And I think we found a proper avenue there, but I cannot — and you ask a good question — I simply cannot tell you with authority who is calling the shots there today."

Bush's inability to reach Deng exacerbated anxieties in the White House that were already at fever pitch. "When the President of the United States calls," Baker noted, "you usually take his call." Deng's failure to do so led to the presumption that he was indeed unable to speak. Any other explanation was almost too insulting to contemplate, though as it turned out, Deng was healthy but merely had nothing he wished to say to his longtime American counterpart, and even less he wanted to hear. "One of the things he [Bush] was disappointed about during the Tiananmen thing," Baker later remarked, "was that Deng Xiaoping did not take his call, and he was really disappointed in that, because he felt that Deng had become a colleague, and a friend." Baker emphasized that last word, "friend," long and slow. This was a word that meant something real to George Bush. When a friend called, especially in a crisis, Bush's value system demanded that the call go through.

Deng was different. He made and kept friends only when they had something to offer. Those without value he discarded, sometimes to house arrest, sometimes worse. He ultimately appeared in public days later, ending speculation about his demise, though his words raised deep concerns for American analysts. "Chinese authorities have been working themselves into a self-righteous rage," Lilley reported home in mid-June. News reports increasingly highlighted evidence of foreign support for the protesters, revealing in lurid detail how American officials had turned honest Chinese students into agents of counter-revolution. "They know what really happened at Tiananmen and they know they made some really bad mistakes, but they cannot admit this" and hope to survive in power, Lilley thus explained. "Instead they are trying to write a new script on the blank page of the average Chinese" by "tapping historic anti-foreign sentiment."

Bush's China policy appeared thoroughly mired down by mid-June. Unable to work directly with his Chinese counterparts, he found himself waging a rearguard action against American legislators intent on ending a relationship that hard-liners in Beijing seemed not to want either. But if those hard-liners believed contact and trade fueled the country's pro-democracy movement,

Bush reasoned, all the more cause to continue them. Deng's cadre would never fully extinguish their people's thirst for freedom, Bush told his staff, especially not after their own reforms had ignited a flame they could now never hope to put out. The desire for freedom was simply too powerful, too universal, and too innate to ever entirely eradicate. It was therefore only a matter of time before the democratic surge they had just witnessed being beaten back arose once again behind the Great Wall. Time was on their side, Bush believed. It was the one thing sure to drive Deng's old guard and acolytes from power.

"The situation in China will remain unsettled at least until Deng Xiaoping and other party elders die," the State Department's Bureau of Intelligence and Research similarly advised. Defense Department analysts predicted they would lose power sooner rather than later. "The current coalition of hardliners is a marriage of convenience that will not last," the Defense Intelligence Agency concluded, "but the transition to a moderate, reform-oriented coalition will take at least five years." At the same time, "current developments have caused a rift between Washington and Beijing that will take 3 to 5 years to overcome."

If he could somehow manage to keep relations from fully rupturing in the short term, Bush believed, Sino-American ties might more easily be rebuilt, ideally while he was still in office — presuming a second term — and thus able, at long last, to capitalize on his long experience with China. "The efforts may look one-sided, and I guess maybe objectively they could be considered one-sided," Scowcroft later replied when asked why Bush felt so determined to maintain the relationship despite opposition from both capitals. "We did it in the context of what we saw as a very difficult period for the Chinese leadership. They were in a panic at what had happened . . . [and] perhaps realized they'd made a terrible mistake in the way they handled Tiananmen Square, but saw no way to admit that, since that, in a sense, would be acknowledging that they were somehow inferior, and that outsiders had a better concept of how the Chinese should manage their affairs than did the Chinese." Bush, therefore, concluded he had to seek some form of reconciliation with the Chinese despite their despicable behavior, rather than standing on formalities or waiting for Beijing to move first. "I'm going to pursue them," he told his deputy. "If they're not willing" to talk, at least "we will have done our best."

On June 20 Bush made good on his promise of further sanctions, canceling a planned trip by Commerce Secretary Robert Mosbacher while authorizing Baker to announce a halt to formal exchanges between the two nations. "I'm sending signals to China that we want the relationship to stay intact, but it's hard when they're executing people, and we have to respond," Bush wrote in his diary. Baker went further than instructed, however, announcing a wholesale curtailment of all high-level contacts.

He simply misspoke, though Bush found in the wake of Baker's testimony that he could neither speak directly to China's top leadership nor publicly deliver or receive any suitable envoy, at least not without seeming to violate his own policy. He found it politically impossible even to arrange a face-to-face meeting with the Chinese ambassador in Washington. An attempt to quietly shuttle the diplomat into the West Wing on June 21 was scuttled when a wandering member of the press pool overheard staffers making arrangements. "A terrible situation," noted Bush, venting in his diary once more. "We immediately cancelled the meeting. We cannot have a meeting like this in public," he complained. It would look as if the administration had backtracked on its latest stance. And still no one in Beijing would answer his calls.

Finally, Bush turned to pad and pen to reach Deng. The extraordinary letter, hand-delivered by Scowcroft to the Chinese ambassador for transmission to Beijing, reveals as much about the author as the situation and is thus worth quoting at length. "I wanted a letter straight from my heart," Bush began, "so I composed it myself."

> I write this letter with a heavy heart. I wish there was some way to discuss this matter in person, but regrettably that is not the case. First, I write in the spirit of genuine friendship, this letter coming as I'm sure you know from one who believes with a passion that good relations between the United States and China are in the fundamental interest of both countries . . .
>
> Second, I write as one who has great respect for what you personally have done for the people of China, and to help your great country move forward. There is enormous irony in the fact that you who yourself have suffered reversals in your quest to bring reform and openness to China are now facing a situation fraught with so much anxiety.
>
> I recall your telling me the last time we met that you were in essence phasing out of the day-to-day management of your great country. But I also recall your unforgettable words about the need for good relations with the West, your concerns about "encirclement" and those who had done great harm to China, and your commitment to keeping China

moving forward. By writing you I am not trying to bypass any individual leader of China. I am simply writing as a friend, a genuine "lao pengyou."

It is with this in mind that I write you asking for your help in preserving this relationship that we both think is very important. I have tried very hard not to inject myself into China's internal affairs. I have tried very hard not to appear to be dictating in any way to China about how it should manage its internal crisis. I am respectful of the differences in our two societies and in our two systems.

I have great reverence for Chinese history, culture and tradition. You have given much to the development of world civilization. But I ask you as well to remember the principles upon which my young country was founded. Those principles are democracy and freedom — freedom of speech, freedom of assemblage, freedom from arbitrary authority. It is reverence for these principles which inevitably affects the way Americans view and react to events in other countries. It is not a reaction of arrogance or a desire to force others to our beliefs but of simple faith in the enduring value of those principles and their universal applicability.

And that leads directly to the fundamental problem. The early days of the student demonstration, and indeed, the early treatment of the students by the Chinese Army, captured the imagination of the entire world. The wonder of TV brought the details of the events in Tiananmen Square into the homes of people not just in "Western" countries but world-wide. The early tolerance that was shown, the restraint and the generous handling of the demonstrations, won world-wide respect for China's leadership. Though people all over the world tried to understand and sympathize with the enormous problems being faced by those required to keep order; and indeed, they saw with admiration the manifestation of policy which reflected the leader's words: "The Army loves the people." The world cheered when Chinese leaders were seen patiently meeting with students, even though there were "sit ins" and even though disorder did interfere with normal functions.

I will leave what followed to the history books, but again, with their own eyes the people of the world saw the turmoil and the bloodshed with which the demonstrations were ended. Various countries reacted in various ways. Based on the principles I have described above, the actions that I took as President of the United States could not have been avoided. As you know, the clamor for stronger action remains intense. I have resisted that clamor, making clear that I did not want to see destroyed this relationship that you and I have worked hard to build. I explained to the American people that I did not want to unfairly burden the Chinese people through economic sanctions.

There is also the matter of Fang Lizhi. The minute I heard Fang was in our Embassy, I knew there would be a high-profile wedge driven between us. Fang was not encouraged to come to our Embassy, but under our widely-accepted interpretation of international law we could not refuse him admittance . . . We cannot now put Fang out of the Embassy without some assurance that he will not be in physical danger. Similar cases elsewhere in the world have been resolved over long periods of time or through the government quietly permitting departure through expulsion. I simply want to assure you that we want this difficult matter resolved in a way which is satisfactory to you and does not violate our commitment to our basic principles. When there are difficulties between friends, as now, we must find a way to talk them out.

Your able Ambassador here represents your country firmly and faithfully. I feel that Jim Lilley does the same for us; but if there is some special channel that you would favor, please let me know.

I have thought of asking you to receive a special emissary who could speak with total candor to you representing my heartfelt convictions on these matters. If you feel such an emissary could be helpful, please let me know and we will work cooperatively to see that his mission is kept in total confidence. I have insisted that all departments of my US government be guided in their statements and actions from my guidance in the White House. Sometimes in an open system such as ours it is impossible to control all leaks, but on this particular letter there are no copies, not one, outside of my own personal file.

. . . I send you this letter with great respect and deep concern. We must not let this important relationship suffer further. Please help me keep it strong. Any statement that could be made from China that drew from earlier statements about peacefully resolving further disputes with protesters would be very well received here. Any clemency that could be shown the student demonstrators would be applauded worldwide. We must not let the aftermath of the tragic events undermine a vital relationship patiently built up over the past seventeen years. I would, of course, welcome a personal reply to this letter. This matter is too important to be left to our bureaucracies.

It was a remarkable letter, interweaving affairs of state with personal remembrances. Even while prudently noting that he was "not trying to bypass any individual leader of China," a reference not only to Deng's unofficial role as first among equals in Chinese politics but also to the real possibility that Deng might not be calling the shots in Beijing for long, Bush nonetheless made it plain that no matter who in the Chinese high command might ultimately read the text, he was "simply writing as a friend, a genuine 'lao pengyou.'" The term by this point had lost much of its meaning, save as a reminder to the Chinese of one inescapable fact: they were unlikely to find a more willing partner in the Oval Office. Carter, Reagan, and most likely Dukakis would not have worked so hard to maintain the relationship. If China's top leaders hoped to resuscitate Sino-American relations in their lifetime, now was their best opportunity. Perhaps their only one.

The letter was a gamble, giving Deng the power to embarrass Bush through its publication or to rile up his own population at the thought of a foreign executive interfering in their affairs. "It's highly sensitive," Bush lamented to his diary late in the evening on June 24. "China is blasting the United States for interfering in their internal affairs, and we are criticizing China, though not as vociferously as most in the Congress would like me to do."

It was this extraordinary effort to maintain the relationship that led political opponents to charge Bush with having coddled the "butchers of Beijing." Bill Clinton used that phrase on the campaign trail in 1992, echoing a common refrain that Bush let his personal feelings for China override what should have been a stronger response. Bush was blinded by feelings of friendship wholly unreciprocated by the other side. In the worst possible reading, he was not merely wrong to have put such faith in Deng; he had been duped by people he considered friends but who used him as a tool.

Such charges contain an element of truth. Bush's inflated sense of his own understanding of China, coupled with his long relationship with Deng in particular, undoubtedly influenced his desire to keep relations from being severed. "I take this whole relationship very personally," he confided to his diary, "and I want to handle it that way." Years later Bush confessed that his personal relationship with Deng colored his every move during the crisis. Believing Deng both a patriot and a reformer, Bush interpreted his crackdown at Tiananmen as heartless and despicable but not irrational. Bush believed that he at least understood why his old acquaintance reacted so violently when threatened with instability and chaos: Deng was simply doing what he thought best to save his country from a worse fate, and could be reasoned with in the aftermath to do what he thought best for China's long-term interests. Once assured of stability, Deng and those around him would reignite their economic reforms and thereby inexorably lead to democracy's rise. "*Had I not met the man*," he conceded years later, "I think I would have been less convinced that we should keep relations with them [China's leaders] going after Tiananmen Square."

In the intervening decades China has indeed become more affluent and internationally integrated, yet democratic reforms have not continued at the same pace. While not unique, that paradox is unusual. Societies typically demand greater civil and political liberties, what Bush would reflexively term "freedom," as they become more prosperous. In continuing to engage China after Tiananmen, he banked on the fact that it, like nearly every other communist country in 1989, would inevitably turn democratic. The only question in his mind was when. Believing that integration and engagement had led China to the brink of democratic revolution, and that China's earlier decades of isolation had led only to spasms of violence and repression, Bush worked hard to keep the "butchers of Beijing" a part of the world. Little good, he reasoned, grew from keeping them apart.

This thinking was logical and consistent, but was it wise? Two questions are embedded in that one: First, was Bush right to think that engagement with China would speed democratic reform faster than the isolation his domestic critics demanded? Second, was it in America's long-term strategic interest for Bush to desire Chinese growth in the first place? In response to both, one is reminded of a quip often employed by historians when asked to project the future, which ironically stems from a key figure in modern Chinese history. "It is too soon to tell," Zhou Enlai once responded when asked about the consequences of the French Revolution nearly two centuries before.

The same argument might be made about China's crackdown of 1989, itself ironically coming exactly two centuries after revolution gripped France. If, on the one hand, China in time becomes fully integrated into the international system, *and* if its people demand greater liberties in turn, which their government eventually permits, *then* Bush will have been right to prioritize engagement over isolation. If, on the other hand, his work to keep China engaged in the world turns out only to have further fueled its economic rise and thus its rise to power, allowing it in time to challenge American dominance in the Pacific without simultaneously instigating a new liberal order in line with American values, then one might well argue that Bush enabled Washington's successor, speeding up the process of imperial decline already confronting the United States.

In truth, either answer, if framed with Bush in mind, gives both too much credit and too much blame for China's fate to outside forces. It reeks of the charge that anticommunist crusaders leveled at Harry Truman's State Department after Mao's 1949 victory, that they had somehow "lost" China. It was never Washington's to begin with, either in 1949 or in 1989. If the story of Tiananmen demonstrates anything, it is that domestic forces invariably dictate events within any country, especially a potentially revolutionary one, far more than foreign influences. Why did Chinese citizens take to the streets after Hu Yaobang's death? For the same reason they had marched in defiance repeatedly throughout the mid-1980s. It was not because of anything Bush did or said, and not in response to Gorbachev, either. They instead rallied in defiance of a wrong they perceived at home after having first enjoyed the fruits of a decade of dramatic economic reform and growth. Bush did not catalyze their protests. Deng's cadre did.

Yet Deng and those around him were deeply committed to economic reform, regardless of what they considered its unfortunate side effects. They had no desire to tear down their work of the prior decade, despite the inconvenience of the interest in political reform it generated, just when it was producing the very prosperity they had long desired. Provided their clique remained in power, China was not going to isolate itself in the long term after 1989. Perhaps in the short term they would indeed, as Lilley put it, close the door to beat the dog, exiling as many foreigners as possible while they reimposed order. But they would in time invite them back. China was going to remain engaged in the world as the twenty-first century neared, with or without the United States.

What Bush did to prevent a break between the two countries after Tiananmen thus accelerated a process of Chinese integration with the world that neither he nor his critics who called out for punishment and revenge could have halted even if they had wanted to. Unable to keep violence from occurring in Beijing, and fearing in particular the descent into nuclear-haunted civil war in Tiananmen's wake, he was at least able to offer a Hippocratic response, doing what he could to keep the situation, and China's potential isolation, from getting worse.

So we know why Bush acted as he did, and that his actions ultimately bore fruit. Deng responded to his personal letter within twenty-four hours, affirming his desire to keep their dialogue alive and his willingness to receive an American emissary. But whom could Bush trust with such a sensitive mission? Baker had pledged to Congress that the administration had sworn off precisely this type of contact. Clearly, he could not go. Nixon and Kissinger were obvious possibilities, as Bush and Scowcroft discussed. Neither would work, however. The former was too high profile for a secret mission, and the latter seemed too likely to leak its details in order to burnish his own image. Lilley wouldn't do either. Bush's message would lose much of its impact if delivered by someone already in Beijing.

Scowcroft himself appeared to be Bush's best option. Anyone more prominent was too easily identified; anyone lower in the administration's hierarchy would fail to impress. He was also Bush's own *lao pengyou*, able to convey the president's deepest wishes. He'd bring along the State Department's Lawrence Eagleburger, useful not only for his diplomatic savvy but also so it wouldn't seem as if the administration was repeating Reagan's mistake of letting the NSC go "operational." The foul-mouthed, chain-smoking Eagleburger was also just the sort of man who could make marathon flights across multiple continents in a cramped jet bearable.

Their journey, which began on June 30, was the stuff of spy novels. An air force cargo plane, stripped of all its markings, took off into the predawn sky from Andrews Air Force Base, supposedly headed toward Okinawa. The destination was a ruse. The pilots and crew changed into civilian clothes once aloft, with Scowcroft, Eagleburger, and Scowcroft's personal secretary Florence Gant tucked away inside what the air force euphemistically dubbed a "comfort pallet," essentially a windowless shipping container with hard cots and chairs loaded into the airplane's belly. It was akin to traveling inside a large metal coffin, but it was at least secretive. Refueling in flight lest ground crews inadvertently identify the passengers, the plane nearly ran headlong into Chinese air defenses, whose radar picked up an unidentified jet screaming toward their coast. "No one thought to tell the military air defense units," Scowcroft later revealed, leaving one to imagine the international fallout if some overzealous Chinese pilot had blown the plane, and its secret cargo, from the sky.

Cooler heads in Beijing authorized the intrusion before a tragedy could occur, and after what Scowcroft described as "an endless time in the air," the

Americans finally landed in Beijing early in the afternoon on July 1. Their plane taxied to the same remote hangar used to house Nixon's Air Force One back in 1972, its passengers and crew spirited directly to the Diaoyutai State Guest House long used to host foreign dignitaries far from potentially prying eyes.

Deng wasted no time coming to the point the next morning. "Sino-U.S. relations are in a very delicate state and you can even say that it is in a very dangerous state," he warned. And it was all Washington's fault. The Americans had catalyzed counterrevolutionary elements while rallying global opinion against Beijing's legitimate government, and even willingly harbored a known fugitive within their embassy walls. "The cause of the issues between China and the United States on this question is not because China has offended or impugned U.S. interests even in the least way," Deng insisted. "The question is that on a considerably larger scale the United States has impugned Chinese interests, [and] has hurt Chinese dignity. That is the crux of the matter.

"To be frank," Deng said, "this could even lead to war."

He let the last word hang ominously in the air. Finally he continued. It was Washington's responsibility to find a solution. "There is a Chinese proverb," he said: "It is up to the one — up to the person who tied the knot to untie the knot. Our hope is that in its future course of action the United States will seek to untie the knot." He wouldn't say precisely what might assuage his regime's anger. Washington would have to make the first move.

That was why he was there, Scowcroft calmly explained. Having shown remarkable restraint throughout Deng's entire harangue, Scowcroft quietly noted that he had been selected for this mission for a reason. The president trusted him, and knew that he understood better than anyone something Deng knew as well: that this was a president who cared more for China than any other person Beijing could expect to hold the office. "I work just down the hall from him. I see him 5–6 times a day and I see his pre-occupation with the problems between our two countries. We both served in the Nixon Administration. We have both been for many years close associates of Henry Kissinger." Indeed, Scowcroft reminded Deng, Bush was "the oldest friend of China still in the United States government." If Deng did not listen to his *lao pengyou* at a moment such as this, Scowcroft implied, then friendship, old or new, meant nothing at all.

And he was there not to criticize, he quietly continued, but rather to explain. What the Chinese did to their own people and within their sovereign borders was their business. It was, he told Deng, "as you have so eloquently said, a wholly internal affair of China." Nevertheless, "how the United States government and the American people view that activity is, equally, an internal affair of the American people." The protesters at Tiananmen Square had proudly pleaded for values Americans held dear, and Americans were outraged to see those protesters so summarily crushed. "That is the crux of the problem President Bush now faces," he said. Bush wanted Sino-American relations to survive, and thus it would help if the Chinese would, at the least, "try to be sensitive" to the way their actions appeared to the rest of the world. Bush was powerful. But no American president was sole master of the nation's affairs. Despite his opposition, Congress had just endorsed even steeper sanctions. "The President will continue to oppose such legislation," Scowcroft promised, "but the magnitude of the vote illustrates the political realities with which he must cope. Even his veto authority is powerless against such unanimity."

Deng retained his defiant air throughout Scowcroft's response, to no one's surprise. "I just hope that United States statesmen and people will understand one point," he offered when his counterpart concluded. "I think that one must understand history; we have won the victory represented by the founding of the People's Republic of China by fighting a 22-year war with the cost of more than 20 million lives, a war fought by the Chinese people under the leadership of the communist party; and if one should add the three-year war to assist Korea against U.S. aggression then it would be a 25-year effort." They'd fought for China's independence, which meant, he insisted, "no interference from foreigners." Ultimately, "so long as these principles are observed, Sino-U.S. relations will continue to develop on the basis of respect for such principles." If, however, "that approach is not adopted," he warned, "then China will not be held responsible for the deterioration of relations."

Their all-morning meeting never moved substantively beyond those initial statements. Anything more would have been a surprise. Its purpose had never been persuasion, nor a breakthrough, nor even a meeting of the minds. The point of the entire exercise was merely for each side to demonstrate that despite their anger, they could still talk. An angrier president, one less convinced that history flowed in his direction, or more easily swayed by public opinion, would never have sent so personal an envoy. The first part of Scowcroft's mission succeeded, therefore, the moment he landed in Beijing. It would prove a complete success if he and Eagleburger could somehow manage to return home without anyone discovering they'd been gone. "We had aired our differences and listened to each other, but we still had a distance to go before we bridged the gap," Scowcroft later concluded. "It was clear to me that the clash of cultures had created a wide divide between us," but at the least, they were still talking. "Let us talk of something else," Li Peng bluntly stated over lunch. The meeting had taken place. That was enough.

Sino-American relations never did warm back up during Bush's remaining three-plus years in office. They were more akin to "treading water," as Baker described it. American sanctions largely remained in place, ultimately because of congressional pressure, indeed outlasting sanctions from any other major power. But relations did not cease. Bush even dispatched Scowcroft and Eagleburger to Beijing again in December in order to brief Deng and his leadership team on the latest happenings in Soviet-American affairs. No other American ally or adversary received such special treatment. By the same token, with Bush due to meet with European leaders, no other member of the United Nations Security Council required its own readout. With Sino-Soviet rapprochement still in the air as 1989 came to a close, and with Gorbachev no less a strategic quandary at year's end than at its beginning, Bush knew he might well need Beijing to recall the virtues of triangular diplomacy and engagement once more, and that the Chinese, for all their flaws and faults, were not going away no matter how much angry critics in Washington longed for their isolation.

Indeed, as we shall see, ties to Beijing proved useful when Bush needed China's acquiescence in United Nations sanctions against Iraq later in his presidency. Ties broken in 1989 might well have precluded even this minimal though crucial cooperation, preventing Bush from claiming the international mandate he sought for war in the Persian Gulf. Chinese leaders unable to forgive a snubbed handshake from decades before were unlikely to forget quickly a president's condemnation the next time he needed their help. Bush could ask for Beijing's support when he needed it in late 1990 for a wholly unexpected crisis only because he had not severed relations the previous summer. As Bush explained, Scowcroft's secret trips, and everything he had done before or since, had one purpose: "It kept the door open."

Tiananmen taught Bush a major lesson as well. Or, rather, it reinforced one. He wielded greater power than any other person on earth. Yet he had been unable to keep violence from occurring when a democratic surge met a recalcitrant regime. Deng had been terrified of chaos and instability, even of civil war. In truth so was Bush, who perceived in Tiananmen's wake just how close they'd come to seeing real chaos descend on one of the world's great powers. When crowds began massing throughout Eastern Europe later that year, chanting for the same freedoms and opportunities that Chinese protesters had also apparently desired, Bush could not help but think of their fate, and could not help but feel that the awful outcome of Tiananmen was about to be replayed in Leipzig, Bucharest, Sophia, or even East Berlin or Moscow. In those cities, citizens marched because Gorbachev, the great reformer, had provided an opening. Then again, Deng had once been called a great reformer too.

"Right now, and especially in light of what has happened in China," Bush wrote Margaret Thatcher in July as he prepared for another trip back across the Atlantic, "there is no global issue more important in determining our future than the outcome of the changes now underway in Eastern Europe." Although he was no less convinced of the power of time and of democracy's inevitable triumph, Tiananmen left Bush more skeptical than before that the path toward that inevitable future would be smooth, or that he or anyone else had the power to keep chaos at bay.

11

EASTERN EUROPE ABOIL

"IT'S NOW JULY 9, and we're three hours out on our way to Poland," Bush penned in his diary on the first leg of a nearly two-week trip that would see him visit both sides of the rapidly disintegrating Iron Curtain. Gorbachev's influence would be everywhere, though he was not on the president's carefully orchestrated itinerary. All of Eastern Europe faced unrest, but as was typical of the region, its various parts faced various problems. In Poland and Hungary, the first two stops on Bush's trip, reformers had taken the first tentative steps toward power. In Romania, Bulgaria, Czechoslovakia, and especially East Germany, conservatives vowed to hold fast. The people must "take up arms" in defense of "the achievements of socialism from external enemies," the wife of East Germany's recalcitrant Erich Honecker implored. The regime "would do all in its power to help defend the socialist order," her husband declared, against enemies near and abroad, including the man currently in charge in Moscow. He invited Chinese police officials to come teach his secret police, the Stasi, what they had learned about "crowd control" following Tiananmen. The "Chinese solution" might do wonders in Germany, Honecker warned.

Poland and Hungary, where market reforms had gone furthest, featured skyrocketing inflation, unemployment, and hunger. Romanians knew hunger too, without any market reforms whatsoever. In East Germany, long the jewel in the Soviet crown and the most prosperous of Moscow's client states, growing pockets of protesters called for change, and an even greater number of citizens plotted escape. The Soviets suffered too: In Siberia and then Ukraine, miners walked off the job demanding greater pay. In Georgia and throughout the Baltics, crowds chanted for the right to secede. Estonia's legislature declared the sovereignty of its laws over those promulgated by the Kremlin, while Lithuania's prepared its own break from the Soviet state. Latvia's pro-nationalist movement was not far behind. "Only three or four years back, a prison camp or insane asylum would have awaited anyone who called openly for secession," Ambassador Matlock noted from his post in Moscow. "Forty years ago, it would have been a bullet at the base of the skull."

"The empire-federation is falling apart," one of Gorbachev's closest advisers confessed in his journal. "Chaos is breaking out." Gorbachev "feels that he is losing the levers of power irreversibly," Anatoly Chernyaev wrote, yet "has no concept of where we are going." China's fate forced a worrisome question. "Is it true that our people can only be brought to order by force?" Chernyaev wondered.

The question ricocheted throughout Washington as well. "Can it happen in the Soviet Union?" The "it" was a violent clash of reformers and hard-liners, another Tiananmen, the query posed by the White House to the National Intelligence Council (NIC) in July. The best minds in Washington's analytical community answered with a resounding . . . maybe. The situations appeared similar, the NIC advised. Each featured authoritarian states undergoing rapid transformation, and "some instability is inherent in the reform process." Yet they were not entirely analogous. China's protesters considered the regime their enemy. Soviet bloc reformers, conversely, largely championed Gorbachev. So long as he remained in power, the American analysts therefore advised Bush, "the probability of such a popularly based challenge to the Soviet regime" as had arisen in Beijing "does not appear to be high for the foreseeable future." Of course, they added, "China experts would have said the same thing prior to the events in China."

So could it happen in Eastern Europe? Could the urge for change swell into a surge of protest followed by repression? The real thing to fear within the Soviet empire wasn't in their considered opinion a revolution against the Communist Party so much as one directed from within, whether by Moscow's conservatives or by the hard-liners still in charge throughout most of the Warsaw Pact countries. "Some are now talking of the possibility of civil war in the next two to three years," Matlock had written in the quiet of his own flight across the Atlantic in late June 1989, en route back to Washington to help brief the president. "Others [warned] of the possibility of famine" if *perestroika* failed. Tapping away on his laptop as his plane flew through the night, Matlock considered his "worst case scenario: prices continue their rise, shortages become more acute. Scattered small protests coalesce and become violent, going after the most convenient local target: a minority nationality, Party or police headquarters — or just on a general rampage."

"What then?" he wondered, staring out into the darkness, unable to sleep. "Could they order a Tiananmen Square? Probably. Would it work? Probably, in the short term at least, but at the cost of the reform process itself." The subsequent turmoil might make Tiananmen Square look like child's play. "By the time I left the plane in Washington," Matlock recalled, "it had dawned on me that the empire's reinforcing rods might soon disintegrate."

Matlock believed in Gorbachev, though his remained the minority opinion among the small group around Bush whose views truly counted. Their skepticism persisted even after the strategic pause, and even after Scowcroft, Baker, Gates, Rice, and crucially Bush too had largely warmed to the idea that Gorbachev's success, and continued tenure in office, offered their best hope of a peaceful evolution of Soviet power. He increasingly appeared their surest bet, but not someone they trusted. "Our residual doubts," the NSC's Peter Rodman wrote in a memo passed from Scowcroft to the president in late June, stem "from the nagging suspicion that, somewhere out there, there is a limit to that tolerance . . . say, a bloc country's attempt to go neutral, or to vote the communist party out of office." Any further revolutionary changes behind the Iron Curtain might yet prompt even Gorbachev to reassert Soviet control, if only to placate critics. "Events may test the new Soviet tolerance some time soon," he concluded. Tell Rodman how much he liked his paper, the president told Scowcroft, which he found "right in its conclusions."

Bush's chief consideration as he prepared for his first visit to Eastern Europe as president was simply to avoid anything that might directly or inadvertently raise the risk of further instability. He had tried not to influence events in Beijing, but Chinese leaders made his words part of their rationale for cracking down. He hoped for a quieter time when visiting Gorbachev's strategic backyard. "Whatever this trip is, it is not a victory tour with me running around over there pounding my chest," Bush instructed his speechwriters. "I don't want to sound inflammatory or provocative," or do anything to "complicate the lives of Gorbachev and the others." Things were precarious enough without his rocking the boat. Above all else, he told them, "I don't want to put a stick in Gorbachev's eye."

The trip was largely "symbolic," he had explained to Polish journalists brought to the Oval Office before his departure. He hoped merely to "salute the changes that have taken place" and to "give our views freely and openly about freedom and things of that nature." He was not, under any circumstance, "trying to complicate matters between the Soviet Union and Poland, or anything else." If asked, he might offer a bit of "free advice" on how to manage a market economy. Then again, he added with a laugh, Warsaw's leaders might "turn around and ask me about our debt—and then I'd be embarrassed."

"So we're going in a constructive vein," Bush continued, waving off his press secretary's theatrical glances at his watch. He knew they were running late, but rather enjoyed speaking to reporters in quiet settings. They didn't interrupt as much as when cameras were rolling. "We want to see perestroika succeed, and I want to see glasnost succeed," and therefore he welcomed Gorbachev's talk of reduced tensions and a reduced Soviet military footprint in Eastern Europe. Perhaps someday, Bush mused, Poland might be free of Soviet troops altogether. "I would like to see continuation of the change that would result in the Soviets feeling comfortable in taking their troops out of there."

He realized immediately that he had gone too far. As presidential proclamations go, Bush's suggestion that Poland might someday be free of Soviet troops was hardly akin to Ronald Reagan directly challenging Gorbachev to "tear down this wall," or JFK demanding the Soviets cease pressuring West Berlin because "*Ich bin ein Berliner.*" Bush merely restated a position held by American administrations since the tail end of World War II. Yet it was precisely the kind of direct advice he'd sworn to avoid.

Words were not always Bush's allies. A good conversationalist and debater, for all his years in public service he had yet to master the interview. His grammar often collapsed in direct relation to his desire for precision, exacerbating his natural tendency to start and stop sentences midstream, smothering one thought with another. "Well, I'd like to see respect for the will of the people," he said, trying to explain his point to the rapidly scribbling journalists. "And I think as we—I don't want to—well, let me start over. I will stand by that, obviously . . . Having said that, I will not be trying to inflame change so that it does what you're talking about. The [Polish] people seem to be handling it very nicely now, with elections and with discussions around a table. And I don't want to do something that would inadvertently do what you're talking about, or that you asked about; and that is, to have some crisis that will compel other answers. And I don't want that, and I'm not going to deliberately do anything that is going to cause a crisis."

He already had. "Bush, in Polish Press Interview, Urges a Pullout of Soviet Troops," blared the *New York Times.* The *Chicago Tribune,* with its large Polish American readership, was no less direct: "Bush Urges Soviets to Pull Troops from Poland." Similar headlines appeared throughout Europe, and especially within the Soviet bloc, where Bush's words came under particular scrutiny. If the main purpose of his trip was to cheer democracy without riling up its opponents, he failed even before leaving Washington.

"Tell the President to please be a bit more considerate," Gorbachev whispered to Ambassador Matlock the next day, pulling him into a private corner of a Moscow performance hall. They were both attending a Van Cliburn concert in celebration of American Independence Day, the first performance on Soviet soil in decades for the American-born pianist who'd won the country's first international Tchaikovsky Competition in 1958 at the tender age of twenty-three. "There is something I'd like you to pass on to the President," Gorbachev continued, especially at this "very complex and difficult period." Opposition was growing, and "what he [Bush] says has an effect here."

Matlock suspected at once that Gorbachev had been "offended" by talk about the withdrawal of Soviet troops from Poland. Of course he would pass along the message, he replied, but asked for "some examples" to be sure to get the message right. "No," the Soviet leader said, waving his hand and releasing the ambassador back into the crowd. "Just let him know that some statements complicate things here. He should just try to be more considerate if he wants to help."

In fact Gorbachev didn't think Bush was helping at all, and increasingly feared he never would. "The hawks [in Washington] were again on the move," he warned his staff. Whom the Poles chose to befriend was hardly Washington's concern, he subsequently told French president Mitterrand. But "to go to Poland and to call for the withdrawal of Soviet troops," he continued, "all this looks very strange to me." He said much the same to Germany's Kohl. "The situation throughout the communist world was precarious" enough, he complained, without ill-considered American demands for how Moscow should manage its own sphere of influence.

The two most powerful men in the world thus entered the summer of 1989 speaking to each other only through intermediaries and the press, and speaking past each other as a result. Mitterrand and Kohl each promised to mediate, each advising Gorbachev to worry less when it came to the still relatively new American president. Kohl stressed that he trusted Bush, and hoped Gorbachev would in time say the same. Bush still "had a long way to go to compete with Reagan," Kohl noted. "He has neither the actor charisma nor the art to communicate with people via TV." But Bush was one thing Reagan was surely not: "he was an intellectual," possessing "a far more European vision of things than Reagan had."

Give Bush time, Kohl continued. The Kremlin once considered Reagan dangerous, "and you, Mr. Gorbachev, were able to find a common language with him." Sighing in response, Gorbachev stressed anew his hope that Bush would tread carefully while visiting Eastern Europe. "I think you should agree with me that you should not stick a pole into an anthill," he said. "Consequences of such an act could be absolutely unpredictable."

With Mitterrand he employed a different analogy. "It is important, Mr. President," Gorbachev told his French counterpart, "to avoid a situation where in these times, which are marked with signs of big changes and common hopes, someone would try to behave like an elephant in a china shop." Mitterrand also advised patience, though he too suggested Gorbachev lower his expectations. Noting that he'd just spent a weekend with Bush in Kennebunkport, he explained that the real issue plaguing the new administration — and thus plaguing them all — was less a desire to see the Soviets laid low than a sheer inability to conceive of doing anything else. "Bush, as a President, has a very big drawback," Mitterrand observed. "He lacks original thinking altogether."

It is more apt to say Bush and Gorbachev — ultimately neither elephants nor ants — circled each other like great cats in the spring and summer of 1989. Each traveled to the same places and met with the same foreign leaders, competing at every stop for international favor and for Europe's soul, but never speaking directly to each other. Gorbachev followed Bush to Beijing. Bush subsequently followed Gorbachev's visits to Germany and France with visits of his own, a few days after having met with Hungarian and Polish leaders whom Gorbachev had engaged with only days before that.

Their trips to Paris, just a week apart, elicited far different receptions. American presidents had been there before, and by 1989 Parisians were well practiced in responding to visiting Americans with resigned apathy (if not worse). Gorbachev they treated like a rock star. "Gorby! Gorby!" the crowds chanted at every turn. On this two hundredth anniversary of the French Revolution, he proudly declared, "Perestroika is also a revolution," with equally ambitious goals. Like France's revolution, *perestroika* too "will know a great destiny that will not be limited to a national context."

Popularity did not equal perfection. "Where's the Bastille?" he naïvely asked France's foreign minister. Hoping for pictures of their revolutionary leader standing before a quintessentially revolutionary symbol, Soviet media officials had not realized the building had been razed centuries before. "It *was* over there," Roland Dumas sheepishly explained, though Mikhail and Raisa Gorbachev found it difficult to see anything over the raucous crowds that followed their every move. Police tried to maintain order. The scrum of photographers proved stronger. Several broke through security lines, rushing toward the Soviet couple with bodyguards on their tails. Cameras flew. Punches followed. An aluminum ladder sailed through the air, turned from pedestal to projectile by a bleeding cameraman. "Gorbachev looked appalled," a British correspondent reported. "Mr. Dumas looked furious." Shevardnadze looked amused, laughing loudly at the sight of it all, and at Gorbachev's difficulty meeting natives within the Paris mob. "He shook my hand but he didn't really say anything," reported a fourteen-year-old pulled from the crowd so the Soviet leader could meet a real Parisian. The boy was a New Yorker on vacation.

Gorbachev's speech at the Sorbonne the next day went equally poorly. Reveling in his academic setting, he situated *perestroika* in the long arc of social advancement, quoting at length Voltaire, Montesquieu, Diderot, Holbach, Mably, and Rousseau, among other famous French philosophers. It was a classic Gorbachev text: flowery, messianic, erudite, yet hard to follow, and long. Very long.

His answer to a professor's question about Tiananmen drew more attention than anything in his prepared remarks, as he seemed to justify Beijing's crackdown. "We cannot interfere directly and give advice" to any nation about matters inside its own borders, he replied. It was not up to Europeans to tell the Chinese how to run their affairs, just as it was foolhardy for foreign zealots to call for the Soviet bloc's outright abandonment of its own values. "Some would like to see the problem of Europe solved by the displacement of socialism," but, he warned, "I think this is unreal and even dangerous."

Such words did little to calm White House skeptics. Was his refusal to condemn the Chinese merely geopolitical realism, they wondered, or a preemptive justification for the day when he might follow their lead? It was hard to know.

He might be trying to signal that the Soviet Union would not invade its Eastern European allies even if they turned away, but to some ears in Washington it could also mean he would crack down on his own people if necessary, and there was nothing anyone outside the Soviet Union could do to stop him.

He thought he was being unambiguous. "I think that what I have just said makes it clear whether there is any 'Brezhnev Doctrine,'" he told Kohl. It was a reference to the Soviet Union's long-standing promise to suppress any antisocialist move by an ally. Whatever his intent, the message failed to resonate across the Atlantic. "I missed completely, really, the revocation of the Brezhnev Doctrine," Condoleezza Rice admitted in 1997. "Later, in conversations with Gorbachev, and particularly his people, like Chernyaev, I think all of us learned that they had been more interested in, and believed that they were making more of a statement about Eastern Europe" during this period. "So this was really an example of not quite getting the message."

Gorbachev kept trying. "Respect for the sovereign right of each people to choose their social system at their own discretion are [sic] the most important prerequisite for a normal European process," he told the Council of Europe in Stroudsburg immediately after his trip to Paris. The continent's inhabitants should therefore strive to construct "a common European home."

Having used that phrase before, Gorbachev seemed to think it was time to offer a full definition. With the Cold War "consigned to oblivion," Gorbachev explained, nations could finally "replace the traditional balance of forces with a balance of interests." Trust would trump threats in the new common European home, and economic partnerships replace fearful rivalries. To be European in this age, he continued, when East-West divides were being dismantled and a new continental union was on the rise, was to trust in Europe's own ability to govern itself, without the influence of foreigners across the seas.

Much like his talk at the United Nations the previous December, the speech was a sensation, even as it revealed anew the great tension of Gorbachev's tenure. He was an agent for change, yet he could not control the changes he unleashed. In such a volatile time, "if someone were to try to destabilize the situation," he warned Kohl, someone like an American president recklessly calling for the sudden withdrawal of Soviet troops from Eastern Europe, "it would disrupt the process of building trust between the East and the West, and destroy everything that has been achieved so far."

Gorbachev should look nearby to find the real source of instability, Kohl responded, rather than raise the bogeyman of American influence. "We understand Moscow much better, and we feel much closer to it than [East] Berlin now," Kohl continued. East German citizens desired the same democratic changes they'd seen occurring elsewhere in the Soviet bloc. They too wanted *perestroika*. Their government, however, refused them the chance. "I just feel sorry for the people," Kohl said, but "it is not a secret to anybody that Erich Honecker is not inclined to undertake any changes or reforms.

"Thus he himself destabilizes the situation," Kohl warned.

IT WAS A PROPHETIC WARNING. East Germany rumbled as spring turned to summer, prompting Honecker's recalcitrant regime to ship off its most vociferous critics before they might begin to gather in the streets as in Beijing. Nearly ten thousand received official notice in May that their emigration applications had finally been approved. Some had submitted the forms decades earlier. Others had never submitted the requisite paperwork in the first place. They were nonetheless being kicked out before they could cause too much trouble, leading many protesters to the logical conclusion that the louder they screamed, the more likely their chances of winning the exodus they desired.

Droves of East Germans made the fateful decision to flee without permission. Close to ten thousand made it to West Germany in May alone. In June the numbers of exiles and escapees totaled 10,646 and 12,428 respectively, their steady arrival overwhelming West German social services. All were welcomed home to their fatherland, even as bureaucrats in Bonn tabulated the cost of long-standing promises of housing, cash, and a passport for any East German who somehow made it out. Such promises had been easier to make when they thought no one was coming.

More emigrants were on the way, and there were more avenues of escape for Honecker's regime to block than merely its own. Each Warsaw Pact nation managed its own borders. On May 2, Hungarian officials announced that they would no longer actively patrol their long boundary with Austria. Their citizens could already cross with relative ease following the previous year's loosening of restrictions, government spokesmen pointed out, yet border surveillance continued to cost a million dollars annually, largely to maintain electronic sensors whose primary accomplishment seemed the detection of stray rabbits and deer. It was money they could better spend elsewhere.

Finances were only part of the reason for Budapest's announcement. The

Hungarians were tired of serving as prison guards for the vast penitentiary known as East Germany. East German citizens had long enjoyed the luxury of relatively carefree travel throughout the Soviet bloc, provided they did not stray farther. It thus fell to East Germany's allies bordering the West to keep them from exiting, an unpleasant task the Hungarians disdained. They hoped the humanitarian gesture might additionally win goodwill and investments from the West. Kohl had already secretly promised as much. "This is the beginning of a new process which we hope will help our international links a lot," the commander of Hungary's border patrol candidly told a reporter.

Not coincidentally, more than 25,000 East Germans chose to visit Hungary that June. By July, vast camps of East German "vacationers" had sprung up within sight of a new, freer life. The border remained officially closed to them, yet more than 150,000 were encamped by August, waiting for an opportune moment to sneak across at an unguarded spot. From Austria they might travel anywhere they liked, including to West Germany. Those unlucky enough to stumble upon a guard or patrol frequently found them looking the other way, or received a gentle suggestion that they might come back in an hour or so, after the unit had moved on.

Word soon spread, prompting even more East Germans to pack their belongings into their sputtering Trabants for a Hungarian summer holiday. Ownership of the tiny auto represented a lifelong dream for many East German citizens, who lusted for its twenty-six-horsepower engine, which could propel four cramped passengers from zero to sixty miles per hour in a lethargic twenty-one seconds—though sixty miles per hour exceeded the car's recommended top speed. Most East Germans had to wait more than a decade for the opportunity to purchase one of the compact proletarian cars. They nonetheless readily abandoned even their precious Trabants for the chance to sneak across the border on foot, though one filmmaker documenting the scene noticed time and again drivers stopping in their tracks on their way to freedom in order to walk back and lock the doors. "A German is still a German," he explained, "even in situations like these."

For all of Hungary's impressive steps beyond Soviet control, Poland had taken even greater strides, making it the other obvious choice for Bush's travel planners. Budapest's reformers were still communists. In Poland the communists were voted out. Concluding that his government no longer had the ability to solve its deepening economic crisis, General Wojciech Jaruzelski had threatened to resign from his post as general secretary the previous December, taking his senior military commanders with him, unless the party's Central Committee authorized direct talks with their adversaries on a power-sharing agreement.

The threat worked. Roundtable discussions began on February 6, 1989, bringing together government, labor, and church leaders as equals. "I would negotiate with the devil himself if it would help Poland," Lech Walesa, leader of the opposition labor union known as Solidarity, proclaimed when asked how he could sit down with his onetime jailers. Negotiations took place, literally, around a twenty-nine-foot round table. Why so big? "Because the world spitting record was twenty-eight feet," participants joked. Weeks of tense negotiations produced plans for a general election on June 4. No one anticipated at the time just how fateful that day would become.

The fact that the Poles planned genuine elections was shocking enough. Citizens routinely voted throughout the communist world, albeit in sham contests between approved candidates. East German leaders, for example, considered elections less referendums than affirmations, or in their words, "a significant political high point in the social life of the GDR [German Democratic Republic] with which the confidence of the citizens in the policies of the party and government can be strengthened anew." They claimed the support of 98 percent of eligible citizens — not just voters but citizens — in their May 1989 municipal elections, results even diehard communists found laughably divorced from reality. Central authorities had commissioned a victory in the 93–95 percent range. Local officials, ever eager to impress their superiors, took it upon themselves to give Berlin an even better return. "They did it out of habit and discipline," Berlin's district secretary later rationalized, "believing that it was willed and blessed from on high."

Poles had their own recent experience with manipulated election results, having the previous year voted on a referendum supporting the government's difficult economic reforms. Jaruzelski's regime wanted popular endorsement of their painful short-term measures designed to produce a better future, and a majority of voters indeed backed the plan. Because fewer than half of the eligible voters had cast ballots following boycott calls by opposition groups, however, the government declared the referendum a failure. The results meant nothing if they failed to coopt Solidarity into sharing responsibility for the inevitable hardships. Such were the confusing times Poles endured, when a besieged government could win its own referendum yet rule against its own economic plan.

In June, with voters able to express real choices, Solidarity candidates captured every one of the 161 contested seats in the lower legislative body known as the Sejm, and ninety-nine out of one hundred available spots in the Senate. The one hundredth, an independent, quickly agreed to join their caucus, making Solidarity's victory complete. The bulk of the regime's handpicked candidates running unopposed for reserved seats failed to win, falling short of the required support of 50 percent of voters and thus calling into question the Communist Party's ability to handpick the next president. The election thus proved a total mandate for change, unlike the false mandates of the past. "Somehow, in the depths of our brains, we were convinced that we would win the election," Prime Minister Mieczyslaw Rakowski lamented, "because, after all, we had always won elections."

The results caused more anxiety than elation, even for the victors. "Total victory, or something close to it" by Solidarity and other opposition groups, the American embassy had warned Bush in advance, "will threaten a sharp defensive reaction from the regime . . . [E]ven possibly military responses cannot be entirely ruled out" if Poland's leaders perceived a sudden end to their rule. Deeply embarrassed, Jaruzelski prepared his resignation. A shocked Central Committee wondered how to avoid "destabilization" similar to the "situation in China," debating if the same tactics might be required. Walesa judged the results "a disaster," immediately reopening negotiations in hope of ensuring space on the next round of ballots for the party's preferred candidates. The point of the election had been to wean hard-liners from power slowly, but now "too much grain has ripened for me and I can't store it all in my granary."

Poland's historic elections thus prompted a political crisis a mere five weeks before Bush's visit. The process assumed Jaruzelski's presidency, for his unquestioned patriotism and fidelity made him trustworthy to all sides, including most especially Gorbachev and the Soviets. Yes, he had initiated martial law in 1981 and was the face of the crackdown to the wider world. But his supporters considered the move Poland's salvation, as it kept the Soviets from intervening. By late 1987 his country had become, according to one respected human rights organization, "the freest country in the Eastern Bloc." This was a compliment only by comparison. Still, it was clear that he shared Gorbachev's view that socialism's deep structural flaws could be overcome only by equally deep structural change. "We need to support the Poles," Gorbachev subsequently told Helmut Kohl after Solidarity's landslide electoral victory. "They do not have anyone who has more authority and respect than Wojciech Jaruzelski now."

Elsewhere in the Eastern bloc, upheaval took different forms. Czechoslovakia's protest movement remained highbrow. While Hungarians marched, Poles struck, and East Germans fled, the Czechs wrote. Leaders of the Charter 77 movement, so named for their initial declaration that appeared in Western newspapers in January 1977, largely criticized the government's human rights violations in essays and plays by such writers as Pavel Kohout and Václav Havel. Censorship at home ensured that most citizens knew of the organization only because of the government's efforts to discredit it. Yet so long as people recalled the sight of Soviet tanks in the streets and retained memories of the crushed Prague uprising of 1968, Czechs were naturally wary of Kremlin promises that each state might truly choose its own path to socialism. Talk of *perestroika* now, a Czech confidant told Gorbachev, was "like telling a person whose legs have been broken: 'Well, come on now, you can go where you want.'"

Little anticommunist fervor seemed to prevail in Romania, either, for entirely different reasons. There the population feared the Soviets far less than they feared their own regime, which, like the one in East Germany, believed that what Gorbachev's acolytes called transformation was in fact counterrevolution designed to undermine their socialist dream. "He [Romanian strongman Nicolae Ceauşescu] told me the measures that we are now undertaking in the USSR in the framework of perestroika had already been implemented in Romania 10 years ago," an incredulous Gorbachev complained to France's Mitterrand in early July. "He told you that?" an equally incredulous Mitterrand responded. Of all Eastern Europe's regimes, Ceauşescu's needed *perestroika* most of all! "I visited Romania several years ago," Mitterrand added. "After that I never went back."

Romania was less a country than a cult by the late 1980s, and its entire political and cultural life orbited the former cobbler and petty criminal at its head. Slight and wholly underwhelming, Ceauşescu assumed power in 1965. Party leaders expected him merely to hold down the office until a real leader emerged. He instead transformed the country into his own satrapy. Schoolchildren sang of his achievements. His birthday became a national holiday. By the 1980s Romanian airwaves overflowed with tales of his triumphs and travels. Cheering crowds emerged everywhere he went (often trucked in for the occasion to ensure a professional level of enthusiasm). American officials reported the ecstatic cheers that erupted every time Ceauşescu entered a tent at one Bucharest trade show, for example, the crowd skillfully collected once he'd passed so they might line up and cheer his arrival at the next tent, and the next, and the next. The people loved him, Ceauşescu consequently thought, and his government was perfectly willing to airbrush away anything counter to his desired reality.

Actual reality mattered far less. Romania's heroic farmers brought in a record harvest of 60 million tons of grain, the government announced in October 1989. The real number was less than 17 million, though few dared tell the emperor. In a world constructed wholly of lies, a Romanian apparatchik's greatest service was enabling his demigod's fantasies. The roads leading to Ceauşescu's country retreat were lined with lush fields of corn, abundant to any passing eye. Four or five rows in, however, well past what Ceauşescu could see from his speeding limousine, crops withered for lack of care. In the words of Newsweek's bureau chief, "There was no fertilizer for ordinary farmers, except those who fed the emperor's ego."

This alternate reality enabled an alternative approach to dealing with the country's foreign debts. Like most of Eastern Europe, Romania ran up huge bills in the West during the 1970s on loans intended for industrial modernization. Total Soviet bloc debt skyrocketed from $6 billion in 1970 to $56 billion by 1980, topping out at over $90 billion by the decade's end. Poles were paying more than $2 billion annually by 1989 on interest payments alone. In Hungary the figure was $1 billion. Worse yet, industrialization never occurred. Most hard-currency loans were instead diverted directly to the purchase of consumer goods in order to appease increasingly agitated populations. Unable to repay their principal, much of Eastern Europe was therefore effectively putting food on its credit card by the time Gorbachev took office, falling further behind every month.

Not so in Romania. Whereas most bloc leaders refused to break the cycle of debt and interest payments lest the flow of Western goods dry up and unrest ensue, in 1981 Ceauşescu vowed to pay off Romania's entire $10.1 billion national debt by the close of the decade. Deprivation ensued as the regime squeezed every possible dollar for repayment. Electric power was severely rationed, and Ceauşescu's planners developed a "program of scientific nourishment," calculating the true (by which they meant minimum) number of calories a worker required for maximum productivity. For the personally abstemious Ceauşescus, the right number seemed very low indeed. "If only we had a little more food," Bucharest residents ruefully joked. "Then it would be just like wartime!"

The pervasive secret police known as the Securitate kept everyone in line. Foreign books and magazines were banned, and typewriters were individually registered so critiques could be matched with their doomed author. Political criticism meant arrest. Arrest meant torture. Or worse. Sometimes a body was returned. Families were more often left to wonder. The police enforced not only Ceauşescu's absolute control but his morality as well. Abortions were criminalized and fertility enforced throughout the 1980s, thus creating a population boom in the midst of a shrinking economy. "Breed, comrade women," Ceauşescu exhorted. "It is your patriotic duty!" Women of childbearing age were examined monthly by state doctors. If a woman had not conceived, the Securitate wanted to know why. If she was no longer pregnant, they wanted to know whom to jail.

Few could afford children, however. Cavernous orphanages with medieval sanitation consequently housed more than 100,000 young Romanians, the revelation of their existence at the Cold War's end as shocking to Western sensibilities as any indictment of communism since the admission of Stalin's crimes in the 1950s. Infants were crammed ten or more to a crib, receiving on average a mere five or six minutes of personal caregiver attention a day. Children so malnourished for basic human comfort often suffer deep psychological scarring. More than three thousand were infected with HIV as well. Visiting in 1990, one American doctor thought "the closest thing that I can relate it to is a concentration camp."

Gorbachev loathed the man he called the "Romanian führer," thinking Ceauşescu "in a class of his own" when it came to vanity and hubris, always with his "arrogant smirk, treating others with apparent contempt, everyone from retainers to equal partners." The feeling was mutual. Neither would recognize the other's self-proclaimed historic transcendence. Gorbachev had little time for messianic competitors, noting that Ceauşescu had for decades agitated for independence whenever Moscow demanded fealty. Now that the Kremlin had embraced reform, he was suddenly screaming for orthodoxy. "It gave me a strange feeling to watch Ceauşescu going out of his way to demonstrate the independence of his positions," Gorbachev later wrote in his memoirs. "Anyone with the slightest political experience could see his delusions of grandeur as well as his psychological instability."

Needless to say, the pair did not fare well in person. A loud argument erupted over dinner during the first night of a 1987 visit by the Soviet leader to Bucharest, for example, leaving their wives to stare silently at their plates in discomfort. "I began *my* perestroika in 1966," the Romanian said. "You are running a dictatorship here!" Gorbachev countered. Soviet staffers witnessed a food riot the next day, when the sight of fresh fruits and vegetables stocked in markets along Gorbachev's route caused residents who'd not seen such luxuries in years to descend like locusts the moment his heavily armed entourage passed. "Human dignity has absolutely no value" in Romania, Gorbachev complained to the Politburo upon his return, subsequently wondering aloud to Kohl how "this kind of family clan would be established in the center of civilized Europe." He "could imagine something like that to emerge somewhere else, like it has in [North] Korea, but here, right next to us — it is such a primitive phenomenon."

The Eastern Europe Bush visited in July was therefore beset by competing forces pitting agents of change against communist hard-liners determined to hold fast. Disdainful of what they'd witnessed in Budapest and Warsaw, and practically terrified of the images from Beijing, hard-liners went on the offensive immediately after Gorbachev's trip to the Council of Europe, when Ceauşescu and Honecker conspired to place the issue of Hungary on the agenda for the Warsaw Pact's annual meeting in Bucharest. "Of course, he [Honecker] did not intend to just 'discuss' the Hungarians," a member of East Germany's Politburo later recalled. "He wanted to stop them. As he saw it, it was time for the entire bloc to hold together," holding the line while there was still time to save socialism, by force if that is what it took. Force had worked in the past. It had just worked in Beijing as well.

The Bucharest counterattack began the moment the Warsaw Pact leaders arrived on the tarmac. Ceauşescu welcomed each with the region's traditional kisses and cries of "comrade." All except Hungary's Miklós Németh, to whom he extended only a cold hand and an even colder "Mr.," signaling that he no longer considered the Hungarian leader part of their fraternal pact. Unhappy to be in Romania in the first place, and fully aware of the impending opposition to all that *perestroika* had wrought, Gorbachev planned a cool welcome of his own, making Ceauşescu wait in the sweltering summer heat for nearly an hour before finally descending from his air-conditioned plane.

The attack continued behind closed doors. Hungary was "destroying socialism," Ceaușescu raged. Its "dangerous experiments" would bring down the entire Soviet bloc. "The unity of socialist countries is a historic necessity," he screamed, demanding the counterrevolution be put down before it was too late. Honecker chimed in. The "forces of imperialism" were trying to "destabilize socialism." Czechoslovakia's Miloš Jakeš called for recognizing the "complexity" of the situation. Poland's Jaruzelski largely sat mute. Embarrassed by his electoral defeat, he'd not wanted to attend in the first place. Gorbachev asked him to reconsider. He needed all the allies he could muster. "I can only speak about Poland," Jaruzelski finally said when prodded. "But I have to admit that we have behaved like an absolute monarch who was always right."

Németh squirmed while all around him the bloc's supreme leaders, titans of the communist world, debated his country's fate. All eyes turned to Gorbachev, the man who had started it all, and still by far the most powerful leader in the room. His continued support was key to the region's entire transformation. By the same token, he was the only man alive who might possibly still have the power to reseal Pandora's box.

Gorbachev in turn gazed at Németh. Their eyes met, and to Németh's great astonishment, the Soviet leader, the world's most popular man with enough nuclear firepower at his command to destroy the globe multiple times over . . . winked. "This happened at least four or five times," Németh recalled. "Each time he smiled at me, with his eyes. I don't know how else to describe it. But I clearly saw he was trying to tell me that he did not share these views." He was trying to tell the Hungarian that, so long as his cadre held sway in the Kremlin, Budapest's quest for greater autonomy was secure.

With that, Németh excused himself for a cigarette. There was no need to hear any more tirades, and it was not as if the troglodytes would be convinced by anything he might say in response. Honecker departed soon after for East Berlin, complaining of illness and exhaustion but in truth felled by anxiety, while Gorbachev delivered his own lengthy discourse on the necessity of change. When he'd finished, he gave the Hungarian delegation the honor of an hour-long private session, while Ceaușescu paced and ranted alone.

Gorbachev had made his choice. It was neither new nor unexpected, but he had once more, and in a more formal fashion than ever before, warned Eastern Europe's communist diehards that he would never condone their use of force. They would be on their own if they imposed a Chinese solution. But having once more made his choice known, the real question was how he, and an American president who still refused his company and counsel, might somehow keep others from choosing Deng's.

THE HUNGARIANS wanted everyone to know. The Americans especially. Thus Larry Napper, in Bucharest for only a month, found himself standing beneath a dim streetlamp, late at night, in the rain, waiting. "It was like a Graham Greene novel," Napper recalled thinking as his contact from the Hungarian embassy arrived, and certainly like nothing he'd ever experienced during nearly two decades in the Foreign Service. Turning up the collars of their topcoats against the rain, the pair walked the city's boulevards, always on the move to complicate life for Securitate eavesdroppers. "They pleaded with Gorbachev," the Hungarian told him. "Ceaușescu and Honecker each made a pitch for Soviet intervention. They wanted the tanks to roll in, just like in '56 and '68." Ceaușescu was "absolutely determined" never to cede power or "become a victim" of popular dissent. He knew what happened to fallen gods and fake messiahs.

Gorbachev had clearly sided with the region's reformers but worried that because of it, his own days were numbered. He barely had time to sleep and felt under "constant pressure," he complained to France's Mitterrand, and Bush certainly wasn't making his job any easier. He "would not last a week" in office, Gorbachev worriedly told Mitterrand, if he pulled Soviet troops entirely out of Poland as President Bush recently suggested. What if he "disappeared?" a reporter had pointedly asked immediately before his flight to Bucharest to meet his seething allies. What then would happen to *perestroika*? "Why the concern?" Gorbachev shot back, feigning anger. "Are you worried about my health? Do you have some terrible news?"

This was the maelstrom into which Bush flew in mid-July 1989, and nowhere was the tension greater than in Poland, the first stop on his trip. More than a month after the election, voters were still wondering who would emerge as their president. American officials did not even know for certain until hours before Bush's arrival who'd be waiting to receive their man as he stepped off Air Force One. Matters of protocol, however, were their least concern. "Most Solidarity leaders are apparently convinced that [General Wojciech] Jaruzelski must be elected president if the country is to avoid civil war," Ambassador John Davis warned from Warsaw. "At the same time, most are reluctant to cast votes for the man who more than any other represented the regime's brutalities."

Much of his own party wanted him gone. Chastened and angry, a growing number of the regime's delegates to the Sejm conspired to give Jaruzelski the early retirement he'd once held over their heads, repudiating change and the man who'd made it possible by simply withholding their votes. Poland's officer corps, conversely, would "feel personally threatened if Jaruzelksi was not President," Davis was warned, "and would [therefore] move to overturn the Round Table and election results." Talk of a coup was in the air.

Jaruzelksi's main base of support came from people he'd once imprisoned, and from foreigners. Untrained in parliamentary procedures, Solidarity leaders turned to Davis and his staff for a crash course in the art of vote counting. "There is much hand-wringing going on," he reported back to Washington, "with some Solidarity Senators saying they will vote for Jaruzelski if they must to save the country, even if it means ending their political careers virtually before they have begun." Communist Party delegates were lobbied to at least abstain from voting rather than cast a negative ballot. An exhausted Jaruzelski was not even sure he still wanted the job. "The General is determined that he will not 'creep' into the Presidency," Davis reported. "He is understandably reluctant to face another public humiliation."

Pause to consider these ironies. Solidarity supported the man who'd once banned the union. Avowed communists refused to back their own party boss. American officials were wading as never before into electoral politics behind the Iron Curtain in order to ensure the survival of a general whose own people had just effectively rejected him and who had once initiated martial law. "If Jaruzelksi is not elected president," Davis warned, "there is a genuine danger of civil war ending, in most scenarios, with a reluctant but brutal Soviet intervention." In his assessment, "There have probably been few occasions when an American president has arrived for an official visit in a more fluid and fast-moving political situation than the present one in Poland."

Bush privately endorsed Jaruzelksi, but only in private, wishing for a way to back the general without riling up his opponents, knowing that his every gesture and comment on his trip would be scrutinized. As Scowcroft privately advised, there was "a political vacuum at the time when Poland needs strong leadership: Solidarity is legitimate but not in power and the regime is in power but not legitimate . . . The message you bring to the Poles will be watched throughout Eastern Europe."

He'd be scrutinized at home, too, especially by the politically powerful Polish American community, which largely considered Jaruzelski the face of their homeland's oppression, and who openly lobbied against the White House's doing anything to support his continued reign. Jaruzelski had a "terrible image in America," Bush had lamented to Margaret Thatcher before his visit. He also suffered from a "cosmetic matter," Bush mused aloud, wondering if perhaps the general might be persuaded to discard his omnipresent archaic dark glasses for something more fashionable. The dry transcript of their conversation does not reveal if Thatcher paused or sighed. It does record her deadpan reply: "Thatcher agreed, and commented on the medical condition that forced him to wear those glasses."

No matter who led Poland one thing was certain: they would be asking for money. "The necktie of indebtedness" constricted everything Solidarity hoped to achieve, Walesa told American officials, and it was therefore "in the West's own self-interest" to reduce the country's debt payments in order to bolster its democratic transition. He and Jaruzelski each independently pleaded for Bush to announce a significant new aid package during his visit, noting in particular a promise Bush had made during his last visit to Poland, in 1987 while still vice president, that American aid would flow once political change occurred. They needed something like a Marshall Plan, Jaruzelski told Honecker, the massive American aid program Stalin denied his empire in 1947, but the total achieved by Washington's reduction in sanctions during Reagan's last year amounted to less than $20 million, or barely enough to cover three days of interest payments on Poland's foreign debt. "That's nothing," Jaruzelski complained. Such a paltry sum "doesn't deserve comment."

A mere $10 billion should do the trick, Polish analysts calculated, or approximately one fifth the total value of the entire Marshall Plan in constant dollars, the number so often cited in Solidarity and government circles as to become less aspiration than expectation. Jaruzelski formally requested an initial package of $1 billion in food aid, $2 billion from the International Monetary Fund, a broad rescheduling of Polish debt, American export credits, and American support for a World Bank loan in the range of $800 million. These would be good first steps, he argued, without which "there exists a real threat that the economic crisis will deepen further with all its consequences." Polish-born former national security adviser Zbigniew Brzezinski privately issued the same warning, telling Bush that without immediate debt relief, "Poland faces, quite literally, the prospect of a major domestic explosion that could in turn precipitate even a Chinese-type repression."

Bush simply did not have a Marshall Plan to offer. Recession loomed, and in the midst of negotiations with Congress over how to limit Washington's own expanding debt, there was simply nothing remotely close to $10 billion available. "The days were over when the United States could pick up the tab for everything," Scowcroft later noted. "The funds that we actually had available to provide to Poland, Hungary, and to others were pathetically small." Hoping to reward democratization though economic incentives, he lamented that the nation's fiscal woes made it "the worst time in the world to implement the policy we had."

Treasury Secretary Nicholas Brady and his aides thought neither Poland nor Hungary capable of digesting anything like even the reduced levels of aid contemplated by the State Department and the NSC. Each country lacked the basic financial infrastructure to process Western funds effectively, they reasoned, and money sent before market reforms took hold would largely go to waste. "We can't throw good money down a Polish rat-hole," Brady argued, and certainly shouldn't merely forgive Polish debts at the very moment when his department was struggling to enforce rigid accountability for loans throughout Latin America. "Look, in 1976 we gave money to the Poles," he told Bush. "They squandered it." This time Washington should "wait until they have a program. Wait until they can use the money."

"It was a fight that never ended," Scowcroft recalled, pitting the administration's strategists against its economists in a bureaucratic tug-of-war that lasted throughout the spring. "No amount of economic assistance reasonably available would have made a substantial difference to the Polish economy," but that was neither the point nor the true spirit of Bush's 1987 promise. "What the Polish people did need was the sense that the United States was standing beside them."

Purely political arguments carried little weight with Treasury officials. By late spring they were arguing that Bush had never meant to promise aid in exchange for political reforms, only for market reforms. That disgusted the NSC's Robert Blackwill. "Expectations in Warsaw and Budapest were ignored and strategic considerations did not come up," he reported to Scowcroft after yet another unproductive interdepartmental meeting. All they cared for were ledgers and numbers. "Harry Truman, who was in office when disputes in Eastern Europe began the Cold War, would have been appalled."

Let's just ask the president what he meant back in 1987, Gates declared in a huff during another frustrating meeting, walking out the door and down the hall to the Oval Office. He returned with confirmation that Bush had indeed meant political reform, and moreover that he wanted economic aid to flow Poland's way without any further delay. No one else in the room could call on the president without an appointment. "I'll tell you, it really makes an impact on the bureaucracy," Gates later joked of his Machiavellian moment. "That kind of thing gets around."

Even direct presidential decrees can become mired in bureaucratic molasses, however, and debate within his administration over the size of the aid package for Poland continued until mere days before Bush was scheduled to announce it to the Sejm. They eventually agreed to a deal whose highlights included creation of a $100 million "Polish-American Enterprise Fund" designed to aid development of private enterprise, $15 million for environmental programs, and the promise of support for both $325 million in new loans pending before the World Bank and a broader rescheduling of Poland's debt. The Poles sought $10 billion, with $3 billion in immediate aid. The total American commitment announced during Bush's trip amounted to less than $500 million, most in the form of loans. "Emotions run high regarding Poland," Bush explained to Germany's Kohl. They wanted to do whatever was possible to shore up the shaky regime following its uncertain roundtable results, yet it was "important to act carefully and to avoid pouring money down a rat-hole."

"I thought our package was inadequate," Scowcroft concluded of the final deal, endangering American prestige not only behind the Iron Curtain but within the broader Western community as well, where policy makers looked to the United States to take the lead. But it was at least something. The Poles were sure to be disappointed, Bush admitted to Canada's Brian Mulroney, but given the fiscal realities, the Americans could only hope to "show interest and concern without making new financial commitments." No matter what the Poles expected or desired, "this encouragement could not take the form of a new Marshall Plan." White House chief of staff John Sununu caused a minor political stir during their trip when he described their quandary in far less diplomatic terms. The Poles were like kids "in a candy store" when it came to loans, he told reporters. "They don't know which direction to take and don't have the discipline to take the right steps." Pundits and Polish American leaders were outraged at the analogy, forcing the routinely acerbic Sununu to backtrack, though only barely. "I apologize for the metaphor," he said. "But I think the point was correct."

JARUZELSKI WAS WAITING when Bush landed in Warsaw amidst sweltering heat. Nighttime brought little relief. The hotel hosting the American delegation couldn't supply enough power to keep the air conditioning running, and temperatures inside soared. Scowcroft ultimately forced his sealed window open just a crack, sleeping on the floor nearby in the hope of catching a passing breeze. The Secret Service refused Bush the same luxury. "It was a painful reminder of how backward the Polish economy was," Scowcroft recalled, "and how far it had to go."

"The world has watched with wonder" Poland's remarkable changes, Bush told Jaruzelski the next morning when the pair met privately, asking the general to "advise him on ways that the United States could help without interfering" while apologizing anew for his ill-advised comments about Soviet troops. "Gorbachev was quite concerned," Jaruzelksi conceded, but Bush needed to keep in mind that Soviet forces were deployed to defend not only against NATO but also against a graver threat with a far longer history. "Because of the German question, the Poles would feel more comfortable, even in the future, with one or two Soviet divisions in East Germany," he said. The Cold War might indeed be waning, Jaruzelksi said, but Polish pain predated Yalta.

Bush wanted to talk instead about Poland's future, and found his counterpart more personally engaging and more serious about real economic reform than he had expected. The general still hoped for large-scale American aid, yet he also implored Bush to push for tough austerity measures as the price for Western assistance, reasoning that only outside pressure could force Solidarity to swallow the bitter economic medicine required. Having never ruled, it had yet to experience the tyranny of impossible choices, Jaruzelksi observed, or face the consequences of disappointing constituents. Solidarity was "at once a trade union and a political party," he said, and its leaders were unwilling to approve any cutbacks, because "workers favor distribution of profits rather than surplus." Poland needed surpluses if it was ever to repay its foreign debt, he continued, a view Bush shared. Though "aware that some in labor unrealistically expected the West to come up with some $10 billion in assistance," Bush responded, he felt that such largesse "would be wasted without fundamental reform first." Scheduled to see Walesa the next day, Bush promised to speak to Solidarity's leaders about ways they might be more reasonable in their demands, and about "the dangers" of "levying heavy demands on the economy."

Bush's enthusiasm gave Jaruzelski pause. The general had expected to explain himself, not enlist the American in his cause. It would create a world of political trouble for them both if he were perceived to have criticized Solidarity in their private conversation. This would be precisely the sort of appearance of outside interference they needed to avoid, Jaruzelski cautioned, prompting Bush's promise "to discuss reform without representing the issues as Jaruzelksi's views." Poland was in a "delicate period," Bush said, its "tremendous accomplishments" standing in stark contrast to the "tragic" situation in China.

Their conversation quickly turned to Gorbachev, subject of so many conversations in Europe that summer. The Soviet leader hoped to meet Bush soon, Jaruzelski told him, and the world was too much in flux for its superpowers to be deaf to each other. Echoing Thatcher and Kohl, and soon thereafter Mitterrand, Jaruzelski urged Bush to sit down with Gorbachev, the sooner the better.

Bush responded cautiously. He wanted to meet with Gorbachev, but it wasn't quite that simple. Would it be worthwhile for them to meet if they proved unable to forge an arms agreement during their time together? Wouldn't such a disappointment derail whatever positive momentum a summit might produce?

Meet with him anyway, Jaruzelksi advised. Merely talking carried benefits of its own. "It should be remembered that Reykjavik looked like a lost cause initially but turned out well because the two sides' positions drew closer," he reminded Bush, recalling Reagan's famous session with Gorbachev in 1986. That summit had ended in failure when the pair proved unable to trade SDI for real nuclear disarmament. But cooperation ultimately grew from its ashes.

"What was scheduled to be a ten-minute cup of coffee turned into a conversation lasting two hours," Bush subsequently wrote in his memoirs, exaggerating considerably, during which "Jaruzelski opened his heart and asked me what role I thought he should now play." The general was tired and humiliated, and wanted to retire. Bush recalled urging him to reconsider and run for president as planned. To hear Bush tell the tale, theirs was a deeply personal and lengthy session.

The official record of their conversation describes it as having lasted forty-five minutes, longer than the brief chat initially scheduled but hardly the bulk of the morning. The discrepancies do not end there. In the official American transcript of their meeting, prepared from Scowcroft's own notes, the subject of Jaruzelksi's presidential ambitions never comes up. The pair discussed Poland's political and economic straits at some length but at no time did they turn to Jaruzelksi's personal role in resolving his country's political quandary. The only thing Bush offered that approached encouragement can be found in the Polish version of their conversation, which records Bush noting that "the personal position and popularity of General Jaruzelski in the United States has never been as high as it is right now."

The point of a memory is more important than its details, in this case the point that Bush backed Jaruzelski and thus a role for reform-minded communist leaders to democratize at a steady pace. He told him so in ways both explicit and subtle, going out of his way to praise the general at every public event during his trip. His body language expressed the same, as he visibly shared laughs and asides with Jaruzelski whenever cameras might be present, while along with his staff he repeatedly endorsed Poland's steady and consistent path toward democratization. "Mr. Chairman," he said in a formal toast to his host, "today you have the goodwill of an expectant and hopeful world."

It was a different dynamic, and a different atmosphere, when Bush left Warsaw for Gdansk, traveling to Solidarity's home and to Walesa's. Crowds in Warsaw had been positive but not ecstatic, giving Bush a polite reception but far less than the rock star treatment Gorbachev frequently received in the West. "Everyone feels uncertain, what to say, how to act," press secretary Marlin Fitzwater observed. After so much downplaying of expectations over Bush's visit and treading carefully with every word and gesture, that hesitancy "spreads to the population as well."

Gdansk was different. Crowds began to gather well before daybreak for Bush's afternoon speech at the Lenin Shipyard, site of the labor union's birth and some of its most bloody battles for survival. The heat had finally broken, and more than a quarter-million residents turned out to watch Bush stand arm in arm with their union's leader. "Your time has come," Bush told them to enthusiastic applause. "It is Poland's time of destiny, a time when dreams can live again!" Once more the crowd exploded with cheers, though as Bush later noted, they "were very much in the mood and would have cheered anything."

Walesa was clearly impressed, Bush thought. "Oh my God, oh my God," he kept saying (in English), in Bush's recollection, adding that it was the largest crowd he had ever seen. Walesa himself recalled perhaps only forty thousand people in attendance, but nonetheless Gdansk "gave him [Bush] a somewhat different reception" than the capital.

The real meat of their conversations was served up in a far quieter venue, over an intimate lunch for the Bushes and Walesas in the latter's modest home at 54 Polanki Street, an address well known to Poland's police. Smoked eel and salmon, hard-boiled eggs, crayfish, salami, pork chops with prunes, several types of cheese and sausages, and two kinds of soup — a cream of mushroom and a cold borscht — greeted the first couple upon their arrival. These were only the starters. Roast turkey and veal, pork loin with mushrooms, and a loin of beef followed, accompanied by potatoes, vegetables, and morels. Cake with fresh strawberries completed the fare. He'd weigh three hundred pounds if he ate this much every day, Bush exclaimed to reporters afterward. At least he'd passed on the beer, washing it all down with vodka instead.

The conversation quickly turned from Poland's bounty to its needs. "I stressed the urgency of our hope for foreign investment of about . . . $10 billion," Walesa recalled, handing over papers detailing his plan. "President Bush listened attentively and seemed receptive to the idea," Walesa continued, noting the president's appreciation for having concrete proposals to share with his staff. Walesa considered this an "encouraging" reply and the meal a success, writing that Bush was "somewhat surprised by my powers of persuasion."

Bush was too prudent and polite to say otherwise. "I wondered whether he [Walesa] understood all of the details of what he was showing me," he skeptically thought while reviewing the proposals. Walesa's loosely sketched plan had little precedent, offering to sell shares in Poland's farms and factories to outsiders rather than rely on another round of loans and grants. At least Walesa proposed to sell something to Western investors, rather than merely taking out yet another mortgage on the country's future. The overall message was the same, however, no matter its form. Without American help, Walesa warned, Jaruzelski's government would fall, the roundtable process would fail, strikes would resume, and the military would ultimately restore order, either alone or with Soviet backing. If Poland failed, Walesa bluntly predicted, "there would be a second Tiananmen in the middle of Europe."

The session revealed Walesa's sense of the stakes at play but also displayed once more Bush's particular diplomatic style. He neither let the conversation deteriorate into argument nor in any way implicated Poland's leader in the discussion. "I gently asked him about Solidarity's demands," Bush recalled. "He told me that the union was pragmatic and understood that it would not have its way on some tough issues. I was not so certain, although I did not say so." He calmly opined instead that the opposition's plans did not fully comport with the country's dire economic state. "Some of their demands, such as a five-year maternity leave, were so far-reaching that no government could fill them and undertake economic reforms at the same time," Bush remembered explaining. Still he listened more than he spoke, repeatedly praised the Walesas' hospitality, and undoubtedly left the labor leader convinced that he had an attentive partner in the White House. "We parted like two brothers," Walesa wrote. "Though maybe half-brothers is more like it, since he is a good deal taller than me."

"Not bad" was how Scowcroft assessed their performance in Poland. Despite receiving far less American aid than he'd hoped, Jaruzelski was practically "beaming" by the time they left, Baker reported, and within days declared he would, in fact, seek the presidency. Poland needed a leader capable of ensuring stability during such tumultuous times, Jaruzelksi explained when asked what finally forged his decision, citing both his widespread support at home and, in particular, Bush's. "After Bush's visit, during which at every occasion he [Bush] spoke about the contributions of [Jaruzelski]," one Central Committee member recalled, "Wojciech's efforts greatly increased." Of perhaps greatest importance, having seen where Bush's preference lay, Walesa formally endorsed Jaruzelski's candidacy as well, with Solidarity's membership quickly falling in line.

Poland thus solidified its first major step toward democracy on July 19, when Jaruzelksi was elected by the Assembly to serve as the nation's first president of the new democratic era. Solidarity activists armed with binoculars could be seen in the gallery above the Assembly's chambers, noting faces and tabulating votes, just as the American embassy had instructed. He won, with their aid, by a single vote. Walesa was one of the first to send Poland's new president his congratulations, even if the warm wishes carried a barb from the onetime inmate whose ex-jailer's political fortunes now hung on his goodwill. "For your sake, sir, and for Poland's sake, I hope that the next president elected by the legislature owes his job to the voices of all the Poles." It was hardly the ringing endorsement Jaruzelksi wanted, but neither was the vote a clarion call for chaos, and there can be no doubt that American influence played a critical role in the outcome.

The rest of Bush's journey behind the Iron Curtain was a walk in the park by comparison. Crowds cheered ecstatically in Budapest, celebrating their new-found sense of national sovereignty along with the first visit ever by a sitting American president. Even torrential storms could not subdue their enthusi-asm. They listened patiently as a lineup of Hungarian officials plowed through their prepared remarks despite the awful weather conditions. Bush, conversely, made a grand show of tearing up his prepared remarks to the wild cheers of the shivering crowd, offering only a few extemporaneous words in their stead after acknowledging their misery.

He then won hearts in a way words never could. Spotting an elderly woman standing near the podium, shaking from the wet and cold and without an um-brella or jacket, he spontaneously placed his own raincoat around her shoul-ders. No aide told him to. No one pointed her out. It was simply something a Yankee gentleman was supposed to notice and do. The crowd erupted anew at the sight. "It was an incredible scene, one I will never forget," Scowcroft said. "The empathy between him and the crowd was total."

There was only one problem: it wasn't actually Bush's coat. He'd borrowed it from a Secret Service agent, who'd insisted he wear it when the skies showed no sign of clearing up. "When we got back to the car, Barbara scolded me," Bush recalled, good-naturedly ordering him to apologize to its rightful owner with a promise that a replacement would be waiting when they returned to Wash-ington.

Hungary offered few of Poland's political land mines. Its economic reforms far outpaced Warsaw's and Moscow's, and its reform-minded regime had yet to take the critical step of opening itself up to full democratic elections. A greater sense of stability reigned, as did a greater sense of optimism so long as hard-lin-ers from abroad could be kept from intervening, leading his hosts to present Bush with a highly symbolic gift: a piece of barbed wire from the defunct bor-der with Austria. "Let Berlin be next," he spontaneously enthused.

Bush's trip left its mark on his thinking. It was one thing to read policy pa-pers or diplomatic cables discussing a region in flux. It was another entirely to see it with his own eyes, witnessing regime and opposition leaders sitting with each other for the first time in generations, sharing glasses of wine rather than recriminations. "I've been to what, 77, 79 countries, or something, as Vice President," he told reporters during a quiet moment on the flight to Paris, their next destination, "and these talks were more than just diplomatic. I mean, you didn't rely on the printed card, and they didn't. I mean, they spoke right from the heart. They said what they thought; they made clear the difficulties they were facing. And I tried to do the same thing. There was something very special and warm and personal about the meetings in Poland and Hungary."

But isn't it "Gorbachev's revolutionary approach to the East-West relationship that has given these people license to move forward?" a reporter pressed.

"I would think so," Bush abruptly responded. "And let me repeat for the umpteenth time: I want to see it succeed."

"When are you going to tell him that personally?" another asked.

"I don't know."

"This year?" another probed. "Geneva?" another query flew.

There were "no plans for that right now," Bush answered. Winding up his impromptu midair news conference, he turned unusually philosophical, searching yet again for the right words to convey his conviction that democracy's inevitable triumph meant time was on the side of the reformers, and of democracy. They'd win, he said again, so long as they could keep from losing. "This is their business; this is their life, their country," he repeated. "To the degree we can encourage change without intervention in internal affairs — why, I'm all for that. But that is not our role. This is too fundamental: The people's aspiration for liberty and for free choice is too fundamental. And they can look to our system, look to our country, as a beacon for all these good values. But it's not our role to go in and dictate to any of these countries how they're going to run their business." The best support America could provide aspiring democracies was as an example, and as a bulwark of stability.

What the reporters didn't know was that Bush had already reached the most important decision of his young presidency. It was time to meet with Gorbachev, face-to-face and without any formal agenda, just as so many global leaders had advised. He needed to understand what the man truly wanted, Bush told Scowcroft and Baker before Air Force One took them home from Europe. He needed to hear it from Gorbachev's own lips, and as Scowcroft recalled, Bush "put it in that way he has when his mind is made up."

But how should they meet, and where? "I'm thinking in the back of my mind what we should do about meeting with Gorbachev," Bush wrote in his journal. "I want to do it; but I don't want to get it bogged down on arms control, and I don't want the meeting to be judged by arms control. I would like to find some cataclysmic world event that we could [work on] together with the Soviets; or helping in some major international catastrophe; or something that shows cooperation and gives me and Gorbachev a chance to talk quietly, though not raising the expectations that we're going to solve the strategic arms problem in one high-level meeting."

Winging his way back across the Atlantic, he decided he could wait no longer. "I am writing this letter to you on my way back from Europe to the United States," he wrote to Gorbachev.

> Let me get quickly to the point of this letter. I would like very much to sit down soon and talk to you . . . I want to do it without thousands of assistants hovering over our shoulders, without the ever-present briefing papers and certainly without the press yelling at us every 5 minutes about "who's winning," "what agreements have been reached," or "has our meeting succeeded or failed."
>
> Up until now I have felt that a meeting would have to produce major agreements so as not to disappoint the watching world. Now my thinking is changing.
>
> Perhaps it was my visit to Poland and Hungary, or perhaps it is what I heard about your recent visits to France—whatever the cause—I just want to reduce the chances that there could be misunderstandings between us. I want to get our relationship on a more personal basis . . . I would propose no more than a handful of advisers on each side. I would visualize long chats between you and me alone and, also, good discussions with my Secretary of State, NSC Adviser, and Chief of Staff present, along with their counterparts, of course.

Bush suggested perhaps a quick visit to Camp David should Gorbachev decide to address the United Nations again. Or perhaps he and Barbara might open their house in Maine. "Late September can be very nice there," he added. "It would also give me a chance to take you for a ride in my speed boat and maybe catch a fish."

He was open to any and all suggestions. Most of all, he wanted time for candid conversations. "General Brent Scowcroft and Secretary Jim Baker, plus my Chief of Staff John Sununu, are the only ones who know of this letter, which I have personally written. I hope I can demonstrate to you that some things need not 'leak.'"

Finishing his letter to Gorbachev, he immediately began a separate note:

> Dear Chairman Deng, Dear Friend: I use this unique form of conversation because General Scowcroft told me that if I would continue to treat you as a friend, you would welcome that — no matter the outcome of the difficulties now between us.
>
> Of course, I, too, want it to be that way.

It was an emotional letter, firmer than the open tone of the note to Gorbachev:

> Brent Scowcroft told me of your reference to the Chinese proverb: "it is up to the person who tied the knot to untie the knot." Herein lies our major dilemma. You feel we "tied the knot" with our actions . . . We feel that those actions taken against peacefully demonstrating (non-violent) students and the nationwide crackdown against those simply speaking for reform "tied the knot" . . . It is what happened next that we believe "tied the knot" . . .
>
> Please understand that this letter has been personally written, and is coming to you from one who wants to see us go forward together. Please do not be angry with me if I have crossed the invisible threshold lying between constructive suggestion and "internal interference" . . . We can both do more for world peace and for the welfare of our own people if we can get our relationship back on track. I have given you my unsolicited advice. Now I earnestly solicit your advice. If there is to be a period of darkness, so be it; but let us try to light some candles.

What are we to make of these twin letters, one written to someone who was effectively a stranger and the other to an old friend, each calling for an opportunity to quietly and calmly move global relations forward? Their quiet tone is the key. Bush's critics had hammered at him for six months, arguing that he was doing too little to meet the new dynamics of the Gorbachev revolution, and too little to condemn the way revolution had been destroyed in China.

"It is sad and embarrassing that Mr. Bush feels he has to choose his words so cautiously, praise democracy so delicately, discuss the past so gingerly, step so politely around Mr. Gorbachev's dismissal of Mr. Bush's suggestion that Soviet troops be withdrawn from Europe," *New York Times* columnist A. M. Rosenthal opined. "Mr. Bush does not have to worry about 'inciting' Eastern Europeans," he said. "Mr. Gorbachev knows communism already has done that." Bush should do more in response, Rosenthal chided.

Unbeknownst to nearly everyone, he already had.

12

ANOTHER BORDER OPENS

"I CANNOT ESCAPE THE FEELING . . . that we are not taking advantage of the chance which history has given us." Bush's invitation took a week to find its way into Gorbachev's hands. The reply, hand-carried for extra security as in days of old, took a week more. Neither side trusted the usual bureaucratic channels to keep word of their budding conversation quiet. They barely trusted each other. "Let me say right away that I welcome your proposal that you and I should meet as early as this September," Gorbachev finally responded. He appreciated the opportunity "to leave behind the cumbersome retinue of aides and, without haste or protocol ceremony, to talk about everything that we must."

Selecting a site proved nearly as difficult as deciding to meet. Superpower summits typically took months to choreograph, and the frenetic nature of each man's schedule left little space for spontaneity or sequestration. "A secluded place?" Gorbachev's personal translator asked. "There can be no such place when heads of state meet. Few aides? This also turned out to be a very elastic formula."

The conversation begun in July lingered throughout August. September started no more promisingly. "I must now admit that the chances of developing a scenario which would permit a spontaneous get-together appear increasingly remote," Bush confessed to Gorbachev in another private note. Perhaps planned spontaneity was the problem. Gorbachev could simply spend a weekend at Camp David, Bush suggested. "The press is not admitted" to the presidential retreat, he assured Gorbachev, thereby making it impossible for the media to "play their usual games of who won, who lost, and who made the most dramatic proposal. In short, you and I could engage in a serious, substantive, and unstructured dialogue that would permit us better to understand each other and to establish a closer relationship."

That would not do either, Eduard Shevardnadze objected during an Oval Office visit later that month. Gorbachev had made the last two visits to the United States. A third would make him look like a supplicant. It was Bush's turn to travel. En route to a long weekend of private conversations with James Baker in Wyoming, Shevardnadze looked forward to hiking, fishing, and, most important of all, unscripted conversations. Similar luxuries simply did not exist for their masters, however. Perhaps they might find a quiet spot in a third country, symbolically halfway between the two capitals. Alaska? Bush offered. Too close to North Korea for Gorbachev to avoid a politically awkward stopover. Maybe Spain? Shevardnadze countered. This time it was Bush who said no. "There has to be some rationale to keep the fact of a meeting from getting out of hand," he replied, and "there was no reason for me to go there absent a crisis."

Scowcroft feared the worst, as usual. "I was beginning to wonder whether the Soviets really wanted a meeting," he recalled. Perhaps Gorbachev preferred the public relations boost he was receiving from accusing the Americans of lethargy. That was certainly the message emanating from Moscow. "A peculiar lull has set in," Shevardnadze had publicly complained the previous week, and only one nation was to blame: "Because of the restrained, indecisive position of the American administration, both the U.S.A. and the U.S.S.R. have lost a lot." To the world, at least, it looked like Bush was stalling.

Yet only days later Shevardnadze was sitting in the Oval Office and, to Bush's eyes at any rate, appeared to be stonewalling the very meeting Gorbachev supposedly had long desired. If the two sides could not even agree on a suitable venue for quiet sessions, Shevardnadze conceded, "we should [instead] prepare for a real summit." It would be sadly lacking in the personal touch both leaders desired, but "holding no summit at all would be inconsistent with the situation in our relationship."

Bush responded with remarkable candor, and not a small amount of frustration, believing he'd already bent over backwards to accommodate Soviet demands. Perhaps deep in his marrow he also believed a gentleman should not refuse another's hospitality. "Just so you hear it from me, let me go back a bit," he thus began to lecture Shevardnadze. "I have been officially opposed to an informal meeting. But, because of rapid changes around the globe, the problems you face and we face, I have changed my position 180 degrees." Presidential

U-turns were difficult, he explained. The Soviets shouldn't make things more difficult still.

"We heard our allies indicate their desire that I meet with President Gorbachev," Bush told him. "If we did, I would have to explain to the press why I changed." But he was willing to have that uncomfortable conversation with the media because, he told Shevardnadze, "we need to get comfortable with each other. So that we can talk really frankly — pick up the phone, for example. My predecessor did not have a close relationship with President Mitterrand," he continued. "I wanted a better one. He [Mitterrand] can be a difficult man — we all can. I invited him to Maine. It was the best darn thing I ever did. Because of this recent step, our relationship is on a completely different plane. The chemistry is different. It could be the same with us."

The two sides finally settled, weeks later, on a shipboard conference off the island of Malta for their non-summit, timed to follow Gorbachev's late November state visit to Italy. The president's brother William "Bucky" Bush suggested the site, having recently visited for business. It was warm and secluded, he reported. None of the men who negotiated the meeting had ever set foot on the small island nation in the middle of the Mediterranean, though it seemed to offer everything required. As it was neutral territory, neither side would officially host. Scowcroft liked the fact that meetings at sea alleviated the need for cumbersome sessions onshore with local officials, while Bush relished the notion of replicating Franklin Roosevelt's legendary 1941 meeting with Winston Churchill off the coast of Newfoundland, where both arrived aboard armed cruisers to produce an Atlantic Charter that cemented their wartime alliance. Only Baker found the plan disquieting. Roosevelt and Churchill had considered themselves sailors. Bush and Scowcroft boasted the iron stomachs of former pilots. Baker got seasick.

While the White House and Kremlin parried, conditions throughout the Soviet bloc steadily deteriorated. Russian miners remained on strike, threatening the nation's energy supply. Prices soared, followed by hoarding. Moscow residents were told to expect food rationing by winter. Only vodka seemed plentiful, the government's anti-alcoholism program having proved an absolute failure. Outlying regions fared no better. Separatists and nationalists continued their cries for independence, their anti-Russian pleas tinged with the specter of ethnic conflict. The Baltics were "a ticking time bomb," Condoleezza Rice warned inside the White House. "Any sense that the rights of Russians are being trampled can only serve to inflame Russian nationalist sentiment," she told Scowcroft, "giving a natural platform to Gorbachev's conservative opposition" while forcing local leaders to choose between "loyalty to Moscow and effectiveness at home." She thought it easy to see which side was winning, zeroing in on a single telling piece of intelligence: Latvia's independence movement already claimed more members than its Communist Party.

Conditions were no better in Eastern Europe, with Czechoslovakia finally joining the list of Soviet client states beset by demonstrations. Though Czech rallies were smaller than those in Poland or Hungary, individual acts of civil disobedience flourished throughout Prague as summer transitioned to fall. Street performers and political activists openly flouted authorities. Some marched with watermelon helmets; another performed magical tricks with pyrotechnics; others carried blank banners through the town. What did watermelons, fire, or the farce of advocating nothing at all have to do with revolution? The mere fact that authorities had to ask that question, and that others saw their fruitless attempts to demand compliance, was proof enough that something was afoot. A regime willing to jail someone for wearing fruit revealed its underlying absurdity.

Hundreds were jailed nonetheless. Food was plentiful in Czechoslovakia, at least compared to the rest of Eastern Europe, yet disquiet grew. "We see what is happening next door to us," Václav Havel confided to Western reporters, "in Poland and in Hungary. We are frustrated and want change too. People know it is necessary to do something. But they do not know what or how." The regime had little idea either. One day the communists will go too far, Havel predicted. "The police would beat some student protesters and forty thousand people will turn out in Wenceslas Square." All it would take to start the conflagration was a guard pushed too far, a new recruit's unsteady hands on a gun, or a protester, felled by a baton, wrapping his fingers around a rock.

Prague was calmer than Warsaw. "Events in Poland have moved much farther and faster than anyone anticipated," and not for the better, Scowcroft warned Bush by the late summer. Poland's economy cratered by August, forcing yet another political crisis and selection of yet another prime minister, this one drawn from Solidarity's ranks. Having focused for months on financial relief, Polish leaders were soon pleading instead for food. Hungary meanwhile longed for relief from East German refugees, who continued to stream into the country seeking passage west. Back in East Berlin, Erich Honecker's government estimated at least 150,000 citizens remained on extended vacation in the resort area surrounding Lake Balaton, refusing to return home yet uncertain where to go next. Hungarians put the number at 200,000. With available housing long since exhausted, local officials no sooner set up shelters than they found them filled. Autumn rains and winter snow loomed.

Honecker was on vacation as well, taking an extended medical leave to combat exhaustion and the ongoing stomach distress he just couldn't seem to overcome. Some said it was his gallbladder; perhaps his kidneys. The former was removed following his collapse at the Warsaw Pact summit in July, along with a cancerous polyp on his liver. His absence left a vacuum within the government's highest echelons. His longtime aide Egon Krenz rushed home from his own vacation, only to find that Honecker had left the entirely unremarkable Günter Mittag in charge in his absence. Increasingly distrustful even of his disciples, Honecker feared anyone with enough talent or ambition to take his place. "Don't take yourself so seriously," he admonished Krenz before ordering him out of Berlin. "You're not indispensable around here."

Honecker had good reason to be worried. Party elites indeed were conspiring against him, though finding potential co-conspirators meant listening for who sighed perhaps a bit too openly at meetings, or whose vote came a second too slowly. The consequence of guessing wrong was jail, or worse, yet as party powerbroker Gerhard Schürer told Krenz once the two finally broke their silence, "I'm prepared to do this because otherwise the GDR will go kaput." Krenz listened, though he was not yet ready to act. "I can't do it," he said. Honecker was his mentor, his "foster father, his political teacher." Revealingly, however, Krenz did not turn Schürer over to the police for his treasonous talk, advising instead that they merely wait for "a biological solution to this problem." Honecker had cancer. Krenz knew. His doctors knew. The West Germans knew as well, passing the secret along to the Americans. No one risked telling the patient himself. The regime was not just sick but in denial.

And still more people gathered at the border. Some became unstuck with a little help. Several hundred East Germans made a successful mad dash to freedom on August 19, seizing the opportunity provided by a well-advertised "pan-European picnic" arranged on their behalf. Hoping to create a spectacle — and a distraction — organizers invited Austrians and Hungarians to celebrate their growing ties. Posters circulated throughout the refugee encampments highlighting the festival's proximity to the border, providing detailed maps and directions as well as the times when crowds, and the cover of welcome chaos, were expected to be the largest. What happened next was no accident. Hungarian border police scrupulously checked and rechecked the documents of Austrians streaming in, bending their heads intently over their paperwork so as not to notice the rush of East Germans streaming out. Carrying only the barest minimum of luggage, many with children in tow, they walked at first, ever wary of a Stasi trap. Some began to trot as the unguarded gate neared. Soon hundreds were running across in a mad dash. Hungarian state television reported on the festival that night. West German programmers did the same, beaming images throughout the East. "This was the Big Bang, the real test of Moscow's tolerance" for change, Hungarian prime minister Németh later recalled. "Will we get a bang on the door from the Russians? If something went wrong, or if Moscow protested, then we would have learned something. But nothing happened."

Officials in Bonn helped facilitate such flights to freedom but fretted over them as well, fearing a sudden influx of immigrants or, increasingly, the sudden collapse of the East German regime and a complete humanitarian crisis unlike anything seen in Europe since 1945. Kohl's government formally closed its consulate in East Berlin on August 8, when the crush of asylum seekers made everyday operations impossible. One hundred thirty-eight East Germans inside refused to leave, knowing the Stasi waited outside. Hungary's embassy in East Berlin similarly hosted 158 uninvited guests. Fearing the same, American officials reduced public services at their own East Berlin installation, instructing marine guards to bar the doors to potential invaders, though not to eject anyone who managed to make it onto American property. Don't let them in, in other words; but don't kick them out, either. "There is no question of turning anyone over to police and Stasi," the embassy's political counselor made clear, and it would be unseemly for the great champion of human rights to throw would-be refugees to the curb. Into the hands of nineteen- and twenty-year-old marines went the power to execute and interpret these potentially contradictory orders, and thus the power to decide a family's fate in the blink of an eye.

In Warsaw, Budapest, and especially Prague, West Germany's embassies filled with a steady supply of would-be immigrants. The Budapest outpost shuttered its doors in mid-August when the rush of refugees proved too great. Hundreds more steadily arrived in Prague, where West German diplomats struggled to supply basic sanitation for the families encamped on the embassy grounds. No one knew how many more would join them or how long they might stay. Police blocked the streets leading to the compound's gate. Hopeful asylum seekers merely scaled the fences or had passers-by boost them through open windows, crowding ever tighter. Mattresses soon covered the floors. Tents crammed every available square foot of grass. Trash accumulated. Nearly three thousand people shared two outdoor toilets. The place smelled.

The desperation was clear: leaving home and family and friends in order to climb behind walls deep within the capital of a foreign country, with no real plan for escape and no idea how or when you might ever leave, knowing only that police stood mere yards away, capable of sending you to prison, and of further inflicting punishment on family and friends for your temerity. And every day more showed up.

West German newspapers filled with criticism of Kohl's handling of the crisis, alternatively chastising him for doing too much or too little. Even typically supportive editors complained that the Foreign Ministry should have preemptively closed its embassies sooner. The long-standing open invitation to East Germans came under fire. As noted earlier, all GDR residents who arrived on West German soil garnered a lucrative arrival package. They also gained immediate citizenship. Bonn had never formally sanctioned the country's partition. West German officials could therefore simply print at will a passport for any East German resident, providing the immediate freedom to travel, after handing over a packet of deutsche marks as part of the formal welcome.

Many thought the time for such charity had passed. Kohl's government retained hope that the East Germans might still seize the chance to initiate genuine reforms, thereby affording citizens a better life at home rather than reason to flee. "If the East Germans tore down the Wall today," a ranking West German diplomat confided to Western reporters, "we would have to build another one tomorrow." A former deputy mayor of Berlin was blunter, revealing an important truth about postwar German politics: "We may say we want reunification and the Wall to come down . . . but not really." Reunification would be expensive, and would force all of Europe to confront difficult questions better left unexplored: Germany's responsibility for the twentieth century's worst wars and the underlying fear that a renewed German state might somehow lead to more. West German politicians routinely and dogmatically expressed a yearning for a single German state, but talk was cheap. Few considered it possible in their lifetime. "I do not write futuristic novels," Helmut Kohl had responded the previous October when asked about unification. "What you ask now, that is the realm of fantasy." When the wall was high and impermeable, they could say whatever they wanted about reconciling East and West and reuniting the fatherland. "We are not," the editor of the influential daily *Die Zeit* stated at the end of September, "an inch closer to reunification than a year ago, or five or ten years ago."

In early 1989 one half of West Germans polled believed unification so far-fetched an idea that their government should no longer even keep it as a long-term goal. In September only one quarter thought they might see reunification in their lifetime. By late October it was nine out of ten, as East Germany continued to hollow out from the inside. "I speak of German unity before the United Nations year after year," said Foreign Minister Hans-Dietrich Genscher, hoping to calm continent-wide fears, but "embedded in the development of Europe."

One will meet today Germans who boast of the unbridled welcome their country extended in 1989 to their fellow citizens from the East. Don't believe them. One quarter of West Germans polled in September reported fearing increased competition from Easterners for scarce jobs and housing. Anti-for-eigner and anti-immigrant platforms fueled a rise in the extreme right's for-tunes during local and European Parliament elections that year, with the ultra-conservative Republican Party capturing nearly 10 percent of the German vote. "One must reckon with us now," crowed the former Waffen-SS officer who led the party, promising "Germany for the Germans," and not necessarily those whom fate had consigned to the East. Kohl's coalition survived the electoral test, albeit barely, leaving him with the lowest approval ratings of his tenure as he faced what was increasingly shaping up to be his nation's most pressing postwar political crisis.

Desperate for at least the appearance of change, Kohl sacked his party's long-serving general secretary. "Those who vote for the radical right must be taken seriously," he declared, speaking not only to his reassure own elector-ate but to calm the worried glances from those throughout Europe who re-membered all too well the last time German nationalism seemed on the rise. "Britain and Western Europe are not interested in the unification of Germany," Margaret Thatcher told Gorbachev in September. "The words written in the NATO communiqué may sound different, but disregard them! We do not want the unification of Germany." The Soviets most certainly did not disagree. "Ev-eryone says in a single voice," Gorbachev's close adviser Anatoly Chernyaev wrote in his diary on October 9, "nobody needs one Germany."

Kohl's coalition government nonetheless pressed the Hungarians to release even more East Germans. From their perspective these were 200,000 of their own citizens trapped in a foreign country. Yet the matter was also personal, especially for Foreign Minister Genscher, standard-bearer of the liberal Free Democratic Party, whose support gave Kohl's Christian Democratic Union its mandate. The pair formed an uncomfortable partnership. Kohl trusted his chief national security aide Horst Teltschik more, but Genscher held the most important title and post, and the power to make or break the government coalition.

The two men at the top of Germany's ruling coalition had taken very different paths to power. Born into relative comfort in a conservative and predominantly Catholic region, Kohl was too young to have served in the war, save for a few weeks in the chaotic final days of the conflict, when desperate Nazi officials dragooned every able body into the nation's defense. Thanks to "the mercy of a late birth," as he often put it, Kohl instead came of age during his nation's postwar economic miracle, pursuing a doctorate in political science and a business career while rising steadily through the ranks of the Christian Democratic Union. Never considered a brilliant orator or visionary, Kohl instead boasted a common touch and a particular penchant for finding safe passage through political minefields. "Kohl's instincts were almost always for the middle ground on an issue even before others could discern where the middle ground might be," according to one set of American diplomats.

He also instinctively understood the lengths to which he'd need to go to fully accept responsibility for his country's horrific past so it might yet safely find acceptance within the new Europe. Kohl made the first pilgrimage to Israel by a sitting German chancellor, for example, bowing his head in shame at the vast Holocaust memorial Yad Vashem. He shared a symbolic handshake with French president Mitterrand at Verdun, site of more than a million German and French casualties during World War I. Kohl understood the power of an image, allowing his more nationalist side to emerge only behind closed doors. "The future belongs to Germany," he'd once been overheard to say, thinking he was beyond earshot. Kohl was a bear of a man at six foot four and nearly three hundred pounds, fleshy and boisterous, his very mass offering a stark if symbolic comparison with the gaunt and sickly Honecker: the West robust and confident, the East visibly withering away.

Genscher had not been born so lucky. Only a few years older than Kohl, he

had served in the war and in the Hitler Youth. He was from the East. Returning home from his Luftwaffe unit in 1945 after a brief stint in an American
prisoner of war camp, he pursued a law degree in Halle, a historic university
town deep within the Soviet occupation zone. Life under communism chafed,
however, and he ultimately fled west in 1952. To Kohl's left, and also more tone-
deaf to history's echo, by 1989 Genscher had been Germany's foreign minister
for an unprecedented eighteen years, having helped construct his nation's *Ost-
politik* policy favoring reconciliation with the communist East while cautioning against too close an alliance with either superpower. He was, in the words
of one American ambassador, a "slippery man," and a constant reminder for
American officials that whatever frustration they might suffer in dealing with
Kohl's bluster, a far worse option waited in the wings.

Nothing in either man's long political career could have fully prepared them
for the secret request they received in late August: Németh wanted to meet.
With images of bursting East German refugee camps in Hungary on the airwaves nightly, the topic was not hard to deduce. "Rarely had I been so filled
with anticipation before a meeting," Genscher later wrote. Indeed, having suffered his second heart attack only weeks before, he arrived at the meeting directly from his hospital bed, refusing to let a little thing like cardiac arrest keep
him from the potentially climactic moment, or allow Kohl to take sole credit
for whatever happened next. With Gyula Horn, Hungary's foreign minister,
filling out the party, the foursome met in a secluded German castle on August
24, there to decide the fate of millions.

Németh wanted to open the border. To everyone. He refused to hand another
East German back to Honecker's repressive regime, yet he needed to avoid doing anything that might endanger "the success of Gorbachev's policies." West
Germany should thus prepare for a rush of refugees unlike anything seen before. "What do you want" in exchange? Kohl repeatedly asked. "What do you
want as a gesture from me?" Nothing directly, Németh said. It was merely the
right thing to do; the humane thing to do; the Western thing to do.

Nevertheless, Németh continued, Hungary would have needs during its unprecedented economic transition. In times like these, "one has to ask oneself to
whom can we turn for help? To Gorbachev? No. To the COMECON [the Soviet
bloc's Council for Mutual Economic Assistance]? No. One can only turn to the
West." It would be unseemly to trade families directly for financial considerations, he acknowledged. But perhaps Kohl could put in a good word when

the Hungarians turned to the International Monetary Fund and to Western bankers in the weeks and months to come. Of course it would be wrong to open the border for a price, but "I ask you, Herr Bundeskanzler, can we count on your support?"

Kohl understood. By October, Budapest held new lines of credit worth 500 million deutsche marks. "We did not do this for the money," Németh later insisted. "But we really needed this money." Certainly the money helped. For the Hungarians, though, more seemed at stake. "Do you realize," the country's interior minister asked Horn when told of the decision to open the border, "that of the two German states we are choosing the West German one?" No, Horn replied. "We are choosing Europe!"

"Everything that followed was a consequence of this decision," Genscher in time concluded. The Hungarians flew home on August 25. The Kremlin was quietly notified of what to expect. One week later Horn was in East Berlin to deliver the news that the border would open on September 11. "That is treason!" his East German counterpart exploded. "You are leaving the GDR in the lurch and joining the other side." That, of course, was the Hungarians' entire point. "This will have grave consequences for you," the East Germans warned. But their bark had no bite. Calls for another Warsaw Pact summit brought no response. The Poles refused to attend. Neither would the Soviets. Honecker remained bedridden. "Hungary is betraying socialism," the Stasi chief roared to the Politburo. The head of the nation's Communist Party was more resigned — and more troubled. "More and more people are asking," Horst Dohlus told his worried colleagues, "how is socialism going to survive here?"

The gates swung open at midnight. Over eight thousand East Germans streamed into Austria the first day. Within three days the number rose to eighteen thousand. Forty thousand crossed in the first week. Austrian and West German bankers prepared to travel in the opposite direction. Bush, seemingly following up on Kohl's promise as well, announced Hungary's promotion to most favored nation trading status. Asked how many East Germans might eventually choose to emigrate, Shevardnadze shrugged and responded, "One million . . . maybe two million." There were only 16 million people in the entire country. By the end of September everyone knew someone who had gone. Workers would close up shop one day and simply fail to show up the next. One hospital reported losing every one of its forty-two doctors on staff. Bus and train service broke down throughout the country, as managers never knew from day to day how many drivers might show up for their shift. Apartments sat empty. Those who left were typically the youngest and most ambitious. Those who remained were older, more tired. "You're still here?" replaced "Good day" as the common greeting.

Hungary's opening sent reverberations throughout Europe. Delegates from Kohl's Christian Democratic Union cheered when the news broke during their annual meeting, prompting calls for immediate reunification and restoration of Germany's 1937 borders. Standing for reelection as party chair, Kohl did little to temper their enthusiasm, calling instead for "realization of this vision" in which *all* Germans could enjoy "freedom and unity." Surging nationalism alone could not shore up the support of his own party, however. Nearly 20 percent of the delegates voted against him, even though his was the only name on the ballot.

Sensitive to anything that even remotely suggested a rise in German nationalism, the Soviets in particular rebelled at Kohl's remarks. The current German chancellor's sentiments sounded "very similar to statements made by German leaders in the 1930s," Shevardnadze privately told Baker. "It is to be deplored that fifty years after World War II," he subsequently declared, "some politicians have begun to forget its lessons . . . Now that the forces of revanchism are again becoming active and are seeking to revise and destroy postwar realities in Europe, it is our duty to warn those who, willingly or unwillingly, encourage these forces." Faced with the sudden erosion of its union, its empire, and in East Germany the very symbol of victory in World War II that bound Soviet society together, Gorbachev's regime seemed to endorse self-determination and sovereignty for everyone, save for its historic foe.

"Moscow has had a rocky summer," Baker told Bush at nearly the same moment the NSC began debating why Gorbachev was permitting this. Why, they asked, at the moment when his own political future seemed imperiled by problems within his borders, would he permit client states to undermine his position among conservatives even further? "For all the ferment we see in Eastern Europe, I simply do not believe the Soviet Union has yet evolved to the point where it will permit the disintegration of its security buffer which has depended for four decades on Communist Party control," the NSC's Peter Rodman wrote. Scowcroft scribbled bluntly in the margin "I agree" before asking in reply, "So what do we do?"

His staff had no easy answer. "We should restrain the euphoria," Rodman offered. Eastern Europe's liberalization, and its liberation, remained "very much at the mercy . . . of how far the Soviet Union has evolved. I remain concerned about a potential clash of expectations — i.e., East Europeans miscalculating." Rice similarly argued that their best course of action, given their inability to control spiraling events overseas, was to concentrate on their messaging at home. "Too many people on the [Capitol] Hill and in the press expect reform to proceed in a straight line," she lamented. By late September, with Hungary open to the West, and East Germany rapidly emptying, Rodman predicted Gorbachev would be out of office within six months, "and that the Democrats [would] assault George Bush for the historic blunder of not doing enough to save him and all the positive trends he represented." Rodman feared the question of "who lost Russia" might ultimately damn Bush and Republicans for years, as had been the case when Harry Truman's administration took the blame after 1949 for "losing" China. "We need, clearly, to educate people to the reality that the Soviet system is so inherently rotten that no Western help could save him [Gorbachev]."

Senate minority leader Bob Dole, fresh from an extended overseas tour, personally delivered the same message. "The Democrats are already drooling over this 'freedom' issue," he warned Bush. If in six months Gorbachev was out and Eastern Europe's democratization reversed, Republicans would bear the blame on Capitol Hill. Scowcroft believed the same, and was worried, telling Bush, "We also need to stay ahead of pressures in Congress and the public at large for a more generous U.S. response," though there was little they could do to change the public's impression of their inaction until their secret negotiations with Gorbachev on a meeting bore fruit.

Events on the ground in Eastern Europe prevented them from ever keeping pace. Honecker returned from his convalescence in mid-September with a solution to the increasingly worrisome problem of overcrowding at West Germany's embassy in Prague. The refugees could return home without fear of punishment, he promised, and would subsequently be allowed to emigrate within six months. They only needed to fill out the proper forms and go through the proper channels. Fewer than two hundred accepted his offer, while thousands more took their place. "Your solution isn't working," Czechoslovakia's Miloš Jakeš pointedly complained, prompting Honecker to suggest stanching the flow of refugees at its source by sealing East Germany's own borders with the rest of the Warsaw Pact. Those holed up in West German embassies would be afforded the opportunity to immigrate immediately to the West. He had only one caveat: they first had to travel back through East Germany so officials could at least properly document their criminal departures. After all, not only were the traitors turning their back on their homeland by haphazardly fleeing, but also they were screwing up its paperwork.

The move was both boneheaded and tone-deaf. Announcing it at a reception for a delegation from Beijing, in town to formally thank the East Germans for their support in the aftermath of Tiananmen, Honecker appeared unaware of how the image of sealed trains transporting undesirables across the German countryside would appear to the rest of the world. He appeared unaware of any global sentiment at all. While the majority of citizens tuned in each night to see images of their fleeing compatriots on Western TV, the party's official newspaper greeted readers the day after the reception with a seven-column-wide picture of Honecker embracing his Chinese guests. "In the Struggles of Our Time," the headline read, "the GDR and China Stand Side by Side." The message, with its embedded threat, was not subtle.

The first trains rolled west on September 30. Every night for a week they ran, filled with apprehensive yet excited East Germans on their way to a new life. Most refused to speak to the Stasi guards who boarded to check their documents, silently handing their papers over for inspection. Others dropped them to the floor, forcing the regime to bend down at least this once before them. The silence continued until the authorities left, replaced by laughter and cheers that peaked as the West German border neared. Hands and handkerchiefs waved from the windows at wistful-eyed onlookers. Keys soon flew from the windows. Why bother keeping them when the houses and apartments they unlocked would never be seen again? East German ID cards soon followed. Soon a cascade of worthless East German cash blew from open windows into the countryside. There were deutsche marks just on the other side! At times the trains seemed encased in their own cloud of paper. Back in Prague, Jakeš breathed a sigh of relief before pausing to ask, "What are we going to do with all these abandoned cars?"

THE FIRST TRAINS passed without incident, but panic swept East Germany once the border closed for good on October 3. Anyone still dithering over the decision to leave suddenly realized the chance had passed. "I was on the last train from Dresden," a young man told an American reporter. "I knew we had to get out now or we never would. I fear a crackdown is coming." With crowds gathering nightly to march and to plead for reform in Dresden, in Leipzig, and even in East Berlin, he feared "a catastrophe as in China. This regime would fire on its own people." Asylum seekers inundated the American compound in East Berlin, pushing past the marine guards. "All had been trying to reach the West German Embassy in Prague when their trains were turned back," recalled a Foreign Service officer, explaining, "They decided during the return ride to come to us because they feared arrest at home and were convinced the GDR was closing the border in order to crack down. It was now or never, they said."

Dresden's rail station soon became a battleground. At least twenty thousand gathered where trains from Prague connected to westbound tracks. Unable to board the full carriages, eight hundred would-be immigrants climbed aboard an empty train on an adjacent platform, refusing to disembark unless they were also taken west. The police were called out, followed by soldiers. Water cannons were loosed on the crowd. The first malfunctioned, spewing only a trickle. Laughter rippled through the ranks of protesters, who were soon doused when other cannons worked. Rocks flew in response, then bottles. Cars were set ablaze. Hundreds were arrested. More were beaten. One unfortunate rioter fell onto the tracks, where the departing freedom train amputated both his legs. "There was a real danger," Stasi officers reported, "that the entire station area would come under complete occupation."

East Berlin's officials faced down crowds estimated in the tens of thousands, but in Leipzig far larger groups gathered with every successive Monday. First two thousand, then eight thousand by September 15, then nearly fifty thousand marched in protest a week later. By the end of October more than 100,000 had gathered to chant and cheer, warily though steadily, around the city's ring road. Stasi chief Erich Mielke wanted them stopped. So did Honecker and Krenz, the latter just back from a much-publicized visit to Beijing, lest anyone miss the regime's point that Deng's response to crowds offered a model worth emulating. No one would order triggers pulled, at least not yet. Gorbachev would be arriving on October 6, joined by representatives from the entire communist world, gathering to celebrate the East German regime's fortieth anniversary. Huge parades were planned, filled with the loyal youth of the party's elite. Surely they would show the regime's true spirit, the Politburo reasoned.

Then something vital and wholly unexpected occurred, something that fundamentally altered the political upheaval unraveling all of East Germany: the crowds stopped chanting for the freedom to leave and instead began crying out for the freedoms they required in order to remain. "We are the people," they chanted in Berlin, in Leipzig, and around the country, and "we are staying." This was the moment when the regime truly died. Others might abandon the sinking ship, but those who remained were now telling their overlords that the time had come for a real change. For the first time in the GDR's troubled existence, the people were telling the government what to do, not the other way around.

They even appealed directly to the man who had started it all. "Save us, Gorby," torch-bearing paraders shouted when passing the reviewing stand for the official anniversary celebration. "Save us! Help us! Perestroika!" It was a moment of profound embarrassment for Honecker, who had already endured two days of difficult meetings with the Soviet leader. "It looks as if they want you to liberate them again," Polish leader Mieczyslaw Rakowski wryly whispered to Gorbachev, "and these are supposed to be the cream of Party activists."

Gorbachev had not wanted to make the trip, despising East Germany's leadership and believing them too intransigent and dull-minded to understand the necessity for reform. Honecker acted "as if he were in a trance," Gorbachev later admitted. The pair talked, but only past each other. At dinner that night Honecker proposed a toast to the GDR's *fiftieth* birthday and to the hope that they might all gather again in a decade to once more drink to the regime's good health. Gorbachev openly snorted at the prospect, convinced that nothing would change the sclerotic minds of the men in charge of East Germany save their own downfall. He was long past the point of caring for diplomatic niceties when confronted with such ostrich-like behavior. "Life punishes those who delay," he warned his hosts. He left unsure if any in Honecker's circle would take the hint, or had even noticed it. Whatever happened now was out of his hands, Gorbachev reasoned. With more than 400,000 Soviet troops stationed throughout the GDR, he could have forced the issue one way or the other. He instead ordered his commanders to confine their soldiers to their barracks, no matter what. Whether the regime fell or survived, it would have to be on its own. The time of the Brezhnev doctrine was well and truly past.

Not everyone was resigned to the regime's collapse, or proved willing to simply cede power to the crowds. Police beat back protesters in Plauen, Magdeburg, Karl-Marx-Stadt, Potsdam, and Arnstadt. Another crowd gathered in Dresden the night of October 7, this one numbering more than thirty thousand. Once more batons and barricades led to their bloody dispersal. "The situation is now comparable" to the crisis that afflicted Beijing two months earlier, Stasi chief Mielke warned his colleagues, "and must be countered with all means and methods. The Chinese comrades must be lauded," he added, because "they were able to smother the protest before the situation got out of hand." He readied his troops. Officers were given double the usual rounds of live ammunition. Soldiers and police throughout the country were called up for service and placed in a communications vacuum: television sets were unplugged, radios confiscated, calls home forbidden. Party officials distributed photos of burned and maimed policemen throughout the barracks, presumably from the Dresden riot, so everyone understood the stakes. "The motto was, it's us or them," one soldier recalled. Having learned from the Chinese how difficult it was to provoke local police to fire on crowds potentially containing their own friends and relatives, officials called in units from outside Leipzig to confront the massive march scheduled for October 9. At least 100,000 were expected in the streets. "Fear," a Leipzig church administrator recorded in his diary that morning. "What will happen? Chinese solution."

And back in Washington they continued to write memos, to debate, and to wait secretly for the moment when their president might sit down with the man who'd started it all.

13

"IT HAS HAPPENED"

THEY ALWAYS SEEMED to want more. For him to scream louder at democracy's enemies and to praise its proponents more heartily. For him to sound more like Reagan: eloquent, assertive, simple. No one ever had to doubt which side his predecessor stood on. Reagan had told Gorbachev to "tear down this wall" even though everyone knew it was impossible. Now it *was* possible, and Bush seemed muted. "Even as the walls of the modern Jericho come tumbling down," House majority leader Richard Gephardt charged in mid-November, "we have a president who is inadequate to the moment."

Bush was no Reagan, and he was fine with that. The potential bumps on the road to democracy worried him more. He'd seen enough bloodshed during his navy days to last a lifetime, and felt a level of impotence while watching blood spill in China that Reagan never had to endure while president. He knew the price of indiscreet words. "I think there is reason to be cautious," he told reporters. "Substitute the word 'prudent' if you like . . . [W]e're seeing dramatic change, and I want to handle it properly." Properly meant calmly, without surprises, and certainly without provocations. "We are trying to conduct our administration with a certain rhetorical restraint," he assured Yevgeny Primakov, chairman of the Supreme Soviet, on a Washington visit in late October. "We recognize some of the changes you are undertaking are difficult," Bush said. "We do not want to appear to exacerbate your problems."

No amount of explanation could satisfy his critics. "President Carter says you've really been slow on the uptake on the most transforming political event of our time," said veteran reporter Helen Thomas, addressing him at a November 7 news conference. "You have failed to show leadership. You have failed to put the U.S. ahead of the curve on these things that are happening . . . [W]hy don't you have any new ideas?"

"Now Helen," Bush chided, "that is not a kinder and gentler way to phrase your question." He'd done "plenty," he said, citing ongoing aid and trade discussions with Poland and Hungary. But "the fact that some critics are out there equating progress with spending money doesn't bother me in the least." More important, he reminded the press corps, "things were going our way" in Eastern Europe. It was increasingly becoming his stock phrase. We were winning, he continued, so why change what was already working? While "I don't think you can ever say [events have] gone too far" for the region's communist regimes to regain control, "I mean, who predicted with certainty what would happen in Tiananmen Square . . . I think it's gone too far to set back these fledgling — I won't say democracies, but steps towards democracy."

The public had no idea of the lengths to which he'd gone to ensure the continuation of Sino-American relations, and similarly no idea that all the while he'd been criticized for failing to speak to Gorbachev, they'd actually been secretly arranging to meet. Legislators and reporters alike were amazed upon first learning about those conversations more than three months after they'd begun. "Frankly," the Soviet ambassador teased Scowcroft once news of the Malta meeting broke in late October, "we were surprised that you kept it quiet as long as you did." The previous White House had been leakier.

"You say the summit is not meant to bail out Mr. Gorbachev politically," another reporter noted at Bush's Halloween press conference. "How about yourself? You've been criticized by the Democrats as being too timid toward Eastern Europe and toward Gorbachev . . . [D]o you think it will help you?" Reporters repeated the same question in various forms fifty-six times that morning. That number is no exaggeration. "After forty years of calling for free markets and an open society," one proposed, "you have a chance to perhaps cement some of these changes in the Eastern Bloc — in Europe and in the Soviet Union. Do you have some kind of plan or vision for getting that accomplished?"

Transcripts of the press conference do not capture Bush's audible sigh. "We're seeing it move, aren't we?" he quickly retorted, a glint of anger in his eye. "We're seeing dynamic change, and I want to handle it properly." Pundits and legislators could say almost anything without consequence. But he knew "the United States can't wave a wand and say how fast change is going to come to Czechoslovakia or to the GDR." Anger finally burst through when another query began "Your administration has been criticized . . ." The questioner was quickly cut off. "I knew exactly what I wanted to do, and I knew how I wanted to go about doing it," Bush shot back with presidential bluster. "And that's why I didn't need the advice of others in this particular subject matter. I knew how I wanted to do it."

The barbs nonetheless took their toll. "I keep hearing critics saying we're not doing enough on Eastern Europe," he lamented in his diary. "Here the changes are dramatically coming our way, and if any one event — Poland, Hungary, or East Germany — had taken place, people would say this is great. But it's all moving fast — moving our way — and [yet] you've got a bunch of critics jumping around saying we ought to be doing more. What they mean is, double spending. It doesn't matter what, just send money, and I think it's crazy." It had been five months since Tiananmen, which continued to weigh heavily on his mind. "If we mishandle it [Eastern Europe], and get way out looking like [promoting dissent is] an American project," he continued dictating into his tape recorder, "you would invite crackdown, and . . . that could result in bloodshed."

There were also world affairs beyond the Soviet bloc to manage. Eastern Europe was not the only situation he was accused of handling with excessive timidity. Panama took center stage in early October. Relations between the two countries had been steadily declining since the middle of the decade, as Manuel Noriega bullied potential political opponents of his military junta, blasted Washington's management of the strategically critical canal that ran through the isthmus, and bemoaned the presence of nearly twenty thousand U.S. troops. Facing indictment on drug trafficking charges in Washington, he added election fraud to his résumé in May after international observers estimated that votes had run three to one against his regime in the country's general election before the ratio miraculously switched once the ballot boxes fell into government hands.

Vote tampering made the election results null and void, Noriega complained, reveling in the irony. The Organization of American States, the Western Hemisphere's key diplomatic body, authorized negotiators to seek his removal. Yet too many Latin American states had their own memories of what ensued when one of their own got caught in Washington's crosshairs to endorse anything more. Even Noriega's opponents understood. "What will happen if another Latin American country also nullifies its elections?" an anti-Noriega negotiator argued. "Will they want the OAS to come in?" Or, often unstated but always in the back of their minds, the Yanquis to the north?

They had done so more than a hundred times before since 1789, employing sometimes dozens but oftentimes thousands of marines and sailors to enforce Washington's will (and safeguard U.S. businesses) throughout the Western Hemisphere. "Today, the United States is practically sovereign on this continent," an American secretary of state pronounced in 1895, and "its fiat is law." Exceptional cases like Cuba or Iran-contra notwithstanding, the United States retained no less power and influence within its own strategic backyard a century later. Bush knew this history too, and the power at his command, coupling his country's public vote in the OAS with a secret presidential finding authorizing covert action against Noriega's regime. "I intend to start ratcheting up the pressure," he privately told congressional leaders, until Noriega stepped down or someone else stood up to him.

Someone soon did. Or at least tried. In late October, plotters of an impending coup asked U.S. commanders to blockade key roads once they seized power in order to prevent Noriega from calling in reinforcements. The coup worked . . . at first. Major Moisés Giroldi successfully captured Noriega in the early morning of October 3, as planned. "If the major had any brains at all there was no way Noriega would walk out of that room alive," Robert Gates subsequently lamented, but Giroldi could not bring himself to shoot the godfather to his children in cold blood. Noriega had no such qualms. Once freed by loyal forces who arrived at the palace unheeded, he ordered his captor tortured — extensively and slowly — before his inevitable execution. More executions followed in the ensuing weeks as Noriega not only eliminated those he thought involved in the plot but also seized the opportunity to purge his military of any whose loyalty he had reason to question.

Giroldi's forces never stood a chance because United States forces had never left their barracks. None of the requested roadblocks went up to hinder Noriega's troops, leading to the coup's inevitable failure. Critics decried the lost opportunity and derided anew Bush's timidity, though more than anything else his administration's reaction revealed deep flaws in the way it shared critical information during a crisis. Put simply, the system failed. Press secretary Marlin Fitzwater learned of the coup from CNN rather than from military or intelligence channels, for example. The Pentagon's Southern Command, based in

Panama, just a few short miles from Noriega's presidential palace, was equally unable to grasp what was happening. Local commanders doubted Giroldi's odds and democratic credentials, leading Defense Secretary Cheney and newly appointed chairman of the Joint Chiefs Colin Powell to balk at recommending that Bush authorize the requested support. They never told Giroldi, however. "Our reporting was spotty," Cheney later admitted, "and we did not want to fall into some kind of trap Noriega might be setting, trying to get us to take the first step militarily only to find out later there hadn't been a coup." Cheney ultimately withheld from the White House any ensuing information that did not comport with his opinion, leaving Bush's spokesmen to offer the press information contradicted by their colleagues elsewhere in the administration.

The coup lasted less than five hours. The decisive factor, according to Southern Command, "was US inaction." By 2:30 in the afternoon Cheney learned that Noriega was free and the plotters doomed. Illustrative of the entire system's breakdown, however, Bush remained unsure of any of those things hours later. "Here it is six P.M.," he told his diary, "and we don't know where Noriega is; we don't know whether he is wounded or not; and we don't know whether he is dead or alive; but that is the nature of this business."

"Amateur Hour," a *Newsweek* headline blared, while *U.S. News and World Report* called Bush's team "The Gang That Wouldn't Shoot." Senator Jesse Helms labeled them Keystone Kops, questioning how they could have failed to prepare contingency plans for the very coup they'd encouraged. One congressman simply described it as "a resurgence of the wimp factor." Such criticism struck home within Bush's inner circle. The usually mild-mannered Scowcroft nearly came to blows with Senator David Boren, chairman of the Select Committee on Intelligence, as the two argued just before jointly appearing on ABC's Sunday morning news panel. "I don't think it's right," Boren said once cameras were rolling, for our "great country to go out and encourage people to take action, to put their lives on the line, to imply by encouraging them to take action that we're going to help, and [then] not be there." There were echoes of Hungary in 1956 in the statement, or more ominously perhaps of Eastern Europe in their own day.

It was "a very candid discussion," Boren's spokesman said of their off-air tiff, though the senator "accept[ed] General Scowcroft's apology" for the confrontation. Boren wasn't done, however, telling reporters in Chicago later that week that the White House had "blood on its hands, for the lack of courage to reach out to the officers who were doing what they thought America wanted."

"I want your ass back here!" Bush screamed through the telephone at Boren after learning of his remarks, organizing military transport when commercial flights proved too slow. The senator found an angry president the next morning in the Oval Office, surrounded by much of his national security team. Powell looked distinctly "uncomfortable," he recalled, clearly fearing his tenure as Joint Chiefs chairman would prove the shortest on record. Cheney squirmed in his chair as well. Boren sensed he was interrupting an already tense conversation. How could he say they had blood on their hands? Bush demanded.

He changed his tone once Boren detailed what he had learned from reviewing American cable traffic, including the key fact that it would have been easy for American forces to aid the plotters had they only received permission to do so. After a long pause, Bush apologized. He'd been wrong to call Boren onto the carpet, he acknowledged, escorting him to the door. Glancing back at the room full of worried faces, he added as the senator departed, "These gentlemen and I have something to discuss."

"We did not act very decisively," Scowcroft later admitted. Refusing Giroldi's plea for assistance was not the real problem. What signed the plotters' death warrants was Washington's failure to let them know they were on their own. "We were sort of Keystone Kops," he added, and "it was probably my fault." Bush ordered his deputy to review anew their crisis management plans and, more important, ordered the Pentagon to begin planning its own full-scale military operation against Noriega. "All of us," declared James Baker, "vowed never to let another such opportunity pass."

IT WAS ONE THING to debate intervening in Latin America. There was precedent for that. A prospective American intervention in Eastern Europe was quite another matter, though real violence seemed imminent. Weekly protest marches in Leipzig grew steadily in size and intensity. Something had to give. "We should expect there will be further riots," Honecker announced on October 8. "Hostile actions should be prevented offensively," he ordered, instructing Krenz to take command of the situation. Troops readied their weapons, as an estimated 100,000 protesters prepared to make their way methodically around the city's ring road. Hospitals were notified to be ready for casualties. Parents were instructed to retrieve their children from day care and school no later than 3 PM. Extra jail cells were prepared. The interior minister authorized police to employ "all measures" to halt the crowd. Everyone understood what he meant.

Krenz refused to go to Leipzig as ordered, though less out of compassion than to save his own skin. He knew Honecker's cadre wanted to fire on the marchers and feared he was being set up to take the blame. He thus called the Soviet ambassador for advice and for cover. "I understood why he [Krenz] called me," Vyacheslav Kochemassov later remembered. "The main blow was being aimed at him," and he wanted the Kremlin either to halt Honecker on its own or at least to understand that he, Krenz, was not to blame for what was about to occur. Gorbachev neither condemned nor endorsed the impending violence, however, choosing only to reissue his standing order that Soviet soldiers should not get involved. The policy of noninterference was meaningful only if consistently applied, he'd explained to the Politburo earlier that week. "We must be realists. They, like us, have to hold on." Told that Tiananmen had produced more than three thousand deaths, Gorbachev finally let his exhaustion show through. "Three thousand . . . So what?"

Dusk found Leipzig filled with tension. As had now become routine, crowds began gathering in the city's central churches in the early afternoon. "People, No Senseless Violence," a large banner hanging from Nikolai Church's front door implored. "Leave the Stones on the Ground." Prayer services began at 5 PM. An hour later the march began. Police readied their barricades as the protesters neared. "Comrades, today the class war begins," officers told one battalion. "It's them or us. If truncheons prove inadequate, arms are to be used." The police had tear gas, machine guns, dogs, and water cannons at their disposal. "Fight them with no compromises," the interior minister ordered.

Police commanders frantically phoned Berlin for final instructions as the mass of protesters neared the first barricades. No one answered. They tried the Interior Ministry, to no avail. At the Politburo the phone just rang and rang. They finally reached Krenz, who told Leipzig's police chief that he was "unable to confirm" what to do next but would call back immediately with news. Immediately meant more than forty-five minutes later, or long enough to ensure that whatever happened would occur without his input or intervention. Still the crowd surged closer. "We are the people!" they chanted in unison. "Join our ranks!" Above the din, voices could be heard shouting, "Not another China!" It was, in the words of one witness, "a river of people to which you could see no end, a river that nothing more could stop."

Except perhaps a hail of gunfire. "We had already received the order to start running in the direction of the demonstrators, and we had gotten to about thirty meters in front of them," a twenty-five-year-old police officer told interviewers a week later. "There could not have been more adrenaline," but "then all of a sudden: Company halt! Turn around!"

It was a near thing. At the critical moment, local officers came to the same realization as had Krenz: if no one was willing to answer the phone or take responsibility in Berlin, they would be the scapegoats for whatever came next. By refusing their call, the spineless bureaucrats back in the Politburo could claim that their euphemistic command to "halt the demonstrations" had been wildly misinterpreted. Realizing his own neck was on the line, the local police commander ordered a general retreat. "Now they don't need to call back anymore," he told his deputy with a sigh.

It is not hard to imagine all that might have occurred had someone pulled a trigger that night, or if the order to stand down had arrived only a few moments too late. Dozens might have fallen in the first volley. More would have followed. The troops carried extra ammunition, clubs, and water cannons. Rocks would surely have been thrown in retaliation. The marchers also carried torches. In Berlin and Dresden, protesters had already set cars afire. There were children in the crowd. It might easily have been Tiananmen Square in the heart of East Germany. And if Soviet commanders defied orders and intervened, whether on behalf of the crowd or the regime? The ramifications of that sequence of events are nearly too awful to consider. The Pentagon had multiple plans in place for how to deal with a mass uprising in East Germany. None ended well.

All of which raises the question: Why not? Why did violence occur in Beijing but not in East Germany, where the same basic conditions existed? In both cases police had already clashed with protesters. Both involved citizens questioning their government's legitimacy and regimes calculating that inaction meant their demise. In both cases the order to fire was given, and the GDR had never lacked for soldiers willing to fire on their fellow citizens in the past. The lifeless bodies of those shot trying to scale the Berlin Wall were typically returned to their families still covered in blood, and with a bill for the government's expenses, including for the bullet

The answer resides in three places. The first was East Germany's deeply ingrained socialist culture, wherein individual initiative proved dangerous. Trained over decades to avoid independent thought, the men in the field and in East Berlin were unwilling to take responsibility for bloodshed. The second reason dovetails with the first: the men in charge were different. Deng Xiaoping's cadre had no qualms about deploying force, believing that history justified their decision, and that history would vindicate them. They took responsibility, individually and collectively. Honecker's crew, conversely, buckled. For all their bluster, they proved as hollow as their regime, unwilling to take the steps necessary to ensure its survival. History can be made by evil men in power. Or saved by their qualms.

Luck mattered as well. East Germany was lucky that night: lucky that no soldier fired, that no protesters lost their composure, that no one slipped and no one broke. The night, the entire month, indeed the entire fall contained all the combustible ingredients of an explosion of epic proportions. If violence had arisen in Leipzig, it most likely would have spread — to Berlin, to Dresden, perhaps then to the border, to Prague, Warsaw, the Baltics, maybe all the way to Moscow. No one can say with precision, but given the circumstances, with crowds and police pitted against each other and the regime's fate at stake, and with ethnic and nationalistic tensions forever bubbling beneath the surface, we should not expect that the same experiment would again produce the same peaceful outcome, nor that the crowd in Leipzig would again be permitted to complete its trek around the city center. The world was lucky.

Honecker lasted barely another week in office. "We can't beat up hundreds of thousands of people," Stasi chief Erich Mielke ruefully told the Politburo, speaking practically rather than from any sudden change of heart, before the motion was put to the group that it was time for their chairman to retire. No one spoke on his behalf. A loyal communist to the end, Honecker raised his own hand in support of the group's decision before silently walking out of the room. Quietly entering his office for the last time, he first called the Soviet ambassador with the news: "It has been decided to relieve me of my duties." His next call was to his wife. "Well," he said, sighing in exhaustion, "it has happened."

KRENZ TOOK HIS PLACE, finally winning the post he'd long coveted, though it hardly came with the public's affection. "What is the difference between Erich Honecker and Egon Krenz?" The joke making the rounds in East Germany by

October's end boded poorly for its new leader. "Krenz has a gallbladder." Having spent a lifetime backing his mentor, indeed having been the menacing face of the regime's threat to follow China's example, Krenz found there was nothing he could do to win the crowd's favor. He tried, immediately announcing the reopening of the GDR's borders with its socialist allies, a full review of all travel policies, and his hope for an "earnest political dialogue" with protesters. The crowds only grew in response. No longer demanding merely the freedom to travel or speak, by the third week of October marchers openly clamored for full democratic elections. Three hundred thousand gathered in Leipzig for the weekly march. Berlin would soon see half a million rally in the city's center. Hoping Gorbachev's endorsement (and a new round of aid) might improve his image, Krenz flew to Moscow on November 1. "After all, the GDR is in a sense the child of the Soviet Union," he implored the Soviet leader, "and one must acknowledge paternity of one's children."

Gorbachev could not be moved. He'd warned them to act before history decided matters. Now perhaps it was too late. Krenz should try being more open with his people, he suggested, endorsing anything short of either the GDR's dissolution or the even more unthinkable option of full-scale absorption by the West. "You must know," Gorbachev told the East German, "no serious political thinker, not Thatcher, Mitterrand, Andreotti, or Jaruzelski, not even the Americans, though their position has recently exhibited some nuances — are looking forward to German unification." Krenz must therefore do everything possible to maintain his regime's integrity, even if it meant loosening the GDR's borders. He must keep his state alive. "In today's situation," Gorbachev advised him, unification "would probably be explosive."

Krenz thus left Moscow with little more than Soviet suggestions, though Gorbachev took time to call his own military commanders throughout the GDR to stress once more that they should not intervene. The fact that he felt compelled to reissue those orders is telling. "This is an absolute priority," he said. "I don't want anything to start there by accident involving our soldiers."

Bush's advisers continued to fear the consequences of Soviet involvement more than the GDR's outright collapse. "Nothing save the U.S.-Soviet strategic relationship is more central to our national security" than what happens in the very heart of Europe, Scowcroft's staff at the NSC reminded him. "Our overriding objective should be to prevent a Soviet military intervention" to quell unrest, "which could and probably would reverse the positive course of East-West relations" and "raise the risk of direct U.S.-Soviet military confrontation."

They developed a series of escalating responses attuned to the level of disorder and danger should the East German regime continue to fall, noting ominously that the mere presence of hundreds of thousands of Soviet troops meant that any political crisis could swiftly escalate beyond control. A general Soviet mobilization against civilian unrest in East Germany was "among the World War III scenarios for which U.S. and NATO planners have been preparing for decades," they ominously noted. Once military planners looked to their operations manuals for instructions, the NSC warned, and "as pressures grow to alert the entire range of U.S. and Allied forces, including strategic nuclear forces, the political margin for maneuver will begin to close rapidly."

Wars had started over less. Having long aspired to defeat communism in Europe, and in Germany especially, American policy makers realized that their very dreams contained the seeds of nightmares. Renewal of Soviet-American tensions seemed dangerously possible if things turned out poorly in the weeks to come. Nuclear war was not out of the question. None of these things was what the NSC expected to happen. There were, however, more realistic contingencies the White House prepared to confront, including the increasingly real possibility that the GDR's collapse would lead in time, perhaps in a remarkably short amount of time, to German reunification. "We should be poised to exert such leverage as we have — including direct discussion of Germany's future . . . before the opportunity for a political settlement is lost," the NSC concluded. Of course, knowing they should do something before it was too late and knowing what to do were two entirely separate things. "While no one wants to confront the German question just yet — before democratic change is secure in Poland and Hungary, before Czechoslovakia has followed suit, and before a wholesale drawdown of Soviet forces forward-deployed," the NSC offered, "we have little capacity to influence the pace of change in the GDR."

Another president might have given fiery speeches. Or put troops on alert. Or provided assistance or even arms to opposition groups. Another president might have seen political points to be gained by siding openly with pro-democracy protesters, especially if he was already under fire for saying and doing too little on their behalf. "We are getting criticism in the Congress from liberal Democrats that we ought to do more to foster change," Bush lamented to Kohl as October neared its end. Another president might have listened, believing the lineup of expatriates who promised, if given arms and aid, to liberate their homelands behind the Iron Curtain. Another president might have picked up the phone to put Gorbachev or other bloc leaders on notice, perhaps threatening retaliation if violence erupted, thus drawing a line from which there could be no retreat, while providing hard-liners with confirmation that the West in fact sought their demise.

Bush did none of these things. Then again, doing nothing when there was no clear choice suited his general approach, though his problems were unlike those that continued to beset Krenz, whose efforts to shore up his regime constantly backfired. "Every idea is needed, and no one is excluded from an exchange of ideas," he proclaimed on state television. He would create a constitutional court, offer alternatives to compulsory military service, and reduce party privileges. Education and economic reforms would follow. The Politburo would even accept term limits, all with an eye toward making a "thorough change," and in desperate hope of appeasing the protesters.

No one believed him. Long terrified into submission, voices of dissent sprang to life once cracks appeared in the government's armor. Hand-lettered signs flourished throughout East Berlin as November commenced which only a month earlier would have landed their authors in jail. "We Demand Free Elections" read one propped against the entrance to parliament. "I do not have the impression that we have unfree elections," the tone-deaf Krenz retorted. Outside the Council of State another demanded "Pluralism Instead of Party Monarchy." Krenz muttered some words about sharing power in response. There was little he could say when faced with the most ominous sign of all: "We Are the People — and There Are MILLIONS of Us!"

Hard-liners continued to demand a crackdown, raising the prospect of outright civil war. Thousands of loyalists paraded past his office bearing their own signs. "We are the party!" they chanted. Krenz "has a two-front fight," the American embassy's chief political officer warned, between reformers and conservatives. West Germany's Helmut Kohl perceived three dominant groups fighting for power within the East German Politburo. There were the "cement heads," unable to change, whose time had passed. A second group, led by Krenz and rapidly fragmenting, hoped to chart a new path even if they lacked any real plan for doing so. Members of a third faction within the communist regime, this one steadily growing, were focused by early November on preparing for whatever might come next, be it a life on the run, perhaps in hiding, maybe in exile, or trapped behind the very iron bars they'd once used to jail an entire country.

Less than three weeks into his reign, Krenz was largely reduced to stalling for time. "We decided something in the morning," he wrote in his memoir, "and had to change it by evening." They tried openness, broadcasting videos of protest marches in Berlin in the hope that their new transparency might calm emotions. The crowds only grew larger, sometimes topping a million participants, according to American and Soviet observers. Perhaps new travel regulations might prove enough? The desire to journey west had started it all, he reasoned; giving loyal citizens what he presumed they truly wanted — the chance to see the other side and then come home — might bring calm back to the streets. An estimated 23,000 East Germans fled through Czechoslovakia the first weekend of November alone. His staff debated the duration of exit visas and the amount of currency East Germans might take with them while his citizens simply left.

Some lines he would never cross. "I can see no circumstances under which the wall will be removed," Krenz insisted. "It is a bulwark against Western sub-version." The other great bulwark of East German society — the Stasi — mean-while quietly began executing plans for its own end. Mielke ordered the sys-tematic burning of their most sensitive files lest they fall into the wrong hands. There was simply "not enough time!" Krenz repeatedly lamented, first to his journal and then out loud, and "too much pressure!"

Günter Schabowski found a room full of mentally and physically exhausted bureaucrats when he entered the Politburo's meeting room on the afternoon of November 9. They'd been arguing for weeks, to little effect. The party's chief spokesman, Schabowski had provided daily news conferences for the growing gaggle of Western reporters in East Berlin in hope of further demonstrating the regime's new transparency. Most were dreadfully dull. With no real media experience and little sense of drama, he'd mostly read official pronouncements and the names of those appointed to midlevel party positions, answering a few perfunctory questions before retreating to the safety of his office. "I can speak German and I can read a text out loud without mistakes," he later said, explain-ing his qualifications. He was the government's mouthpiece, not its brain.

"Take this," Krenz said, thrusting a piece of paper into his hands. "It will do us a power of good." It was a copy of the new travel regulations, their release embargoed until the weekend. Citizens with passports would thereafter be allowed to apply through their local authorities for a thirty-day exit visa. Those without passports could apply for one. Legitimate requests would be approved, provided everyone followed the rules. Schabowski thought little of the news. Tucking the paper into his briefcase, he did not even pause to read it while en route to his 6 PM press conference.

He proceeded to bore the room nearly to tears. The Associated Press's correspondent actually fell asleep. NBC's Tom Brokaw began to question his decision to rush to East Berlin in search of a story, dreading the interview his producers had lined up with Schabowski for later that evening. This man seemed incapable of making good television. Nearly an hour into the proceedings, an Italian journalist asked if there was anything new to report on the government's long-promised travel reforms. Remembering Krenz's memo, and forgetting its embargo, Schabowski began riffling through his papers, rambling as he looked for the answer. "We know about this tendency in the population, about this need of the population, to travel or to leave . . . and . . . uh . . . we have the intent . . . to implement a complex renewal of society . . . uh . . . in that way to achieve, through many of these elements . . . uh . . . that people not see themselves obliged to master their personal problems in this way . . . Anyway," he droned on, "today, as far as I know . . . a decision has been made . . . Private travel to foreign countries can be applied for without presentation of existing visa requirements, or proving the need to travel or familial relationships. The travel authorizations will be issued within a short period of time . . . the responsible departments for passports and registration control for the People's Police district authorities . . . are instructed to issue visas for permanent exit without delay."

Proud of his answer, Schabowski peered over his glasses to ask if anyone had any further questions. The room had perked up. When does this new policy go into effect? one asked. "As far as I am aware," he said, riffling once more through his papers, "immediately. Without delay."

Bedlam ensued. Questions rained down from all directions. Reporters sprinted to find telephone booths. Only the first in the queue would secure an outside line. Brokaw's crew had a hefty mobile phone in their car. It rarely worked. His producer sent a staffer racing to tell the news. They had a planned satellite broadcast window in six hours; if what they had heard was true, New York would have to do the impossible and get them on the air sooner. Each ensuing question landed on Schabowski like a body blow. "Is that also valid for

West Berlin?" a reporter asked. Shuffling his papers yet again, hoping for the answer to suddenly appear, Schabowski weakly offered, "Yes, Berlin." A dozen more questions flew, a British reporter's voice rising above the rest: "What will happen with the Berlin Wall now?"

Realizing he was generating far more excitement than he'd intended, Schabowski looked for an escape. "It has been drawn to my attention that it is 7:00 p.m.," he announced. "This is the last question, yes, please understand!" His final answer, however, did little to clarify matters. "Surely the debate about these questions, um, will be positively influenced if the FDR [West Germany] and NATO also agree to and implement disarmament measures in a similar manner to that of the GDR and other socialist countries. Thank you very much."

What had he done? The BBC carried the news a few moments later. West German television led with the same bulletin moments after that. "The GDR is opening its borders," the nation's most respected news anchor declared. "The gates in the Berlin Wall stand open." The Soviet news agency broadcast the same without comment. "This is a historic night," Brokaw soon told American audiences. "The East German government has just declared that East German citizens will be able to cross the wall . . . without restrictions." This surely was not what Schabowski had intended, nor even what the regulations had pre-scribed. But the news traveled faster than the truth. All anyone with a television set or radio could know for certain was that something had changed, that it had something to do with travel, but with what documents precisely, and how to proceed?

Puzzled crowds began to gather at Berlin's main crossing points. Nearly two hundred massed within the hour at Checkpoint Charlie, long the site of East-West drama and intrigue, only to find scoffing border guards. They'd heard nothing from their superiors about any new travel regulations. No one at headquarters or at the other crossing posts seemed able to tell them any-thing, though reinforcements were dispatched just in case. Across the fence American soldiers watched silently as the crowd grew larger and louder. Two generations of GIs had watched with similar passivity as East German authori-ties tyrannized and even shot their citizens. What might they do now if things turned ugly right before their eyes? Their commander had his orders barring intervention. He also had dozens of heavily armed troops, a fully loaded tank, and a conscience. There had not been a significant American military presence in Beijing back in July. Berlin was different. "Open the gate; open the gate!" the crowd chanted. Hadn't the guards heard the news? A flashbulb popped, leading one of the East German policemen to recoil, reflexively reaching for his rifle.

The scene repeated itself to the north at the Bornholmer Strasse border crossing. Within ten minutes of the news conference's end, twenty would-be crossers showed up, demanding to enter West Berlin. "Nerves are stretched taut," the wife of an American diplomat breathlessly reported. Coincidentally making her way through the checkpoint at the same time, she saw crowds of young men she presumed were drunk for all the noise they were making, until she realized she couldn't spot a single bottle. They chanted and waved their arms at the guards, who in recent weeks had become increasingly used to such requests from those derisively dubbed "wild pigs." Such miscreants were typically shunted away brusquely yet without incident. More than the usual number of wild pigs began to materialize, however. By 7:30 there were more than a hundred in the crowd, with more arriving each minute. The streets were filled with cars, the glare of their headlights making it nearly impossible for the complex's commanders to effectively discern how many people they were facing. An hour later there appeared to be a thousand, each demanding to cross, each saying the same thing: They had heard the news. Why hadn't the guards?

Harald Jäger had seen the news but didn't know what to make of it. His troops badly outnumbered and increasingly frightened, the lieutenant colonel in charge of the Bornholmer station did what any East German officer would have done in that situation: he called for orders. He'd been in the border guards for twenty-five years. His father had served before him. His spangled uniform made it plain that this highly decorated officer was a man of the regime. More than halfway through a twenty-four-hour shift, he'd watched Schabowski's news conference from a break room as it unfolded. "Bullshit!" he spat at the screen. He was typically the model of calm and control, but the explosion was longer in coming than his colleagues realized. Unwell for weeks, he was scheduled to receive test results from his doctors the next morning. He should expect the worst, they'd cautioned. It looked like cancer. (It wasn't. Jäger's health scare proved merely a timely false alarm.)

Jäger was unprepared not only for what he was witnessing outside his command post but also for what he was hearing from his superiors. No one would confirm anything. Stasi Operation Command reported no change in policy.

Everything looked as it should from their desks. Thirty calls later, still with no useful reply, he finally snapped when asked if he was "capable of assessing the situation" or was "simply a coward."

At this moment, as historian Mary Sarotte aptly put it, "a man who had not disobeyed an order in nearly three decades had, with that insult, been pushed too far." Fearing the edgy crowd would soon lose its composure, perhaps Jäger realized he would eventually have to answer to God for whatever he decided in the next few minutes. Given what he expected to hear the next morning, that conversation might be coming sooner than he'd hoped. "All I was thinking about now was how to avoid bloodshed. If a panic started, people would have been crushed. We had pistols. I had given instructions not to use them, but what if one of the men had lost his nerve? Even a shot in the air . . . I cannot imagine what reaction that would have provoked."

He gave the order to open up. The first border-crossers stepped tentatively across, their documents marked for later punishment by the nervous guards. Soon this proved too slow, and the crush of people too great, as dozens streamed past, then hundreds. By 10 PM the gates were fully opened at Bornholmer, at Checkpoint Charlie, and across the city. Once one opened, the rest followed suit. "Trips have to be applied for!" East German newscasters impotently implored. It was too late, their words overwhelmed by images on television screens of East and West Berliners embracing, then drinking, and then finally dancing atop the despised wall.

The import of the moment simply cannot be overstated. Europe's entire postwar structure, enmeshed in and codified by years of protocol and promises, came crashing down in large part because of a spokesman's mistake. More than once global attention had focused on the lines cutting through and across Berlin. World War II effectively ended there. World War III had nearly begun, more than once, over the city's fate. Whole armies squared off on either side. Built by the Soviets in 1961 and continuously reinforced by East German governments ever since, the Berlin Wall was solid, built of concrete, steel, and barbed wire. It was defended by guards, guns, dogs, and above all else fear. Honecker vowed it would last for a hundred years.

And then it was gone, or at least traversed and trampled, with ordinary citizens chipping away at this scar across Germany's heart with shovels, axes, hammers, and even their bare hands. Strangers embraced. Cameras flashed. Champagne flowed. It was, an NBC correspondent declared, "the largest block party in history." Every human emotion was present. For some it was a rebirth. A young East German chemistry student named Angela Merkel found the experience so enthralling she decided on a career in politics. For others it was a wake. "Why have I been standing here for the last twenty years?" one of Jäger's fellow officers asked over and over again as jubilant crowds rushed by.

No one had had any inkling that the Berlin Wall would fall on November 9, 1989. The Soviet ambassador to East Germany had gone to bed. Once roused, he hesitated before waking anyone in the Kremlin, fearing inexperienced staffers on duty at that hour might overreact and order "measures that we would all bitterly regret later." There was still "a real danger of bloodshed," the KGB's senior officer on his staff warned. "Even a single shot on this night," the deputy ambassador cautioned, would cause "a worldwide catastrophe" in which the involvement of Soviet troops could not be ruled out. Surely, he reasoned, not everyone in Krenz's government had endorsed this radical step. In fact, no one had.

Kohl was in Warsaw, finally undertaking a long-planned reconciliation trip fifty years after the Nazi invasion that formally started World War II. His absence was proof enough that no one, least of all in his government, expected the events of that night. He found Polish leaders consumed by the fear that history would soon repeat itself. The GDR's transformation was happening too fast, Lech Walesa told him earlier in the day, well before the drama in Berlin began to unfold. Its disintegration would lead inevitably to unification, and with a single Germany inevitably came "the widespread fear of German aggression, [and] German tanks." The Poles had hoped East Germany might prove the fifth or sixth Soviet satellite to embrace *perestroika,* after enough time had passed that their own reforms could not easily be reversed, not the third so soon after Poland and Hungary, and before even the Soviet Union fully transitioned to democracy. Europe worked best when Germany was surrounded by powerful states, and there had not yet been time for Germany's neighbors to build the solid foundations they would need if they were to feel comfortable with unrestrained German power. "You know, the Wall will come down soon," Walesa ruefully predicted. "I don't know when, but I really think very soon, maybe weeks." Unification would naturally follow.

Kohl laughed in response. Even on a trip timed to a painful anniversary, he grew weary of Poles forever bringing up the past. "You're young and don't understand some things," he said, reminding his hosts how little they liked being lectured to by Germans. "There are long historical processes going on and this will take many years." All this talk in Warsaw of Germany's past and future took place before the news from Berlin reached Kohl's staff. "I'm at the wrong party," Kohl immediately told his hosts when he learned what was happening to their west. Leaving Genscher to make his apologies, he beat a hasty retreat to the airport in search of a flight to Berlin.

It was nearing dinnertime in London. Margaret Thatcher's office issued an immediate statement of praise for the GDR's lifting of travel restrictions, though inside 10 Downing Street the mood proved less joyous. While others fixated on the image of Germans dancing on the wall, another image stuck in her mind, this one from Bonn. "The Prime Minister was horrified by the sight of the Bundestag [the West German parliament] rising to sing *Deutschland über Alles* when the news of the developments on the Berlin Wall came in," one of her intimate aides reported. The nationalist song favored by the Nazis had officially received less jingoistic lyrics in 1952. The tune remained unchanged, however, and for Thatcher at least, so did its foreboding meaning. "I think it is a great day for freedom," she said, stone-faced, when finally emerging to meet with reporters. Asked if German unity was something she could live with in her lifetime, she replied only, "I think you are going much too fast."

A similar scene played out among the French. François Mitterrand publicly praised the news out of Berlin—"These are happy events," he told reporters—but his mood turned to one of "grave concern" in private. "Gorbachev will never agree to go further," he feared, "or else he will be replaced by a hawk. These people [the demonstrators in Berlin] play with world war without understanding what they are doing!"

It was lunchtime in Washington, where James Baker was hosting Filipino president Corazon Aquino at the State Department. An aide quietly handed him a note: "Mr. Secretary: The East German government has just announced that it is fully opening its borders to the West. The implication from the announcement is full freedom of travel via current East German/West German links between borders. We are asking EUR to give you an analysis." Baker didn't need further analysis to feel comfortable sharing the news with his guests and proposing a champagne toast to the moment before quickly excusing himself to rush over to the White House.

Bush had already heard, and the news rattled him. For all the months he had been proclaiming his support for *perestroika*, and for all the weeks he had publicly and privately been telling his staff that Gorbachev seemed truly intent on seeing his reforms succeed, he had retained until this moment significant doubt about his adversary's sincerity. No longer. This had to be part of Gorbachev's plan, he reasoned. There was no way the East Germans could pull off something this momentous without Kremlin approval. He was only half-right. Gorbachev had not known. At the same time, the Soviet leader had both set the conditions for the GDR's ultimate reform and, perhaps more important, tolerated it. Watching the images beamed from Berlin on a small television set in a study adjacent to the Oval Office, Bush was hit hard by the real meaning of what he was seeing. "If they are going to let the Communists fall in East Germany," he said, turning to Scowcroft, "they've got to be really serious."

Baker soon joined them, the three Cold Warriors transfixed by the images until Marlin Fitzwater interrupted. The media needed a statement. Pundits were already filling the airwaves, proclaiming the triumph of a half century of American policy. Bob Dole and George Mitchell both planned to issue congratulations on the Senate floor, claiming a share of the victory and celebration. The president had to say something. Bush at first demurred. Overexuberance from the Oval Office might yet prompt rethinking in the Kremlin, or a change in occupants. He finally conceded to Fitzwater's insistence, and the reporters filed in.

For the ensuing half hour, Bush succeeded in saying remarkably little. "Is this the end of the Iron Curtain, sir?" he was asked. "Well, I don't think any single event is what you'd call the end of the Iron Curtain," he responded, choosing his words deliberately, almost lethargically, repeatedly peering down at his papers and glancing at Baker for confirmation. "What's the danger here of events just spinning out of control?" another pressed. "I'm not going to hypothecate that anything goes too fast," Bush said. "But we are handling it in a way where we are not trying to give anybody a hard time." Overall he appeared unimpressed by the news from Berlin. Perhaps even uninterested.

Crowded in next to Baker at the edge of Bush's desk, kneeling so her camera crew could secure a clear shot, Lesley Stahl of CBS grew visibly impatient. "This is sort of a great victory for our side in the East-West battle," she finally blurted out. "But you don't seem elated." Bush visibly shrugged, offering, "I am not an emotional kind of guy." Yes, he was pleased. But "the fact that I'm not bubbling over — maybe it's getting along towards evening, because I feel very good about it."

He was more candid in his diary. Democratic leaders, and many from his own party, were urging him to travel to Berlin to celebrate, to declare communism dead and buried once and for all. Democrats Mitchell and Gephardt in particular wanted Bush to convene an immediate European summit and to travel directly to Eastern Europe after his meeting with Gorbachev at Malta, or better yet before, to demonstrate unbridled American support for the region's democratization. Leaders of the EU were already scheduled to meet in Paris on November 17. Bush should be there too, the legislators argued. If the continent's map was about to be redrawn, then the United States needed a seat at the table, one only the president could properly fill.

Bush refused, however, in his words, "to dance on the wall," rejecting anything he thought might in any way complicate matters further in Europe, and for Gorbachev. "The big question I ask myself is," he admitted to his journal, "how do we capitalize on these changes? And what does the Soviet Union have to do before we make dramatic changes in our defense structure? The bureaucracy answer will be, do nothing big, and wait to see what happens. But I don't want to miss an opportunity." If nothing else, his public inaction gave further ammunition to his congressional opponents. "At the very time freedom and democracy are receiving standing ovations in Europe," Gephardt protested, "our President is sitting politely in the audience with little to say and even less to contribute."

Bush thought others in Europe were already complicating matters enough. Kohl flew to Berlin, requiring transport aboard an American plane because of lingering regulations left over from the World War II division of the capital. The symbolism of this detail should not be overlooked: for the West German chancellor even to speak in Berlin, let alone to speak for one unified country, demanded American consent. As Kohl took the stage in front of a teeming crowd, his aide Horst Teltschik barely had time to pass a note with a plea just received from Gorbachev, who asked for calm in the midst of this "chaos." The Soviet leader specifically implored Kohl to refrain from saying anything "to stimulate a denial of the existence of two German states," which "can have no other goal than destabilizing the situation."

Whether actively ignored or simply lost to the adrenaline of the moment,

Gorbachev's message failed to have an effect. The raucous crowd was intoxicated, and not merely by the moment. "We Germans are now the happiest people in the world," West Berlin's mayor proclaimed to wild applause. Kohl received a less enthusiastic welcome when he approached the podium. Berliners rarely supported his party, and the gregarious West German was personally unpopular in gritty Prussia. Hearing scattered hissing from those who did not want a far-off politician sharing in their local triumph, Kohl nonetheless blustered ahead, speaking beyond the crowd to the global audience he knew was watching. Thanking the allies and the Soviets alike for their support of self-determination, he embraced the moment. "The division of our Fatherland is unnatural," he said. The road ahead led to "unity and right and freedom. We demand this right for all in Europe! We demand it for all Germans!" At this moment "we have less reason than ever for resignation or permanent acceptance of the two German states," he added, before climactically proclaiming, "A free German Fatherland lives."

He could hardly control his excitement when speaking to Bush later that night. "I've just arrived from Berlin. It is like witnessing an enormous fair. It has the atmosphere of a festival. The frontiers are definitely open." Both men agreed on the need to keep in touch before Bush's long-scheduled meeting with Gorbachev in three weeks' time, with Bush adding in direct contrast to Kohl's enthusiasm his desire "to see our people continue to avoid especially hot rhetoric that might by mistake cause a problem." It would seem, he told Kohl, that his "meeting with Gorbachev in early December has become even more important."

Gorbachev had not ordered the wall breached, contrary to what Bush presumed. He'd not been told of East Germany's plans, since there hadn't been any. The rapidly deteriorating situation there was not even on the Politburo's agenda for their five-hour meeting on November 9, their session concluding before Schabowski's unexpectedly momentous news conference. The brewing separatist crisis in the Baltics dominated discussion instead. "Everything is aimed at preparations for secession," Prime Minister Nikolai Ryzhkov warned. "As soon as they [the region's nationalist parties] win elections, they will adopt a decision to leave. What should be done? . . . What we should fear is not the Baltics, but Russia and Ukraine. I smell an overall collapse."

Preoccupied by his own domestic troubles, Gorbachev took the news from Berlin badly. Apart from pleas for calm, the common refrain for policy makers in the first hectic hours, his initial public reaction was therefore muted. He said more when writing to other world leaders, beginning with Kohl. He'd been painfully clear in recent weeks on what changes the Soviets could tolerate, with unification topping his list of intolerable outcomes, and believed both Thatcher and Mitterrand agreed with him. With, in his calculus, three of the Four Powers holding legal authority over Germany's fate on the same side, he thought there was little prospect for any swift change in Germany's status. His larger concern was therefore to keep the situation from getting out of hand on the ground, and to remind the Germans — and the world — that for all their problems, the Soviets still held important cards when it came to determining Europe's future.

Immediately after dispatching his plea for calm to Kohl, Gorbachev sent identical messages to Mitterrand, Thatcher, and Bush to inform but also to warn. "I have appealed to Chancellor Kohl to take the extremely pressing steps necessary to prevent a complication and destabilization of the situation," he wrote. "If statements are made in the FRG, however, that seek to generate emotional denials of the postwar realities, meaning the existence of two German states, the appearance of such political extremism cannot be viewed as anything other than attempts to destabilize the situation in the GDR and subvert the ongoing processes of democratization and the renewal of all areas of society.

"Looking forward," Gorbachev continued, "this would bring about not only the destabilization of the situation in Central Europe, but also in other parts of the world." Thatcher "clearly understood" Gorbachev's position, she said, trying to comfort an anxious Soviet ambassador who arrived on her doorstep at 10 PM on the evening of November 10. "While the countries of Eastern Europe could choose their own course in their domestic affairs, the borders of the Warsaw Pact must remain intact." The GDR was a member of the Warsaw Pact; for the alliance's boundaries to remain intact meant nothing could happen to alter its borders, its status, or, frankly, its side.

Amidst the celebration and tumult in the street, therefore, leaders of each of the major powers reacted to the news with trepidation. Only Kohl was truly happy that night. For Thatcher, Mitterrand, and a host of European leaders such as Walesa, the fall of the wall meant an end to an era that for all its troubles at least featured stability and restraints, on Germany in particular. They feared the repetition of history, the return of German power, and perhaps most dangerous of all the collapse of Soviet influence. "Gorbachev's policy in Eastern Europe is being overrun by events," the British ambassador cabled home from Moscow. "Gorbachev's problem is now to control the forces he has unleashed."

Bush, through it all, simply could not shake memories of Beijing. Indeed, his handwritten notes from his conversations with both Kohl and James Baker from those nights, only declassified in 2015, reveal his darkest fear. "So far no conflicts e[ven] though in E[ast] Berlin, Dresden, Leipzig hundreds of thousands in streets," he scribbled as the German talked. The Soviets could still disperse the crowd, he jotted while hearing the latest from Baker, but so far at least, they'd promised "no Tiananmen here." The name appears repeatedly. He privately pledged to push for a responsible conclusion to whatever came next, which at the least had to include one vital component: a continued American presence in Europe. "I'm telling our people we must challenge the defense system, and go back to demand new studies" on how to "keep the West secure," he wrote in his diary the evening of November 10, without undermining either America's strength or its vital European presence. "As the changes happen, I'm absolutely convinced there will be declining support for defense all around Europe."

Gorbachev feared the ominous specter of chaos as well. In the sudden collapse of the GDR and potential rush for unification, he saw the erosion of whatever slim measure of political support he retained from conservatives within the Soviet Union. Already stressed by declining supplies and growing calls for secession, they would be unlikely to stomach the loss of not only their country's most valuable ally but also the security that came from a divided Germany, and the power and prestige that accrued to the Soviets as one of the four legal guarantors of Germany's (and thus Europe's) future. Gorbachev's vision of a common European home had never been particularly specific, but it had taken as its central premise a melding of Europe across the traditional East-West divide represented by the Iron Curtain and the Cold War. If the Cold War's central questions were resolved before he managed to solve the Soviet Union's, the rest of Europe might well build its own common home without him. Soviet socialism would then truly be doomed.

The mood was therefore dark in the Kremlin as the broad significance of what they had done settled in. "The Berlin Wall has fallen," Gorbachev's chief foreign policy adviser wrote in his diary. "This entire era in the history of socialism is over . . . This is the end of Yalta . . . the Stalinist legacy . . . This is what Gorbachev has done. He has indeed turned out to be a great leader. He has sensed the pace of history and helped history to find a natural channel."

But what would come next? "A meeting with Bush is approaching," Chernyaev mused. "Will we witness an historic conversation? There are two main ideas in the instructions M. S. [Gorbachev] gave me to prepare materials: the role of two superpowers in leading the world to a civilized state, and the balance of interests. But Bush might disregard our arguments." In that case, he confessed, "we do not really have anything to show except for the fear that we could return to totalitarianism."

14

GERMANS PAUSE . . . AND ACT

THE BERLIN WALL'S COLLAPSE did not make Germany's reunification in-evitable. With thousands of East German citizens daily streaming west and the GDR's Politburo in disarray it is easy to assume the opposite. That narra-tive makes sense and feels good. Long-suffering citizens rose up in defiance of their totalitarian state, choosing first with their voices, then with their feet, and ultimately with their votes to join the democratic alternative that had long beckoned just beyond barbed wire. Once free, nothing could keep the German people apart.

That is an inspiring story, though not the way it happened. The forging of one singular German state from two was never a certainty. Neither was it uni-versally desired, even among Germans. The critical weeks that followed the Berlin Wall's breach instead brought vigorous debates over the country's (and the continent's) future, punctuated by continued efforts to save the flailing East German regime, amid repeated admonitions for calm from each of the great powers. Europe's future geography was at stake, and, some feared, a return to its terrible past. "All sides should think through their actions very carefully," Gor-bachev warned Kohl two days after the wall fell. Anything done in haste might lead to "chaos." Margaret Thatcher recoiled at the idea of a speedy reunification as well. "Much too fast!" she protested when pressed on the matter immediately after the wall fell. "You have to take these things step by step, and handle them very wisely."

Kohl promised to do nothing that might further inflame passions, keeping his word for a few weeks, until the moment in late November when he feared losing his opportunity to father a new Germany. His sudden shift upset his allies yet also set the conditions for Bush, in the new year, to make the most meaningful contribution of his presidency. Bush too wanted calm in the heart of Europe. But he also saw in Germany's unification an opportunity to cement an American place across the Atlantic for the foreseeable future, no matter what Gorbachev's revolution might ultimately bring.

Hanging over it all was the impending superpower conference in Malta, set for early December. "This one isn't a summit," Bush insisted repeatedly, denying speculation that he and Gorbachev might strike some grand bargain for Europe. "Summits take on a definition" and "an expectation of grand design and grand agreements." This was merely a chance to talk, "a non-summit summit," he called it. It was an "intermediate summit," Foreign Minister Shevardnadze helpfully offered, with nothing on the agenda save conversation.

History suggested otherwise, being replete with examples of great powers convening to resettle international systems in flux. The 1815 Congress of Vienna codified Europe's boundaries after Napoleon's rampages. A century later, World War I's victors similarly met with pens and maps at Versailles. Sessions at Yalta and Potsdam in 1945 largely defined World War II's aftermath and the ensuing line between democracy and communism. For those who interpreted the Berlin Wall's breach as proof of the Cold War's end, the coincidental Malta sessions gave every appearance of being yet another moment when great powers might take it upon themselves to seal the fates of millions.

They might even decide about Germany, whose occupiers after 1945 (Britain, France, the United States, and the USSR) retained a legal right to oversee its sovereignty. The leaders of three of these four openly feared its unification, and with it creation of a new economic and political colossus in their midst. A unified Germany might bring about "1913," François Mitterrand lamented to Gorbachev in late November, returning Europe to the geopolitical rivalries that had made the twentieth century the bloodiest on record. Unrestrained German energy had started two world wars, Mitterrand similarly lamented to Thatcher. If they did nothing to impede its growth, "he and the Prime Minister would find themselves in the situation of their predecessors in the 1930s who had failed to react in the face of constant pressing forward by the Germans." Thatcher immediately agreed. "If we were not careful reunification would just come about," she worried. "If that were to happen all the fixed points in Europe would collapse: the NATO front-line; the structure of NATO and the Warsaw Pact; Mr. Gorbachev's hopes for reform."

History itself seemed determined to reinforce the magnitude of the issues at hand. The Berlin Wall fell on November 9, 1989, the fifty-first anniversary of Kristallnacht, when Nazi hordes had rampaged through Germany and Austria in an orchestrated pogrom considered a key milestone on the road to the Holocaust. Hundreds had died, with thousands more incarcerated, many ultimately exterminated, solely for the crime of staining German soil with their blood. Millions more followed.

Crimes like these can never be fully forgiven or forgotten. "I don't begrudge their exuberance," the Nobel Prize–winning author (and Holocaust survivor) Elie Wiesel wrote of the German youth who danced atop the Berlin Wall a half century later. "They deserve the chance to begin again." And yet "I cannot hide the fact that the Jew in me is troubled, even worried. Whenever Germany was too powerful, it fell prey to perilous temptations of ultranationalism." West Berlin's mayor proudly declared that "Nov. 9 will enter history." But as Wiesel noted of the Germans, "They forgot that Nov. 9 had already entered history —51 years earlier."

Respected voices throughout the international community echoed his concerns. "Auschwitz speaks against every trend born of manipulation of public opinion," German novelist Günter Grass wrote of his country. "Even against the right to self-determination granted without hesitation to other peoples." It could never be a normal country again, not only because of what it had done but also because of the dark stain at the heart of German culture that its crimes revealed. Perhaps geography was to blame as much as culture. Set at the crossroads of civilizations, Germans were forever cursed by being "too big for Europe, too small for the world," Henry Kissinger noted, a sentiment Mitterrand privately voiced to Gorbachev as well in mid-November. "Germany had never found its true frontiers," the French leader said. Never content, the Germans "were a people in constant motion and flux," driven to dominate others.

Nothing could change Germany's location or, some feared, its drive for dominance, which division had seemed to mute. "Did not the two earlier German unifications lead to war?" the editor of the *New York Times* consequently asked soon after the Berlin Wall's breach. "Is there not a terror in millions of minds and hearts that the nightmare visage of the past may be the face of the future?" Surveying the strategic landscape, his newspaper concluded that "neither Europeans, Russians, Americans, nor indeed German leaders themselves are ready or eager to see a Fourth Reich." The "same can be said of almost all Europeans, East and West," the German-born Kissinger counseled, while the Polish-born Zbigniew Brzezinski similarly advised that "to end the division of Germany you need to end the fear that it generates."

The pervasiveness of such fears cannot be overstated. It infused even the jokes that circulated throughout the Pentagon.

Question: "What was the French response to German unification?"

Answer: "They put up speed bumps to slow down the panzers."

Its message could not have been clearer: Germans (history taught), when unbridled, were uncontrollable. In time the same cries of *Lebensraum!* (living space) that animated Hitler's followers would once more curse the continent. Airwaves filled with similar pointed quips. No less a cultural icon than *The Tonight Show*'s Johnny Carson told viewers, "The Berlin Wall is down. That means that all Germans are now free to go wherever they want in Europe. Hey, wasn't that the problem back in 1939?"

Many of Carson's viewers no doubt remembered 1939. Younger generations did not. Thus their historical sensibility — that is, why people with no personal memory of the events in question knew to laugh — better demonstrates their society's consensus view of what happened in the past and why. As *Saturday Night Live*'s Dennis Miller (born in 1953) put it: "You know, I'm trepidatious about a unified Germany in much the same way I am about Dean Martin and Jerry Lewis getting together . . . I haven't really enjoyed any of their previous collaborations, and I'm not sure I need to see any of their new stuff." *Late Night*'s David Letterman (born in 1947) was blunter. German unification will have three phases, he told his studio audience and the millions more watching on TV. "One: Political unification; Two: Economic unification; Three: France surrenders."

The message behind all these jokes was no laughing matter in the end, and it echoed across the Atlantic. "Unshackled once more," the editors of one British tabloid warned on November 11, 1989, "the Germans might well decide to seek, in the time-worn phrase of their rulers for the past century, 'a place in the sun.'" They'd done so before. "Where these ambitions will take them we cannot

Witness to the war that seared his generation, Bush served as one of the youngest aviators in the Pacific.

Memories of the war never faded, and neither did those of Bush's beloved Robin.

Bush knew presidents and powerbrokers from a young age, including one of his father's favorite golfing partners, Dwight Eisenhower.

Posted to the United Nations when his Senate bid fell short, Bush discovered a new passion for diplomacy that ultimately guided his presidency.

His diplomatic education continued in China, where he first met Deng Xiaoping and learned to work with experienced Washington insiders such as Brent Scowcroft (shaking Deng's hand), Henry Kissinger, and President Gerald Ford.

Eager to impress the world's most famous man, Bush instead learned firsthand of Mikhail Gorbachev's mercurial style during a 1987 limousine ride neither would soon forget.

A president-elect, an outgoing president, and the man who hoped to shift his friendship from one to the other, in December 1988 in New York.

An "old friend" of China's, Bush made a visit to Beijing — and to Tiananmen Square — a priority for his first months in office.

Initially organized to protest corruption, Beijing's crowds in time threatened the very fabric of Communist Party rule.

Peter Charlesworth/LightRocket/Getty Images

Bicycles proved no match for the armored vehicles sent to dispel demonstrators once Deng Xiaoping's government feared an end to its rule and the potential for civil war.

David Turnley/Corbis Historical/Getty Images

In the wake of the Tiananmen violence, Bush found raucous crowds in Poland and Hungary eager to show their enthusiasm for democracy by cheering an American president.

Meanwhile, crowds of protesters gathered throughout the Soviet bloc, such as in Leipzig, where the order was given to fire on crowds — but not obeyed. *Christian Günther/Ullstein Bild/Getty Images*

When the Berlin Wall finally fell near the end of 1989, lines of cars and refugees fled the German Democratic Republic, including through this opening in the Iron Curtain near Ulitz, Germany. *Sven Creutzmann, Mambo Photo/Hulton Archive/Getty Images*

With Baker to his left and Scowcroft and chief of staff John Sununu to his right, Bush gave an impromptu Oval Office press conference on the day the wall came down.

Bush would not dance atop the wall, but he did receive a piece chiseled from it, handed to him in the Oval Office by West Germany's foreign minister, November 21, 1989.

Crowds continued to surge after the Berlin Wall fell, their enthusiasm spreading to Czechoslovakia, where longtime dissident Václav Havel addressed a pro-democracy rally in the midst of the country's "Velvet Revolution" in November 1989. *Sovfoto/Universal Images Group/Getty Images*

Bush and Gorbachev at Malta, December 1989 — the first time that an American president and a Soviet general secretary held a simultaneous joint press conference.

Two weeks after Malta, recently deposed Romanian despot Nicolae Ceaușescu was in custody following his flight from Bucharest. He'd be dead days later. *David Turnley/Corbis Historical/Getty Images*

Not every revolution ended bloodlessly. Here, civilians in Bucharest huddle behind a tank after violence overtook their revolution in December 1989. *David Turnley/Corbis Historical/Getty Images*

Moments of personal diplomacy and German unification. Bush and Helmut Kohl, February 25, 1990.

Bush and Gorbachev, alone with an interpreter at the White House, June 1, 1990, the weekend German unification became real.

Forced by fate to work together, the pair also found time to work on their friendship at Camp David that same weekend.

Bush speaks to the press before the start of the first National Security Council meeting following Iraq's invasion of Kuwait, August 2, 1990.

Bush at Walker's Point during the Gulf crisis, with (from left) Saudi Prince Saud, the NSC's Richard Haass, Baker, deputy national security adviser Robert Gates, Sununu, and Saudi Prince Bandar, August 16, 1990.

A quiet moment for granddaughter Ellie LeBlond in the midst of the Persian Gulf crisis, Walker's Point, August 26, 1990.

Partners for decades, on the tennis court and in government. Bush and Baker confer in Paris, November 19, 1990.

Recalling his own wartime holidays away from home, Bush spent Thanksgiving with American troops in Saudi Arabia, November 22, 1990.

A telephone to his ear, Bush nonetheless learned much of what transpired in the Gulf War by watching the same CNN broadcasts as other leaders around the world, including this speech by Iraqi foreign minister Tariq Aziz, January 9, 1991.

A solitary moment on the eve of war, January 16, 1991.

American soldiers in liberated Kuwait sit beneath the gaze of the man whose invasion brought them to the desert. *Peter Turnley/Corbis Historical/Getty Images*

The "new world order" also meant a new power dynamic within the Soviet Union, exemplified by Russia's new president, Boris Yeltsin, who visited the White House soon after winning election in order to boost his own credibility at home and abroad.

In the wake of Desert Storm, Americans at last enjoyed victory parades denied the Vietnam generation. Here Bush receives a salute from General Norman Schwarzkopf and staff, July 8, 1991.

The G7 met in mid-July 1991 in London, where Gorbachev's plea for aid failed to produce the largesse he thought necessary to stay afloat and in power.

Gorbachev and Yeltsin, often at odds but soon to need each other more than ever, welcome George and Barbara Bush in Moscow, July 1991.

Bush's first conversation with Gorbachev following his release from Crimea during the August 1991 coup.

Youth perched on Moscow barricades during the coup, mere days after Bush's departure from the city. *Igor Gavrilov/The LIFE Images Collection/Getty Images*

By June 1992, the Soviet Union was no more, and American officials found a new ally in Moscow, who, like the previous one, insisted on visiting Washington to prove his political standing at home.

Successful abroad, Bush ultimately fell to charges of domestic inaction launched by Bill Clinton and Ross Perot during the 1992 presidential campaign. The trio are seen here in their October 19 debate.

guess. More captive markets? More living space? This much is certain: they will not set out to spread joy and benefit humanity."

Deep in their hearts, beyond where reason and logic held sway, Gorbachev, Mitterrand, and Thatcher each believed this trope, and feared a new unchained Reich. Each had known the pain of German power, and considered it their duty to spare future generations the same. Too young to have fought but not too young to have suffered under German occupation, Gorbachev recalled watching his village's men depart for the front lines in the Great Patriotic War against the Nazis. Most never returned. He thought his own father was dead for more than a year before learning of his survival. Once, while searching for food, he stumbled upon a pile of half-decayed corpses revealed by the spring thaw, each riddled with German bullets. He was ten. Such experiences are not easily forgotten, but neither were they unique. One of his principal advisers owed his painful limp to a German hand grenade. His foreign minister lost a brother in the war. Everyone lost someone. "We were wartime children who survived," Gorbachev once observed, adding, "Nothing of the life and deeds of our generation is understandable unless we take this into consideration."

Their nation having suffered more deaths during World War II than any other, Soviet leaders were not shy about voicing their concerns over their historic enemy's revival. "Where are the political, jurisdictional, and material guarantees that German unity, in the long term, will not create a threat to the national security of other states and to peace in Europe?" Shevardnadze publicly asked little more than a month after the wall fell. That he made these remarks at NATO headquarters, the first Soviet official ever to speak from that podium, lent added weight to his message: an organization designed to repel the Red Army had far more to fear from one of its own. "History itself," he warned, "prescribes heightened circumspection of Europe."

Mitterrand sang a different tune when speaking publicly, however, or directly to Kohl. "I am not afraid of reunification," he'd declared a week before the wall fell when standing shoulder to shoulder with his German counterpart. "If they want it, and if they can get it." He was less sanguine in private, realizing that France would have to live next door to the country that emerged, be it pleased or embittered by whatever came next. "There is a certain equilibrium that exists in Europe," he thus cautioned Gorbachev when pressed on the question of German unification. "We should not disturb it." Unwilling to speak more forcefully in public, he thus quietly encouraged his allies to disrupt Germany's plans on France's behalf. "The French may be with us in their hearts," Thatcher's staff thus advised, "but we must expect them to play a very canny game in talking to the Germans."

Thatcher was more openly recalcitrant, having never fully shaken her own childhood memories of scurrying from Nazi bombers, or the stories of close friends who had fled Nazi persecution, and afterwards barely disguised her contempt whenever Germans spoke of peace. "We've been through the war and know perfectly well what the Germans are like," she reportedly told one colleague after the wall fell. "National character basically doesn't change," she said on another occasion, reminding her entire cabinet that "a single European currency was no answer." Yes, "Western governments had taken a formal position since 1955 in favour of East German self-determination," but that was easy to say when no one thought it would actually happen. Therefore "German re-unification should not be treated as an immediate issue," she counseled, before ominously adding, "Although events were moving in a favourable direction . . . Europe faced a difficult decade ahead."

She said much the same to Bush after the wall came down, hoping to win his support in stymieing the march toward unification. "Talk about that [unification] here has led to some strong emotions," she told him. "History here is living history." Bush subsequently wrote "Coventry" alongside notes used during one of their conversations, a reminder of the city's infamous firebombing at German hands, and in turn of the touchstone for her fury.

Bush alone, among all the big four leaders, embraced unification from the start. He was also the only one who openly rebuffed the basic assumption of Germany's innate aggressiveness. "I think there is in some quarters a feeling — well, a reunified Germany would be detrimental to the peace of Europe, of Western Europe, in some way," he'd told reporters while on a brief trip to Montana earlier that fall. "I don't accept that at all, simply don't." West Germany had come a long way since World War II, he insisted. It was a trusted ally whose forty-plus years of democracy could not easily be overturned. If unification "was worked out between the two Germanys, I do not think we should view it as bad for Western interests."

Bush's vehemence, as well as his timing, took even his closest advisers by surprise. "I called Brent [Scowcroft] right away," Robert Gates recalled, having witnessed the impromptu press conference, "and said, 'Brent, we now have a

policy on German reunification.'" NSC staffers had been reexamining the issue for weeks, evaluating Washington's habitual support for a single Germany,
but had yet to return a formal conclusion. As far as Scowcroft was concerned,
therefore, they didn't yet have a current policy. "He said 'what is it?'" Gates
continued.

> I said "we're for it."
> He said "Who says so?"
> I said "The President."
> He said, "Oh, shit."

There were two reasons for Bush's stance. The first is how lightly he wore the
past. Ready to forgive Japan for World War II, he afforded Germans a parallel
opportunity. "I prefer to look at Germany's 45 years of contribution to democracy and to the security of the West," he told an interviewer on November 21,
"and that's what we are focusing on." How would he respond to Europeans who
remained afraid of German power? the reporter asked. "I say to them: that's a
matter for the German people to decide. And there are some that worry about
it. I understand Mr. Gorbachev has some understandable constraints, because
he looks at borders, he looks at history — he's concerned. But . . . this is 1989.
And we can learn from history, but we can also look to the future."

His second rationale for backing unification was, ironically, based very much
on history. The Germans alone were not to blame for Europe's long history
of wars, he believed. Europe itself was at fault, having shown no ability over
the course of millennia for managing its affairs without resorting to violence.
It had never known real peace, Bush believed, without American oversight.
Drawn into Europe's conflagration in 1917, Americans tipped the Great War
in democracy's favor. Leaving Europe once more to its own devices after Versailles, they'd been forced to return within a generation, sacrificing hundreds of
thousands of lives by 1945 in order to bring Europeans once more a peace they
could not achieve on their own.

This time they stayed, and so came decades of peace. It was a nuclear-tinged
and frequently terrifying peace, but the presence of American power in NATO,
an organization designed to keep the Soviets from overrunning the continent,
also served as a way to keep the rest of the continent in check as well. This was
the history that mattered to Bush, to his inner circle, and indeed to an entire

generation of policy makers in power at the Cold War's end. American force remained Europe's only real gyroscope, one ranking State Department official testified to Congress in early 1990. Without American stabilization of the kind present since 1945, it seemed inevitable that "European squabbling would undermine political and economic structures like the EC," leading to resumption of "historic conflicts."

Leaving Europeans in charge of their own affairs was not an option, Bush therefore told his staff soon after the collapse of the Berlin Wall, and not just because the continent would shudder under German leadership. The Americans had not been there over the previous four decades exclusively to hold off communism. "In a new Europe, the American role may change in form," he pledged, "but not in fundamentals . . . [W]e will remain in Europe."

Yet just as there was no guarantee that Germany would inevitably reunify after the dramatic events of November 9, it was even less assured that whatever security arrangement emerged after 1989 would include either German or American participation. New countries in new situations typically reassess their alliances, and a newly unified Germany, hoping to be the center pole of a broad new European tent that encompassed both East and West, might well choose to forgo any association with either the Warsaw Pact or NATO. The former appeared to be on the ropes; and the Soviets, meanwhile, were equally unlikely to warmly embrace East Germany's wholesale absorption by the latter. German popular enthusiasm for NATO — and for American leadership — had plummeted in recent years thanks to the ongoing dissent over short-range nuclear missile deployment. Perhaps the country's leaders might acquiesce to the popular will in the wake of unification and join Europe without retaining its membership in either of its dominant military blocs.

A neutralized Germany would have been disastrous from Bush's perspective. The Western alliance could not long survive without Europe's strongest power, American analysts calculated. Without NATO, what reason did the United States have to justify its continued involvement in European affairs? While his counterparts fretted and schemed to retard German unification, Bush therefore did the opposite, but not for their sake. "I've got to look after the U.S. interest in all of this," he told his diary, "without reverting to a kind of isolationist or stupid peace-nik view of where we stand in the world." After all, "there are all kinds of events that we can't foresee that require a strong NATO, and there's all kinds of potential instability that requires a strong U.S. presence" in Europe.

Unification faced two additional obstacles, both German: the lingering hope within the GDR that the state might somehow survive its current trials, and the equally strong hope among West Germans that they might somehow welcome close ties with their fellow Germans without assuming responsibility for their welfare. Modern states do not easily disappear, and the GDR was no trifling state. Notwithstanding its financial debt and moral bankruptcy, the country boasted a powerful military, a seat in the United Nations, and by some measures the world's tenth-largest economy. East Germany had problems as 1989 came to an end, but so too did other socialist states such as Poland and Hungary, where protests had produced reforms, not extinction.

Hans Modrow became prime minister a week after the wall fell, promising precisely that. Born to a family of bakers in a far eastern province that became Polish territory after 1945, he had a quick mind and a mild temperament that fueled his rise. "In the past several weeks, Comrade Modrow has made no negative comments regarding our informational activities," the Stasi reported to the Politburo in late October 1989, for example. The same could rarely be said of anyone else during that tumultuous period.

This was a man who knew how to keep his mouth shut, making him all the more persuasive when he finally rose to defend the GDR. Honecker had not been seen in public in weeks. Krenz retained control of the state's security portfolio, but his days seemed numbered. Modrow was therefore the new face of East Germany and its last hope for real reform. "I know that I have accepted a difficult legacy," he acknowledged. Anyone who thought they could simply go back to life as it had been before November 9 "is either blind or malicious." The only reasonable path forward was instead a fundamental restructuring capable of achieving "a better socialism," rooted in equality and designed to "create for ourselves a life that is manifold and meaningful, that allows individuality to thrive and makes comradery in the collective possible." Socialism still had value, he preached, especially when juxtaposed against capitalism, in which "one person is the other person's devil."

These were not the words of a man ready to throw in the towel. His appeals to the GDR's national mission found receptive ears within a population weaned on its message. According to one survey, conducted before the wall's unexpected fall, only 1 percent of those who marched in Leipzig listed unification among their ultimate goals. Most marched instead for greater access to the West and a greater say over their own lives, in effect, in the hope of renewing their state rather than tearing it down. Some were just following the crowd. "Those Who Cried 'Stalin Hooray' Now Shout 'Reform, Today!'" a banner at the Leipzig demonstrations had sagely pointed out.

"Away with the *Wendehalsen*," the banner further read. The word might be translated as "turncoats" or, more dramatically, "traitors," and its message should not be quickly dismissed. Socialism had not failed; East Germany's treasonous and corrupt leaders had, a point reinforced by a flood of news reports printed under the GDR's new policy of transparency in October and November, revealing the depths of the government's material hypocrisy. While the proletariat struggled, the investigations revealed, a privileged few enjoyed all the trappings of power: cars, vacation homes, and, most damning of all, access to Western products. Such sensationalistic stories offended a citizenry raised to think in egalitarian terms, but simultaneously reinforced the idea that their national mission might live on once purged of those who had strayed. Asked what type of economic reforms Modrow's new government should develop, 86 percent of East Germans in one late November poll supported "socialist" reforms. Only 5 percent endorsed wholesale adoption of Western-style capitalism. "Poland minus Communism is Poland," one popular formulation of the time warned. "East Germany minus Communism is . . . West Germany."

In late November an influential group of thirty-one writers, church leaders, and reformers published an appeal to save "our country." East Germany had its own history, they argued, and "either we can insist on GDR independence" and strive to "develop a society of solidarity, offering peace, social justice, individual liberty, free movement, and ecological conservation," or "we must suffer . . . a sell-out of our material and moral values and have the GDR eventually taken over by the Federal Republic" solely on the basis of its economic might. More than 200,000 signed their petition.

Confederation, rather than unification, became Modrow's goal. East and West Germans needed to find some new means of living integrated lives, he told reporters, but without one side's capitulation, a position that gained support in West Germany as well once euphoria over the breach of the wall subsided and the prospective toll of absorption grew clearer. By the third week of November, 70 percent of West Germans polled endorsed some form of closer political ties with the East. Some wanted unification, but nearly an equal number thought it more likely that the two countries might integrate without one side's capitulation, forming ties similar to those between Austria and German-speaking regions of Switzerland. Confederation would be less disruptive, they reasoned. "The existence of two German states [is] the basis for stability in Europe," Modrow declared. He'd just spoken to Gorbachev, who'd told him no other option was open for discussion.

LITTLE NOTICED amidst the nonstop television coverage from Berlin was startling news out of Bulgaria. Todor Zhivkov had been in power since 1954. By November 10 he was under arrest. Inspired by *perestroika*, and fearful of replicating the East German regime's inaction, leaders of the coup that removed him promised reforms, roundtable negotiations, and elections by spring, all with the Kremlin's explicit endorsement. Whereas the opening of the Berlin Wall had been entirely unplanned, Gorbachev knew full well about the coup in Sofia ahead of time, and his refusal to intervene served as tacit approval. "Somebody had to start, and you took that responsibility on yourself," he praised Bulgaria's new leader a month later. "I can only note that if such events — the change of leadership in the GDR — had taken place earlier, everything would have been much calmer," he ruefully told Mitterrand.

Bush's White House was taken completely by surprise, having been assured by its embassy in Sofia only hours before Zhivkov's ouster that there would not be "any major personnel changes" anytime soon within the Bulgarian regime. Few staffers were paying attention to Bulgaria anyway. "When word reached us at the White House of Zhivkov's resignation, my NSC colleague Condi [Condoleezza] Rice and I looked at each other in utter bemusement," Robert Hutchings later recalled. "We had been so preoccupied with Germany — this being the day after the night the Berlin Wall fell — that we had not even thought about Bulgaria for weeks."

Czechoslovakia was at least on their radar. Prague remained tense, though once more American analysts predicted little prospect for wholesale change. "Widespread popular pressure remains muted," Ambassador Shirley Temple Black wrote Washington as the wall fell barely two hundred miles away. Still scarred by memories of 1968, her staff concluded that "the average man, distinct from dissident and intellectual circles, has become more, not less, cautious about change in the face of the GDR developments." Opposition groups, yearning to wake their fellow citizens from their stupor, organized a series of increasingly provocative marches throughout the fall of 1989 in the hope of inciting a violent government response capable of catalyzing popular opposition against the regime. All they needed was some police or militia member to lose his cool.

Miloš Jakeš's regime largely refused to take the bait. Officials lambasted the crowds, calling them agitators, punks, and worst of all, writers. Yet they held fast at words, prudently letting the demonstrations peacefully run their course. As Václav Havel had earlier predicted, all the elements of a bonfire were in place. But would it burn?

Police finally provided the spark on November 17, after an otherwise peaceful demonstration commemorating the fiftieth anniversary of a Nazi crackdown on Czechoslovakia's universities took a wrong turn. Government officials initially endorsed the march, deeming this a particularly good time to recall the costs of unbridled German power. Batons met skulls, however, once demonstrators mistakenly veered off their prearranged course and toward the Politburo, the object of their chants simultaneously switching from Nazis to communists. Police surged, rushing the protesters' lines. Bones were broken. Hundreds were chased down and beaten, including women and foreign journalists, the melee caught on camera. Videotapes spread rapidly through the country, couriered by preestablished activist networks. Radio Free Europe erroneously reported a young protester's death, driving yet more angry citizens into the streets.

The several thousand who'd marched on November 17 morphed in response into several hundred thousand by the twentieth, overflowing Prague's Wenceslas Square, the city's spiritual heart. Havel rushed to join them. Some ninety miles away when the spark was struck, he'd been planning his own provocative march timed for December's international Human Rights Day. "Milos, it's over," the crowd chanted as Havel addressed them. They saved their loudest applause for an old man's surprise appearance on the dais. "*Dubček na hrad!*" some shouted as they recognized the Prague Spring's ousted leader. "Dubček to the castle; Dubček for president!"

Their cheers quieted once Alexander Dubček began to speak. His calls for renewing his generation's dream of socialism with a human face held little purchase twenty years later with a crowd bent on wholesale change. East Germans retained affection for their socialist dream; Czechs did not. "Poor Dubček was an anachronism," historian Tony Judt has written. "His vocabulary, his style, even his gestures were those of the reform Communists of the Sixties. He had learned nothing, it seemed, from his bitter experiences, but spoke still of resurrecting a kinder, gentler, Czechoslovak path to Socialism." An aged relic, he was revered but no longer worshiped.

What did they want instead? A document titled "Programmatic Principles of Civic Reform," enumerated by allied opposition groups, offers some insight. "1. A state of law. 2. Free elections. 3. Social justice. 4. A clean environment. 5. An educated people. 6. Prosperity." These demands led to a solitary conclusion: they wanted more, and better, from their government and for their lives. Nothing specific: just more. Their final demand revealed something even more profound, specifically the feeling that generations forced to live under Nazi and then Soviet rule had been deprived of a birthright that was finally within their grasp: "7: Return to Europe."

Jakeš promised to crush the mob but found no one willing to carry out his orders. Militia refused to march on the capital, and his own defense minister vowed on television to disregard any command to break up the crowds. Resigning on November 24, Jakeš took the rest of his party leadership with him, leaving in charge a cadre of midlevel officials with neither the talent to have risen to the party's elite nor the internal drive to join its opposition. Their new general secretary was an entirely unremarkable railway manager described by one contemporary as "possibly the only man dumb enough to accept the job."

The crowds continued their protests, culminating in a countrywide strike on November 27. It was a very Czech affair, with protests occurring over lunchtime and after five in the evening, lest a revolution disrupt the workday too much. Officials got the point nonetheless. Roundtable negotiations that had taken months in Poland now commenced and concluded in a mere two days, with opposition leaders agreeing to join the government once its communist members collectively resigned on December 10. By year's end, Dubček was named to the ceremonial post of chairman of the new Federal Assembly, whose first act was the unanimous election of Havel as the new nation's first president, elevating to power a man who only weeks before had openly lamented that his compatriots "might not live to see the day" when their opposition bore fruit. After years of struggle, the transition from communist dictatorship into budding democracy occurred so smoothly it became thereafter known as the "Velvet Revolution."

White House staffers could only watch in amazement as the world they knew unraveled, preparing Bush for his early December sit-down with Gorbachev while simultaneously disputing claims that the two sides intended to negotiate the terms of socialism's collapse. Every bit of news reinforced Bush's sense that Malta shouldn't be a negotiation at all. "Things are coming our way," Bush told his diary once again, "so why do we have to jump up and down, risk those things turning around and going in the wrong direction" either by celebrating the West's victories too loudly or by giving Gorbachev the opportunity to arrest the Soviet empire's unraveling before it was too late?

Each day's news nonetheless fueled concern that Gorbachev might come prepared to offer a deal. "We were getting hints from Moscow that one of Gorbachev's objectives at Malta was to gain some sort of 'understanding' for this situation and for the measures he might take to crack down," Bush later recounted. "I could not give him that, and if I did, it would have a lasting historical, political, and moral price." All he wanted was a conversation, and to make it out of Malta without Gorbachev's yet again capturing the initiative with a surprise proposal.

Bush's cadre quietly schemed to find a way of turning the tables, preparing a lengthy agenda designed to keep Gorbachev off guard. "Brent offered me about twenty topics to choose from," Bush recalled. "I took them all. I wanted to be prepared for anything." Memos flowed in from throughout the national security bureaucracy, far more than were typically prepared even before major summits. The White House called in outside experts for discussion: Soviet specialists, regional authorities, arms negotiators, and human rights advocates. "Whatever more he's got in his bag of tricks," Robert Blackwill boasted, referring to Gorbachev, the president would be "ready for him."

To Gorbachev, they sent only bland signals, with Bush dispatching yet another handwritten personal letter in late November to stress anew his hope that their upcoming meeting would bring nothing more than a much-needed chance to talk. "I want this meeting to be useful in advancing our mutual understanding and in laying the groundwork for a good relationship," he emphasized. He wished only that they might discuss "your vision and mine of the world 10 years from now," to discern "the differences when you say 'common Europe home' and I say 'Europe whole and free.'"

He closed his letter to Gorbachev with a highly personal sentiment. "I am writing this on the eve of our Thanksgiving Day," Bush noted, "a day in which all Americans thank God for our blessings." This year he planned to "give thanks for the fact that we are living through times of enormous promise. I will give thanks that our 5 kids and our 11 grandchildren might have a real chance to grow up in a more peaceful, less scary world." He also planned to give thanks "that you are pressing forward with *glasnost* and *perestroika*" and by extension for the man who'd done more than anyone else to make such once unfathomable events in Europe possible. "For, you see, the fate of my own family and yours is dependent on *perestroika*'s success."

Many aspire to influence the course of events, leading great nations at pivotal moments in history. Few are granted the opportunity. This was Bush's first such moment. For all the years he'd spent preparing to lead, and for all the time during his career he'd spent in the presence of power, even after he'd been president for nearly a year in his own right, Malta was the moment when the full magnitude of his responsibilities took hold.

Gorbachev had felt this weight before, most clearly at Reykjavik with Reagan. Indeed, he'd of late been drowning under the weight of his country's problems. Fully absorbed by seemingly insurmountable domestic issues, he'd only lately come to realize that his fate was now intrinsically intertwined with that of Germany, where events had moved further than he'd ever imagined. He'd wanted the GDR to reform, not to disappear, and as Ambassador Matlock warned from Moscow, while "the Soviets stress that they will not intervene militarily" to halt the pace of political change, they too had limits, and loss of their military alliance's most valuable member — and the centerpiece of their wartime triumph over fascism — was beyond easy comprehension. Backed too far into a corner, Matlock cautioned, hard-liners might choose to fight their way out, pressing Gorbachev to crack down in a manner he'd previously rejected, or pushing him aside in favor of someone who would. The CIA's leading Soviet specialist echoed Matlock's concern. "If it were to appear that Soviet troops were being forced to retreat from the GDR," he told the White House, that "he [Gorbachev] had lost Germany, and the security environment for the Soviet Union was more threatening, the domestic fallout — when combined with other complaints — could threaten his position."

Having already declared unification off the table, calling it "no issue of current policy," Gorbachev told Mitterrand that he "would support the direction in which the situation is moving now, with the exception of one aspect. I have in mind all the excitement that has been raised in the FRG around the issue of German reunification." The GDR held a special place in Soviet thinking. Lest there be any doubt, he stressed to Mitterrand, the day Germany unified, "a Soviet marshal will be sitting in my chair."

Bush feared the same. "It is this that really concerns me," he wrote in his diary after hearing reports of a potential crackdown on protesters in the Soviet republic of Moldova. "Some violence coming out in the Soviet Republic, where the Soviets use force, and that tightens up what is happening in Eastern Europe. [It] turns world opinion around against Gorbachev, [and] pushes us to take military action which, of course, we couldn't take way over there, and wouldn't take, and sets back, for proper reasons, the relationship that's moving in the right direction." *Perestroika* would wither, perhaps even taking Gorbachev with it. "This," Bush wrote, "is the big dilemma."

Kohl promised both leaders that he understood. "We will do nothing that will destabilize the situation in the GDR," he told Bush, saying much the same to Gorbachev. "We are prepared to honor the German people's choices about their future," Scowcroft thus told Bush on November 24, "including a choice for reunification," and "are comfortable with the way Bonn is meeting the challenge posed by recent events." Marking the memo in his typical fashion, Bush wrote an emphatic "yes," and underlined the word.

They were therefore all shocked when Kohl announced a ten-point plan for rapid reunification on November 28. He called for further democratization in the East, leading to gradual development of a federation of the two states embedded in the broader European process. His ultimate goal was not in doubt: "a state of peace in Europe in which the German nation can recover its unity in free self-determination."

Kohl's bold announcement took nearly everyone by surprise, including his own foreign minister. Domestic politics had forced his hand, as he believed German voters would ultimately favor unification, rewarding whoever provided the sovereignty they'd sought for generations. He also feared the way resurgent right-wing parties were taking control of the nationalism issue, and the growing popularity of Modrow's suggestion for an intra-German confederation. He thus moved the instant he perceived the slightest softening of the Soviet position. "We are thinking about all possible alternatives for the German question," one of Gorbachev's advisers told Kohl's national security adviser on November 21. "Even things that were practically unthinkable."

The Soviet diplomat did not mean what Kohl chose to hear, having only days before denounced the existence of a single Germany as "incompatible with the geopolitical and geostrategic requirements of stability." Yet Kohl secluded himself and his staff to work out the details of their rash proposal in the aftermath of the Soviet's inadvertent admission, specifically excluding anyone from outside their own party from the discussions. By speaking so boldly of a real plan for unification on November 28, he had effectively broken his word to the British, the Americans, the French, and the Soviets, and outflanked his own coalition partner at home, but Kohl succeeded in seizing the initiative while others dithered, setting the terms for future negotiations while the rest merely waited for negotiations to be scheduled. "Great speech," Genscher reluctantly admitted when Kohl finished his address. There was nothing else to say.

Every other major power was furious, albeit in different ways that reveal much about each nation's perspective on the issue. French policy makers felt insulted, having scarcely imagined that Kohl would make such a dramatic move without discussing it with them first. They had little choice but to acquiesce, however. The British too complained behind closed doors. Thatcher privately told Kohl that no matter what he might unilaterally declare, she still refused to consider unification "on the agenda." But publicly, Foreign Secretary Hurd would go only so far as to suggest that Kohl's plan lacked an eleventh essential point: "that nothing will be done to destroy the balance and stability of Europe or create anxiety in the minds of people who have a right to be worried." The other ten points could stay as well.

Kohl ultimately reasoned that he could win Thatcher's and Mitterrand's grudging approval because he felt confident in Bush's. The Four Powers held legal claim to decide Germany's future, but not all four votes were equal. Thatcher might lament the resurgence of German power, but she dared not directly contradict the Americans lest they form a new special relationship with the Germans instead. Bush had made his position on unification clear, and "when the tide of history is bringing a chance of freedom and democracy to Eastern Europe, with the prospect of reducing or even eliminating the Soviet presence," Hurd's principal political adviser warned, "the Americans would have no sympathy with a policy which put all that at risk in order to maintain the division of Germany." Indeed, the British might find themselves wholly isolated on the issue, and thus in the future as well. Given their proximity to Germany and their real sense of German power, and of its potential consequences, "the French would not stand against that tide either," he added, "whatever they may say in public."

The opinions that mattered most were in Washington and Moscow, and the latter's reaction was predictable. Shock turned to fury when news of the ten-point plan reached the Kremlin. "Maybe the federal chancellor no longer needs to cooperate with us?" Gorbachev indignantly told Genscher. "How can we talk about 'building a new Europe' if you act this way? Kohl assured me that the FRG doesn't want to destabilize the situation in the GDR, that he would act responsibly," Gorbachev raged. "But the chancellor's actual steps contradict his assurances.

"An all-European process is under way," Gorbachev continued, growing redder in the face; "we want to build a new Europe, a common European home. We need trust to do that." Kohl had broken whatever level of trust the Germans and Soviets had achieved. What does confederation mean? Gorbachev asked. "Where will the FRG be, in NATO or the Warsaw Pact? Or maybe it will become neutral? And what would NATO mean without the FRG? And, in general, what will happen next? Have you considered everything?" His rhetorical question offered its own answer: the Germans had not, at least not to his satisfaction. "If we wanted to sour or even break off our relations with the FRG," Gorbachev thus warned Genscher, Kohl could not have devised a better reason.

No less taken aback by Kohl's unexpected remarks — he'd not been told the chancellor planned to address reunification until thirty minutes before the speech — Bush reacted far more calmly, realizing that Kohl's move presented him with a choice: to back the Soviets or the Germans. It was a choice, in effect, between *perestroika* and NATO, or between a process he still did not trust or fully understand, and the basic foundation of American-led peace and security for the past two generations. Clearly, when framed in those terms, it was no real choice at all. Swift unification threatened to undermine Gorbachev, risking the real possibility of a broader Soviet crackdown in Eastern Europe. Yet Germany mattered more to Bush's overall strategy for ensuring Washington's place in Europe. He could break with Gorbachev yet, so long as NATO survived, still remain a vital presence on the continent. Indeed, anything that slowed *perestroika* also retarded calls for the American withdrawal. He needed Germany on his side to ensure America's staying power.

Bush chose the Germans. "I will respond negatively to any Soviet request for special US-USSR arrangements on Germany," his briefing points for a late November conversation with Thatcher thus made plain. "Or Four Power Talks, or other devices that shift the focus at this time from the pressing need for reform and democracy in the GDR." He'd reject anything the Soviets, or anyone else, might say or do to thwart Germany's ability to decide its future, because

Germany's fate was so deeply interwoven with his own. Even with the Malta summit looming and the ink barely dry on his personal letter to Gorbachev pledging support for *perestroika*, when forced to decide between favoring a vital democratic ally or the man whose democratic push depended in large measure on keeping Germany divided, Bush knew whom he had to choose. "We shared Kohl's reasons," one NSC staffer explained, "while the others did not."

Baker offered the administration's official response the day after Kohl revealed the ten-point plan, offering four principles on which unification should proceed: First, there must be respect for German self-determination. Second, any unified state should take shape "within the context of Germany's continuing alignment within NATO and an increasingly integrated European community." Third, unification should be peaceful and gradual. And finally, whatever form a unified Germany took, it could not involve a redrawing of the state's borders. (This was not the moment to renegotiate the historically contentious German-Polish border, for example.)

None of those four American principles in any way contradicted or constrained Kohl's agenda, a point Bush reinforced when speaking to the German chancellor the next day. Their conversation featured nothing like the venom infusing Gorbachev's conversation with Genscher. Indeed, Bush let Kohl speak first, hearing anew the German's promise, however hollow-sounding at present, that "we should try to avoid doing anything that would destabilize him [Gorbachev]." He did not even laugh when Kohl promised to "act with reason and caution," because he heard Kohl say something of greater import: once his plan came together, once Germany was unified, it would remain "in NATO."

"Stability is the key word," Bush finally said after Kohl's long monologue concluded. "We have tried to do nothing that would force a reaction by the USSR." Ultimately he planned to do nothing more at Malta than talk; he certainly had no plans to scheme with Gorbachev for some new division of Europe. "We will go forward cautiously," he told Kohl, "as part of a strong alliance. I've been criticized for being too cautious, but, first, things are going our way, and, second, stability should be the byword. We don't want inadvertently to create instability."

15

MALTA

PILLS WOULDN'T HELP. Nor a drink. He couldn't sleep. Gorbachev awaited him at the end of the flight. The Soviet Foreign Ministry's spokesman had built up expectations with talk of traveling from "Yalta to Malta," from the Cold War's beginning to what increasingly seemed like its end. The same thought permeated Washington. "The basis of the communist doctrine is dead," longtime insider Paul Nitze enthused in anticipation. "The *West* has won an ideological victory of tremendous dimensions." It was "the end of the status quo that has existed in most of this region for four decades," George Kennan, the architect of containment, declared, "and in Russia for a full seven." Triumphalism infected even the sports pages. "A funny thing happened on the way to the Mediterranean: The Cold War ended. We won," the *Washington Post*'s Tony Kornheiser wrote. What more could Bush possibly demand: "an Arby's in Romania?"

Such arrogant talk meant the Americans didn't understand *perestroika* at all, Gorbachev complained to his staff. The West had not "won" what was everyone's to share. More ominously, if they thought themselves triumphant, they would seek to dictate terms to their onetime foes. Any peace built on such a shaky foundation would never last. The Cold War was indeed ending, Gorbachev told an audience at the Vatican while en route to Malta, but "not because there are victors and vanquished, but precisely because there are none of either." *Perestroika* meant abandoning the idea that any single ideology, or any nation, held the ideal formula for all. None had the right to dictate another's affairs. "We no longer think that we are the best and that we are always right, and that those who disagree with us are our enemies," he declared, tossing aside

Marxism's central tenet of inevitable victory over the capitalist world. In its place, "we have now decided, firmly and irrevocably, to base our policy on the principle of freedom of choice . . . [T]his is the essence of perestroika's philosophy: we are getting to know ourselves, revealing ourselves to the world, and discovering the world."

The Americans needed to do some discovering of their own, he told Canada's Brian Mulroney, because unless they too learned some humility, the world would never fully advance into the brighter future that beckoned. "It is not easy for the Americans to comprehend the essence of the new world, of the new values," he said. They thought themselves perpetually right, that they owned a "hotline to God," and thus felt no need to listen when others talked. An annoyance during ordinary times, this was downright dangerous amidst such wholesale change. "As far as Eastern Europe is concerned," Gorbachev argued, "it is hard for the United States to give up the habit of teaching others." It was worse than a habit, he continued. "This for Americans is like an illness — AIDS — so far there is no treatment for it."

"We plan to work with the current administration on the same principles" while in Malta, he told Pope John Paul II. "If we want to talk about politics" with the Americans, "about how to change the world for the better, then we have to do it as equals." Indeed, he intended to do a bit of teaching of his own, instructing Bush on why talk of recent events in Eastern Europe as a triumph of "Western values" was insulting to many around the world who considered themselves Western already. I am a "Western Slav" myself, the Polish-born pope noted in agreement, adding, "It would be wrong to claim that changes in Europe and the world should follow the Western model." Europe, he said, "should breathe with two lungs."

It was an anxious and increasingly angry Soviet leader who arrived at Malta, worried over the fate of his empire and fearful that Bush cared more about obtaining communism's surrender than in partnering for the peace to come. "We felt the Soviet Union was in free fall, that our superpower status would go up in smoke unless it was reaffirmed by the Americans," one of his advisers recalled. "With the avalanche of 1989 almost behind us, we wanted to reach some kind of plateau that would give us time to catch our breath and look around." They therefore sought something hard to define: Bush's continued endorsement and practical means of supporting the Soviet bloc's transition to a free market economy without letting that support seem paternalistic. They needed help but wouldn't grovel for it. Neither would they negotiate away their empire. As Gorbachev had told Mulroney, the notion of a Polish or Hungarian departure from the Warsaw Pact was "not on," adding, "Any change in the alliances would be seriously destabilizing." He was even clearer on the most destabilizing issue of all. Asked about German reunification, Gorbachev answered simply, "Not in my lifetime."

"They want to be like us," the Canadian leader briefed Bush when the two met in advance of the latter's departure for Malta, "but not have it lorded over them." Mulroney had flown down from Ottawa just for dinner. Bush found him an engaging sounding board and messenger. If it looked as if the Americans intended to capitalize on the Soviet Union's troubles, he reported Gorbachev arguing, expanding their influence rather than ensuring stability or helping *perestroika* succeed, the consequences could be cataclysmic. But there remained a sense of optimism amidst the Soviet leader's gloom, Mulroney ultimately told Bush. "If you take care, yet lend a hand, you and Gorbachev will walk together into the history books."

Bush had been hearing reports of Gorbachev's precarious position for months, but Mulroney's conversation proved newly troubling, revealing his opponent's deep sense of insecurity. It was the same basic insecurity that had long driven Russian leaders: a sense that they did not fully belong to Europe, or rather that Europe would never fully accept them as equals. He already intended to avoid anything that resembled a "we won, you lost" attitude, Bush made clear, with Baker adding that they must pay particular attention to how they phrased any discussion, employing terms like "technical cooperation" rather than "technical assistance," given the latter's connotation of charity. The Americans would continue to make sure "Gorbachev understands that we will not conduct ourselves in ways to make life more difficult for him," Bush promised.

Yet what was all this talk of values? Gorbachev seemed to want something more, Bush mused, something he could not quite put his finger on: some form of American help that neither man could as yet fully define. Bush had already voiced his support for *perestroika* and had tried to avoid gloating about Soviet problems. Why couldn't he simultaneously, though in his own cautious manner, celebrate Eastern Europe's long-awaited turn to democracy? America stood for democracy. He stood for democracy. Hadn't his side and its values in fact won? How could Gorbachev in one breath praise self-determination yet in the next deny Eastern Europeans the right to choose their own alliances? In short, what kind of world were they inhabiting where a Soviet leader could laud an entire region's turn toward freedom but an American president was criticized for doing the same?

None of Bush's advisers could answer these questions or predict which version of Gorbachev might show up in Malta: the visionary who inspired crowds, the anxious politician who could not keep shelves stocked at home, or the angry Kremlin autocrat determined to safeguard Soviet interests. Gorbachev's own history of arriving at conferences with surprise proposals contributed to their anxiety.

This time, though, Bush planned a surprise of his own. He'd promised there would be no agenda to their meeting. On his desk lay the opposite, the fruit of his White House's work over the previous month: a full-scale response to Gorbachev's long-standing plea that Bush do more than just talk about his support for *perestroika*. "We're going to beat the timidity rap once and for all," chief of staff John Sununu crowed, his staff massaging Bush's presentation right up until the moment they placed it in his briefcase.

None of this was why Bush couldn't sleep while flying toward Malta, however. News from the Philippines kept him awake. Corazon Aquino was facing an armed coup, her fifth mutiny in the past three years. Early reports filtering into Air Force One suggested this was the most serious yet. Opposition units were driving toward the presidential palace. Rebel-piloted planes circled above. Aquino was pleading for help. Desirous of supporting the fledgling democracy yet reluctant to intervene — and aware that his inaction in Panama had led to disaster — Bush ordered his air force to scramble fighters over Manila, seizing control of the skies by sheer intimidation while demonstrating which side the Americans backed.

The show of force worked. With rebel pilots unwilling to risk flying, the coup fizzled, though not before leaving Bush with bloodshot eyes and a palace intrigue of his own brewing at home. Vice President Dan Quayle had called a full meeting of the National Security Council in Bush's absence in order to deal with Aquino's request. Dick Cheney refused to attend. Quayle lacked the authority, Cheney said. The vice president wasn't even in the chain of command. Lawrence Eagleburger shared Cheney's qualms yet drove to the White House to represent the State Department nonetheless. Bob Gates was already there. He rarely left the building anyway, and thus was hard-pressed to explain why he shouldn't consult with the vice president in a crisis. "I don't actually know what happened back in Washington," Scowcroft later recalled. "I had phones up to each ear. Everyone was calling with their version of events." He frankly didn't care who had the authority to call what meeting so long as his boss had all the information he needed. "It was a tempest in a teapot," Scowcroft later concluded, but one whose revelation might further undermine the administration's already spotty crisis management record, and whose adjudication consumed precious time that would otherwise have been spent on last-minute preparations for Gorbachev. Or merely resting.

The real tempest raged outside. Dark clouds built in the distance as the American delegation made perfunctory calls on Maltese officials after landing. Steadily strengthening gales ripped flags from their staffs. Torrential rains soon followed, then flooding. Locals called it the worst storm in recent memory. Soviet and American warships, anchored and ready to host their leaders, bobbed violently in the heaving seas. Most miserable were the teams of Secret Service and security personnel assigned to patrol around the vessels in flimsy rubber dinghies. These were the toughest waves he'd witnessed in his twenty-five years in the navy, the captain of the USS *Belknap* confirmed as he welcomed his commander in chief aboard. Bush had been looking forward to reliving some shipboard memories, meeting the cruiser's crew, and perhaps even fishing off its bow before hosting the Soviet delegation for dinner and their second day of scheduled talks. Yet so far Malta was shaping up to provide nothing like the sun-drenched Mediterranean paradise they'd all envisioned when Bush's brother had first floated the locale as the site of a "shipboard summit."

Reporters were calling it the "seasick summit," their cameras onshore barely able to make out the rain-soaked warships in the harbor. White House staffers vigorously denied rumors that Baker had been seen vomiting. The Soviets heard it was Bush who had been sick. Neither would admit it if true, though Bush did place a seasickness patch behind his ear as a precaution. Gorbachev fared worst. Uneasy on the water in the best of circumstances, he took one look at the heavy seas and refused to leave the relative stability of the docked Soviet cruise liner *Maxim Gorky*. If the storm persisted, they might have to cancel the meetings altogether, the ship's captain warned, or at least hold the sessions on land. He couldn't guarantee their safety otherwise. The Soviets quickly readied the *Maxim Gorky* as a substitute, scrambling in particular to locate a table long enough to accommodate both delegations. When none could be found, they tossed a single green tablecloth over two adjoining tables in the ship's card room, hoping no one would notice that one half stood several inches shorter than the other.

Few of the participants cared. They'd come too far — not just in miles but in months — to see their meeting washed out. Mid-morning on December 2, Bush, Baker, Scowcroft, and staff clambered into a small launch perilously tied to the *Belknap* for the short trip across the bay. Dipping out of sight of land with every swell, the boat nearly capsized more than once. Baker looked pale. Sununu grasped the launch's rails with both hands. Everyone gripped tightly to the metal stairs as they finally staggered aboard to be met by Gorbachev, Shevardnadze, and a bustle of Soviet aides. "It was the damnedest weather you've ever seen," Bush recorded that night in his diary. Shaking the rain from his overcoat, he realized with satisfaction that he'd already met every senior Soviet official in the room. There was thus little need for introductions, and with a sweep of his arm, Gorbachev invited both sides to sit, their papers overlapping across the small table. "It's so narrow that if we don't have enough arguments," he quipped, "we'll kick each other."

Bush was their guest, Gorbachev offered, and had also proposed their meeting. Perhaps he'd like to speak first? "With your permission," Bush began, his voice cracking and hands clearly shaking, "I'd like to put some ideas on the table." Scowcroft couldn't remember ever having seen him so nervous. There is something important to be said, Bush stammered, before finding his voice and stating: "The world will be a better place if *perestroika* succeeds. I know you had some doubts in New York" the last time the two met. "But you are dealing with an administration and, for the most part, a Congress that want to see you succeed."

The man who came to Malta boasting "no agenda" then proceeded to speak nearly uninterrupted for more than an hour, outlining seventeen discrete proposals designed to reinforce his opening statement. They included revision of trade restraints, with a promise to waive the congressionally mandated link between Soviet emigration policy and bestowal of most favored nation trading status; expansion of U.S. technical economic cooperation (not, it should be noted, assistance); support for Soviet observer status in the General Agreement on Tariffs and Trade (GATT) talks; and the promise to explore expansion of export-import credits to Soviet firms.

"Gorbachev's face lit up like sunshine," Marlin Fitzwater reported. It was precisely what he'd wanted to hear, and his response was precisely the reaction the Americans had hoped to achieve. "It's the end of economic warfare," Gorbachev's senior expert on the United States recalled thinking. Others on his staff thought they'd reached a milestone. "For the first time the Americans made a commitment to give economic support to perestroika, to our reforms," Chernyaev wrote.

It wasn't much of a package in strictly financial terms, but the sum total of Bush's proposals mattered far less than their collective symbolism. "We all knew that none of them significantly amounted to much," Robert Gates recalled. "But, taken together, we thought they represented a significant political investment in Gorbachev and perestroika." It wasn't charity, Bush carefully stressed. "These steps will not be presented as the superiority of one system," he told the Soviets. "I am not making these suggestions as a bailing out. That is not the spirit I came here with."

Scribbling in a small orange notepad throughout Bush's presentation, Gorbachev mostly nodded and smiled, but not to everything. American support came with strings attached. In exchange, Bush continued, the Soviets must further demonstrate their commitment to reform, improving human rights, working toward new arms control agreements, and ending what he termed "the single most disruptive factor" in Soviet-American relations: Moscow's continued support for Cuba and other communist forces in Central America. "The poor guy [Cuban leader Fidel Castro] is practically broke," Bush exclaimed. "The best thing would be if you gave him a signal that it would no longer be business as usual."

"I am going to finish, not filibuster," Bush promised. Their conversation was rapidly becoming a monologue. "No problem," Gorbachev nonchalantly replied. "You are doing it in a businesslike, direct, American way." Continuing farther down his list, Bush touched on the need for shared environmental efforts and more student exchanges, and suggested they jointly promote Berlin as site of the 2004 Olympics as a sign of their newfound solidarity. Drawing a deep breath, he finally deadpanned, "This is the end of my non-agenda."

Gorbachev was clearly pleased by what he'd heard. "It shows that the Bush administration has already decided what to do," he loudly proclaimed, and he was glad to hear it. "The emphasis on confrontation based on our different ideologies is wrong," he said. "We had reached a dangerous breaking point, and it is good that we stopped to reach an understanding."

"I want to begin my remarks," he opened, warming up for an extended monologue of his own, by saying that "in the American political community there is still one idea very present. It is this . . . The policies of the Cold War were right; those policies should not change. The only thing the U.S. needs to do is to keep its baskets ready to gather the fruit." What he had heard already that day convinced him that Bush did not share this triumphalist view, Gorbachev said, pleased in particular that Bush had brought "steps," rather than mere words, to demonstrate his goodwill.

But while steps mattered, the Americans had to change their attitude as well. "The U.S. has not entirely abandoned old approaches," he noted. "I cannot say we have entirely abandoned ours," though through *perestroika* they at least were trying. "Sometimes we feel the U.S. wants to teach, to put pressure on others." Maybe this one meeting would not be enough to ensure that American leaders learned this lesson, he said, while slapping the table for emphasis, "but we must understand this fundamental point." There would be plenty of time over the next two days for "details to be filled in." First they had to agree that the Cold War's end was everyone's triumph, not just Washington's. He was not there to surrender; neither were they there to proclaim an American victory. Their conversation could not proceed otherwise.

The meeting could have quickly deteriorated at that moment. Participants described the mood as friendly but not warm. Nerves remained taut. Gorbachev made plain his intent to refuse any overtures cloaked in American superiority. The scene offered a remarkable contrast: the one who harangued held the weaker hand, while the more powerful of the two proffered olive branches.

Bluster was Gorbachev's best option. The Kremlin still retained awesome power, including the ability to bring Europe's democratic transformation to a screeching halt. Bush offered ways to help, but what Gorbachev really gave in return was the promise not to hinder. He could backtrack on his nonintervention promise, and still had plenty of troops in East Germany. Gorbachev did not know precisely where he hoped European affairs might lead next, but by raising the specter of further conflict, he made sure the Americans knew that the Soviets intended to remain relevant.

Bush moved quickly to defuse the tension, citing his own behavior as proof that he understood Soviet sensibilities and Gorbachev's plight. Holding the stronger hand, he had the luxury of conceding on matters of pride. "We have not responded with flamboyance or arrogance that would complicate USSR relations," he pointed out. "I have been called cautious or timid," Bush said, "but

I have conducted myself in ways not to complicate your life. That's why I have not jumped up and down on the Berlin Wall."

"Yes, we have seen that, and appreciate that," Gorbachev responded, though he hadn't yet finished making his point. Just yesterday Bush had displayed the typical American arrogance that so troubled the rest of the world. He'd intervened in the Philippines, unilaterally deciding another country's fate. Imagine the outrage if Moscow had done the same! Bush even criticized the Soviets for aiding allies in Central America but said nothing of Washington's heavy-handed domination of the region. This was precisely the sort of hypocrisy the Americans tolerated only in themselves. "If your position doesn't change," Gorbachev warned, "ours won't" either.

Gorbachev didn't see why Nicaragua was so important to Washington in the first place. It will have elections, he said, and then form a new government. The United Nations could monitor the vote. What happened there shouldn't be a concern for either superpower if they truly cared about self-determination. And as for Cuba — the Soviet leader laughed — well, no one can really give orders to Castro, "absolutely no one," as the Cuban had "his own views on perestroika."

Their conversation continued in this vein for the rest of the day, alternating between moments of tension and levity, through a group lunch and then a lengthy private session for the two leaders and their closest aides (Scowcroft and Chernyaev). The weather might not have been perfect or the venue as planned, but the day ultimately offered precisely what each desired: an opportunity to speak candidly, and to ensure that even if they did not agree on every point, they at least had listened.

It also gave each the opportunity to spring surprises. Castro wanted to normalize relations with the United States, Gorbachev revealed. "His very words: find a way to make the President aware of my interest in normalization."

Bush wasn't expecting this. He'd hoped to put Gorbachev on the spot with a discussion of ongoing Soviet support for Cuba, not to have his country painted as the intransigent one. "Let's put all our cards on the table about Castro," Bush responded. America's allies couldn't understand why the administration cared so much about Latin America, but it was a "gut issue" for conservatives in his party and for the influential Cuban American lobby. Moreover, "I am convinced he [Castro] is exporting revolution," said Bush, disrupting fledgling democracies throughout the region. A region, he need not mention, that Americans typically considered their own strategic backyard. "Castro is like an anchor as you move forward and the Western Hemisphere moves," Bush continued. "He is against all of this — Eastern Europe and the Western Hemisphere." The same could be said for Daniel Ortega in Nicaragua, while Manuel Noriega in Panama was a problem of a different sort, an "open cancer" on the region. "I'm seeking to get him out," Bush mentioned casually.

"Let me tell you how your steps are perceived in the Soviet Union," Gorbachev quickly interrupted. "People ask are there no barrier [sic] to the U.S. action in independent countries?" If the Soviets shouldn't intervene in Eastern Europe, their sphere of influence, why didn't the Americans have to abide by the same rules? "The U.S. passes judgement and executes that judgement," Gorbachev said. This was the very arrogance he was there to rail against.

Bush was taken aback, not fully appreciating that he had fallen into his counterpart's logical trap. Having committed to the freedom of choice within his own empire, Gorbachev demanded the Americans do the same within theirs. But here the truth of Bush's position revealed itself, as did the difference between their positions and their worldviews, and the disparity of hard power between them all boiled down to a single point: Gorbachev yearned for all sides and perspectives to be equally respected and openly questioned everything his state stood for and believed. Bush already knew his side was right. "If we are asked for help against the scourge of drugs," Bush replied, explaining his position on Panama and the region more broadly, "we will help." The same could be said for the Philippines. "Aquino is democratically elected and asked for our help," he said. "It never occurred to me that this would cause problems in the Soviet Union, though I probably would have done it anyway." At that moment Bush had little interest in issues of irony or hypocrisy. Allies asked for help, and he had provided it. That is what one does for one's friends.

This was also, however, the same argument used over the previous two generations to justify Soviet interventions in Eastern Europe. Fraternal allies needed help against insurgents; the Kremlin provided it. "In the Soviet Union some are saying the Brezhnev Doctrine is being replaced by the Bush Doctrine," Gorbachev quipped, leaving Bush somewhat flustered in search of a response. "I want to understand," the president pleaded. "Here is a democracy saying that it needs help against the rebels." Why wouldn't the United States help?

Gorbachev was happy to explain, though it was still unclear where he expected a victory in this debate to lead. "In Eastern Europe there are governments, legitimately elected, that are now being replaced," he offered. "The question is in Eastern Europe it is prohibited for Soviet troops to intervene . . . Some now are seeing that we are not performing our duty to *our* friends." If the doctrine of nonintervention was good enough for the Soviets, why couldn't the Americans pledge the same? "All is now interrelated," he said. "Our position is non-interference. The process of change can be painful but we believe in non-interference."

Bush showed signs of rapidly losing patience with the conversation. If both sides held fast to the notion of peaceful change, he said, his ire rising, superpower tensions would no doubt diminish. But both men knew the world was ultimately not that simple. "I would never give advice to a senior office holder like you," Bush said, bluntly implying that Gorbachev should afford him the same courtesy. "I can accept your criticism, but not on this issue. Aquino is struggling to bring democracy. I would hope your criticism could be muted. I can accept it but I think your criticism would cut the wrong way."

But Gorbachev wouldn't let go. He loved a good debate. Latin America was only the example. Eastern Europe was where this point really mattered. "What I dislike is when some U.S. politicians say unity of Europe should be on the basis of Western values," argued Gorbachev, especially as this celebratory sense of its unbridled expansion cut to the heart of the potential East-West divide. "Mr. Kohl is in too much of a hurry on the German question," Gorbachev said. "His approach could damage things." There were still too many unknowns to make any good decisions on Germany, he continued. "For example, would a united Germany be outside alliances or within NATO?" You and I are not responsible for the division of Germany, Gorbachev told Bush. Ultimately, they should "let history decide what should happen."

They should keep things just as they were. When Gorbachev and his cadre said "let history decide," a phrase they returned to frequently over the ensuing weeks and months, they meant retain what history had already provided. They had inherited two Germanys. Two was the number they should pass along.

That was his last major word on Europe for the day. Lunch featured a broad discussion of Soviet economic problems. Bush and Baker strove not to lecture, holding back even when Gorbachev contended that there was no real private property in the United States. "Perhaps only on a family farm," he said. Stockholders owned the rest. The group broke for the afternoon, with the Americans departing for a planned respite on the *Belknap* despite Soviet pleas that they use the horrific conditions outside as a reason to stay and continue discussions within. They would return at 4:30 for another session, Bush promised, followed by dinner for all back on the *Belknap*.

It was the last they saw of each other that day. Tossed by the waves, the small launch carrying Bush and his staff required more than a dozen passes before successfully tying up alongside the American flagship, breaking the ship's starboard landing platform in the process. Neither the *Belknap*'s captain nor the Secret Service would sanction another trip out into the storm, leaving the leaders of the world's two great superpowers a mere few hundred yards apart yet unable to meet. Deep in thought and frustration, Bush stared out through the rain at the *Maxim Gorky* for more than an hour, walking the deck despite the torrent, his only means of calming his nerves. Even the cruiser's communications system faltered in the storm. "We were truly isolated," Scowcroft recalled, and "I said a small prayer that no crisis would explode that night" before joining the president's staff for the luxurious dinner originally planned to show off American hospitality. The ship's crew ultimately enjoyed the swordfish and lobster intended for Soviet palates. Was the weather ruining the summit? reporters asked the next day. "Hell no," Bush yelled back, perhaps a bit too forcefully to be believed. "Hell no!"

TALKS RESUMED the next morning on the *Maxim Gorky*, the seas still too rough for Gorbachev to venture from the dock. He nonetheless took the offensive from the start. Bush asked about nationalist sentiment in the Baltics. There was little to discuss, Gorbachev replied. He was sympathetic to the desire of any people for autonomy, but not at the cost of his nation's integrity. Secession was simply not an option. "I must not create a danger to perestroika," he told Bush, and would do what he had to in order to keep his movement, and his union, alive. "This is a sensitive issue for us," he said. "I hope you understand our position . . . [I]f the U.S. has no understanding it would spoil relations with the U.S. more than anything else."

It was very much like Deng's words of warning before Tiananmen. The Chinese would do whatever proved necessary to maintain public order, Bush had been told repeatedly, and would brook no interference in their internal affairs. There was nothing he could do to stop any ensuing carnage. This was the essence of Gorbachev's message as well, that *perestroika* demanded respect for Eastern Europe's autonomy but did not require him to disband his own country or to heed anyone's advice about how to operate within it. "But if you use force," Bush protested, "that would create a firestorm." Gorbachev responded nonchalantly. "We are committed to a democratic process and we hope you understand," he said, once more echoing Deng, but "destructive forces should not be allowed to undermine" *perestroika* and all that had been achieved. Some might disagree with his agenda, he acknowledged. But "we are right."

The second day featured few of the rhetorical fireworks or potential flash-points of the first. When talk turned once more to Gorbachev's critique of "Western values," Baker ultimately defused the entire potential argument with simple wordplay. "You have said governments should choose their form," he offered. "We agree as long as people can choose their governments. This is what we mean by Western values — not that there should be specific forms involved." Let us thus simply refer to everything both sides desired as "democratic values."

Gorbachev accepted the clarification, though nomenclature mattered less in the end to the Soviet side than their having raised the issue as a signal of their remaining power even in their moment of material weakness. "Western values — Western strength," Shevardnadze interjected. "Some are saying it is because of Western strength" that Europe was changing, but might should never be confused with right. "Eastern Europe is changing to be more open, democratic, and to respect human values," Gorbachev added. This was all to the good, but "it is dangerous here to try to force the issues."

The winds eventually subsided, and the meeting closed with Bush and Gorbachev jointly fielding reporters' questions. It was the first time an American president and a Soviet general secretary had conducted a joint press conference. The image of the two leaders sitting together was to both of them well worth the difficulty of the previous two rain-drenched days. They had gained "a deeper understanding of each other's views," Bush said, adding that there was "virtually no problem in the world, and certainly no problem in Europe, that improvement in the U.S.-Soviet relationship will not help ameliorate." Gorbachev nodded. "I share the view voiced by President Bush that we are satisfied, in general, with the results of the meeting." It was the best way to describe what had happened. They were not suddenly friends as a result of their hours spent together. Neither had they solved all that potentially divided them. But both were satisfied.

Little that can be measured changed in Soviet-American relations as a result of the Malta talks. There were no dramatic breakthroughs on arms or human rights, and certainly no shared path forward for Germany. Gorbachev didn't want to change the status quo. Bush believed the opposite. "We cannot be asked to disapprove of German reunification," he argued. "I realize it is a highly sensitive subject" for the Soviets, he conceded, but ultimately it was not a matter for anyone but the Germans to decide. This was, Bush stressed, the essence of the "democratic values" they'd both endorsed.

Even if little was resolved, Malta mattered. It gave each man a truer sense of the other. Upon returning to Moscow, Gorbachev told his staff that back in 1987 he'd thought he would be able to "trust" Bush. The pause had raised doubts, but "in Malta I became firmly convinced that my original instinct did not deceive me." Bush felt much the same way. He did not always agree with Gorbachev's conclusions, but at least now he better understood his thinking. "He's really got a problem with anything that makes it look as though we're on the march while he's on the retreat," Bush told his staff when he got back to Washington. "I sort of knew that" before, "but hearing it from *him* was somehow still different."

What each man had heard most clearly was the other's limits. For Bush it was the prospect of violence in the Baltics or beyond, as well as continued Soviet aid to Latin America's communists. Either would derail Soviet-American relations. For Gorbachev it was German unification and anything that reeked of a dictated peace in Europe. Indeed, because there were no winners or losers in his view, there was no reason for either side to retreat from its current sphere of influence. "It would be a mistake" in particular for the Americans to withdraw back across the Atlantic, Bush felt, especially given the uncertainty over both Germany and the European Union. But that did not mean he wanted the Americans to dominate, either.

"I think the answer is self-determination" for Germany, "and then let things work out," Bush told Kohl over dinner on December 3. "Then avoid things which would make the situation impossible for Gorbachev." He'd flown to Europe to brief his NATO allies on his Malta discussions, though Kohl was the only leader granted a private session. Together they needed to bring Britain and France along, Bush explained, so Kohl could achieve the unification he'd long desired, and so he could save NATO and thus Washington's entree into Europe. It wouldn't be easy. "Great Britain is rather reticent," Kohl noted. "That is the understatement of the year," Bush agreed.

The anxiety of the previous week, indeed of the previous tumultuous months, finally burst through the next day when Bush once more faced the press. "Why do the West Europeans need us once the military threat recedes?" an American reporter asked. More profoundly, absent the Soviet threat, "why should there have to be a NATO?"

"You mean why will there always have to be a U.S. presence?" Bush responded, trying to understand the query. It wasn't what the reporter had asked, but his every thought as to Europe's future circled back to the same question. "Well, if you want to project out 100 years, or take some years off of that, you can look to a utopian day where there might not be. But as I pointed out" to both Gorbachev and to his NATO allies, he continued, "that day hasn't arrived; and they agree with me. And so, the United States must stay involved . . . and Mr. Gorbachev understood that. He made that point to me."

Does this mean the Cold War is really over, then? another reporter asked. Gorbachev had said as much. "Do you agree with him that the Cold War is over?" he probed.

"We're fooling around with semantics here," Bush replied, gradually losing his composure as exhaustion finally began to take its toll. "I don't want to give you a headline. I've told you the areas where I think we have progress. Why do we need to resort to these code words that send different signals to different people? I'm not going to answer it!"

Gorbachev did, the reporter interrupted.

"Well, good," Bush responded, his temper flaring. "He can speak for himself in a very eloquent way. But in terms of if you want me to define it, is the Cold War the same — I mean, is it raging like it was before in the times of the Berlin Blockade? Absolutely not. Things have moved dramatically. But if I signal to you there's no Cold War, then it's 'what are you doing with the troops in Europe?' I mean, come on!"

BEFORE THE MIRACULOUS YEAR of 1989 came to an end, Bush had two fewer dictators to contend with. The first he deposed. The second met the fate he'd always feared. The demise of Manuel Noriega's rule began on December 16. A U.S. marine lay dead, shot at a Panamanian checkpoint. A young U.S. Navy lieutenant and his wife fell prey to Noriega's henchmen as well. He was shackled and beaten. She was harassed and, according to reports given to Bush the next day, most likely raped. It could have been Barbara from fifty years before. He'd been a young navy lieutenant once. "We've had enough," he told his diary in response. "We cannot let a military officer be killed, and certainly not a lieutenant and his wife brutalized."

Twenty-six thousand American troops landed in Panama in response, striking twenty-seven targets simultaneously on December 20. Noriega had survived coups, assassination attempts, and years in the drug trade. He fell quickly before overwhelming military force intent upon his capture. The men who'd bled and died as young officers in Vietnam and who now held the reins of power at the Pentagon were not about to waste time with piecemeal efforts when they had firepower on their side. "My conviction — that you go in to win — was shaped in small encounters," Joint Chiefs chairman Colin Powell later wrote. "Go in with a clear purpose, prepared to win — or don't go in at all."

Noriega sought sanctuary in the Vatican embassy. The Americans could not follow. But they could subject the building to a massive audio assault. Psychological warfare experts believed no one could withstand forever hard rock music blasted at unbearable levels, denying anyone for blocks the ability to think or sleep. They clearly enjoyed their work, or at least its comedic potential. "Wanted Dead or Alive" by Bon Jovi led directly to Jimi Hendrix's "Voodoo Child," to be followed by Guns N' Roses crying for "Patience," all played at volumes capable of drowning out a 747 on takeoff. Noriega lasted ten days, far longer than most of his troops and subordinates, though his capture ultimately cost twenty-three American military lives, over five hundred Panamanian fatalities, and the widespread destruction of residential areas that left tens of thousands displaced and homeless.

On January 3 he surrendered. Unwilling to violate the church's commitment to sanctuary, yet wholly uninterested in hosting a despot indefinitely, Vatican officials simply explained to him their intention to move everyone in their official embassy to the Catholic high school across the street, making that new building their home and legal territory. Noriega was not invited. He could wait for the Americans to arrive once the Vatican's sovereignty over the old building evaporated. Or he could take his chances with the crowds that had gathered just beyond the Americans' perimeter, chanting, "Death to our Hitler." Seeing no escape, Noriega attended mass, took communion, and walked to the front gate with a Bible in his hands. Moments later he was in a helicopter bound for an American prison. Most of what the Vatican officials had told him was untrue. Noriega had other options. But as the local monsignor put it, "I'm better [than him] at psychology." Better, he implied, than the U.S. Army, too.

For Bush, Noriega's capture was a relief. It was the first time he had ordered troops into harm's way. Forty-five years before, he'd been the one flying the mission. Now he was back in Washington while younger men wondered if they'd ever see home again. "I am thinking about the kids," he told his diary late at night on December 20. "Those young 19 year olds who will be dropped in tonight . . . So the tension mounts. They asked whether I would sleep; but there's no way I will be able to sleep; during an operation of this nature where the lives of American kids were at risk."

Baker was more sanguine, at least in retrospect, realizing that his president — and his friend — had just crossed a Rubicon. So too had the nation. "In breaking the mind-set of the American people about the use of force in the post-Vietnam era, Panama established an emotional predicate," he later explained, "that permitted us to build the public support so essential to the success of Operation Desert Storm thirteen months later."

Nineteen eighty-nine had been a year of revolutions and crushed dreams like no other in living memory. Force was used to quell disorder and impose stability in Beijing and in Panama City, albeit under very different circumstances. Force was not employed in Berlin or Budapest. Its use typically led to heartbreak. But perhaps, Baker mused aloud as Christmas approached, not always. There was indeed one instance in which he could conceive of Soviet military force in particular doing some good. "If the Warsaw Pact felt it necessary to intervene" in Romania, he told reporters the morning before Christmas, the United States and its allies, and this president in particular, would understand. Rioters had taken to the streets. Civil war beckoned, and "the Soviets have got the incentive and the capability to do something to stop the bloodshed." After decades of decrying the Brezhnev doctrine and fearing that Moscow might crack down with military force to quell change within its own empire, Baker was virtually pleading with Gorbachev to give just that order.

Nicolae Ceauşescu, at long last, had fallen. Romania's revolt began, innocuously enough, the week before Noriega's thugs shot and harassed their way into an American invasion. It began with Bucharest's order to remove and relocate a pesky priest from his parish in Timişoara, a small outlying city with a large Hungarian minority. Parishioners gathered to say no, something unheard of in Romania. Their defiant spirit spread. A dozen elderly people prayed inside the church on December 15, sending the Securitate away. Even hardened operatives balked at shoving aside grandmothers. Forty people gathered that night upon hearing the news. By the next day there were several hundred, all ready to offer something they'd thought impossible until that week: their own voices. By the time the authorities realized the extent of the problem, two thousand were clashing, chanting for something other Soviet vassals already seemed to have received: "Freedom!"

The cries were infectious and, in some circles, terrifying. "Kill [the] hooligans, not just beat them," Ceaușescu ordered. His wife insisted, "Not even one should see the daylight again." Promising serious consequences if his police chiefs and generals did not halt the rioting, the pair left as planned for a state visit to Iran on December 18. He needed Tehran's oil and gas, which Gorbachev refused to export. Romania had plenty of both beneath the ground. It had once largely fueled Hitler's panzers. But that was before the communists took over the refineries.

The protests only grew in his absence, however, erupting throughout the country. Crowds in Cluj offered a textbook example of what might go wrong, and of what had been so narrowly avoided elsewhere in Eastern Europe thus far. Protesters chanted anti-Ceaușescu slogans. Police appeared. A young man stepped toward their line, daring them to shoot. Someone did. The first shot was most likely an accident, a misfire from one of the officers in charge whose sidearm went off in a scuffle. Panicking, he ordered the rest to fire. By morning, twenty-six were dead and hundreds wounded, all because someone yelled and someone else slipped.

The protests swept into Bucharest by December 20, prompting Ceaușescu's premature return from Iran. He addressed the nation that night, calling upon loyalists to gather the next day in front of the presidential palace to defend their beloved regime. He planned to appear to cheers, as he'd ordered so many times before, the image beamed to his supporters throughout the country. As before, the first ten rows would be filled with plainclothes Securitate officers, ready to applaud and sing on cue, all of them disarmed lest any get the wrong idea.

Only this time . . . the crowd booed. First it was a whistle. Then a hiss. A firecracker went off, startling the despot and the throng alike. A look of fear passed across his eyes, simultaneously flashed through tens of thousands of television sets into homes and bars in every province. He was, at long last, afraid. "Be calm! Be calm!" his wife shouted. Then the screen went blank. The picture returned two minutes later showing Ceaușescu standing alone, but the seed of doubt had already been sown. As historian Gail Stokes noted, "Romanians all over the country, some of whom had begun to demonstrate in their own cities, saw the 'genius of the Carpathians' waver, and they smelled blood."

Nicolae and Elena retreated into their palace. Troops manned barricades throughout the city. Protesters erected their own. By evening, police with riot shields and sticks roamed the streets, covered by army snipers who fired freely and without discrimination. Forty-nine people lay dead by morning. Nearly five hundred more were wounded, even more arrested. News reports the next morning told of the defense minister's death. He had refused Ceaușescu's order to fire on the city. Rumors swirled that he'd committed suicide. Or had been shot in cold blood for disobedience. The army largely stood down in confusion, with some units ordered to their barracks lest they join the crowds, others simply uninterested in being Ceaușescu's pawns any longer. Nicolae and Elena soon fled toward waiting helicopters. Sporadic fighting nonetheless continued for days, featuring loyalists and fanatics unwilling to concede to each other, and personal vendettas masquerading as justice for those who'd long felt oppressed. Unconfirmed reports told of massive bloodletting, largely exaggerated but widely feared at the time, with casualties rumored to reach into the tens of thousands.

The spark that had not caught fire in Germany, Poland, or Hungary seemed to have finally ignited within one of Eastern Europe's darkest states. Six thousand miles away, Baker could do little more than wonder if perhaps Moscow might finally have cause enough to impose order. "We would support efforts to assist the Popular Front," he stated while appearing on *Meet the Press*. His words echoed Dean Acheson's from two generations before. "They are attempting to put off the yoke of a very, very oppressive and repressive regime," Baker emphasized. Would he support Soviet intervention? the host asked. "That would be my view, yes."

It was neither an idle concern nor an isolated idea. In Paris, Foreign Minister Roland Dumas told his Soviet counterpart that if the USSR "thought it useful to intervene, France would help." A Dutch spokesman echoed the endorsement, though noting the "enormous dilemma" the issue caused, given bitter memories of past Soviet occupations in Eastern Europe, and Romania's historic fanaticism over its own independence. Romanians faced "action orchestrated by circles . . . who want to halt socialist construction and again put our people under foreign domination," Ceaușescu had argued in one of his last statements to the crowd. "Better to die in full glory than again be slaves in the ancient land."

Intervention might be the only thing that could save him, forcing his minions to choose between subservience either to their own despot or to foreigners. Closer to the action in Moscow, though still hundreds of miles away from the bloodshed, Ambassador Matlock asked his Soviet contact on December 24 if perhaps it was indeed time for the Kremlin to send in its own troops before the fighting spread. For humanitarian purposes, of course. "We do not visualize, even theoretically, such a scenario," Matlock was bluntly told.

Shevardnadze was even clearer, calling the idea "stupid." *Perestroika* relied on the principle of noninterference. It could not be violated, even for righteous causes. "We stand against any interference in the domestic affairs of other states and we intend to pursue this line firmly and without deviation," Soviet deputy foreign minister Ivan Aboimov insisted. Having just removed their last bloodied and wearied troops from Afghanistan, Soviet officials also had little desire to engage nationalistic fanatics defending their own land again.

The Ceauşescus were already effectively dead in any event. Captured by opposition forces, they were executed on Christmas Day. Their "trial" lasted less than an hour, its verdict certain from the start; hundreds volunteered to serve on their firing squad. The paratroopers eventually afforded the honor pumped more than 120 bullets into the couple the instant their commander gave the order. Rebel leaders had hoped to televise the entire execution, but the deed was done before cameras could be erected. Photos of their lifeless corpses began circulating within hours. The execution "left a bad impression on me," Shevardnadze later coldly noted.

Halfway around the world, American troops were storming a Latin American nation ruled with another iron hand, albeit intent on capture rather than assassination, and the irony of the moment was lost on no one. Soviet troops stood pat, refusing to violate a despot's sovereignty. Bush, meanwhile, the great advocate for democracy, ordered thousands into battle because, after months of frustration, he had finally heard enough when a young American officer's wife was brutalized. It was, to be sure, the straw that broke the camel's back, and tensions between Washington and Panama City had been brewing for years. Yet what became known as "Operation Just Cause" was also in its execution, and provocation, an assault ripped straight from nineteenth-century imperialism's playbook, not one designed for the cooperative spirit of the eve of the twenty-first.

The invasion of Panama was "a special case," Matlock explained to Shevardnadze after the dust had cleared from both conflicts. Clearly uncomfortable with justifying the entire operation, he read directly from prepared talking points. Encyclopedic in almost all other respects, Matlock's memoirs fail to recall the incident or the ensuing discussion. Merely because the Americans had taken matters into their own hands to restore order and to remove an abrasive irritant, he argued, and even though they had hinted (strongly) that the Soviets do the same in their own sphere of influence, they should not take from Panama the lesson that force was ultimately a hegemon's right, even within its own strategic backyard.

Shevardnadze replied calmly: "We disagree with what you have done, but we would like to go on with our relationship. But how do we explain the situation to those who say that we are being taken advantage of?" A crisis would invariably develop in one of the republics, he later lamented to his deputy. Perhaps even in his native Georgia. Crowds would gather, chanting for independence and wildly employing phrases like "freedom" and "democracy," and no matter how many times Gorbachev promised to forswear violence, Bush had just shown that violence could work, even for democracies. Perhaps especially for democracies. Gorbachev's critics "would immediately say, 'why don't we use force to restore order?'" he predicted. "See what the Americans did in Panama? And that's a foreign country!"

"It seems," one of Shevardnadze's deputies told Matlock, "that we've turned the Brezhnev Doctrine over to you."

16

NOT ONE INCH EASTWARD

THE FOCAL POINT OF EUROPEAN POLITICS, and of European history, shifted fully to Germany's reunification as 1990 began, and in turn from the streets to the negotiating table. Nine months of laborious diplomacy ensued, though four moments effectively tell the tale: one in January, two in February, the last in June. By their end, each great power believed it had secured what it needed most. The Germans had one country; Europe had its path toward unity; the Soviets retained the prospect of fuller East-West integration; and the Americans held fast to their strategic foothold in Europe. One party ultimately felt betrayed, however, and this sowed the seeds of future conflict.

At the heart of these negotiations were competing visions for Europe's post–Cold War strategic architecture. Might it be Gorbachev's "common European home," in which cooperation proved more powerful than brute force? If so, there would be no need for military blocs and no reason for Germany to remain in NATO. The entire continent might look inward instead for its security, and finally toward Moscow just as a reinvigorated Soviet Union stood poised to exert its influence once more.

Or might George Bush's vision of a "Europe whole and free" prevail, in which the stream of history continued its inexorable flow toward democracy, safeguarded by revitalized American-led institutions and overseen by on-site American power? "There is almost euphoria about the declining threat and talk of a $60 billion peace dividend," he lamented to Margaret Thatcher in late January. Withdrawing in triumph might seem a good idea to many. Bush considered it naïve. America had to stay. "There are a lot of weirdos over here who have all sorts of crazy ideas," he complained. They talked of a peace dividend but sounded too much like isolationists of old, using the same tired phrases that had left his generation no choice but to sacrifice for their shortsighted mistakes. "Many of them don't want our troops in Europe at all."

Hoping to get ahead of the calls for retrenchment, Bush used his January State of the Union address to propose dramatic cuts of his own. The Red Army currently had 650,000 troops in the region, the Americans 300,000. They'd each retain 195,000 under his new proposal, "a level that will reduce the Soviet threat," he told Thatcher, but "not cause the Alliance to unravel." It wasn't a step he relished taking, but "I want to get in front," he said, "so the charge is not made that we are oblivious to the exciting changes that are occurring in Central Europe." Indeed, it didn't matter if the Soviets accepted his proposal or went further, he wrote Gorbachev. The United States planned "to retain a substantial military presence in Europe for the foreseeable future, regardless of the decisions you take about your own forces." The past year had brought unprecedented and wholly unexpected changes, most of them wonderful, he told Thatcher. But "the jury is still out on the changes we may see" in the future.

Perhaps neither Bush's nor Gorbachev's vision would come to pass. German unification might instead bolster Europe's integration along its current Franco-German axis, leading to the ejection of both Soviet and American power. France's Mitterrand certainly hoped this might prove the case. "The system of alliances as it has functioned will be outmoded, outstripped in a few years," he told Brazil's president in early February. Even if Europe's ultimate liberation from outside influences took decades — and he considered both the Americans and the Russians outsiders — this could be the moment the continent's real emancipation commenced. "The Americans want to extend the Alliance to other subjects," one of Mitterrand's closest advisers argued, "confining to the Atlantic Alliance — where the American preeminence is a given — the global mission of leading the evolution of Europe." The French, by contrast, "should assert Europe."

A fourth vision hovered above the negotiations as well: the nightmare scenario. Failure to come to terms over Germany might send Europe back to the dangerous balance-of-power trap it had faced in 1913. East-West tensions would certainly reignite as a result, forcing Germans to exchange neutrality for the right to unify, or perhaps to have neutrality thrust upon them, leaving them unmoored and bitter. "We wanted to avoid singling out Germany," Baker's aide Robert Zoellick subsequently remarked, "having Germany unified in a way that some future generation of Germans would say, 'we weren't treated fairly and we have objections we have to rectify.'" In this scenario all the twentieth century's suffering would have been for naught, the continent returned right back to where it had begun.

One thing was certain: any settlement that seemed overly dictated by the West, and in reality by the Americans, might cut short the post–Cold War peace before it had ever truly begun. Gorbachev had warned as much at Malta. His own survival rested on the results. His "fate depends more on you" than on the radical critics or conservative hard-liners he faced every day, Mitterrand reminded Kohl at year's end. Only Kohl had the power to unify Gorbachev's opponents behind the one thing they all feared more than poverty or an outright end to socialism: a resurgent Germany. "Gorbachev knows this too," Mitterrand added.

The stakes therefore could hardly have been higher as the euphoria of 1989 gave way to the practical problems of managing an empire's collapse. "We are entering the end-game of the Cold War and your own role will be decisive," Scowcroft advised Bush. The finish line was in sight, but victory was far from assured. "When the end-game is over, the North Atlantic Alliance and the U.S. position in Europe [must] remain the vital instruments of peace and stability that we inherited from our predecessors." Anything less spelled disaster.

These issues came to a head in early 1990, as German unification finally became inevitable. More to the point, it became a necessity. The flow of refugees that had slowed at year's end, as East Germans paused to weigh Hans Modrow's promise of a new brand of socialism, became a torrent once more. Two thousand a day had departed by mid-January, leaving the state increasingly populated only by those too stubborn, aged, or tired to move. Modrow found he couldn't even fill key positions within his own government, which left him unable to carry out the regime's most basic functions or even pay its day-to-day bills. He hoped only to hold on until national elections in May. By mid-January, however, few thought the GDR would last that long. He moved the vote to March. That soon seemed far away.

There simply weren't enough qualified people left. Public services disintegrated. Hospitals could not staff their shifts nor schools their classrooms. Medical students were conscripted to see patients and soldiers to keep factories open. Garbage piled up in the streets. "Only an orientation to West Germany was a realistic alternative for us," Modrow reluctantly decided. The sole question that remained was whether he'd be able to negotiate confederation or give in to absorption.

Kohl reached the same conclusion. In December he'd promised that unification would take years, pledging to "reduce the speed" of his ten-point plan as long as the final destination was assured. A month later his foot was off the brake. "If the DM [deutsche mark] does not come to us," East German crowds chanted, "we will go to the DM." Kohl feared that the refugee flow might turn into a full-scale human crisis if calm was not quickly restored. "Everything leads the representatives of the Federal Chancellery to consider the evolution towards German unity as ineluctable," the head of the French Foreign Ministry's policy planning unit told Mitterrand on January 16, "and more rapid than initially foreseen."

Outright chaos seemed a real possibility, as did bloodshed. Infused with a spirit of radical democracy and disheartened by the broken hierarchy around them, army conscripts near Potsdam seized control of their barracks, demanding the right to elect their own officers. Armed maritime cadets forcefully ejected their commanders as well. Government workers went on strike throughout East Germany's smaller cities, some demanding higher pay, others insisting that Stasi members be removed from the payroll. Having ebbed in December, Leipzig's weekly rallies grew anew in the new year. "The majority in the crowd no longer presses for reform and democracy but for unification," Britain's foreign secretary warned Thatcher.

The once-peaceful crowds soon turned violent after newspapers uncovered plans for a counterrevolutionary coup. Exhorting the "armed organs of our common homeland" to "unmask and paralyze the hate-filled machinations against the organs of the state power," the call to arms — circulated by diehard loyalists in desperate hope that a silent majority of citizens still backed the GDR's core mission — seemed proof that the communist elite would not yet give up without a fight. Rioters stormed Stasi headquarters in East Berlin in response. Rumors swirled that they'd been let in by someone inside who'd had enough or, more ominously, was hoping to trigger a full-scale retaliation. Perhaps it was not too late to follow the Chinese model after all. Things seemed ready "to go completely out of control," Modrow openly lamented, with the masses eager to "become a law unto themselves."

KGB stations came under assault as well, raising the dire prospect that Soviet troops, unwilling to fire on crowds in defense of a decrepit regime, might ultimately be forced to do so in self-defense. "I have no doubt the Russians would obey an order to fire on the Germans," one of Baker's top aides had warned earlier. The sentiment echoed across the Atlantic. "We were prepared to defend ourselves against the crowd," a KGB colonel later confirmed. "And we would have been within our rights to do so."

Colonel Vladimir Putin meant what he said. Stationed in Dresden since 1985 and tasked with recruiting would-be operatives, the future Russian leader had hoped to be posted farther abroad. More promising officers won those coveted posts instead. Lacking enough resources to tempt even impoverished students to betray their country, he largely spent his days filling out reports. Exercise fell by the wayside. An extra twenty pounds settled around his midsection. Friends worried he was depressed.

The last few months had at the least proved exciting. He'd witnessed the previous summer's freedom trains passing through Dresden, and commiserated with local police charged with restraining the crowds that wished to join the exodus. In January he saw the local Stasi headquarters ransacked, before returning to oversee the preemptive burning of the KGB's prodigious files within his own. Heat from the ensuing fire broke the building's furnace, even as an angry mob gathered outside. Calling for reinforcements and ordering his troops to the rear of the building, Colonel Putin donned a sidearm and threw open the front door. "My comrades are armed," he told the would-be attackers, doing his best to make all of his five feet seven inches fill the doorway. "Don't try to force your way into the building."

It was one against hundreds. Yet something in his eyes made the mob doubt its odds. Dissipating in search of easier prey, they left Putin's station unscathed, though his faith was shattered. His calls for reinforcements had gone unheeded. "We cannot do anything without orders from Moscow," the voice at the other end of the line had replied. "And Moscow is silent." This unwillingness to come even to the defense of Soviet soldiers left Putin, already angered by the Kremlin's abandonment of a fraternal socialist regime, convinced that perestroika's proponents had lost their way and forgotten the Soviet Union's real security needs. Gorbachev, he concluded, simply wasn't strong enough. Not for Russia, anyway.

"It was clear the [Soviet] Union was ailing," he'd later write, but until that moment he'd not fully appreciated its "paralysis of power." Putin, his wife, and two young daughters would soon join the exodus of Soviet families heading back home, a used car, a twenty-year-old washing machine, and an intense sense of abandonment all they had to show for their five years in East Germany. "We had the horrible feeling that the country that had almost become our home would no longer exist," Lyudmila Putin recalled. "My neighbour, who was my friend, cried for a week. It was the collapse of everything" for the East Germans, "their lives, their careers."

The GDR's collapse in January happened too quickly for Germany's former enemies and victims to reconcile their fears with their new reality. Freely admitting that unification was "inevitable" by this point, Britain's normally controlled ambassador to Bonn nonetheless laughed in surprise when Kohl predicted a single Germany by 1995. That was still far too soon for London's liking. From Moscow, Gorbachev similarly warned Bush that "any haste, leaping over stages," or "overly categorical conclusions and assessments" on the question of German unity "can only result in chaos." Such sentiments broke out in public as well. "It is difficult to imagine an undivided Europe with a divided Germany," Czechoslovakia's Václav Havel told the Polish Sjem in January. "Just as it is difficult to imagine a united Germany in a divided Europe."

Havel said the same to Baker in early February, albeit in stronger terms. "Germany could not be neutral," the playwright turned president argued. But he also "couldn't imagine a united Germany in NATO." To his thinking, the very existence of military alliances, "remnants of the Cold War that kept them [Europeans] oppressed," meant the continent's ongoing division. Yes, the United States had saved Europe twice, Havel conceded, and its support had made the democratic revolutions possible. But as Baker informed Bush, "the Czech reformers want Europe to develop its own security protections over time," a vision that cut to the heart of the White House's worry that communism's collapse would ultimately force the Americans' expulsion from Europe as well.

In the end it was the opinion of the four occupying powers — the British, French, Soviets, and Americans, who retained legal control over Germany as no formal peace treaty had ever fully ended World War II — that mattered most. They alone could potentially veto unification. Bush was already on board, having early on offered support in exchange for a German pledge to remain firmly embedded in the Western alliance. "Germany's NATO membership" was the president's "unequivocal prerequisite," as one of Genscher's chief aides put it. So long as German leaders maintained this position, the Americans would champion their cause within any Four Powers discussion. "If America had so much as hesitated," recalled Genscher, by then foreign minister of a reunified Germany, "we could have stood on our heads and gotten nowhere."

The GDR's rapid collapse forced Kohl and those around him to wonder if they'd have the luxury of following the American lead, and what they expected would be Bush's cautious, even plodding leadership on the issue. To prevent real chaos, they might have to act sooner and give up more, especially if German membership in the alliance appeared to be the final barrier to unification. Soviet leaders had thus far refused to condone either unification or the GDR's wholesale absorption by a military organization whose sole purpose was to repel Moscow's advance, and would undoubtedly perceive its acquisition as a redrawing of Europe's strategic divide some three hundred miles closer to their own border. "I asked how one could reconcile German unity within NATO," Britain's ambassador in Bonn reported home on January 25 after a lengthy session with the German chancellor. "Kohl frowned and paused and said he could not answer this question now."

He'd promised Bush something quite different in December, but that was before the GDR's full-blown collapse. Kohl similarly wavered on the issue of NATO membership in mid-January when queried by reporters, leaving American policy makers desperate to shore up his commitment. Time, it seemed, was no longer on their side. "Kohl tended to take the line that we need not worry," Baker recalled, that "Germany's home was in the West." But he'd not feel comfortable until Kohl's promises had a bit more than mere words at their back. "We needed a framework approach," Baker argued, a structure to negotiations in order to keep the Germans both on schedule and in line. "We had a really strong and seamless relationship" with Kohl, he added. "He gave us his word. We gave him ours." When asked if the administration had a "plan B," however, in case Kohl made a sudden dash for unification without NATO, or lost out to another politician who would, Baker was blunt but honest: "Not for the United States."

None of the other Four Powers shared Washington's enthusiasm for a unified Germany, despite the quickening pace of East Germany's collapse. Thatcher continued to express her disquiet in the new year, publicly urging the Germans to place a "long view of Europe's needs" above their own "more narrow, nationalistic goals." Bush expected her grudging support in the end, reasoning that for all her dislike of the Germans, she valued the Anglo-American special relationship more and would do nothing to scuttle an outcome his White House clearly backed. "Mrs. Thatcher will undoubtedly support our position regarding Germany's full membership in NATO," Scowcroft's NSC informed Bush. She might hope to "slow down" unification but would undoubtedly come around. Indeed, Thatcher's own staff urged the same. "I am not in favor of using (and thus tying the PM to) the phrase 'slowing down,'" Foreign Secretary Hurd cautioned. "It puts us in the position of the ineffective brake," unable to thwart unification despite having poisoned both Anglo-German and Anglo-American relations, which "we should avoid as offering the worst of all worlds."

Bush made a point of calming her fears whenever possible, speaking to her frequently throughout the winter, stressing in particular that under his plan, no matter how strong the Germans grew, the Americans would remain there to watch over them. "You and I think the same," she finally conceded in February, at least on this crucial point. "Germany must stay in NATO, or all our security goes . . . [M]y fundamental philosophy is that you never know where the next threat will come from, and when it comes it will be too late to prepare for it, so you should always keep your defenses strong" and your primary ally close by. "When I am asked who our enemy is now," the president replied, "I tell them apathy, complacency." For him they were synonyms for American disengagement.

Thatcher also failed to build a coalition in opposition to unification, though not for lack of trying. She hoped initially to coordinate with the French, who had their own special relationships in mind when dealing with Germany. Mitterrand rebuffed her overtures. "I will not say 'no' to reunification," he declared. "This would be stupid and unrealistic," solving little but scuttling decades of shared Franco-German work on reconciliation and toward the new Europe. However distasteful, unification was a reality, he cautioned her, and the Germans were keeping close watch on who was on board and who opposed.

Mitterrand instead traded his support of unification for something he valued more: securing Kohl's backing for speedier monetary integration and thus tighter European bonds. For French policy makers, unification was a gift that kept on giving. They repeatedly bargained their continued support whenever Kohl appeared to hit a potential speed bump. "The government of West Germany would now have to agree to practically any French initiative for Europe," Kohl's personal foreign policy aide conceded. "If I were French, I would take advantage of that," which France most certainly did.

The Soviets ultimately proved the only one of the big four willing to mount a sustained objection to unification. They were, after all, the only ones for whom it represented an outright loss. "The GDR is our strategic ally and a member of the Warsaw Pact," Shevardnadze had said at year's end, and its survival was of paramount importance. Anything else would "destabilize Europe." If legalities and treaties were not enough to keep the Germans from defying the continent's political will, he then ominously added, the Four Powers also had "at their disposal a considerable contingent of armed forces equipped with nuclear weapons on the territory of the GDR and the FRG."

These are not the words of a great power ready to give in. Yet unbeknownst to anyone outside his inner circle, by January's end Gorbachev had reluctantly conceded unification's inevitability. "We don't have the moral right to oppose it," he told his closest advisers as they gathered in the Kremlin to brainstorm and strategize. Neither did they wield the strength to prevent it. For all of Modrow's efforts, the GDR had little hope of survival. "He behaves on the basis of concessions," one of Gorbachev's advisers pointed out, "but soon there'll be nothing for him to concede." It was time to accept the awful truth.

But on what terms? The Soviet Union's potential veto in any Four Power negotiation over Germany's future provided Gorbachev with considerable leverage, and he might yet trade acquiescence for something of use. All four would have to sign any treaty granting Germans full sovereignty. The French, British, and Americans each had their price for support: the common European market, the special relationship, and NATO, respectively. Unsure specifically what to demand in exchange for his endorsement, but fully intending to exact a steep price, Gorbachev insisted that he could never condone that which would jeopardize his revolution's survival: "The main thing on which no one should count is that a united Germany will join NATO," he told his advisers. "The presence of our troops alone will not allow this." It was one thing for the GDR to disappear. It would be something else entirely, and quite intolerable, for it to actively join the other side. "We have to defend the interest of our country to the utmost," Gorbachev thus declared, including "security and recognition of borders, [and] a peace treaty with the departure of the Federal Republic of Germany from NATO — or at least with the withdrawal of foreign troops and the demilitarization of all Germany."

What Gorbachev needed most as 1990 progressed was what always seemed to elude him: time. Time first to gain the West's trust so that whatever he bargained in exchange for allowing unification might stick. Time to bolster his government's standing at home so it could withstand the inevitable political backlash from losing the crown jewel of the Soviet empire. Time might even undermine West German enthusiasm for unification once the costs of integrating 16 million impoverished new citizens became more apparent. "The main thing right now is to draw out the process," Gorbachev told his aides. "This is our strategy."

They could gain the time needed only if the East Germans could first restore order. "It would be good if Modrow came out with a reunification platform," for once seizing the initiative from Kohl's regime, Alexander Yakovlev advised at Gorbachev's late January session in the Kremlin. "We could actively support it" and "win the sympathies of the German people with this." Indeed, having finally endorsed the principle of unification, the Soviets might still keep it from occurring wholly on Western terms if they made it plain that the only thing standing between the Germans and a unified fatherland was their remaining in NATO, and thus their continued subservience to the Americans. Then "we can sit on a hill and look down on the fight from above" while the Western alliance tore itself apart in search of a solution. The key would be finding a wedge to insert between these seemingly tight allies, or exploiting one that already existed. Time was what they needed, even as Kohl looked to speed ahead, and Bush pondered if he would have to move far quicker than he'd ever imagined.

ALL THIS TALK of vetoes and strategies raises a crucial question: Might the Four Powers, singly or collectively, have prevented unification once Germans both East and West desired it? Collectively the four retained legal rights, more than half a million troops on German soil, and, as Shevardnadze pointed out, nuclear leverage as well. Given such power, might they simply have said no?

Not really. Not without stirring up the same rabid German nationalism each desperately hoped to avoid, halting the process of European integration and undermining every semblance of trust that had brought the continent to this optimistic moment. In the extreme, not without ordering the military occupation of Europe's largest economy and most populous democracy. Not, in other words, without destroying any hope of an enduring peace. Legal and political rights matter. So too do troops, guns, and the ominous prospect of mushroom clouds.

But let us be practical: once Germans decided to unify, the four occupying powers had no good options other than to acquiesce. Anxiety over German might notwithstanding, the three democracies among the four could hardly have rallied support for the forceful denial of German popular sovereignty, much less the imposition of martial law that might ultimately prove necessary to defy German popular will. Bush in particular was pilloried by domestic opponents for inadequately celebrating the triumph of democracy in Europe, and for failing to prevent its suppression in Beijing. Imagine the outcry, and inevitable calls for his impeachment, if he subsequently thwarted a democracy's expression of popular will at gunpoint. He had sent troops into Panama to depose a dictator. He would hardly have been cheered at home if he had sent American soldiers into Germany's cities and towns to put down a duly elected government, and it perhaps goes without saying that German military forces would have been unlikely to take kindly to such harsh treatment. He had the legal right to do so, under a narrow reading of the occupation rules drafted nearly a half century before, but little practical or moral backing for such draconian action. At best in that scenario NATO would have been finished. At worst, it would have meant war.

Any unilateral Soviet military move to derail unification or to finally salvage the GDR by force, resembling the crackdowns of 1956 and 1968, would likely have triggered a full NATO response as well. In such an eventuality, conflict would have been precariously possible, *perestroika*'s demise certain. The Kremlin "could only have stopped the course of events by the force of arms," Baker later pointed out. But "under his premise of no military force Gorbachev had no other choice," unless he was willing to forsake his revolution and destroy any hope of building trust with the West.

Soviet bluster or deployment might also backfire, demonstrating weakness if the Germans called their bluff. "Three months ago, the great Soviet power only had to frown and the whole world submitted," Mitterrand told his Defense Council. "Now the Germans are beginning to think that the Russian threat no longer exists." American analysts reached much the same conclusion. "I believe (and this is a hunch and I guess if we did this that I would spend a lot of time in church praying that I was right)," Condoleezza Rice advised at the end of January, "that the Soviets would not even threaten the Germans" if they moved without Four Power consent. "Within six months, if events continue as they are going, no one would believe them anyway." The time Gorbachev counted on to safeguard Moscow's interests, in other words, appeared to the other great powers the thing that would most likely prove just how little effect the Soviets would have over Europe's final disposition.

And the Four Powers' legal rights? Like all aspects of international law, they were effective only to the extent they could be enforced. If the powers were unwilling or unable to compel compliance, then "what purpose would be served by meetings of the four" or any decision they might reach on their own? Mitterrand asked Thatcher in late January. The Germans could simply make unification a "fait accompli." They could still "maintain order" and "watch over the problem of unification," she replied. "That is judicial," he skeptically retorted, and "legal," determining nothing in the real world.

By the same token the Germans could not simply unite in the face of concerted opposition without so disrupting their international standing as to make sovereignty practically worthless. Just as it is implausible to suggest that American or Soviet troops might have imposed martial law to prevent unification, Kohl likewise knew that the cost of unilateral action in its pursuit would be his nation's productive relationships, its foreign trade, and ultimately any hope for a more peaceful Europe. "This is our breakthrough," Konrad Adenauer had declared four decades before, after completing the first tentative moves toward continental reconciliation and consolidation. Kohl did not want to be known for saying, in effect, "This is where the breakthrough ends." Germans already carried enough historical baggage.

The Four Powers, and the two Germanys, needed to work together as the optimistic zeal of 1989 transformed into the more complicated realities of 1990. The story of their negotiations begins in late January, in Moscow, with yet another desperate trip by an East German leader in search of salvation. There Hans Modrow found leaders preoccupied with troubles of their own. International issues like Germany's fate, while monumental, nonetheless ranked lower on Gorbachev's list of immediate problems than the real possibility of Soviet disintegration. The new year had done nothing to improve economic or political conditions. Prices continued to soar. Shelves remained empty. Once unthinkable antigovernment demonstrations were increasingly routine. Lithuanians, among others, were threatening outright secession, forcing Gorbachev to go to Vilnius in order to personally oversee a resolution to the crisis. None was forthcoming, though he at least managed to forestall a full break in relations or the outbreak of violence. The same could not be said of Azerbaijan, however. Gorbachev's handpicked delegate retaliated with force to a pogrom in Baku and the massacre of sixty ethnic Armenians. At least eighty died in that crackdown. Unofficial sources put the number at several hundred. Pro-independence sentiment soared from the Baltics to the Caucasus in the wake of the violence, with once trusted voices like Boris Yeltsin even calling for Russian secession.

"These days are very difficult," one of the Kremlin's top strategists confided to Condoleezza Rice. "I come to work every day to see what new disaster has befallen us." CNN even reported — citing sources within the Kremlin, something new for an American news agency — that Gorbachev was seriously weighing resignation in advance of the Communist Party meetings in early February. The report proved erroneous. That it seemed plausible spoke volumes.

Modrow hoped amidst this cacophony to win Moscow's endorsement of a gradual confederation of the two German states, which the Soviets had not yet publicly conceded, in exchange for something they clearly desired: Germany's eventual neutralization. He had no way of knowing that Gorbachev already ranked a neutralized Germany among the best potential outcomes for him and was thus pleasantly surprised when the Soviet leader quickly gave his consent. "It is essential to act responsibly and not seek a solution to this important issue in the streets," Gorbachev told him, allowing Modrow to think himself remarkably persuasive. "Among the Germans in East and West as well as the four power representatives there is a certain agreement that German unification has never been doubted by anyone."

This statement was blatantly untrue — indeed, Gorbachev had done as much as anyone else to sow seeds of doubt over unification's prospects — yet the import of his concession mattered far more than its accuracy. "GORBACHEV GIVES IN," German newspapers and media proclaimed. "UNITY INESCAPABLE." American observers reached the same conclusion. "Soviet statements [were] now recognizing unification will happen," Baker's notes of a private meeting with Bush on January 31 record, "but making clear the *terms* will be at issue." Modrow returned triumphantly to East Berlin, calling now for a "gradual . . . and calculable coming together of both German states." So long as it was neutral.

West Germany's response further catalyzed the process even as it once more revealed divisions within Kohl's ruling coalition. Delighted by Modrow's pliability and, more important, by Gorbachev's apparent endorsement, Foreign

Minister Genscher announced his own plan the following day in the resort city of Tutzing. It began with unification, and offered instead of full neutralization a compromise designed to appease both the Western allies and the Soviets: a united Germany would remain in NATO, but the territory of the former GDR would not. "Any proposals for incorporating the part of Germany at present forming the GDR in NATO's military structure would block intra-German rapprochement," he declared. And since rapprochement was paramount, "what NATO must do is state unequivocally that whatever happens in the Warsaw Pact there will be no expansion of NATO's territories eastward, that is to say, closer to the borders of the Soviet Union."

The Warsaw Pact would lose the GDR; there was no preventing that now. But under this Tutzing formulation, the Soviets would at least be spared the indignity of losing their dearly won security buffer or of seeing their former adversary take up onetime Warsaw Pact positions. "We should give special attention to the security interests of the Soviet Union," Genscher urged. The Soviets deserved at least that. Indeed, Gorbachev's survival might require it.

Genscher's plan took Kohl wholly by surprise, being sprung without consultation in much the same way the chancellor had announced his own ten-point plan the previous November. And it was bizarre: NATO would maintain a rank of troops deployed right down the middle of a single Germany, defending a line that was no longer a border. Now it was Genscher's turn to persuade voters that his was the party that could guarantee national unity, something increasingly important, given the swiftly approaching East German elections. By incorporating the sovereignty of East Germany but not its security, he argued, the two nations could unify in the truest spirit of the European process, continuing the continent's "move away from confrontation to cooperation," so that both NATO and the Warsaw Pact might eventually become "elements of cooperative security structures throughout Europe."

"How should that work in practice?" Kohl's national security adviser wondered. "One united Germany, of which two-thirds [remains] within NATO, and one-third outside?" The alliance's most basic plank was each member's commitment to mutual defense, the agreement that an attack on one was an assault on all. What then would it mean to have part of a key NATO country outside this guarantee? Put plainly, if Soviet forces entered the former GDR — or refused to leave — would NATO act? Under this plan, could NATO forces even traverse the eastern half of one of its own member states?

Genscher batted all such questions away like so many pesky flies. It "sounded more difficult in theory than it would work in practice," he insisted, being more interested in staking out a political position than in working out its details. Yes, the formulation of a country only partly within a security alliance seemed odd and without precedent. But everything they were doing these days lacked a good precedent. "To think that the borders of NATO could be moved 300 kilometers eastward, via German unification, would be an illusion," he told a visiting U.S. senator. "No reasonable person could expect the Soviet Union to accept such an outcome."

More unreasonable, and troubling, to American eyes was the increasingly frequent German habit of uncoordinated diplomatic pronouncements, and the potential rift these revealed within the country's ruling coalition. Such a rift might all too easily be exploited should Moscow promise unity — and thus likely electoral victory — to whichever German party first offered to guarantee neutrality in exchange. "If the Germans work out unification with the Soviets," the State Department's Robert Zoellick warned, "NATO will be dumped," and with it American strategic influence across the Atlantic.

"We cannot leave it to the Europeans," the NSC warned Bush the day of Genscher's Tutzing speech. Doing so risked "preemptive West German actions and the isolation [of] the Federal Republic of Germany if it were forced by its allies to go it alone." Scowcroft was even blunter. Germany "was like a pressure cooker," he told Bush, and it would take swift American leadership to "keep the lid from blowing off in the months ahead," and more specifically to keep the West Germans, or the Soviets, from cutting whatever deal they thought necessary to stave off real chaos in the East. Earlier they'd feared Kohl might retreat from his strong commitment to NATO. These latest developments made it appear likely that he or some other German leader might willingly bargain away what Washington required most. Scowcroft "doubted whether Kohl wanted to leave NATO," Britain's foreign minister reported home after meetings at the White House. "But if Gorbachev made the offer of unification in return for neutrality, he [Kohl] would be very tempted."

American officials realized they needed more than just one policy for dealing with the West Germans, requiring instead one for the country's chancellor and another for its foreign minister. Each necessitated separate coddling and persuasion. They'd also need to keep negotiations over Germany's future from devolving into a series of bilateral exchanges, making the State Department's desire for a more "structured" approach all the more imperative. Baker's aides consequently floated a "2 + 4" process. The two Germanys would negotiate terms of their union, with the Four Powers in turn joining the discussion over its subsequent international role. Superficially obvious, the formulation mattered. Anything resembling a "4 + 2" — that is, any system that denied Germans a sense that they ultimately controlled their own future — seemed destined to fail. Gorbachev similarly spoke during this period of six-power talks, though that too denied Germans a vital sense of sovereignty. As Baker reminded Genscher, the three formulas meant the same thing mathematically, but not politically.

Development of the 2 + 4 framework in early February did not guarantee a successful outcome, but it at least provided a mechanism to begin discussion, while ensuring an American voice within every key international negotiation. Hoping to win support for the plan, Baker thus greeted Genscher's proposal warmly, if cautiously, when the pair met two days after the latter's Tutzing speech. They were in "full agreement," the German subsequently told reporters, with "no intention of extending the NATO area of defense and security toward the East." Pressed for further details, Genscher obliquely insisted that there would be "no halfway membership this way or that. What I said is there is no intention to extend NATO to the East." What Baker had said quietly resounded loudly in Bonn, where he instructed the American ambassador to present Kohl with a full description of his discussions with Genscher, lest the chancellor hear a different report from his own foreign minister.

The first key moment of unification was therefore the apparent American agreement with the Tutzing formula and its ensuing connection to the American 2 + 4 agenda. Its next took place the following week, when Baker and Kohl each journeyed to Moscow, finding the Soviet capital, and its leadership, abuzz with anxiety. Moscow had not seen political turmoil like this since Nikita Khrushchev denounced Stalin in the mid-1950s. Now, a raucous Communist Party meeting was debating a fundamental change to Article 6 of the Soviet constitution, which designated the party as the "leading and guiding force" in Soviet society. Its entire power derived from this one statement. Perestroika's most radical proponents wanted it gone. Conservative critics considered that move an unfathomable step back from socialism into a democratic abyss.

Nearly a quarter-million Muscovites marched that week in support of the revision, the largest such gathering in Soviet history. "Resign! Resign!" the crowd chanted. "Communist Party, We're Tired of You" one sign read; "72 Years on the Road to Nowhere" said another. "The time for half measures is up!" Yeltsin shouted outside the closed-door Central Committee meeting. "We are sitting on a volcano. This is the last chance for the party!" Inside, Gorbachev faced equally vigorous opposition from those who thought the party should seize more power, not less. Initially unsure which side to support, he finally lent his weight to the revisionists. "The Party will be able to fulfill its mission as a po-litical vanguard only if it drastically restructures itself," he said. Conservatives howled in response. "These kinds of people are just not used to having to work to get everyone's vote," an exhausted Shevardnadze lamented to Baker when he was finally able to break away from the plenum. "It's madness."

"Welcome to democracy," the American replied, though he was less san-guine when reporting back to Bush later that night. Ever the negotiator, Baker sensed an opportunity amidst Moscow's general dysfunction. "Can they sort it out and implement reforms that may finally promise to revolutionize the system?" Baker asked rhetorically. "I'm not sure . . . [T]hough I don't doubt the Shevardnadze or Gorbachev commitment, they've got a long way to go and a very steep hill to climb."

Their crisis provided him room to operate, however. "What it all means for us," Baker wrote, "is that their preoccupation with their own situation is great and their need for a quiet, stable, and cooperative international environment is also great." In his opinion, "this isn't going to change in the near-term, and we have a stake in trying to lock them into a very new and more stabilizing set of security arrangements" while they were otherwise focused on problems at home.

Baker therefore reaffirmed Genscher's offer. "We understand that it is im-portant not only for the USSR but also for other European countries to have guarantees that — if the United States maintains her military presence in Ger-many within the NATO framework," he told Gorbachev and Shevardnadze, "there will be no extension of NATO's jurisdiction or military presence one inch to the east."

Of such things international vendettas are forged. The promise that NATO would not move "one inch" to the east would echo throughout Russian-Ameri-can relations for decades, ultimately forming the root cause of twenty-first-cen-tury resentments and antagonism. But what precisely did it mean? By "juris-diction" did Baker mean NATO's mutual defense guarantee? Did "military presence" mean the actual physical presence of NATO troops on the ground? Did German troops count as well? Baker's handwritten notes offer little clarity, though they reinforce the offer's broad outlines. "End result: Unified Ger. Anchored in a *changed (polit.) NATO — whose juris. would not move *eastward."

Constitutional crises rarely produce succinct diplomacy, and Gorbachev might be forgiven if his mind wandered to his never-ending litany of problems. He offered Baker merely a tepid response. "A broadening of the NATO zone is not acceptable," he said, to which Baker quickly replied, "We agree with that." Seemingly feeling that they had reached some semblance of an understanding, despite ambiguous language, Gorbachev summed up what he believed they had decided: "What you [Baker] have said to me about your approach and your preference is very realistic. So let's think about that."

The two men departed with an overall framework of an agreement, vaguely put at best. Certainly nothing one would call a formal accord. They had not agreed so much as brainstormed. Baker nonetheless felt confident enough in his performance to publicize his offer, telling journalists that in exchange for Germans' freely exercising their right to unify, "there should be no extension of NATO forces eastward in order to assuage the security concerns of those of the East of Germany." It was, he explained, "the position of the United States" that German participation in NATO was "not likely to happen without there being some sort of security guarantees with respect to NATO's forces . . . or the jurisdiction of NATO moving eastward."

Baker's was not the only offer Moscow received that day. Robert Gates was also in town, though records of his conversations with Soviet officials remained classified long after versions of Baker's and Kohl's became available. They go largely undiscussed in Gates's memoir as well, in particular his discussion with Vladimir Kryuchkov, chairman of the State Committee for Security.

The discussion between a CIA and a KGB man took on all the air of mutual professional respect and wariness one might expect. Little in this meeting was said by chance, lest an inadvertent word reveal too much. Each proffered barbs, probing for a reaction. The introduction of public debate to Soviet politics made it "more difficult for foreign analysts to understand what was happening," Gates said. "In the old days it was simpler — only one version." Kryuchkov thanked his counterpart for giving him an idea. "He would use it next time he had to justify a request for personnel or budget increases — the more information was available, the harder it was [for foreigners] to understand." Kryuchkov initiated as well as received. When talk turned to the introduction of new political parties, the Soviet suggested perhaps they would divide into two sparring cliques with identical platforms. "Then we could be like the U.S." Gates wondered in retort "whether one of the parties could be capitalist."

Their discussion offered a perfect snapshot of superpower relations at that moment. One country was clearly on the ascent yet filled with trepidation, the other in disarray and jealously protective of its pride. "Perhaps the U.S. could help" offset their current financial woes, the Soviet asked. "We are already helping," Gates replied, which generated a prompt "not very much" from Kryuchkov. The two sides had no choice but to work together, but years of competition were not easily forgotten. Neither was the habit of making threats. Yes, they had problems, Kryuchkov admitted, and would welcome American aid. But, he warned, if anyone "tried to corner us, to exploit our current difficulties, or put us in awkward situations," they'd find the Soviet government and its people adopting a new attitude indeed.

The meeting also displayed signs of their embryonic partnership, their talk concluding with what Gates termed "professional matters." These included his inquiring into the current whereabouts, or more accurately the final disposition, of captured American agents, and an offer that he'd always be available for a "special meeting" should Gorbachev or anyone else in his government wish to send Bush a message too sensitive to put through normal channels. Kryuchkov in turn handed Gates a list of individuals the KGB believed were involved in the European and American drug trade. All happened to be Soviet opponents from Afghanistan. "It was a rare opportunity in which he could kill two birds with one stone," the Soviet told him with a smile, fully expecting that the Americans would, in his words, "end that supply channel." He clearly assumed he'd never have to worry about the names on that list again.

The real heart of their discussion had to do with Germany. "Events are moving faster than anticipated," Gates observed, and under the circumstances, "we support the Kohl-Genscher idea of a united Germany belonging to NATO but with no expansion of military presence to the GDR." He wanted to know what his counterpart thought "of the Kohl/Genscher proposal under which a united Germany would be associated with NATO, but in which NATO troops would move no further east than they were now? It seemed to us to be a sound proposal."

Kryuchkov let the question hang in the air. The USSR had lost 20 million people in World War II alone, he told Gates, who in the course of obtaining his doctorate in Soviet history had undoubtedly run across this fact. "We can't exclude that a reborn, united Germany might become a threat to Europe," Kryuchkov said. "We would hate to see the US and USSR have to become allies again against a resurgent Germany."

As for the specifics of Gates's question, the Soviet spymaster finally meandered toward an answer: "Kohl and Genscher had interesting ideas — but even those points in their proposals with which we agree would have to have guarantees." The Soviets, he said, "could have no enthusiasm about a united Germany in NATO. We should look to other options." Most important of all, he concluded, was the way "trust between the US and USSR is growing," but "that trust still had to be materialized."

Like Gorbachev, Kryuchkov did not formally accept or even endorse what the Americans offered. Yet both of Bush's most trusted lieutenants employed the same language in separate conversations held at the same time, making it extremely unlikely that either misspoke. In fact, Kohl offered the same basic deal the next day, arriving in Moscow just as Baker's delegation was departing, lest it appear that the United States and Germany were conspiring too formally against the Soviets.

They did, of course, coordinate with each other. "We were consulting continuously," Baker later noted, having provided Kohl with advance warning of Gorbachev's likely response. "And then I put the following question to him [Gorbachev]," a departing Baker wrote: "Would you prefer to see a unified Germany outside of NATO, independent and with no US forces[,] or would you prefer a unified Germany to be tied to NATO, with assurances that NATO's jurisdiction would not shift one inch eastward from its present position." According to Baker, Gorbachev answered, "Certainly an extension of the zone of NATO would be unacceptable." In the end the results were promising, but in Baker's words, the Soviets were not yet "locked in." That would be up to Kohl.

"Naturally NATO could not extend its territory to the current territory of the GDR," the German leader told Gorbachev. Genscher likewise told Shevardnadze the same: "NATO will not expand itself to the East." This assurance was offered in exchange for Soviet acquiescence both to unification itself and to Germany's continued membership in the organization. Or as Kohl put it, "In parallel with the unification process in Germany, satisfactory solutions would have to be found regarding the question of alliances."

Gorbachev nodded. "This is our breakthrough!" Kohl's aide Horst Teltschik wrote in his diary. The Soviet leader repeated the same pledge later in the meeting: that he was prepared to accept unification, provided NATO did not expand east. Seeking to publicize the news as quickly as possible, thereby further "locking in" the Soviet leader through the weight of international opinion, Kohl hastily called a news conference for ten o'clock that night. "Secretary Gorbachev has promised me clearly that the Soviet Union will respect the decision of Germans to live in one state," he said as flashbulbs popped, "and that it will be a matter for Germans themselves to decide the path to, and the timing of, their unification." Too thrilled by the accomplishment to sleep, Kohl took a long predawn walk around Red Square. It was not until the next day, when the Soviet news agency confirmed Kohl's announcement, that the German delegation could effectively exhale and then fully exult.

The Americans were not yet ready for self-congratulation. While Baker briefed Kohl on how best to "lock in" Soviet support, Bush simultaneously moved to lock the Germans in as well. "If events are moving faster than we expected," he wrote in a personal letter to the chancellor, timed to arrive before his departure for Moscow, "it just means that our common goal for all these years of German unity will be realized even sooner than we had hoped." Bush pledged to stand by the Germans no matter the cost, and with greater speed than he had once thought possible. "In no event will we allow the Soviet Union to use the Four Power mechanism as an instrument to try to force you to create the kind of Germany Moscow might want, at the pace Moscow might prefer."

Provided, of course, that Germany remained firmly within the Western alliance. "Naturally, this is something for the German people, and its elected representatives, to decide," Bush wrote. "So I was deeply gratified by your rejection of proposals for neutrality and your firm statement that a unified Germany would stay in the North Atlantic Alliance." It might require "a special military status for what is now the territory of the GDR," but ultimately "we believe that such a commitment could be made compatible with the security of Germany."

This was not an offer given for free. "In support of your position," Bush continued, "I have said I expect that Germany would remain as a member of NATO." He'd stand by Germany, in other words, so long as Kohl stood behind him. In addition, "I know we also agree that the presence of American forces on your territory and the continuation of nuclear deterrence are critical to assuring stability in this time of change and uncertainty."

His White House was, in a sense, right back where they had started: concerned with ensuring Germany's reliability and making certain that nuclear-armed American forces remained where they were most needed to keep the peace, whether the adversary be the Warsaw Pact, Europe's general proclivity for war, or what Bush called the new common enemies of "apathy and complacency." There was, Bush assured Kohl in conclusion, "nothing Mr. Gorbachev can say to Jim Baker or to you" that "can change the fundamental fact of our deep and enduring partnership."

The fundamental fact of the German-American partnership, as Bush made plain, was dominant American power. The Germans could have what they truly desired only if they consented to Washington's continued oversight, on their soil, and with whatever weaponry the Americans chose. These things were not negotiable. Without them, Washington would not stand up to Moscow, Germany would remain divided, and Kohl would lose the opportunity to go down in history as the father of his new nation. Sovereignty, in other words, would not entail real independence, because the Germans would forever know who had made that sovereignty possible, and who remained dominant.

This was power politics in the extreme. Bush, as much as anyone else, and certainly more than any other foreigner, can lay claim to being the father of modern Germany. If so, he was a parent with unbreakable rules, with a firm commitment to the Fifth Commandment in particular. Which Kohl understood. "I do believe the letter you sent to me before I left for Moscow will one day be considered one of the great documents in German-American history," he told Bush a week later by phone. "I told Gorbachev again that the neutralization of Germany is out of the question for me." He'd be traveling to Camp David later in the month to shore up all they had discussed. "Without our American friends this would not have been possible," he said. "I will await your visit," Bush responded. "We have been supporting your stated position that NATO membership would be appropriate. We won't move away from that."

He would instead move closer, pulling Kohl nearer while resolving two fundamental questions: First, what precisely had been agreed to in Moscow? And second, could Kohl in fact be trusted? "There is no reason to doubt Kohl's desire to keep his commitment to membership in NATO or his willingness to stand with us on the key issues of Western security," Scowcroft advised Bush in preparation for their upcoming talks at Camp David. "But Kohl will do what he must — even at the expense of NATO and the U.S. link — to become the Chancellor who united Germany. With history beckoning, all else will become for him secondary and negotiable." At Camp David they would have to, in Baker's words, ensure that the Germans were "locked in" indeed.

"We are about to enter into the most critical period for American diplomacy toward Europe since the formation of NATO in 1949," Scowcroft advised. And with unification increasingly appearing to be "wholly on Western terms," this "places us on a probable collision course with the Soviets."

17

CAMP DAVID

JAMES BAKER FELT distinctly unappreciated, but not surprised. Over the past ten days he'd traveled across two continents and fourteen time zones, and come closer to solving Europe's greatest strategic quandary than anyone else in generations. He'd opened relations with newly empowered democratic reformers in Eastern Europe and secured Moscow's tentative acquiescence in East Germany's unification with the West. He'd coordinated with Kohl and Genscher, whenever possible ensuring that the uncomfortable pair coordinated between themselves as well, and then corralled a foreign ministers' conference in Ottawa in support of the clearest road map to unification ever produced. "We have never been so close to our goal," Kohl crowed upon hearing the news. Talks would begin as soon as possible, Genscher added in his own separate comments. "We don't want to lose any time."

The 2 + 4 framework was a start, but only that, with unification's third key moment—Kohl's impending visit to Camp David—mere days away. "If it's this hard just to get agreement on who is in the game," Baker wondered as exhausted aides distributed the accord's final text in Ottawa, "what's the game going to be like when we start playing with substantive issues?" Getting this far had required hours of cajoling, persuading, assuring, and haggling over specific language, punctuated by a single vague promise forswearing NATO's expansion. "Why do they take so long arguing about words?" Czechoslovakia's new foreign minister exclaimed in wonder. He'd been a janitor two months before —albeit a journalist before the Prague Spring—until Havel gave him a promotion and a passport.

Baker, even if feeling unappreciated, was nonetheless in his element, which

meant in control. On February 13 alone he'd spoken to Genscher and Shevardnadze five times each, and privately with his British and French counterparts.
The Western allies hammered out the basics of the 2 + 4 the next day over
breakfast at the German embassy, the site subtly suggested by the Americans
in the hope of further bolstering Genscher's psyche. Shevardnadze was not invited. Baker intended to negotiate with him separately once he had the rest in
line. The Soviets had not seen fit to send any German specialists to Ottawa in
any case. "We have no policy," a Kremlin arms specialist sheepishly confessed.
"It is good that German experts are not around, they would only complicate
things," perhaps even forcing detailed discussions when all Gorbachev really
wanted was delay.

Baker managed all this while constantly coordinating with the White House
and simultaneously orchestrating a joint Warsaw Pact–NATO conference on
the topic — international aviation — that was the original purpose of their
long-scheduled Ottawa meeting. It was a record of rapid-fire accomplishment
rarely seen in the annals of American diplomacy, reinforced by not one but two
personal calls from Bush to Kohl to ensure that Germany's chancellor agreed
with what his foreign minister negotiated in Canada. Baker kept Genscher occupied in a hotel suite until an aide signaled the White House's transatlantic call
was complete. "I don't want to get crosswise with you," Bush told Kohl, even as
Baker sat with his foreign minister. "I am talking about Helmut Kohl, as well
as the FRG."

There were details still to hammer out as Baker's tired team returned to
Washington. But look at what they'd done. The Germans remained in line,
the Soviets remained engaged, and the Americans remained the keystone of
it all, just as Bush required. "I prefer to call two plus four the 'two-by-four,'"
one American diplomat quipped, "because it represents in fact a lever to insert
a united Germany in NATO whether the Soviets like it or not." Bob Gates was
blunter. What the scheme really offered was a way to "give the Russians a feeling that they're a participant in this and it's not done over their heads," even as
the Americans prepared to "roll them."

Not everyone accepted the deal so enthusiastically. It included a mechanism for talks, but nothing more than a vague and perhaps even unworkable
arrangement for how Germany's two halves might actually fuse into Europe's
existing security structures. "An otherwise demilitarized East Germany that is
part of a unified Germany that is part of NATO?" a skeptical reporter asked

the State Department's spokesperson, Margaret Tutwiler. "That sounds a little surreal to most of us." She could only nod in agreement. "I know. But German unification used to sound surreal to a lot of people too."

Sam Donaldson cast his lot with the skeptics. He had not built his journalistic reputation by making politicians comfortable, a trait he shared with his colleague George Will. Together the bombastic reporter and the curmudgeonly columnist strove each Sunday morning to make interviews on ABC's *This Week with David Brinkley* the most contentious in American politics, especially whenever they were deprived of their senior colleague's moderating voice. His name adorned the show. They offered its primary entertainment.

Ten days after Baker's meeting with Gorbachev in Moscow, less than a week removed from Ottawa, and a mere four days before Kohl's scheduled arrival in Washington, the pair had the secretary of state on their set and in their sights. And Brinkley was out sick. "It is said that the pledge will be no NATO troops in what is now East Germany," Donaldson charged, transforming from moderator to inquisitor. "I'm not quite clear on how that works. Does that mean no German troops in addition to any other NATO troops?"

Baker started to explain: "What's been discussed this far—and it's only been discussed to this extent, is that there would be no extension of NATO forces . . ."

"Well, what's a NATO force?" Donaldson interrupted.

". . . in the territory of the German Democratic Republic." With his every word being parsed around the world, Baker was determined to answer only one question at a time.

"The Bundeswehr is a NATO force," Will quickly interjected. Under the agreement just brokered in Moscow, would Germany's own army not be allowed on German soil? "I want to find out what the facts are going to be," Donaldson chimed in.

It was ultimately the best question of all. In the days since Moscow, no one had yet fully explained how such an unprecedented arrangement might work. Even the words used to describe it had changed in subtle though potentially crucial ways. Baker had spoken in Moscow about NATO's "jurisdiction," promising it would not extend "one inch to the east." Heeding warnings from White House staffers, by Ottawa he'd begun referring instead to NATO's "forces," suggesting the GDR's territory might yet fall under the Western alliance's security pledge while remaining devoid of Western troops. Bush was equally obscure, publicly describing East Germany as likely to enjoy a "special military status," at once inside and beyond NATO's protection.

The invented term sounded nice but had no meaning in international law, where full sovereignty typically required full control over territory. Even the Kremlin's spokesmen said they "couldn't understand such a scheme." Indeed, as far as Moscow was concerned, "the idea that a unified Germany be part of NATO was categorically rejected" during Gorbachev's talks with Baker and Kohl. So what had in fact been decided? It was not so much the question of the hour as that of the entire era. Shevardnadze called it "the question of questions."

Baker had not come on the show that morning expecting to be lauded. Neither was he in the habit of openly discussing ongoing negotiations. "Now you're getting — you want to take me beyond where we've already gone," he objected. "Again, this is one of the major questions involved." All would become clear in time.

"Well, what are we going to do?" Donaldson asked anew. "Should a unified Germany have the right to put its own forces on the border?"

"You heard the Secretary General of NATO speak to the question," Baker meekly offered.

"No," Donaldson retorted. "I want to hear the Secretary of State of the United States speak to it."

Scowcroft wanted to hear Baker's response as well. As did Bush. Neither was fundamentally unhappy with what their chief diplomat had achieved in Moscow; but at the same time, they feared he'd negotiated and then publicized an unworkable plan. "We're moving too fast," Scowcroft complained, both on promises for NATO and on the 2 + 4. "It's too late for that," Baker responded, once back in Washington and behind closed doors.

"You can't say no to this," he implored Bush. They had momentum, and had already achieved a level of Soviet acquiescence no one had thought possible only weeks before. Following the Ottawa and Moscow talks, the Soviets seemed taken aback by how far they'd come, and how quickly. Gorbachev, who a year before had ridiculed the American "pauza," now called for a pause of his own so each side might take stock before potentially agreeing to anything they'd later regret. "History has decided that two German states should come into being and it is up to history to decide what, ultimately, the state form of the German nation will be," he said in a lengthy interview on February 22. There were matters of international law and moral responsibility at stake, but most fundamentally, any "violation of the military-strategic balance of these two international organizations [NATO and the Warsaw Pact] is impermissible." No one could yet say what would emerge from Europe's current convulsions, but when it came to the East-West balance, "here there must be complete clarity."

Clarity was in short supply during this hectic yet pivotal mid-February period. The Germans were of little help. Having witnessed their chancellor agree to a plan sprung on them by Genscher, and having seen that plan form the basis of ensuing negotiations, Kohl's deputies subsequently debated the very essence of the Tutzing formulation following Genscher's return from Ottawa. How, specifically, they privately asked the foreign minister, might their sovereignty be extended to include all of East Germany but not their NATO security guarantee? And what of the Soviet troops still encamped on East German soil? There was still no plan or timetable for their withdrawal, indeed no Soviet commitment to that effect, leaving the possibility, however impractical, that Warsaw Pact troops might remain deployed within NATO territory. Without explicit agreement on their removal, Soviet forces might simply stay. On the ground, meanwhile, Soviet conscripts had begun selling their uniforms, equipment, and even their firearms on East Germany's blossoming black market. The Red Army was hollowing out. For an East German society still echoing with calls for civil war and retribution, the flood of weapons did nothing to promote stability.

Genscher remained adamant that his plan offered the only real hope of securing Soviet acquiescence. It was his best electoral strategy as well. East Germans would go to the polls in less than a month to vote for parties aligned with those in the West. Furious at the way the chancellor's minions questioned his motives and methods, and unwilling to cede any political ground over which party could ensure unification faster, Genscher took their spat public, once more suggesting that Kohl lacked the courage to make a clean break with the past, or to stand up to the Americans and make the clean break with NATO that unification increasingly seemed to require. Trailing in the polls, Kohl caved, announcing on February 19 a new compromise: East Germany would indeed join NATO following unification, but "no formations or institutions of the Western Alliance should be moved forward to the present territory of the GDR."

The Solomon-like pronouncement resolved little, and stretched what Gorbachev had seemingly accepted only ten days before beyond anything he might now readily concede. Kohl had not only "'de-NATOed' but demilitarized" the GDR, American diplomats in Bonn cabled home, conveying their fear that whoever won the impending election might yet cut a separate deal with the Soviets. "We had great faith in Kohl," Scowcroft noted. But "we could not rule out the possibility that Gorbachev could tempt — or threaten — him" into breaking his pledge to remain in the Western camp.

The same idea was percolating through Soviet circles as well. The Kremlin should make a "very attractive" offer to both Germanys, the Central Committee's Valentin Falin advised Gorbachev, proposing a "neutral, democratic, and basically demilitarized Germany" in exchange for rapid unity. The Americans would undoubtedly protest. But they were only one vote out of four, and Moscow might yet persuade French and British leaders that neutering their old adversary was wiser than endorsing its revival.

The allure of neutrality grew in mid-February in European circles too, as did Bush's concern for Kohl's fidelity, but he ultimately had no choice but to back the chancellor. Kohl's rivals seemed worse. Genscher was clearly willing to bargain away full German participation in any existing security alliance, while leftist and nationalist party platforms jointly demanded that "a future united Germany should belong neither to NATO nor to the Warsaw Pact." Polls showed 58 percent of West Germans favored neutrality as well. Years of living between nuclear-armed superpowers had apparently diminished either side's appeal among the German electorate. Just a few months after Eastern Europe's optimistic turn toward democracy, Bush faced the very real possibility that he might gain German unity while losing Germany's central role in his entire strategy for Europe.

Václav Havel's appearance in Washington mere days before Kohl's crucial Camp David visit did little to ease the specter of a neutralized Germany. Havel wanted the entire continent freed from the burden of military blocs. Stirred by the possibilities that history presented, and not yet weighed down by the practical difficulties of governing, Europe's optimists hoped not so much for a democratic transformation as for a wholesale revolution touching every aspect of human relations, up to and including the way states interacted. "For those who would rather not touch anything because everything is a monument," Havel once offered, speaking both practically and metaphorically, remember that "if our ancestors thought like this, we would have no castle at all — just some sort of a pagan heart and a hole in the ground."

Havel's buoyancy frightened a White House team for whom optimism was professionally anathema and inexperience the most dangerous quality of all. "Havel presents problems," Scowcroft warned Bush on the eve of their meetings. "In his zeal to break with the immorality of one of the nastiest communist regimes, he is staking out what he considers the high moral ground" with little thought for the long-term consequences. He was "a man on a mission" but also "a man in a hurry." Most of all he was the subject of endless fascination, this playwright turned politician who routinely interspersed formal sessions with cigarette and sausage breaks, and who visibly yearned for solitude rather than the raucous crowds that greeted his every American stop.

Similar to Gorbachev, who'd also fascinated American audiences, the new Czech leader offered increasingly messianic promises, albeit in a halting monotone that demonstrated why he'd flourished behind a typewriter rather than onstage. "The human face of the world is changing so rapidly that none of the political speedometers are adequate," he told a joint session of Congress. Prague could host a grand Soviet-American disarmament conference, he suggested, leading to elimination of all military pacts across the continent. With Europe solved, he'd then mediate the Israeli-Palestinian dispute, he further mused, before turning to other global quagmires. After all, "consciousness precedes being, and not the other way round." For most of the legislators in attendance this hardly explained a thing, though none wanted to be the only one not applauding.

Havel's blue-collar radicalism sharply contrasted with the demeanor of a blue-blooded president determined to preserve a key plank of the status quo. Responsibility required propriety, Bush pointed out upon Havel's departure, and sobriety as well. "Let's get all the Soviet troops and all the U.S. troops out, and life will be beautiful," Bush privately characterized the Czech's naïve thinking, scarcely able to contain his derision. "Everything will be pruning hooks and plowshares." At least by the end of their talks, said Bush, "I think I convinced him that the United States [military] — wanted by Western Europe and, indeed, by some of the countries in Eastern Europe," was necessary and, along with NATO, "there as a stabilizing force."

Bush's reluctance to consider anything more than incremental change offered critics yet again the opportunity to wonder if he was simply unable to think in grandiose terms, and ultimately if he'd ever prove a match for the times. "With the breakup of the Soviet Empire," Helen Thomas asked at a February news conference, "you want Germany to remain in NATO." But why? It was three months since the breach of the Berlin Wall. Ceauşescu was dead, democracy was alive, and Warsaw Pact leaders openly called for tearing down what was left of the Iron Curtain. In such a radically transformed world, she pressed, "who's the enemy?"

Chuckles rippled through the press corps, but Bush did not even crack a smile. He'd answered this question before. Or thought he had. "The U.S. troops are there as a stabilizing factor," he explained anew, clearly struggling to contain his exasperation. "Nobody can predict, Helen, with total certainty, what tomorrow's going to look like. I've been wrong. You've been wrong. He's been wrong. She's been wrong, on how it's going to go, and we don't know—"

"Do you expect the Soviet—" she interrupted, doing nothing to improve the president's mood.

"May I finish, please," he tersely interjected. It wasn't framed as a question.

Regaining his composure, Bush slowly began again. "Our European allies want us there," he said. "I have a feeling that some of the Eastern Europeans want us there, because they know that the United States is there as a stabilizing factor" against unknown threats and uncertainty. The Americans had kept the peace for generations, "and we will be there for a long time to come." Cutting directly to the heart of the matter, answering a question neither Thomas nor anyone else had yet posed that afternoon, he finished bluntly, "I'm not contemplating a neutralized Germany."

No amount of repetition helped. All his work to secure an American presence overseas would mean little if he failed to persuade Congress to invest for another generation—or more—in American troops and bases across the Atlantic. Reagan might have found the words. At least an actor of Reagan's skill might have won over skeptics with an artful turn of phrase. But for Bush, the question, and his frustration over his inability to articulate a strong case, trailed him even back to the privacy of his personal study.

"Who's the enemy?" he dictated aloud to his diary. "I keep getting asked that." Alone with his responsibility, he tried yet again to find the words. "It's apathy; it's the inability to predict accurately; it's dramatic change that can't be foreseen ... there are all kinds of events that we can't foresee that require a strong NATO," he said, "and there's all kinds of potential instability that requires a strong NATO presence."

He lacked the words but not the conviction. Indeed, the depth of his conviction was part of his problem in selling his case. As with democracy, markets, and freedom, the values touted at his inaugural a year before, Bush believed that American oversight simply "worked." It had been a fundamental plank of American foreign policy his entire career. He considered his youth stolen by its absence, and an ensuing war that demanded the death of so many comrades. The necessity of American engagement was something he just did not question. Neither could he ever fully conceive why others might.

But something had changed since he'd taken his oath that brisk January day a year before, not only in the Cold War's temperature but within himself as well. He'd found a purpose. Reagan had entered office determined to crush communism, Gorbachev to save it. Lacking his counterparts' crusading zeal, Bush took until his second year in power to realize the means to achieve a more peaceful and democratic world. He now had his call to action. To date he'd largely stood above the fray when international incidents arose, preaching the virtues of quiet when others chanted in the streets. He had been unable to affect events driven by raucous crowds in Berlin or Prague, unable to prevent regicide in Bucharest or a bloodbath in Beijing, and in every case afraid to speak out against communism's tyrants lest his words topple its reformers.

The moment for such Hippocratic diplomacy was over. It was at last time for statesmanship, fully engaged and without reservation, to ensure Washington's full place in Europe now and into the twenty-first century. Kohl and the Germans had set the pace of the game, and Baker had established its rules. NATO's future, and with it the American presence overseas that he cared about most, was in his court, front and center in the negotiations. "You and Germany have to agree now on German unity within NATO," the alliance's general secretary, Manfred Wörner, implored Bush only a few hours before Kohl arrived at Camp David. Halfway measures like France's associated membership wouldn't work. "Just association with NATO will create temptations for Germany to make diplomatic deals with Russia," he warned. There was thus only one clear path forward for the organization, and one chance to prevent American leadership from becoming a casualty of victory—provided, Wörner ominously added, Bush could also "sell this to the Russians."

The Germans weren't helping. Consumed with their own politics, they appeared incapable of refraining from their stereotypical arrogance. The Americans lobbied and reassured their smaller allies in Ottawa, for example. The German delegation merely made declarations. "We have worked together within the alliance for forty years," the Italian foreign minister lamented after hearing of the proposed deal for NATO's future negotiated without the group's full consent. Genscher cut him off. "You are not a player in this game," he responded coldly, leaving the room gasping at his tactlessness and audacity.

It was not their only misstep. Kohl eagerly broke with precedent when it suited his agenda, yet refused to forswear German claims to disputed territory assigned to Poland after 1945. His argument that he could not negotiate on behalf of a unified Germany until unification had been concluded made legal sense. It was, however, tone-deaf. "Nothing about us without us," Polish diplomats urged in protest over their exclusion from the 2 + 4, reviving painful memories of previous great power decisions regarding Poland's fate. Their prime minister, Tadeusz Mazowiecki, was blunter. "To exclude the voice of the neighboring states concerned and to shut them off from the relevant stage of the discussion on the aspects of the German unification which affect the security of the neighboring states," he warned Bush, "would be tantamount to a repetition of the Yalta formula of 1945." Europe's smaller powers lacked a seat at the table, Mazowiecki noted, but were not powerless. All they required was a champion to make their voices heard.

Their brewing opposition was all the opportunity Thatcher required. She was "not against" unification, her foreign secretary argued in the Oval Office in February. She was merely "reluctant" to see the Germans unbridled. Despite having grudgingly acquiesced in Bush's plans, she'd never fully accepted them, and now seemed ready to backtrack. "Don't you realize what's happening?" she'd recently upbraided her staff when they tried to convince her that times, and Germany, had changed. "I've read my history." To slow this increasingly speeding train, she'd have to do more than bluster, her foreign secretary privately advised. "We must not appear to be a brake on everything. Rather, we should come forward with some positive ideas of our own."

But she had none. Mitterrand rebuffed her last-ditch overture of an alliance to retard unification, and Gorbachev's position on NATO made the Soviet leader an unsuitable ally as well. Grasping by mid-February for anything that might slow the process, she therefore took up Poland's call for settlement of Germany's borders as a precursor to any final deal on its sovereignty, recalling a previous British prime minister's pledge to guarantee Poland's security in 1939. The Poles, moreover, were only the beginning. "Other neighboring states will have their particular views too," she told the House of Commons a week after the 2 + 4 agreement. Each should be addressed, singly and at length, before a final deal was signed. It was "understandable" that "bitter memories of the past should color their view of the present and the future," Thatcher pronounced, leaving Bush to wonder if her support, supposedly locked up weeks before, might not be forthcoming after all.

"Unification was coming fast indeed," she fretted during an early morning phone call with Bush on February 24, the day that would ultimately become unification's third critical moment. "All are worried about the consequences." Kohl was set to arrive at Camp David that afternoon, giving her one more chance to lobby against removing the constraints on Germany, and one more opportunity to remind Bush which special relationship mattered most. Germans were bound to be "dominant" if allowed to unify without restraints, she said, warming up for a lengthy and likely rehearsed monologue. They'd be the "Japan of Europe, but worse than Japan. Japan is an offshore power with a tremendous trade surplus. Germany is in the heart of a continent of countries, most of whom she has attacked and occupied." If "we are not too careful," she told him, "the Germans will get in peace what Hitler couldn't get in the war."

It was more lament than plea. Thatcher described her fears at length yet never specifically asked Bush for help. Neither did she offer any alternatives. Perhaps she accepted her adviser's assessment that Bush's commitment to German unity was unshakable, and merely hoped to fan the embers of Anglo-American ties during a moment when a new Atlantic partnership between Washington and Bonn seemed in the offing. Politically buffeted at home after a decade in power, perhaps she simply longed for the days when a president would heed her advice.

Whatever her intention, Bush had heard this from her before. "This is very helpful, timely, and interesting," he replied once she had expended her breath. But he had an answer. "When I talk to Helmut, I will seek a clear statement from him about full membership in NATO for a united Germany," he insisted, "that would include continued integration of German forces in the NATO command and the continued presence of U.S. troops."

Anything else was simply unacceptable to either of them. "A lot of people here don't understand the need to keep a strong defense," Bush continued, yet no matter how many times he described his reasoning, the same "trick question" kept arising. "When I am asked who our enemy is now," he told her, rehearsing his answer, "I tell them apathy, complacency."

The unknown was a greater threat than a revived Germany, Bush added, and American oversight its best defense. Anything that enabled proponents of withdrawal, such as opposition from trusted allies or a watering down of the 2 + 4's principal issues onto smaller matters, only empowered those who thought the United States forces should leave for home anyway. In other words, her objections were not helping. And she needed to stop. "You and I think the same," she conceded. "Germany must stay in NATO, or all our security goes."

It was just what Bush wanted to hear. "This has been very helpful," he said as their conversation wound to a close, promising to fill her in "when Helmut leaves." He planned to invite her and Mitterrand soon for talks, Bush continued, so the three might strategize in person. "I get many invitations to [American] universities to speak," Thatcher quickly interjected. Anytime Bush could arrange for the French president to travel to the United States, she'd find an excuse to be there too.

Bush's conversation with Thatcher highlights two critical aspects of his style and strategy, especially when set alongside the other call he placed that morning, to Canada's Brian Mulroney. Both conversations demonstrate Bush's commitment to the time-consuming task of personal diplomacy, and the way he listened as well as spoke directly to his counterparts, without a formal agenda but also without intermediaries. In private he could speak freely and hear their candid concerns, ensuring that even if unheeded, his allies would at least know they had been consulted. Privacy also lessened any embarrassment foreign leaders might feel when confronted by the harsh reality of his relative strength and their comparable weakness. The man who refused to dance on the Berlin Wall out of deference for communist diehards knew better than to dress down his allies publicly.

Bush's conversations in advance of Kohl's arrival also reveal that he'd moved beyond what Baker and Kohl had negotiated in Moscow. There the talk had been of NATO jurisdiction. By the time Kohl came to Washington a fortnight later, without any input from the Soviets the terms of their negotiation had changed. Baker "wants to let the Soviets stay in East Germany," Bush told Mulroney. "It gives me heartburn, though, if we suddenly, in an effort to get Germany stability, acquiesce in or advocate Soviet troops remaining in Germany," he said. "That is what we have been against all these years."

Mulroney vehemently agreed. He too believed in the necessity of the New World's oversight of the Old, and believed the Canadians had an equal stake in NATO's survival. "We are not renting our seat in Europe," he added. "We paid for it. If people want to know how Canada paid for its seat in Europe, they should check out the graves in Belgium and France." To his mind the "minimum price" for German unity should be "full membership of Germany in NATO and full membership in all the Western organizations and full support for American leadership in the Alliance." If they failed to keep their alliance intact, which in reality meant if Bush failed to keep the Germans in line and the Soviets out of the mix, Mulroney warned, "we could be back where we were three years ago, three years from now. The linchpin," he said, "is NATO."

The third key moment of German unification, Kohl's visit to Camp David, began for Bush even before the chancellor's arrival, with the time-consuming process of coalition building. The ally who wavered he reminded of American strength. The one already committed he reassured. In both instances Bush ensured that his partners felt consulted, but also that they knew who ultimately would call the shots. "I don't want you to think we are such a big deal that we leave others out," he said as his conversation with Mulroney drew to a close. It was the same sort of self-effacing comment whose virtues his mother had preached his entire life. It was also the kind of statement that became necessary only when reality bespoke otherwise. They'd all be better off if Kohl and Genscher did "what you do so well," Mulroney concluded, "which is to consult with your colleagues."

Bush's conversations with Kohl proceeded in much the same manner. He listened. He commiserated and agreed. And he ultimately wielded American power, tactfully yet without leaving any doubt that Germany would achieve its goals only by holding fast to its NATO pledge, and to Washington's dictates. "We are going to stay involved in Europe," Bush promised. "We have some pressure here to lower the level of US troops and defense spending," and "we are being asked: who is the enemy? The enemy is unpredictability, apathy, and destabilization," he said once again. "But we will stay in Europe."

Kohl did not object, listing instead the issues he required American support to solve, including the tricky question of the current German-Polish border and his relations with both Thatcher and Gorbachev. "Most of the French people are on our side," said Kohl, listing too the Danes, the Dutch, and the Norwegians, all countries the Germans had once occupied, which now supported its reunification. Within Western Europe only one nation, and specifically one person, stood out from the rest in opposition, and "I can't do anything about her," Kohl complained of his British counterpart. "I can't understand her. The Empire declined fighting Germany—she thinks the UK paid this enormous price, and here comes Germany again."

"We don't look at it that way," Bush reassured him. "We don't feel the ghosts of the past; Margaret does." Therefore, "you and I must bend over backwards to consult, recognizing her unique role in history." Both agreed that Genscher's undiplomatic behavior in Ottawa had not helped Germany's current image, but in the end the Americans would guarantee full support from their mutual NATO allies. "You and I must take care to consult," Bush said again. "I called Margaret today just to listen to her, which I did for an hour."

Listening to allies did not entail taking orders from them. Together the Germans and the Americans would make all the critical decisions required on unification, Bush argued repeatedly. Kohl would direct things domestically, dealing with difficult questions that were ultimately only for the Germans to decide, such as monetary union, social benefits, and the myriad reorganizations and syntheses required when two separate societies combined. The Americans would clear the path internationally. "We must have the closest possible consultation between us," Bush told him. "We are going to win the game, but we must be clever while we are doing it."

Indeed, once the Germans and Americans had decided on a course of action, and in particular after Kohl won enough support in the upcoming election to speak with one voice for all of Germany, Bush planned to invite Mitterrand and Thatcher for consultations, turning the 2 + 4 into a "1 + 3." So long as the Germans (the "1") and the Anglo-French-American triad (the "3") agreed, no one in Western Europe could possibly muster the political will to stand in their way. Kohl quickly assented. "The One Plus Three must know what they want," he said. "They must establish a joint position. Then we would go to the Two Plus Four" for its final endorsement.

Camp David's algebra made the thing clearer: the Western allies, led by the United States, agreed to determine among themselves a unified position for Germany and in turn for Europe, inviting the Soviets into negotiations only after achieving a fait accompli. Bush's diplomatic code required consultation, but when it came to the most critical issue of the day, American power afforded the luxury of consulting with potential adversaries only after the fact. The key was to wield power in a gentlemanly manner, forcefully yet tactfully, so hard feelings could be allayed without jeopardizing the outcome. "We need to be low-key about our activities at One Plus Three," he said. "We don't want to give the Soviets the impression that they are being dealt out."

Kohl concurred. "For the next few weeks we need to conduct discreet consultations, so we don't offend people. We need full US-German agreement, then a clear position at One Plus Three. Then we can look toward the Two Plus Four, agreeing on how we will handle the Two Plus Four. You will have to deal bilaterally with the Soviets on this," Kohl added, noting that Moscow's demands would undoubtedly boil down to two things: money and security. "The Soviets are negotiating," he acknowledged. "But this may end up a matter of cash. They need money."

"You have deep pockets," Bush replied, making plain that he expected the Germans to cover the bulk of the cost of whatever aid Gorbachev required to justify his strategic losses. As for the security guarantees, Bush noted, here was where real power would ultimately matter most. "The Soviets are not in a position to dictate Germany's relationship with NATO," he said, temporarily allowing his long-pent-up tension to erupt. "To hell with that. We prevailed and they didn't. We can't let the Soviets clutch victory from the jaws of defeat."

Gorbachev, in absentia, was the meeting's real loser. "We support you," Bush told Kohl, "because we have a big stake in this," and in a united Germany wholly tied to NATO. He meant *all* of a newly unified Germany, despite what they'd tentatively agreed with the Soviets only two weeks before. In future pronouncements they'd forsake the term "jurisdiction" to refer to NATO's reach, employing instead the word "forces." The subtle difference was critical. "Forces" meant NATO or even German military power. "Jurisdiction," conversely, meant a security guarantee. "I used the term 'jurisdiction' before I realized that it would impact upon Articles 4 and 5 of the North Atlantic Treaty," Baker confessed during the second day of talks. Now he knew better. They'd be willing to negotiate with Gorbachev on the final disposition of troops on East German soil. The question would take some time to resolve, given the ongoing Soviet presence and the Red Army's current disarray, but ultimately the West Germans and Americans changed the terms of the deal cut in Moscow simply by changing the terms used to describe it. No matter what the Soviets believed, NATO mattered most. It "is the raison d'etre for keeping US forces in Europe," Baker noted, and "we couldn't have US forces in Europe on the soil of a non–full member of NATO." The Soviets would simply have to accept that reality.

The meetings could not have gone better from Bush's perspective. Wine and laughter flowed each night over dinner. Kohl even survived a chilly hike up a nearby summit, though his labored breath by the end prompted Bush to dub it thereafter the Helmut Kohl Memorial Hill. The weekend's only hiccup: too much profanity in the movie Bush's staff chose for Camp David's screening room. The two principals' wives left early.

A bit of foul language could hardly diminish the sense of accomplishment that flowed through Bush's staff upon Kohl's departure. No better evidence exists for how well the sessions went than a close reading of Bush's final statement to the press after their talks had ended: "We share a common belief that a unified Germany should remain a full member of the North Atlantic Treaty Organization, including participation in its military structure. We agreed that U.S. military forces would remain stationed in a united Germany and elsewhere in Europe as a continuing guarantor of stability. The Chancellor and I were also in agreement that, in a unified state, the former territory of the GDR should enjoy a special military status that would take into account the legitimate security interests of all interested countries, including those of the Soviet Union."

White House staffers had written this statement two days before Kohl arrived.

Once Bush retired for the night, he turned once more to his diary, to celebrate but also to steel himself for the battle with Gorbachev still to come. "We have a disproportionate role for stability," he dictated. "We've got a lot of strong-willed players — large and small in Europe — but only the United States can do this . . . [and] I don't want to see us fettered by a lot of multilateral decisions. We've got to stand, and sometimes we'll be together with them; but sometimes we'll say we differ, and we've got to lead . . . I've got to look after the U.S. interest in all this," he said, drawing his entry to a close, "without reverting to some kind of isolationist or peace-nik view on where we stand in the world."

18

CONCESSION

THE BASIC DYNAMICS of the contest to define the new Europe did not so much change as deepen following late February's Camp David meeting. Kohl's hold on German politics strengthened. Thus too his ability to make good on his NATO pledge. Buoyed by his new sense of mission, Bush worked to win broad support for his vision of Europe's post–Cold War strategic architecture. Some allies he tutored. Others he persuaded. The rest he flat out bribed. By early summer, only the Soviets remained unswayed. Diplomatically isolated, politically battered, fiscally strained, and increasingly exhausted, Gorbachev could ultimately do little more than shrug as he lost the Warsaw Pact, the essential pillar of Soviet security since 1945, in exchange for the vague promise of Western goodwill consistently dangled just beyond his reach.

These were Bush's finest moments in office. Many longtime and potential European allies still harbored profound anxiety over the prospect of unhampered German power, and an infatuation with denuding Europe of both Soviet and American influence in the apparent wake of the Cold War. Countering both impulses required intense effort and great sensitivity. "We should not project an image of larger countries cutting up the world," Bush told France's Mitterrand upon Kohl's departure, and "need to insure that the smaller countries are not left out."

Faced with a need to persuade, Bush instinctively turned to his first lesson in diplomacy: every nation jealously guarded its self-respect. None, not even the smallest, wanted to be in the position of following. Diplomacy worked best, he'd realized at the UN and repeatedly after, when one could "treat people with respect and recognize in diplomatic terms that the sovereignty of Burundi is as important to them as our sovereignty is [to us]."

He and Kohl might simply have forced allies into line. The pair wielded tremendous power. But Bush believed carrots built better long-term relations than sticks. Europe had no choice but to live alongside the Germans. American influence required an invitation, and thus goodwill. NATO "must be consulted," he insisted, even on fundamental issues of security that the 2 + 4 was designed to adjudicate. More important, new and old allies alike needed to be heard as well, so they might sense that their opinion mattered. Bullying might have produced short-term gain, but it was no way to secure enduring allegiance.

Baker simultaneously hoped to educate Eastern Europe's new breed of democratic leaders. Long on optimism yet short on experience, they "tend toward a naïve worldview," he fretted, and thus advised Bush that he could "have an enormous impact on their thinking" merely by taking the time to explain why NATO's survival should be the primary concern even of states still formally aligned with the Warsaw Pact. No matter what dreamers like Havel or others fantasized, Baker told him, "you should particularly reject the idea of a Central European demilitarized zone as destructive of NATO" and "damaging to European security." The lawyer in their partnership, he'd handle the details. Bush, the salesman of the two, would make the case and close the deal.

Dozens of phone calls and face-to-face meetings ensued, part symbolic gesture and part substantive plea, ultimately forming a familiar pattern: listening came first, laced with empathy and respect. "We need to stay in close touch on how you view developments," Bush told Havel. "I would love to have you tell me how you, as a free thinker, see Soviet-US relations," he said to Hungary's president. "There will be no condoning of the exclusion of any ally on German unification," he assured Italy's prime minister, still smarting from Genscher's casual dismissal weeks before. "There was an unfortunate comment at Ottawa," Bush commiserated, but neither Genscher nor Kohl truly ran NATO. "I will insist upon full consultation within the alliance," he promised. "It is not the role of the US to sit around and divide the world."

Explanation came next. Europe needed the stability American troops provided, Bush repeatedly said, especially in such uncertain times. "The press asks me who is the enemy," he told his Italian counterpart, giving his stock response. "I say it is instability and unpredictability. Events have moved in the right direction, but will that continue? I wanted you to know my feelings, because you are such a good and close friend."

Commitment came last: American forces would remain in Europe for as long as their hosts would have them. This was the fundamental difference between the American and Soviet military presence, Bush explained: the first was invited, the second despised. "Post-unification Europe is best served by the U.S. remaining involved," he thus told the Hungarian president. "With troops."

Two points must be stressed about Bush's campaign of consultations: first, that consultation at no point entailed conceding control; and second, that he never relied on words alone. Bush considered listening to allies a mark of respect. Obeying their wishes was another matter altogether. Neither did he believe that mere phone calls or meetings guaranteed an ally's support. As Henry Kissinger had been warning him for nearly two decades, and as Kissinger's protégé Scowcroft frequently echoed, foreign affairs was ultimately about power rather than friendship. Leaders would do what they thought best on the basis of their own strategic calculations, whether they liked their foreign counterparts or not. Bush concurred then and now, but only to a point. "It seems to me that he [Kissinger] is overlooking the trust factor and the factor of style," he'd recorded in his diary during the mid-1970s. Personal ties did not ultimately persuade; they instead ensured that everyone understood one another's position, what they were being offered, and when they were hearing another's bottom line. "Openness is the enemy of mistrust," he'd said earlier in his presidency. Charged with securing support, he personally guaranteed his deals. "This was personal diplomacy in the finest sense of the term," Scowcroft later noted. He was biased, of course, but even if we put aside this assessment, his words provide a sense of the administration's goals: "Coalition building, consensus, understanding, tolerance, and compromise ... There was no Versailles, no residual bitterness."

But there was also power, and the repeated message that compliance brought rewards. Europe's uncertain future demanded choices, Bush said again and again, and specifically the choice for its new leaders to choose which side of history to join: The West, which "worked," and the side most demonstrators in the streets seemed to favor. Or continued separation, perhaps liberated from Moscow's direct control, but still disconnected from the remainder of Europe and the wider world by an invisible cultural and political barrier.

The question, which Bush put directly to each, was if Europe's democracies wished to float down the stream of history in Washington's flotilla, or if they'd rather travel alone. American protection required only the willingness to accept it, exemplified by support for a continuation of NATO and Germany's full place within it. The organization would have to change, Bush granted, finding new purpose in combating the new enemies of uncertainty and instability. But it had to remain, too. "We've arrived at this historic point by maintaining a strong partnership with our European allies," he announced in early May, calling for a new NATO summit designed to look anew at the organization's key missions. To start the bidding, he announced curtailment of the short-range nuclear missile deployment that had been such an irritant to Atlantic relations when he'd first taken office, while pledging an overhaul of the organization's mission. NATO will have to change, Bush told Mitterrand in advance of his formal announcement. "Its role will be different. The organization must be flexible," he said. "It will be guaranteeing against instability," even as it helped former adversaries like the Soviets take comfort that it was no longer designed as a hostile force.

Europe's new democratic leaders had a choice to make as well, Bush made clear during his springtime conversations. Their support for Washington's lead on security would reveal their answer. They could continue to reside beyond the Western orbit, always in Moscow's shadow and bereft of the full economic and cultural benefits American approval conferred. Or they could prosper beneath the protective umbrella of American oversight. The United States wants "to play a stronger role in the new Europe, particularly in supporting the transition to market economies in Central and Eastern Europe," Baker advised Bush to tell East Germany's new leadership. Opportunities abounded for "expanded commercial and cultural contacts across the board." American endorsement would go a long way toward aiding a new state's application to join the World Trade Organization, while direct American aid similarly meant more than merely the dollar amounts negotiated, representing a seal of approval to Western bankers and investors alike.

Securing that blessing was easy. In Baker's description, Eastern Europe's leadership had only to "reciprocate." Which meant one thing: endorsing American leadership on security, and most immediately on Germany's future. Told of Hungary's commitment to democracy, for example, Bush promised further investment. "We have already crossed the border to a new Hungary, with multiparty democracy," the country's acting president remarked during an Oval

Office session, pledging that "the new democratic Hungary will be a reliable and calculable partner."

Bush offered an illuminating response. "Let me assure you that we want to work with you on the economic front in every way we can," he said. "We want to see more private investment and better trade relations. In addition to our governmental activities, we want to see as many leaders as possible from the American private sector be involved." Bush's pledge of economic aid and private investment was not specifically in line with the political transformations the Hungarian had mentioned. But it was precisely what his interlocutor wanted to hear. Eager to speed its difficult market transition, Eastern Europe's new leadership was keen to win American approval in the hope it would further open the door to prosperity. It had been this way when Bush visited Warsaw, Gdansk, and Budapest the previous year, when pleas for aid had dominated his every discussion.

The pressure to secure Western support was even stronger a year later. To gain favor in Paris and Bonn, the new democracies needed to support European integration. The Americans demanded fealty on security. Asked by Gorbachev why he would condone the Warsaw Pact's dissolution, for example, even without the explicit promise of inclusion in a Western alternative, and despite his earlier pleas for a continent devoid of alliances altogether, Václav Havel offered a frank response: "We do not hide that our goal is the inclusion of Czechoslovakia in the West European integration." In the long term he still longed for elimination of all military blocs. But since taking office, and after numerous conversations with Bush and Baker, he'd learn to temper that dream, or at least to see it most likely accomplished in discrete steps, beginning with the Soviets' withdrawal and a turn toward their old adversary. "It seems that NATO, as a more meaningful, more democratic, and more effective structure, could become the seed of a new European security system with less trouble than the Warsaw Pact," Havel finally conceded in May.

Bush had demanded as much, winning from each of Europe's new leaders a promise to push for Soviet, but not American, withdrawal. "What worries me is that the longer the Soviet Union stays" in Central Europe, "the more they will say 'okay, we'll go out if you [the United States] go out,'" he'd told Hungary's leader. "Free countries have to help us avoid that linkage." Bush could not have scripted a more favorable response. "We are in full agreement, Mr. President," the Hungarian responded. Soviet troop withdrawals "cannot be linked to U.S. troops."

Similar negotiations followed with each of the Soviet Union's onetime vassals, some explicit and others vague, but at all times with a clear quid pro quo: American benevolence and influence in exchange for compliance. The Hungarians, as already noted, wanted aid. East Germany's leaders told Bush of their concern that Soviet troops might linger well past unification. He promised to work for the Red Army's speedy departure. Told anew of the Poles' anxieties over their border with Germany, he offered to broker a deal. "Confidentially, what if I could get Kohl to agree with you on a text of a treaty now?" Bush asked Poland's prime minister in March, one that would settle their increasingly bitter dispute before the completion of unification. "Don't answer now," he added, "and please keep it confidential." In the context of a lengthy discussion of European security and the need for the continent's new democracies to back NATO's survival, Poland's part in this bargain was obvious. "Think about my proposal," Bush told the Pole. "Be assured we are playing no games . . . We think Germany should be in NATO."

Bush had every confidence in his ability to deliver on his part of the bargain, having secured that very pledge from Kohl the day before. "I am ready and willing to agree on an eventual text with Mazowiecki," the chancellor promised. "That is something I am not able to make public, but it may be useful to you." Bush found it useful indeed, ultimately cashing in Kohl's pledge to secure Poland's acquiescence on the issue he cared for most. As two of Bush's aides who were engaged in these negotiations later wrote, "Mazowiecki left Washington knowing just what he could expect from the leading power in the West." And, of equal significance, his price.

The pattern of these conversations, and the clear yet deferential manner in which Bush wielded American power, continued even when he spoke with Europe's most influential players, talking to Mitterrand and Thatcher frequently during the spring, listening, reassuring, and then ultimately conveying whose opinion held the most sway. "I need your consultations and advice," he told the British prime minister. "Before each two plus four session, we should carefully make sure that our two countries, the French, and the FRG have identical positions," and could thus maintain a common front against expected Soviet objections. Having asked her opinion, however, Bush then spoke for the bulk of their next hour together. She'd lectured at length when they spoke before Kohl's February arrival at Camp David. This time Bush consumed all their conversation's available oxygen. To counter him she'd have to challenge him, which the Iron Lady, ever worried about losing London's place as Washington's favorite, dared not do. "Britain wasn't a free agent" when it came to European affairs by 1990, one of Thatcher's confidants noted in his diary that spring. "The British choice would be much influenced by what the Americans did."

It was much the same for the French, who planned for their own special relationship with Germany, but who desired, for the foreseeable future at least, American influence in Europe just in case. "I have no agenda," Bush began one April conversation with Mitterrand, and "want to have a clear view of your ideas." By the conversation's close it remained clear, however, which one of the two would be offering advice. "I don't [yet] have a clear answer" on how to corral the Soviets, Bush said. "But I will have to do something."

The pronoun mattered as much as the process, as did each nation's fundamental appreciation of their disparate capabilities. The Americans "do not know well what they want because they feel that Europe is escaping them," Mitterrand told his principal advisers in private. But given their varied needs, "there is no reason to make ourselves unduly disagreeable," he felt, by fighting Washington on something so vital as its ongoing presence on the continent.

Of course, the world is far more than just a series of bilateral relationships emanating from Washington. German leaders spoke to Eastern Europeans, who also huddled with one another. The Soviets confided in the French, commiserated with the Poles, and used former subordinates to convey messages to Washington. "It was important for Americans to understand . . . that the Soviet leadership was not focusing so much on specific issues as on obtaining an overall understanding concerning the new European security structure which would emerge from their assent to German unification and to full NATO membership," East Germany's chief arms negotiator in Europe told Washington's. Since Germany was a test case for what to expect in the future, "the Soviets also considered it very important to know Moscow's role in the new European dispensation."

Eastern Europeans asked Canada's Mulroney to deliver the same message. "I can't hide the fact that I would like Poland's standard of living to equal that of Canada," Walesa told a visiting delegation from Ottawa in April. "But the truth is that we can't do it without your help, or without your investments." The

Canadian representative replied that he'd brought along twenty-nine business representatives for just that purpose, though even the most cursory observer realized the Canadians, like others, intended to follow Washington's lead. Indeed, Mulroney and Bush were that very day enjoying dinner, a ball game, and most important a private conversation of their own. "We have a hell of a selling job to do with the Europeans," Bush remarked as the two looked out at the view from a suite in Toronto's new SkyDome. "Especially [with] the French, but also the Eastern Europeans. However, I think we can do it." Bush could count on Canadian support no matter what, Mulroney replied. "On the subject of European architecture, we do not like structures that do not include the U.S. and North America."

Of the two of them, it was obvious which one followed the other, and obvious to Bush what his neighbors to the north cared for most: that same week he announced the start of joint talks on acid rain and other environmental issues critical to Canadian voters. Preliminary talks were under way as well for a broad North American Free Trade Association including both countries and Mexico, something Mulroney sought. He was therefore particularly keen to deliver Bush's message on the strategic issues the president valued most. "I urge you not to pronounce your opposition" to Germany's full membership, Mulroney dutifully implored Gorbachev in May, following the president's directions. "Bush cannot and will not relent on this question."

By late spring only the Soviets remained as a real obstacle to Bush's plan for extending American leadership through NATO's survival. Gorbachev's position appeared weaker with every passing day. New market reforms were causing the Soviet economy to buckle, eroding the last vestiges of political stability. Prices soared yet higher, driven by a ruthless combination of inflation and scarcity. Radicals once more blamed their crumbling economy on Gorbachev's lethargy. Conservatives and the Red Army's increasingly disgruntled officer corps bemoaned the Kremlin's diminished authority at home and the Soviet Union's faltering fortunes overseas. "It seemed increasingly clear that the left and the right despised Gorbachev equally," one aide later wrote, "and there was no center — certainly no organized center."

Gorbachev's disparate critics found at least one area of agreement: a foreboding sense of distrust of the new European architecture. "There used to be two Germanys," one of Gorbachev's advisers said to Condoleezza Rice in May, explaining why the Velvet Revolutions celebrated elsewhere stung in Moscow. "One was ours and one was yours. Now there will be one and you want it to be yours. That would be an unacceptable strategic shift in the balance of power." Full German membership in NATO was a "deal-breaker," he warned. *Perestroika* had been sold to the Soviet people as a way for everyone in Europe to win. It had no future if they came to perceive themselves as its only losers. Shevardnadze deployed that word specifically. "We understand that a neutral Germany is a problem," he confided to Baker. "But we have a problem with NATO. It's an imagery problem." If Germany united under NATO's banner, "it would look as if you had won and we had lost." The Soviet public would never stand for such a thing, he declared, or forgive any leader who gave up so much.

Gorbachev was blunter. "A unified Germany in NATO will threaten the stability that has existed for the last 45 years," he told Baker in May. More than unacceptable, "it will be the end of *perestroika*." Shevardnadze seconded the point, lest the Americans have any doubt that the Soviet government's current unpopularity made it particularly vulnerable to any further blows. "I have told U.S. reporters in the past that there is no alternative to *perestroika*," he confided to Baker during a private moment. "Well, the truth is, that's a mistake. There is an alternative to *perestroika*. If *perestroika* doesn't succeed, then you are going to have the destabilization of the Soviet Union. And if that happens, there will be a dictator."

Secessionist pressures further complicated Gorbachev's position and fueled his calculated appeals to the Americans for understanding. Calls for independence multiplied in far-off republics and in Moscow's near-abroad, as well as for the first time from within Russia itself. Having earlier overseen the removal of the Communist Party from its formal place at the center of Soviet political life, Gorbachev arranged his own election as president of the USSR through voting conducted by the Congress of People's Deputies. The technicalities of these maneuvers matter less than their impact: by staking his legitimacy on Soviet political rather than ideological power, Gorbachev opened the question of whose political authority ultimately mattered most, the union's or its component republics'. On the one hand, the Politburo could no longer dismiss him, which allowed him greater flexibility. "Having become President, the General Secretary de facto turned into a Tsar," one of his generals warned. "There remains nobody to challenge him."

On the other hand, enhanced control over the Soviet state mattered only so long as the Soviet Union remained relevant. Boris Yeltsin had spent the spring consolidating antigovernment fervor behind a vision of a Russia freed from Soviet bureaucrats, leeching allies, and ungrateful republics. Elected speaker of the republic's new parliament in May, he appealed to the intelligentsia who had made *perestroika* possible. Now they cheered his once heretical assertion that Russia's laws trumped those of the Soviet Union. "I won't bear any grudges," Gorbachev told his aides upon hearing the news. But "if Yeltsin is playing a political game, then we are in for difficult times."

Yeltsin had bigger plans indeed, all of which centered on building Russia into an independent power base of his own. July brought his dramatic — and well-advertised — resignation from the Soviet Union's increasingly impotent Communist Party. It "raised me," he said, explaining his move. But it was time for Russians to strike out on their own. "The center is for Russia today the cruel exploiter, the miserly benefactor and the favorite who doesn't think about the future," Yeltsin told his fellow deputies to the Russian Assembly, describing the central (Soviet) authority that had governed their lives since Lenin's day. "We must put an end to the injustice of these relations. Today it is not the center but Russia that must think about which functions to transfer to the center, and which to keep to itself."

The Baltics were roiling as well. Nationalists won control of Lithuania's legislature in late February. A declaration of independence soon followed. "We cannot ignore the interests of our neighbors, particularly our neighbors to the east," the new president declared. "But we will not be asking for permission to take this or that step." Not a single delegate inside the raucous parliament building voted in opposition. Even Lithuania's Communist Party joined the majority. Outside, boisterous crowds pried Soviet insignia from the building with screwdrivers and crowbars. Estonia and Latvia soon issued their own proclamations of sovereignty, albeit with less gusto, hoping to negotiate Moscow's peaceful withdrawal.

Scenes of joyous revolt within the Soviet Union prompted far less international enthusiasm in Western Europe's halls of power than when similar crowds had gathered in the satellite states. "The Lithuanians will mess up everything," Mitterrand ranted. Gorbachev had enough problems already, and any replacement brought to power in order to bring the Baltics to heel might simultaneously roll back democratic reforms throughout the Soviet empire, perhaps even before negotiations over Germany were complete. "Lamentable people," Mitterrand continued. "I would understand it, if Gorbachev [felt] obliged to react by using force."

Gorbachev faced a nearly impossible situation as the spring of 1990 progressed. Hard-liners called for a forceful response to the secessionist crisis, threatening to take matters into their own hands if he faltered. The Soviet Union could not long survive once it began shedding limbs. Yet Gorbachev also knew *perestroika* could never survive if he violated its central tenet of self-determination, and that he could never condone violence deployed in contravention of his most cherished principle. Abraham Lincoln had not let secessionists leave their union without a fight, Soviet commentators reminded their American counterparts. Why should they expect anything less of Gorbachev, who vowed never to "negotiate" with Lithuania's rebels? "You carry out negotiations with a foreign country," he explained. Not with your own citizens whose leadership claims were "illegal and invalid."

We must "keep Gorbachev in the saddle," Thatcher said, even as news reached London that the Soviet Congress of People's Deputies had just invalidated Lithuania's declaration of independence. Soviet military planes buzzed the Lithuanian capital. Helicopters dropped handbills calling for the secessionist government's immediate dismissal. Soviet authorities began organizing ethnic Russian militias armed with weapons confiscated from native Lithuanians. If the secessionists did not come to heel by March 19, Gorbachev vowed, his military would make them. "If one republic secedes, Gorbachev is through," his own defense minister admitted to a visiting American military delegation. "And if he has to use force to prevent one from leaving, he's out, too."

Mid-March brought two crucial, nearly simultaneous events. East Germans were scheduled to cast their votes the day before Gorbachev's deadline for Lithuanians to recant their demands for independence. Late polling put Kohl's Christian Democratic Union behind by more than twenty points, as East German voters faced West German choices for the first time. His party might still retain power, Kohl told Bush the day before the election, by somehow forming a "reasonable coalition" in the aftermath. Already grieving over his expected defeat, Kohl privately told his deputy that he might instead simply "give up and go home."

March 18 brought instead a stunning victory: 93 percent of eligible East German voters cast ballots, and nearly half of them went for Kohl's coalition.

His main rivals secured a mere 22 percent, the communists 16 percent. Most surprising of all, the New Forum movement, led by the dissidents who'd done the most to topple the GDR, garnered less than 3 percent of the vote. The crowds that had marched under their banner in October and November, in other words, and who'd once heckled Kohl as a bourgeois outsider, now considered him their best shepherd to unity and prosperity. "The sensation is perfect!" Teltschik wrote in his diary. "Who would have expected it?" No one was more pleased than Bush, who had anticipated at best a disputed election result, at worst Kohl's outright defeat. "You are a hell of a campaigner," the president happily congratulated his negotiating partner. Soviet policy makers simultaneously interpreted the stunning news as the defeat it was. "Completely unexpected," one diplomat admitted to his American counterpart, and unnerving for the momentum it now placed at Kohl's command.

Kohl's electoral victory quieted talk of neutralization within Western circles. Even Genscher and Thatcher fell in line, the former largely affirming by April that a unified Germany must have full sovereign rights to determine both its alliances and the disposition of forces within its own borders, the latter receiving Kohl politely, though not warmly, in London. "Very well, very well, I am outnumbered," she conceded to her staff. "I promise you that I shall be sweet to the Germans, sweet to Helmut when he comes next week, but I shall not be defeated. I shall be sweet to him but I will uphold my principles."

There must be no "new discriminatory restraints on German sovereignty," she affirmed when she met Bush in Bermuda two weeks later. They agreed to back Germany's right to choose its alliances. Mitterrand would make a similar statement the following week in Florida. NATO's foreign ministers formally did the same by month's end. "The rapid route," which was initially a German creation but now bore American markings as well, had become "a fact," France's foreign minister observed.

The worsening threat of secession consumed whatever political capital Gorbachev might have deployed in defiance. Western leaders collectively warned against precipitous action in the Baltics, noting the ostracism Beijing's leaders had faced in the wake of Tiananmen. Baker in particular reminded Soviet officials that Bush would refrain from angry public rhetoric lest his words complicate affairs for them, yet they should not take his public restraint as indifference. "If you use force or coercion, there would be all sorts of consequences," he told Shevardnadze.

"We've taken note of the nuances of your public statements, and we appreciate them," his Soviet friend replied. Indeed, while directing his surrogates to remind the Soviets that their actions in Lithuania would directly affect relations with Washington, Bush tried whenever possible to apply as little public pressure as he could. "We rejoice in this concept of self-determination," he told reporters when pressed for comment on the growing crisis, "and it is very important to the people of these Baltic States that the evolution be peaceful. And so, I am, just as in East Germany when the Berlin Wall started down — some of my opponents were saying I am unenthusiastic about it . . . [but] sometimes caution and prudence, I think are right. And I think in this case it proved right because that evolution has moved peacefully, and we did not provoke some kind of outbreak through exhorting there at the Berlin Wall that could have caused other countries to act differently." So once more, he said, "we're watching it closely" but hoping to do nothing more than "encourage the fundamental principle of self-determination" and "the concept of peaceful change."

His surrogates said more when presented with what increasingly appeared like Soviet justifications for an expected crackdown in the Baltics. "The Soviet Union had shown restraint in Eastern Europe," Gorbachev's close adviser Alexander Yakovlev told the American ambassador in Moscow at the end of March, and "would continue to do so because it believed in the right of all peoples to make their own socio-political choices."

Lithuania, however, "was an entirely different matter," according to Yakovlev. It was part of the Soviet Union itself. Believing deeply in the principle of self-determination, Gorbachev was willing to discuss Lithuania's departure. He'd even begun drafting a legal framework for independence. But no Soviet leader could tolerate secession at the point of a gun. At least, not if he hoped to survive. "This is a serious matter," Yakovlev stressed, "not one for dilettantes." By weakening Gorbachev, the Lithuanian leadership had provided "a tremendous gift to the opponents of *perestroika*." Their actions might "please some rightists in Europe and the United States," he concluded, "but it would harm the entire world."

Those were indeed the stakes, Ambassador Matlock responded, and the audience. "The world was now watching." Gorbachev had massed troops on the border (to defend against insurgents, Yakovlev countered) and confiscated Lithuanian firearms ("to avoid any provocation"), but Washington's message was plain: "If the Soviet Union used force or intimidation, the world would begin to doubt the Soviet Union supported democracy and self-determination." Hasty action in Lithuania might keep the Baltics in line, Matlock warned, but would simultaneously preclude the Soviet Union from ever fully joining the West. "The United States should not threaten that the Lithuanian issue will affect US-Soviet relations," Yakovlev retorted. Threats were not the basis of the cooperative friendship they all sought, he said, pointing out that the Kremlin had not threatened Washington over Bush's use of force in the Philippines or Panama.

Matlock had a ready answer. "It is not our policy to threaten, only to explain the situation as it was. If force were used in Lithuania, it would be impossible for the United States to continue relations with the Soviet Union as we wanted." Soviet-American relations had improved more over the past year than anyone might ever have imagined. All that progress, and the trust it engendered, would evaporate the moment anyone ordered troops in the Baltics to fire.

Gorbachev's problems gave Bush and Kohl the upper hand in their ensuing negotiations. He needed the pair's understanding, or at least their tacit acceptance of whatever steps he might be required to take in the Baltics or closer to home in order to keep his union and reforms alive. But he also needed their direct support, exemplified by the conclusion of a broad Soviet-American trade agreement first outlined at Malta and expected to be signed when Gorbachev visited Washington in early June. Soviet coffers required at least $20 billion in new loans and credits over the next few years just to remain solvent, Gorbachev confided to Baker that spring, and needed the trade deal as a green light to signal Western investors that the Soviet Union was worth their time and credit. "The best way to put it is we need some oxygen," Gorbachev said. "We are not asking for a gift. We are asking for a loan; we are asking for specifically targeted loans for specific purposes." The Soviets had no good options left. Shevardnadze, clearly pained by political infighting and a secessionist crisis that touched his heart as a Georgian and as a Soviet, had what Baker termed the look of "a diplomat with a political gun to his head. Any step forward could lead to suicide."

Kohl heard the same pleas from Moscow. Bonn guaranteed $3 billion in mid-May, with the promise that more aid would be in the offing so long as the Soviets continued to demonstrate their commitment to reform. The German aid wouldn't be enough, Treasury Secretary Nicholas Brady advised Bush, not even if supplemented with American funds. In each case the infusions were merely temporary salves, most important for their political effect. "The Soviets seem to lack any sense of the true magnitude of reforms that are required," he wrote. "The sort of piecemeal approach Gorbachev is attempting has proven time and time again to yield nothing but further miseries." But piecemeal was all Gorbachev could politically sustain as spring progressed into summer, along with the hope that his economy would eventually pick up steam before voter patience fully evaporated. "If the Moscow ministerial" between Baker and the Soviet leadership "demonstrated one thing," State Department analysts reasoned, it was that "Gorbachev is . . . prepared to bargain. Despite signs of bureaucratic disarray around him . . . he views the [upcoming Soviet-American summit] as a mark of how far U.S. Soviet relations have come . . . and sees the summit as a means of easing domestic pressures" by winning further American approval, and the funds he needed to give his economic reforms time to succeed. Baker's intelligence unit was clearer: "Despite signs of increased domestic unease over the course of Soviet policy," his analysts argued, "Gorbachev will come to Washington able to deal."

Bush wanted to give Gorbachev room to succeed, but he knew that space was tightening. The Senate voted down a resolution calling for immediate recognition of Lithuania's independence by only a narrow margin, and only after significant White House lobbying. Senators then voted unanimously in condemnation of the Soviet government's actions in the Baltics, their chamber filling with angry calls for immediate retribution if the Kremlin made any further aggressive moves. "We are playing out one of the great moral moments in modern history," William Safire opined in the *New York Times*, while two other influential columnists wrote that Bush's "sweet talk" toward the Soviets, rather than the harsh talk the situation required, was telling Gorbachev that "military force can work." Even moderate Republicans were "beating the drums for Lithuanian independence," Scowcroft worriedly told his staff, while the editors of the *Philadelphia Inquirer* warned that "there must be no hint that the administration would respond to bloodshed in Vilnius as it did to the slaughter in Tiananmen Square, with warm toasts to the Communist officials who ordered the killing."

So far Soviet troops had refrained from opening fire. But they'd hardly been passive. Soviet paratroopers occupied Communist Party buildings inside

Lithuania on March 23. The next day their tanks rolled up to parliament, gun turrets pointed toward the legislators within. "If there is a Tiananmen Square in Vilnius," Senator Edward Kennedy warned during a visit to the Kremlin, Soviet-American relations would undoubtedly suffer. "You don't know what pressure I'm under," Gorbachev replied. "Many in our leadership want us to use force right now."

Bush might well have said the same thing. "Everyone wants us to 'do more,' though nobody is quite clear on what that means," he lamented to his diary after recording the Senate's angry words and near-run resolution. It was easy for legislators to talk and columnists to rail. Their words meant little. His echoed. "So they're hitting me as a wimp unwilling to move," and in particular uninterested in recognizing Lithuania's sovereignty, he complained, venting into his tape recorder in a manner he'd never allow himself to do before rolling cameras or reporters. "But the big thing is to get through this so the Soviets and the Lithuanians can get into negotiations and handle it without bloodshed and force . . . It's funny how these things aren't quite as simple as they seem from the outside."

Gorbachev had other options shy of brute force. The late March deadline passed without further incident, but without resolution either. He thus authorized a full energy blockade on April 19, hoping darkness and cold might bring the nationalists to heel. Mitterrand pleaded for Bush to refrain from any commensurate response, at least until resolution of the German issue, arguing that the West should "not require of Gorbachev what we would not require of the dictator who would succeed him." The Lithuanian "fanatics" were clearly in too much of "an unfortunate hurry," the French leader argued. But without restraint, "this will end in blood."

"I don't want people to look back twenty or forty years from now and say, 'That's where everything went off track'" in potentially ending the Cold War, Bush worriedly told his staff. He didn't want them to say, "'That's where progress stopped.'" Anything he could do to keep Gorbachev from deploying force, and to keep him in power, he reasoned, brought them closer to a peaceful resolution. The truth, no matter how ugly or painful to admit, was that the fate of the Baltics mattered less to him at that moment than the broader fate of the waning Cold War. Eastern Europe had turned toward democracy on Gorbachev's watch. If the Lithuanians and others merely bided their time, democracy would come their way too, especially if Gorbachev remained in power, and if they all somehow managed to ensure German unity and NATO's survival for the long haul.

BUSH BEGGED FOR CALM in Vilnius as in Beijing, a year apart. The similarities and differences between his options in the summer of 1990 and those of a year earlier are nonetheless illuminating. He'd wielded little leverage over China's leadership in 1989, being capable only of threatening long-term political and economic consequences when all the leadership in Beijing cared for was their immediate survival. With Gorbachev and Lithuania, conversely, Bush held powerful cards indeed. Beijing faced economic uncertainty in 1989 but nowhere near the threat of imminent chaos that pervaded Moscow in 1990. Neither did the Chinese have so desperate a need for immediate Western aid, prompting Bush's closest advisers to recommend he directly tie the Soviets' behavior in the Baltics to their financial fate. "Jim Baker and I talked this morning and we believe that suspending the economic agenda that you proposed to Gorbachev at Malta is the cleanest and most coherent response" to the Soviet energy blockade, Scowcroft advised Bush on April 23. They should suspend trade, tax, and investment agreements and withdraw American support for Soviet observer status in the General Agreement on Trade and Tariffs, otherwise known as the GATT, which largely set the rules of international trade within the noncommunist world.

Collectively these measures were something the Kremlin desperately sought, both as a symbol of Western support and as an economic lifeline. "This administration has been restrained in both action and deed," read Bush's notes for his late April session with his national security team; "I know this is a difficult situation for the Soviet leadership." Fundamental issues were nonetheless at stake. "I cannot stand by and watch the repression of the Lithuanian independence movement," he declared. Neither did he want his passivity to once more fuel calls for a more draconian response. "I also want to capture the agenda and not appear to be reacting to Congressional pressure."

Equally important was avoiding the mistakes of the past, and not only of his own time in office. "I'm old enough to remember Hungary in 1956," he remarked yet again to reporters, the example clearly stuck in his mind. "We exhorted people to go to the barricades, and a lot of people were left out there all alone." He'd seen the effect even of inadvertent words at Tiananmen, and knew how close the world had come to real violence the previous fall in Europe. Nothing the Soviets had done was as yet truly irreversible, he feared, especially if someone else seized control in Moscow. "We need to remember Gorbachev's problems with his own right wing," Bush warned Mitterrand. "But we must remember [as well] that it is immoral to accept the Soviets cutting off gas supplies to Lithuania. Economic countermeasures may be the best solution."

The Soviet economy would quickly prove even weaker than feared. New price controls announced in late May led to panic buying. Then hoarding. For the first time in years, Muscovites were forced to queue for basic staples like noodles, rice, and sugar. Yeltsin railed against Gorbachev's "plunder of the people" and called for Russians to stand in solidarity with the Lithuanians against the government they both yearned to discard. Gorbachev took to the national airwaves pleading for calm, but to little avail. His speech contained nothing but the same worn promises, delivered wholly without spark or enthusiasm. He looked "less like a man in control and more [like] an embattled leader," Matlock reported home. More ominously, "signs of crisis are legion: sharply rising crime rates, proliferating anti-regime demonstrations, burgeoning separatist movements, deteriorating economic performance . . . and a slow, uncertain transfer of power from party to state and from the center to the periphery." Having called Gorbachev to offer her support, Thatcher marveled afterward at how tired he sounded. "Like a man who had just lost his father."

Russia's new legislature elected Yeltsin its president on May 31, leaving Gorbachev, already on his way to Washington for his long-awaited summit with Bush, politically embarrassed and, if possible, even more fatigued than before. "You only live once, is all this really worth it?" he lamented to his inner circle. "I don't mind giving my all for something worthwhile, life's not meant for just wine, women, and song. I have no regrets." But "now they're screaming: 'chaos!' 'The shelves are empty!' 'The Party's falling apart!' 'There's no order!' But was there a choice? History's never been made any other way. As a rule, big changes like ours are accompanied by major bloodshed. We've managed to avoid it thus far." But now, he said, his own people "swear at me, curse me . . . they hate me."

Flying to North America in late May, first to Ottawa and then to Washington, Gorbachev needed Bush more than ever, and for the trade deal Bush was suddenly threatening to withhold. "Germany definitely overloads his circuits right now," Baker advised Bush. Gorbachev had the previous week directly accused the Americans of taking advantage of Soviet difficulties. "He almost seemed to be saying that in his hour of need, he didn't need us to complicate his life." Two weeks earlier, Baker had offered nine points of assurance on Germany, ranging from its prohibition on developing nuclear or biological weapons to the granting of a "specified transition period" during which Red Army forces would be able to withdraw peacefully, without pressure from oncoming German forces. NATO, moreover, would rethink its overall priorities in light of Europe's altered strategic map. If the Soviets conceded Germany's right to join whatever military alliance it desired — in other words, NATO — the Americans would make sure the door to East-West cooperation remained open as wide as *perestroika* required.

What if "I say to the President that we want to enter NATO?" Gorbachev asked Baker in response. It was at once a rhetorical question but also an important symbolic query useful for determining just how far the Americans were willing to go in their talk of a new Europe. "After all, you said that NATO wasn't directed against us, you said it was a new Europe, so why shouldn't we apply?"

Baker didn't say no. "I got that question in a news conference in Bonn," he replied, leading Gorbachev to interject: "Well, it's not such a hypothetical question. It's also not so farfetched." Frustrated by Baker's continued evasions, Shevardnadze offered an interjection of his own: "If we are to achieve a new Europe without blocs then it is not such a fantasy to think that the Soviet Union might apply to NATO." The question revealed the heart of Soviet concerns. Neither Kremlin leader truly wished to join the organization. But neither did they want to be the only ones excluded. If the Americans were unable to imagine a day when the Soviets might be their partners, refusing even to accept that possibility no matter how remote, then they would never prove able to treat them as equals, either. "You really should not leave us isolated at this crucial moment," Gorbachev warned. "We may have unusual moves to make."

Gorbachev's stress made any major concession on Germany unlikely during the upcoming summit, Scowcroft's NSC briefed Bush. "We should not expect him to agree to our approach" but should instead focus on winning "practical results" and identifying areas of agreement "where he might be willing to move." It would be a long time before the Soviets conceded Germany's right to unify within NATO and on its own terms, White House strategists predicted. "NATO is associated with the Cold War," Gorbachev told *Time* magazine, "as an organization designed from the start to be hostile to the Soviet Union." The issue hit the newsstands just in advance of his arrival in Washington. "We will never agree to assign it the leading role in building the new Europe. I want us to be understood correctly on this."

Scowcroft's strategists took that admonition to heart. NATO was "our one core objective" for the summit, they reminded Bush. But it was also the one thing Gorbachev could not easily concede, though "economic support for perestroika (now high on Gorbachev's agenda) may also prove the big wild card for bringing the Soviets along to our unification objectives." Bush wanted Gorbachev to "come out feeling he has had a good summit," he told Kohl. The Soviet leader clearly needed a victory to use at home, and might even be willing to pay dearly for one.

Hope of just such a victory, coupled with an appeal to Gorbachev's inherent love of logic, ultimately brought a breakthrough on Germany. Seated in the Oval Office on May 31, Gorbachev came right to the point. The Soviets needed American economic help. He in particular needed to return home with the trade deal currently hamstrung by congressional objections to the situation in the Baltics. He would pursue a peaceful resolution to the crisis, Gorbachev pledged, but he required the trade deal in exchange.

"To the degree that we see a commitment to your own principles of self-determination" in the Baltics, and an end to the embargo most especially, Bush replied, "we can cooperate." Restraint was key, for nations as well as leaders. "I've tried to conduct myself in a constrained way because I know you have big problems," Bush explained. "But I am being hit both on my left and my right by those who say that I am subordinating U.S. dedication to principle" by not pushing harder for Lithuania's freedom. "We need to work together," he concluded.

"The sums are not big," Gorbachev responded before their session came to a close, "but it is important that they be available. The U.S. would not need to provide it all, but you should be supportive. It is important that we not fail in this effort." Determined to ensure Bush understood the full import of his message, he pulled the president aside after dinner. "Gorby got me in the hall and said if we didn't have a trade agreement, it would be a disaster," Bush told his diary that night. "It would be terrible — repeated: a disaster . . . don't embarrass us," the Soviet leader had implored. "Don't humiliate us."

Bush perceived an opportunity. Taken aback by Gorbachev's direct appeal, and his apparent willingness to concede Lithuania's eventual right to self-determination, Bush raised the issue of Germany once their formal session began the next afternoon. It too should have the sovereign rights of any other state, he began, including the right to choose its own alliance.

"To my astonishment," Bush later wrote, "Gorbachev shrugged his shoulders and said yes, that was correct."

The room went silent. Then it erupted. Gorbachev's aides openly recoiled, shooting one another angry and perplexed glances, unable or unwilling to believe what they'd just heard their leader concede. Bush's aides implored him to make Gorbachev repeat the statement. "I'm gratified that you and I seem to agree that nations can choose their own alliances," Bush thus offered again. Once more Gorbachev nodded. "I agree to say so publicly, that the United States and the USSR are in favor of seeing a united Germany, with a final settlement leaving it up to where a united Germany can choose."

Bush pushed further. "I would put it differently," he said. "We support a united Germany in NATO. If they don't want in, we will respect that."

Slowly, deliberately, almost as though suffering under the weight of 20 million dead on his shoulders, Gorbachev nodded again. His staff, meanwhile, were in a near uproar, caught between rescinding the monumental concession their leader had just offered and publicly undermining what was left of his authority in the midst of a high-level negotiating session at the White House. Shevardnadze and Baker would handle further details, Gorbachev proposed, only to have his own foreign minister repeatedly refuse. If Gorbachev was to make this deal, Shevardnadze effectively said, he'd have to do it on his own, and suffer the consequences on his own as well. Soon after their session ended, Soviet officials could be seen screaming at one another on the White House lawn, Gorbachev's personal aides pitted against their military and Foreign Ministry counterparts, the latter demanding to know why their leader had said such outlandish things, the former wondering why the others had not dutifully leaped to his support. "It was an unbelievable scene," Bush later wrote, "the likes of which none of us had ever seen before—virtually open rebellion against a Soviet leader."

It was a rebellion indeed. Gorbachev had clearly exceeded his brief, and the position the Communist Party's Central Committee had authorized only weeks before. "Once again emphasize that it would be politically and psychologically unacceptable for us to see a united Germany in NATO," they'd advised him to say. "We cannot agree to the destruction of the balance of power and stability in Europe that would inevitably result from this step."

Why did Gorbachev rebel, directly contravene his own party's wishes, surrendering in what would become the fourth and final key moment in Germany's unification after so many months? Three intermingled reasons suffice to answer the question. First, he needed American aid, and the broader Western aid he hoped it would signal, more than he needed Germany to remain out of NATO. He'd get the aid he needed, and the symbolically important trade deal, the moment he released Lithuania from his grip, Bush promised. Bush would suffer on Capitol Hill to be sure, but he'd see the deal passed through Congress, provided genuine negotiations commenced with Vilnius. Gorbachev would not get that deal unless he first retreated on Germany.

Second, and this reason should not be discounted merely because it is simple, Gorbachev conceded Germany's right to choose its own alliances because it made sense. He'd lived his life driven by the pursuit of philosophical truths, and he ultimately could not sustain an illogical argument past its breaking point. He believed in self-determination. He'd staked his entire political philosophy on it. How could he publicly state that the fundamental principle he upheld as a universal truth applied only when convenient? History should decide Germany's fate, he had proclaimed repeatedly as a shield. But his entire revolution was predicated on transcending history as well.

Third and most important, Gorbachev conceded Germany's right to join the alliance of its choosing in order to keep alive his slim hope that, with further Western aid and support, he might yet prove his country a worthy partner in Europe's future architecture. He'd seen Bush stand up to public opinion when geopolitics demanded, first in his refusal to cut ties with Beijing immediately after Tiananmen, and more recently through Bush's steady support for China's continued most favored nation trading status despite fierce congressional opposition. He knew therefore that Bush could do the same for Moscow if necessary, and if so inclined. "You have now given MFN treatment to China after Tiananmen Square," Gorbachev noted to congressional leaders during a brief visit to Capitol Hill. "What shall we do" to receive similar treatment? It appeared Americans gave positive attention only to despots capable of pulling triggers. "Maybe we should introduce presidential rule in the Baltics and fire off some rounds in the Baltics?"

"Mr. President, I think already we can state the fact that our meeting became an important event," he told Bush the next day. "There still remain some disagreements between us, but yesterday already allows us to say that we're not just making statements but making practical steps, practical policy . . . Mr. President, I want to assure you that we are ready to do further constructive work."

"As a whole, the work we have done is a big plus for both of us," Bush responded. "Of course, some people in our country will throw little stones at us. In concrete terms, Congress will criticize the trade agreement," he continued. "They will definitely ask me how did it happen that two weeks ago I said that the deterioration of the situation in Lithuania was causing tension in U.S.-Soviet relations, but then we signed a trade agreement at the summit. To this," he finished, "I will reply that the trade agreement was in the interest of both sides."

There remained significant technical issues still to decide, not least who might pay the cost of repatriating Soviet soldiers in East Germany back to their homeland, though Kohl ultimately agreed to pay the bulk of these expenses when meeting with Gorbachev in July. But in Washington, having already lined up his allies through countless hours of persuasion and bargaining over the previous months, Bush won concession on the issue he cared for most, trading Soviet acquiescence on Germany, and thus on America's future in Europe, for the promise of further cooperation.

The trade deal signed in Washington was not ultimately enough to keep the Soviet economy afloat. But it was enough to keep the idea of *perestroika* alive. It was, in the words of one of his advisers, "purely symbolic." Gorbachev himself termed it "a political gesture of support for a market-oriented economic policy." It was also a gesture of hope. "At a moment when perestroika had far greater problems — even the danger of being overturned altogether," one of his aides noted, "Gorbachev still tried to persuade the U.S. president that even a symbolic gesture would be of great significance."

For the chance to continue toward Europe, he gave up what Soviet leaders had vowed for decades never to allow: their German enemy fully embedded in their rival's alliance. He told the Supreme Soviet as much on June 12 — crucially, after his sessions with Bush but before his final meetings with Kohl — admitting that while it might take some time to work out the details, the Soviet Union would ultimately accept Germany's right to choose its own alliances, and thus its right to join NATO fully, without reservations. Two weeks later he lifted the Soviet embargo against Lithuania. Kohl pledged additional credits and loans in response and a limit to the initial deployment of German troops on former GDR soil until the Red Army's complete departure. As thanks, and to keep their agreements and increasing trust snowballing, Gorbachev promised yet again to respect Germany's full sovereign rights. "Gorbachev accepts that Germany remains a NATO member," Teltschik recorded in his diary. "What a sensation!"

The news echoed differently in Moscow. A ranking general raged, "We have lost World War III without a shot being fired."

19

"THIS WILL NOT STAND"

IF IT IS SURPRISING that violence never erupted in Germany, it is unsurprising where it did emerge. No modern-day White House is ever truly taken aback by bad news emanating from the Middle East. "Mr. President, it looks very bad," Scowcroft said on August 1, 1990, finding Bush in the building's basement medical suite, his shoulder wrapped to counteract the effects of the golf balls he'd hit earlier that afternoon. Richard Haass, the NSC's Middle East director, followed his boss into the room, the same worried look on his face. "Iraq may be about to invade Kuwait."

News reports quickly confirmed the worst. More than 1,800 tanks and 140,000 troops were swarming across the border. A mere four ill-trained and ill-equipped brigades opposed them. Outgunned local defenses melted away. Civilians fled, trailing Kuwait's leadership. By nightfall Kuwait City would be under Iraqi control, then soon the entire country, destined, in the words of Baghdad's propagandists, to become their "19th province, an eternal part of Iraq." In little more than a day, Iraqi leader Saddam Hussein fundamentally altered the world's strategic map, placing himself in control of one fifth of known oil reserves. Another 20 percent lay little more than a day's drive away. Whether he'd be sated by his conquest or would still desire more remained anyone's guess. "Saudi Arabia looks like the next target," the State Department's Lawrence Eagleburger warned.

It had otherwise been a largely satisfying summer, at least on the international front. By July, Bush had secured the objectives he valued most: Germany appeared set to remain fully in NATO; the Soviet Union for all its woes remained committed to working with the West; and his own country appeared secure in its self-proclaimed role as Europe's overseer. The stream of history continued to flow in the direction of democracy. "I take pride in the way Europe is moving into this new era of freedom," Bush boasted to reporters. "We've got some very important problems that lie out there ahead of us," he added. "But isn't it exciting when you think back to a year and a half ago to where we stand today?"

Voters appeared largely to agree. More than 60 percent approved of his performance, which, aides gleefully noted, outpaced Reagan's numbers at the same point in his first term. Midterm elections were still months away, and unexpected events would undoubtedly arise. Gorbachev continued to have problems, and thus so did any strategy predicated upon his hold on power; Eastern Europe's new democracies faced a difficult struggle to repair the damage inflicted by generations of communist rule; and those at home whom Bush called "isolationists" in public — and "stupid peace-nik[s]" in the privacy of his diary — might yet try to prematurely curtail Washington's global presence. But the democratic future he'd proclaimed at his inaugural remained intact. "Maybe it's the martini," Bush had dictated weeks before, pausing to reflect after a day's work. "[But I'm] convinced we're on the right track in the big picture."

Domestic perils dampened his enthusiasm. Estimates put the upcoming federal deficit at more than $170 billion, over 4 percent of gross domestic product, an amount scheduled to trigger across-the-board spending cuts if not drastically reduced before October's new fiscal year began. Negotiations between the White House and congressional leaders remained deadlocked, leaving all manner of federal programs on the chopping block, including the defense spending Bush was so keen to preserve. "If I didn't have this budget deficit problem hanging over my head, I would be loving this job," he told his diary in July.

Those negotiations hung on a single issue. Democratic leaders sought to offset spending cuts with higher taxes. House Republicans, conversely, rejected any whiff of taxation as a betrayal of their core conservative message. Bush was on the record as supporting their side. "Read my lips," he'd promised to cheering applause at the Republican National Convention in 1988. "No new taxes." Trailing Dukakis by double digits, he had considered it a prudent promise at the time.

Two years later it felt more like a straitjacket. Conservative stalwarts and his own pollsters urged him to stand by his pledge. Reagan had supported budget compromises and "revenue enhancements" in the past, and the entire system of automatic cuts derived from a bipartisan response to his profligate spending. Yet few doubted Reagan's conservative credentials, even when the facts suggested otherwise. Conversely, nearly two years into his presidency, and after two-plus decades of national service, "you are [still] not known for much," political consultant Roger Ailes advised Bush. "But you are known for character." Voters "identify you as a man of your word," he noted. "Don't break it."

Political reality nonetheless demanded compromise. Having negotiated for months to keep American troops welcome in Europe, Bush loathed the prospect of having to recall them because of draconian budget cuts he should have been able to avoid. Besides, his economists warned, cuts alone would never fully dig the country out of its current fiscal hole. Ideology was one thing. The numbers — including those not reflected in the polls — demanded a more balanced approach. "It was a mistake," Bush later said of the 1988 campaign pledge. "I meant it at the time, and I meant it all through my presidency . . . but when you're faced with the reality, the practical reality, of shutting down the government or dealing with a hostile Congress, you get something done." By late June 1990, he was quietly signaling staffers to begin discussing potential new revenue streams.

"Read My Lips," the New York Post headlined once word of his concession leaked. "I Lied." Conservative hard-liners revolted. Calling Bush's decision "a betrayal of Reaganism," House minority leader Newt Gingrich openly disavowed the deal, while the National Republican Congressional Committee advised candidates to run against their president's record in November. The man they'd accused of being too much of a "wimp" to make hard decisions had made one. "The right-wingers are very upset with me," Bush confided in his diary in response. "I guess this is the biggest test of my presidency. Time will tell."

Iraq's invasion quickly wiped the headlines clear of tax battles. Like a massive snow globe shaken by an invisible hand, the Cold War's prospective end had unsettled every part of the world, forcing strategists to wonder where all the scattered pieces would eventually come to rest. Each superpower had since the 1950s developed its own allies and minions in the strategically vital Middle East, and by the start of the 1990s, they all had to wonder if the end of Soviet-American antagonism might also mean the end of their own strategic importance — and of the largesse that accompanied it. Gorbachev pulled the last Soviet troops out of Afghanistan, freeing Islamists to direct their anger at local apostates rather than toward the atheists in Moscow. The long Iran-Iraq war, meanwhile, which had for the better part of a decade pitted Tehran's desire to expand its religious revolution against Iraq's vision of secular influence, finally ended in 1988. Iranian leaders boasted of nearly 500,000 battlefield and civilian "martyrs." The Iraqis had lost 300,000 troops of their own, with a further million wounded. To put those numbers in context, each represented a greater proportion of Iraq's overall population than the casualties suffered by Britain in World War I.

The long conflict left economic scars as well. Iraq claimed $35 billion in known foreign exchange reserves in 1980. A decade of war left it $80 billion in the red, alongside what American analysts calculated to be an additional $230 billion in immediate reconstruction needs. Almost entirely dependent on oil exports, the country had a gross domestic product of only $38 billion annually, leaving little hope for Saddam Hussein to repay Iraq's creditors or alter his country's fortunes without radically changing its geopolitical position.

His own survival seemed at stake. Weakened by popular frustration over the war's muddled end (a stalemate, with neither side gaining ground despite so many years of conflict) and confounded by the impending reintegration of tens of thousands of wartime veterans into an economy ill prepared to receive them or to pay the pension costs for its widows and orphans, Saddam faced a growing wave of popular dissatisfaction. He knew better than most that unpopular Iraqi leaders rarely left office of their own volition, or alive.

Saddam also felt himself to be in Washington's crosshairs. The Reagan administration had provided $200 million in arms to Iraq during the war while simultaneously selling missiles to the Iranians. The ensuing scandal rattled American politics, though in truth there were some in Washington who were happy to see the two Muslim powers weaken each other. But whereas Washington pundits (and investigators) came to care most about whether Reagan had illegally authorized arms transfers, the missiles themselves were taking Iraqi lives. Arming both sides was low, Saddam complained, even for "conspiring bastards" like the Americans.

He nonetheless took Washington's "stab in the back" as a compliment of sorts, believing that Reagan's decision to arm Iraq's enemies had been fueled at least in part by a desire to keep him from achieving his true destiny: leadership of the entire Arab world. Arabs would throw off colonialism's historical yoke and evict the Zionists in their midst, Saddam frequently argued, only by uniting under a single banner. "Who can carry [out] this role?" he had rhetorically asked. Surely not Egypt, which had made deals with Israel and Washington; not the Lebanese or Palestinians, each in their own way stateless; and surely not the decadent, oil-soaked Persian Gulf states he dubbed "the Arabs of decay, the Arabs of shame." To his mind, "no one else but Iraq can make this [Arab] nation rise and be its center post of its big abode," and no one could lead Iraq but him.

Saddam considered it only a matter of time before the Americans turned their full attention from securing Europe to ensuing their dominance over the greater Middle East, and over him. "The Iraqis never quite trusted us," Reagan's ambassador to Baghdad noted. They considered Bush's administration little better, especially after the Voice of America altered its programming in early 1990 to reflect the profound strategic changes of the previous few months. "The tide of history was running against dictators," broadcasters told Iraqis in February, furthering Saddam's belief that Washington had earmarked him for removal. It "had already swept aside several, such as the Ceaușescus in Romania." The image of their bullet-riddled corpses carried special meaning for an Iraqi regime that similarly ruled through fear. "The Americans want to control the region and we are the only obstacle to them," Saddam's brother-in-law advised. With the war against Iran over and Soviet influence waning, Iraq's "real danger is the United States and its follower, Israel."

Needing cash and feeling pressed for time, Saddam looked nearby for easy prey. Saudi Arabia and other Persian Gulf states forgave or refinanced Baghdad's wartime debts on the generous terms he demanded, conceding his argument that Iraqi blood had protected them all from Tehran's reach. Saddam's massive military, by some measures the world's fourth largest, provided any additional persuasion required. Oil-rich Kuwait proved less pliable. The small principality refused to forgive any of the $8 billion in loans provided during the conflict, and simultaneously opposed Iraq's plea for an increase in its export allotment regulated by the Organization of Petroleum Exporting Countries (OPEC), to which each belonged. "War is fought with soldiers and much harm is done by explosions, killing, and coup attempts," Saddam publicly blustered in response, "but it is also done by economic means . . . [T]his is in fact a kind of war against Iraq."

Saddam promised that his soldiers would have the last word. "It is not reasonable to ask our people to bleed rivers of blood for eight years then to tell them, 'Now you have to accept aggression from Kuwait, the U.A.E., or from the U.S. or from Israel,'" he complained to American ambassador April Glaspie during a private meeting on July 25. "Saddam defends them [Arabs] for ten years [and] they consider his defense a liability," but "the time has come for every person to say . . . I'm Arabian . . . I'm Saddam Hussein . . . Iraq will pay this amount of money to develop the Arab nation and to defend it, [thus] the other Arab countries must pay this amount of money—if they don't, we will fight them."

"We can harm you too," he warned Glaspie, should Washington intervene where it was not wanted. "Everyone can cause harm [according] to their ability and their size. We cannot come all the way to you in the United States," he added ominously. "But individual Arabs may reach you." More profoundly, Americans had no stomach for long wars in far-off places. "Yours is a society which cannot accept 10,000 dead in one battle." His, conversely, had already proved capable of making such sacrifices in his name. And more.

This was Glaspie's first private meeting with Saddam, but she had heard all this before. Two decades of service in the Middle East had rendered her largely immune to the region's theatrics and well versed in its basic dynamics. Saddam had threatened his neighbors before. Each time, they'd paid. She consequently took his bluster in stride, even when his official note taker and his translator both spontaneously, albeit simultaneously, burst into hysterics upon hearing the drastic cuts he'd soon be forced to impose if Iraq's debt crisis continued unresolved. Saddam knew his lines. His underlings knew their cues. Those he intended to extort, everyone presumed, understood their next move as well.

Glaspie understood her role too, which was to serve as messenger rather than arbiter. President Bush appreciated Iraq's sacrifices, she replied once calm had returned to the room. All of Washington desired Iraqi prosperity and the region's overall security. The Americans had recently allotted Iraq the largest agricultural credit in U.S. history to help alleviate food shortages, she reminded him, and had authorized high-level congressional delegations to stress Bush's peaceful intent. Having "lived here for years," Glaspie added on a personal note, "I admire your extraordinary efforts to rebuild your country. I know you need funds. We understand that, and our opinion is that you should have the opportunity to rebuild your country.

"But we have no opinion on the Arab-Arab conflicts, like your border disputes with Kuwait," she said. "We hope you can solve this problem using any suitable methods." While her government hoped for a speedy resolution of the crisis, it also would "never excuse settlement of disputes by other than peaceful means."

Those words essentially ended her career. Subsequent release of the transcript of their conversation made it appear to those uninitiated in the language of diplomacy as though, by urging speedy resolution, she'd given Saddam carte blanche to take Kuwait by force. A rebuttal stronger than a rote statement that Washington "took no position" on the border dispute, critics charged, might well have quelled Saddam's ambitions. "The United States may not have intended to give Iraq a green light," two esteemed foreign policy analysts suggested in 2003, "but that is effectively what it did."

Such condemnations simply ignore the facts. American officials routinely refuse to take formal sides in any of the myriad border disputes ongoing at any time around the globe, lest the world's most powerful state find itself perpetually adjudicating contested lines on maps. Glaspie's words, the product of years of training, were thus the ones employed whenever diplomats confronted similar historic geographic grievances. She "took the American line," as another U.S. ambassador to the region subsequently explained, "which is we do not take positions on border disputes . . . [T]hat's standard. That's what you always say. You would not have said, 'Mr. President, if you really are considering invading Kuwait, by God, we'll bring down the wrath of God on your palaces, and on your country, and you'll all be destroyed.' She wouldn't have said that, nor would I. Neither would any diplomat."

For one thing, the United States lacked any formal defense agreement with Kuwait or its neighbors. Any promise of retaliation for an assault would therefore have been entirely unsupported by recent precedent or obligation, and would thus wholly have lacked credibility. American diplomats routinely warned of their country's willingness to fight on behalf of its friends and allies. In the summer of 1990, however, America's partners in the Persian Gulf — Kuwait and Saudi Arabia especially — were friendly acquaintances, not treaty partners. "We do not have any defense treaties with Kuwait, and there are no special defense or security commitments to Kuwait," State Department spokeswoman Margaret Tutwiler reminded reporters back in Washington who pressed her about how Bush might respond if Saddam made good on his military threats. Indeed, hoping to bolster the credibility of a potential military response, Pentagon planners found only the United Arab Emirates among all the gulf states willing to conduct joint military exercises as a sign of deterrence. The rest found scant reason to put their faith in Washington. Saddam "felt secure" from attack, Glaspie reported home, and "secure in the belief that no Arab government would ever allow us to use their land for that purpose."

Glaspie did not threaten what she could not guarantee, and subsequently declassified documents reveal that her responses also echoed her specific instructions. If there is blame to be laid at the feet of American officials for not warning Saddam off, in other words, it is not hers alone to bear. "We believe that differences are best resolved by peaceful means and not by threats involving military force or conflict," the State Department instructed her to emphasize anew following her session with Saddam. She'd already said as much. "While we take no position on the border delineation issue raised by Iraq with regard to Kuwait, or on any other bilateral disputes," her instructions continued, "Iraqi statements suggest an intention to resolve outstanding disagreements by the use of force, an approach which is contrary to UN-Charter principles."

If Saddam actually did invade Kuwait as a result of a supposed "green light" from Washington, blame would extend all the way to the White House, though real blame for the invasion should lie with the one who ordered it, not with those who attempted to resolve the brewing conflict. Bush planned to say precisely the same as Glaspie during a telephone call scheduled with Saddam for the evening (Washington time) of August 1. His staff hoped that a bit of personal diplomacy might help reduce tensions over the disputed border, though the call was canceled once word of Iraq's aggression spread. "I again want to underscore my strong belief that it is important to find a peaceful solution to this dispute," Bush's talking points for the conversation, prepared by Scowcroft's NSC though never employed, encouraged him to say. "As we have said before, we do not intend to take sides on the issues currently in dispute between you and Kuwait." Such documents validate Glaspie's performance and rebut her critics but ultimately do little to salvage her reputation. The first American woman appointed ambassador in the Middle East, she spent the remainder of her career relegated to less volatile posts, retiring in 2002.

Like everyone else familiar with the region's recent history, she presumed Saddam was merely bluffing. Baghdad's angry threats would soon dissipate "like a summer cloud" once the two sides negotiated a new settlement, Kuwait's emir predicted a week before the invasion. They'd done so before. Both Egypt's Hosni Mubarak and Jordan's King Hussein had worked in the preceding weeks to mediate the dispute, keeping it "an Arab affair." Everyone assumed Saddam's fiery rhetoric was for show. "[I] talked to the Kuwaitis, particularly the oil minister," the American ambassador to the United Nations recalled, "and I asked if they thought there was any possibility the Iraqis would resort to use of force. They said no, absolutely not, that everything is okay, that the Iraqis just wanted to get more money . . . that is what this negotiation is all about."

This was the way Middle Eastern politics worked, King Hussein privately told Bush on July 31. "They [the Iraqis] are a bit angry about this situation, but I believe that hopefully something will be worked out to the benefit of greater cooperation and development in this arena." Bush asked, "Without any fighting?" The king responded confidently, "Oh yes, sir, that will be the case . . . I really implore you, sir, to keep calm," adding, "We want to deal with this in an Arab context, to find a way that gives us a better foundation for the future."

Egypt's Mubarak cautioned the same, reminding Bush of Saddam's ambition to lead the Arab world. Any intervention by the Americans, and consequently a perception that an outside colonial power had forced his capitulation, would prove more embarrassing than helpful and only make matters worse. "We are trying hard to solve this [on our own]," Mubarak pleaded, "to find a good solution for withdrawal and not throwing away the regime." Having no desire to intervene in the matter, and not wanting to discount their advice or poison subsequent relations, Bush consented to let the Arabs take the lead. Not to worry, Mubarak reassured him. "I'm used to disasters in our area."

But Saddam wasn't bluffing, and Iraq's invasion provoked a crisis unlike any other Bush had yet faced. American officials were sent scrambling once news of the aggression broke, working through the night to freeze Iraqi (and Kuwaiti) assets and build support in the United Nations for a resolution of condemnation, with little optimism about the ultimate results. Scowcroft felt a "very instinctive" certainty, an almost "visceral reaction," that the Iraqi leader would likely get away with it, given how difficult it would be to build any coalition against him. The Kuwaitis were, as one esteemed historian of the conflict gently put it, "generally unloved," noting that "there would be more than a few in the Arab world who would be quietly pleased if they got their comeuppance for insufferable arrogance."

The only thing the world loved about Kuwait was its oil, which cash-starved Iraq had every reason to continue pumping once the dust from the invasion settled. Saddam was in fact likely to pump more oil into the global economy than the Kuwaitis had allowed, filling his own coffers but also sending the price per barrel in time plummeting, to the general betterment of international wallets. As long as Saddam kept the great Middle Eastern gas station open for business, some American analysts initially considered, did it really matter who controlled Kuwait? "The fundamental US interest in the security of the Persian Gulf is oil," the Pentagon's Paul Wolfowitz had opined to Defense Secretary Cheney in a private assessment of the Iraqi threat only a week before. It was not sovereignty, ideology, democracy, or liberty. What mattered lay below the earth's surface, and the continued access to the energy it provided. Cheney thought enough of the memorandum to pass it along to Scowcroft and Baker with his endorsement.

This petroleum-based perspective dominated Bush's first harried NSC meeting following the invasion, where advisers openly discussed whether it really mattered, from a purely strategic standpoint, if an Iraqi or a Kuwaiti flag was printed on the side of an exported barrel. Among all of Bush's formal NSC meetings during the war, this was the only one whose transcript remained classified more than a generation after the conflict, and with good reason. Many of its participants went on to play key roles in formulating American policy toward the Middle East in the George W. Bush administration, until 2009, and revelation of their candid views back in 1990 would undoubtedly have complicated contemporary relations, especially during another period of direct American engagement in the region. Released only in 2013, the transcript shows that the administration's initial response was disjointed, unclear, and largely devoid of any high-minded principle of salvation or defense of the Kuwaiti regime.

The administration had already frozen Iraqi assets by the time Bush first formally gathered with his national security team, and would soon secure a 14–0 resolution from the UN Security Council condemning the invasion, but beyond that, his main advisers openly wondered, was there a point to doing more? Protecting Saudi Arabia was a different question. Its proven oil reserves were the world's largest. As Wolfowitz's prior memorandum made plain, it would be "inimical to US interests to permit any power — including Iraq — to gain dominance over Gulf oil supplies. Such dominance by a single country would enable it to dictate prices and production, placing the economies of the US and its allies in an extremely vulnerable position." But control of Kuwait only strengthened Saddam's hand. It did not provide regional dominance. Building the type of international and, in particular, local support necessary to reverse his conquest might therefore prove not only nearly impossible but also more trouble than it was worth.

"The rest of the world badly needs oil," Cheney observed. "They have little interest in poor Kuwait." Bush's budget director, Richard Darman, seconded the point. "There is a distinction between what to do to defend Saudi Arabia and to liberate Kuwait." Returning to the theme later in the conversation, Darman added: "There is a chance to defend Saudi Arabia if we do all that's possible. On liberating Kuwait, I sense it's not viable. Therefore we need an intermediate option if we can't liberate Kuwait, one to limit Iraqi power with an annexed Kuwait."

An Iraqi conquest of Saudi Arabia was something Washington could never condone. Kuwait, by contrast, could cease to exist with little more than a ripple. Even if Iraq were fully embargoed and cut off from global oil markets as punishment, anything Saddam might be barred from selling could easily be replaced by excess Saudi capacity. The Energy Department's representative to the meeting even called the current crisis an "interesting opportunity" to boost overall regional production and thus ultimately further drive down energy costs around the world.

"Should we put out a strong redline on Saudi Arabia as a vital interest?" Colin Powell asked. "I think there is no choice," he volunteered, stepping well beyond the invisible line that typically kept military officers from offering political advice. "The question is[,] do you lay it out to the public." Securing the world's largest oil fields seemed clearly worth the risk, Powell initially reasoned, but he doubted the United States should "go to war over Kuwait." After all, "it's Kuwait," he'd earlier told Cheney and others in the Pentagon's inner circle. "Does anybody really care about Kuwait?"

Once they had both returned to the Pentagon, Cheney dressed down Powell for the general's blunt input on political matters, in this meeting and others. But Bush appeared to agree with his chief military adviser's overall strategic assessment. So too did most who spoke up during that first NSC meeting. "We should be careful drawing the redlines we need to and throwing Kuwait to Iraq," Bush said. Yet he was not yet ready to promise Kuwait's liberation, or to guarantee Saudi Arabia's continued independence, especially as no one in Riyadh had yet formally asked for American protection. Indeed, Saudi king Fahd had yet to make plain that he'd accept American troops on his soil, even if deployed for his own defense. "The Arabs have been disappointing," the State Department's Robert Kimmit noted, and "the more they shrink, the tougher it will be for us."

Thinking globally during that first chaotic meeting, indeed revealing anew that Europe and the Cold War constituted the strategic arena that really mattered most to his mind, Bush noted his desire to bring the Soviets further into the discussion. The crisis could help strengthen Soviet-American ties and Moscow's peaceful connection to the broader world. "We can get them to kick in," Bush said, in a way they never had before. "No adventurism, but get them to agree to some action" in the hope that the symbolic value of both superpowers condemning the same international affair would carry tremendous weight throughout the globe. "US/Soviet relations are good," Bush added, though as was so often the case, his trust in Moscow's designs remained shallow. "We don't want to overlook the Soviet desire for access to warm water ports," he noted. "We don't want to resurrect that" by putting Soviet forces directly into the Persian Gulf. "Maybe something positive like a joint statement" would be enough to demonstrate their newfound cooperative spirit. At the least, strong words from Moscow *and* from Washington might "give some spine to Saudi Arabia and others to take difficult actions."

On the other side of the world, Baker was already working that angle. "We hope you'll try to restrain these guys," he'd implored Soviet foreign minister Shevardnadze upon hearing rumblings of potential Iraqi aggression in late July 1990. The pair were together in Irkutsk for long-scheduled meetings, where, between sessions on arms control and trade, Shevardnadze hoped to repay his newfound friend's hospitality for the hunting they'd done the previous year in Wyoming. The meeting had been months in the making. Shevardnadze admitted at its start to a strong "premonition" that the end of the Cold War "might create dangerous power vacuums in various regions." Pulling Baker aside, he quietly warned that "local dictators would be tempted to fill" any ensuing voids, suggesting the two should consider ways to work together in response. New Soviet thinking, as he'd stressed repeatedly before, enabled newfound East-West cooperation, over big issues like Germany and over others yet to arise.

Neither man realized how quickly Shevardnadze's vague fear would come to pass, or how soon they would have their opportunity to demonstrate solidarity. They were sitting together when word of the gulf crisis first broke. As Baker's chief aide Dennis Ross noted, that proved how "in life and in foreign policy" a healthy dose of "serendipity counts for more than planning." Shevardnadze had initially sniffed away the rumors of potential conflict between Iraq and Kuwait. Long a Soviet client, Saddam would never dare move without consulting Moscow first, he assured Baker, adding the next morning that Soviet officials had met with Saddam's and that "the crisis is being defused."

Interrupted hours later with news that Iraqi tanks had crossed the border, Shevardnadze was incredulous, if not amazed. "I can't believe that's true," he exclaimed, questioning not so much the report as the reality of the new world it represented. "We haven't heard anything about that. There's simply no logic to it!"

In truth the logic of Saddam's action was all too clear: it was a direct response to Moscow's diminished influence. Billions in Soviet arms and aid delivered over the prior decades had only rented his loyalty. "The present circumstances in the world today have given us the opportunity of a lifetime" to redraw the region's map, one of his advisers counseled. The Soviets were down, the Americans not yet fully ready to take their place, and for the time being, the world seemed remarkably unrestrained. In his words, "This opportunity will not happen again for fifty years."

Saddam and those around him overestimated their degree of freedom, and in turn misjudged the extent to which diminished Cold War tensions had provided each superpower with a newfound sense of freedom as well. The last time a superpower had moved aggressively in the region was in 1979, when the Soviets invaded Afghanistan. Every region mattered to Cold War policy makers, who frequently saw the world as their own zero-sum competition. The oil-rich Persian Gulf, however, mattered more than most. President Carter had thus threatened a nuclear response, if need be, to any subsequent Soviet move that threatened the region's oil routes.

But times had changed. Where once Soviet and American strategists had threatened war to keep each other from unilaterally acting in the gulf, in response to Saddam's invasion the two superpowers must "send a signal that together we have entered a new era," Baker told Shevardnadze. Jointly condemning the invasion would "demonstrate that when a crisis develops, we're prepared to act swiftly and affirmatively in a meaningful way." If the Soviets and the Americans could not agree to oppose a despot's violent takeover of its neighbor, Baker reasoned, a use of force of the kind antithetical to Gorbachev's vision for a new diplomacy if there ever was one, they'd never agree on anything.

His plea worked. "Today, we take the unusual step of jointly calling upon the rest of the international community to join with us in an international cutoff of arms supplies to Iraq," the pair announced in Moscow two days after the invasion. Both had slept little over the preceding hours. Baker struggled to communicate with his colleagues back in Washington. Shevardnadze struggled with his colleagues too, though not because of distance. "Let me tell you that it was a rather difficult decision for us . . . because of the long-standing relations we have with Iraq," Shevardnadze told the assembled reporters. But "this aggression is inconsistent with the principles of new political thinking and in fact with the civilized relations between nations."

Forging that joint statement had not been easy. Baker knew that Bush would condemn the Iraqi incursion, at least verbally, even if the president had not yet decided what actions to take in response. Soviet Middle Eastern experts could not even predict how their leaders would ultimately react to the invasion. Many were loath to cast aside their long-standing ties to Baghdad, and unwilling to move past the overarching zero-sum thinking of superpower politics that infused their traditional approach to the region. Old habits are hard to break, especially those formed over a lifetime of ideological struggle. "We could play a good game in this crisis" and perhaps even profit from the ensuing maelstrom, one Soviet diplomat mused to a Shevardnadze aide, "if only we did not let ourselves be had by the Americans."

Gorbachev initially stalled, fearing from the first that the crisis would result in the kind of military action he considered at odds with the underlying principles of *perestroika,* and knowing that his weakened political position at home offered little room to fend off domestic critics. He ultimately endorsed Shevardnadze's desire to stand side by side with Baker for the good of Soviet-American relations. "If the world community could not stop the aggression against Kuwait," Shevardnadze argued within the Kremlin, echoing Baker's plea, "then it would have gained nothing from the end of the Cold War."

More selfish reasons also drove Gorbachev's decision. Just as with the negotiations over Germany, the embattled Soviet leader and those closest to him perceived a desperate need to show Washington and the West that the transformation in Soviet foreign affairs was real. They would be a responsible partner for international stability, he'd repeatedly said — in New York, Malta, Washington, and at every diplomatic stop in between. As a responsible partner, the Soviets stood a far better chance of securing Western rewards of their own, and hope for such rewards had underlain Gorbachev's grudging acceptance of Bush's vision for Germany. Similar hopes now played a part in shaping Moscow's response to the Iraqi crisis as well. Iraq "was one of the few countries with which we were not just close, but also [had] mutually beneficial economic relations," Deputy Foreign Minister A. M. Belonogov stressed in Kremlin discussions. "Naturally we were not pleased at the prospect of breaking off Soviet-Iraqi relations. But there was no other way out." Inform Moscow's remaining friends in the region, Gorbachev subsequently ordered, that they too would be wise to follow Washington's lead. "Tell them that we will not sacrifice our alliance with the U.S." for Saddam — or, by implication, for them, either.

American leaders noted the change. "I think the Cold War ended . . . when the foreign minister of the Soviet Union stood side by side with the American Secretary of State on August the 3rd in 1990 at an airport in Moscow and condemned the action of a Soviet client state Iraq in marching over Kuwait, in occupying Kuwait," Baker later recalled. "It was quite obvious then that the Cold War was over."

The ensuing Gulf War became the world's first real post–Cold War crisis, and Bush's vigorous response, so contrary to his usual Hippocratic character, can be understood only in those terms. It forced him to realize he'd be not only a wartime president but also the nation's last Cold War president, one charged with outlining with greater precision than ever before what the future of the world might hold. And he had new options, unlike any predecessor since Truman. "If we were worried about the Soviets supplying weapons" to Iraq, he told Canada's Mulroney, "it would be a different situation." That would make it Korea, or Vietnam, or Cuba, or any of the other Cold War hotspots that threatened in time to turn into an open superpower conflict. That possibility was off the table with Iraq from the start.

Faced with a genuine superpower conflict — that is, if the Iraqis were believed to be operating on behalf of the Kremlin — Bush might have responded with threats, or with force, as previous presidents had done in similar situations. Given Soviet acquiescence and support, however, Bush faced instead a debate among his own advisers over whether or not to merely accept Saddam's conquest on its own terms, and whether or not to push Saudi Arabia and other gulf states to accept American aid they had in no way shown they desired. He thus felt the freedom to consider Iraq's invasion purely in Middle Eastern terms, and ultimately in terms of what it, and his reaction would mean as a precedent for crises still to come.

The Gulf War can thus be truly understood only as the Cold War's coda, though it took Bush's White House the better part of a week to fully appreciate the real measure of the situation before them. Scowcroft later wrote of being "frankly appalled at the undertone of the [initial NSC] discussion, which suggested resignation to the invasion and even adoption of a fait accompli." It seemed to many in the room, veterans of Tiananmen Square, Panama, the fall of the Berlin Wall, and the Velvet Revolutions, merely "the crisis du jour" rather than "the major crisis of our time."

Scowcroft typically didn't mind prudent responses. But he did not believe the U.S. should be somnolent either. He thus tried to move the discussion from questions of oil and inter-Arab relations to their broader meaning, thinking about the signal that America's response would send to the wider world. "It would be a significant event if the US were to say that this small fracas is of little concern," he warned Bush and those gathered around the table. "That signal would send shock waves throughout the Middle East" and beyond. "We don't have the option to be inactive in reversing this."

"No one here would disagree," Bush said, yet nobody else in the room pursued Scowcroft's reasoning. Alone with Bush in the Oval Office once that first discombobulated NSC meeting had disbanded, he argued his case more forcefully. "This was a case of naked aggression," Scowcroft insisted, "as clear as you could find." Emerging from their private discussion, he quickly found additional sympathetic ears. "That was one of the worst meetings I've ever sat through," Haass complained, no doubt rightly reading his boss's frustration. They'd need to marshal their arguments carefully, Scowcroft told him. "Write me a memo about why we have to act."

Given time to collect their thoughts, Bush's White House team, and other powerful elements of his administration, slowly began to recalculate the full consequences of inaction. "We would be setting a terrible precedent" if Saddam were allowed to profit from his aggression, Haass wrote, "one that would only accelerate violent centrifugal tendencies — in this emerging 'post–Cold War' era . . . That also raises the issue of US reliability in a most serious way." Watching over his shoulder as a fatigued Haass hunted and pecked his way slowly through the document, Condoleezza Rice finally pushed him aside, taking the keyboard so he could merely dictate. An accomplished pianist, she was also by far the better typist, and as Haass conceded, minutes mattered in a crisis.

Across the Potomac at the Pentagon, with time to reflect rather than just to react, Dick Cheney was slowly reaching similar conclusions, though his medium was a yellow legal pad and pen. "Shouldn't our objective be to get him [Saddam] out of Kuwait?" he wrote in longhand notes to himself. "Isn't that the best short and long term strategy?" He didn't think it would be easy; neither did he believe that Saddam was apt to be persuaded by anything less than a show of force. "No non-military option is likely to produce any positive result," Cheney wrote. "The key" was "U.S. military power — the only thing Saddam fears."

While his advisers scrambled and considered, Bush left chaotic Washington for a previously scheduled trip to Aspen and a meeting with Margaret Thatcher. Canceling was not an option. It would make the crisis appear worse, and global markets and opinions were already rattled enough. His usual 707 unable to land on Aspen's short high-altitude runway, a lighter Gulfstream became Air Force One for the day. The smaller venue enabling Scowcroft to further plead his case to Bush in private, the two men crammed into the small jet, seated two by two alongside the president's deputy press secretary and chief of staff, papers and memos overflowing across their laps. The shortest of the four men leaned forward toward the tallest, cutting the air with his finger, reiterating over and over the consequences of inaction.

Bush meanwhile heard telephoned pleas from abroad. "I really implore you, sir, to keep calm," Jordan's King Hussein pleaded, urging him once more to let the Arab world police its own. "George, give us two days to find a solution," Egypt's Mubarak similarly begged. Saudi Arabia's King Fahd was aghast yet noncommittal, even with Iraqi tanks reconnoitering his border. Saddam "doesn't realize that the implications of his actions are upsetting the world order," Fahd told Bush. "He is following Hitler in creating world problems." Nothing would ultimately persuade Saddam to retreat, said the king, "but the use of force."

Unable to defend Saudi Arabia against the Iraqi military juggernaut, and certainly lacking the means to eject him from Kuwait, King Fahd nonetheless hesitated to accept Bush's offer to station a squadron of fighter jets on his soil. History gave him pause. President Carter had deployed fighter aircraft to Saudi Arabia back in 1979 in response to the Iranian Revolution, the deterrent wholly undermined by the revelation that the planes were flying unarmed. Reagan's withdrawal of American forces from Lebanon in 1983 left similar concerns that American commitments weren't hard and fast. The last two American presidents had demonstrated scant resolve when it came to defending Arab interests. King Fahd "held back a little bit," Bush told his diary that night, worrying that the Saudis would "somehow . . . try to buy a solution, or they will accept the status quo with guarantees." As he later confessed, "I had to wonder if, under pressure, they might be inclined to strike some kind of behind-the-scenes arrangement with Saddam."

Despite the combination of Arab reluctance and potential Saudi resignation, Bush took Scowcroft's admonition to heart, realizing that whatever he did next would set a precedent. "The enormity of Iraq is upon me now," he told his diary the next day, having returned to Washington from Colorado. "The status quo is intolerable." His ensuing meeting with his National Security Council featured a far different tone from the first. Offering to lay out the case for intervention on his own, Bush yielded to Scowcroft's prudent suggestion that he instead make what Haass termed "the Churchill speech." No one would seriously demur if the president made his decision plain from the start, and Bush needed to hear each of his advisers' unvarnished opinion, even if it contradicted his own. "My personal judgement is that the stakes in this for the United States are such that to accommodate Iraq should not be a policy option," Scowcroft began instead, his sentiment quickly echoed by the State Department's Eagleburger, sitting in for Baker, who remained overseas. "I couldn't agree more," he said, pounding the table for effect. "This is the first test of the postwar system. If he [Saddam] succeeds, others may try the same thing. It would be a bad lesson."

Bush ultimately weighed in, settling the debate. The Cold War was over, he said. "At stake is the shape of the world to come." By day's end he was telling foreign leaders that "the status quo is totally unacceptable," while telling King Fahd in particular that they "must be viewing this with more urgency." He would not send a high-level delegation to Riyadh to discuss the terms of an American military mission until and unless the Saudis first agreed to accept American troops. As Scowcroft simultaneously put the situation to Fahd's ambassador in Washington, there was no way the Americans would allow Saddam to proceed farther down the Arabian Peninsula, capturing more and more of the world's oil every step along the way. The Saudis, he bluntly told their envoy, ultimately had a simple choice: they could be defended or liberated.

With Kuwait's royal family camped out in Saudi hotel rooms while their people increasingly flowed across the border as refugees, the argument struck a chord. "The Amir of Kuwait asked us for military intervention," Bush told Fahd, "but Iraqi troops had already occupied Kuwait City and there was nothing the US could do. It takes a long time to deploy troops . . . Maybe we can help him [the Amir] down the line but now we can help Saudi Arabia, but we have long lines of supply and we have to move forces long distances. Saddam lied. He is unpredictable and ruthless. He told you, my Ambassador, [and] Mubarak that he wouldn't attack Kuwait. He told them all and he lied."

Two days later, striding from his helicopter across the White House's South Lawn, Bush made what ultimately would become the most famous statement of his entire presidency. "Are you disappointed in the failure of the Arab nations" to secure Saddam's withdrawal from Kuwait? a reporter asked. "Well, I was told by one leader that I respect enormously — I believe this was back on Friday — that they needed 48 hours to find what was called an Arab solution," Bush replied. "That obviously has failed. And of course, I'm disappointed that the matter hasn't been resolved before now. It's a very serious matter." Then came the statement that revealed more than merely his intent to respond to the crisis at hand but additionally his sense that, amidst the growing crisis, after struggling for more than a year in office to ensure Washington's continued ability to play the world's leading role, he'd finally decided to be not just a global leader but the first world leader of the new era: "I view it very seriously, not just that but any threat to any other countries, as well as I view very seriously our determination to reverse out this aggression . . . This will not stand, this aggression against Kuwait.

"I've got to go," Bush continued. "I have to go to work. I've got to go to work."

20

WITH US, OR NOT AGAINST US

THE ENSUING SIX MONTHS defined Bush's legacy. He became a wartime president, orchestrating the largest American military expedition in a generation. He also, at long last, laid out a full vision for the post–Cold War world beyond mere defense against uncertainty, using the crisis in the gulf to model how the international system might respond to future threats in a manner that was at once multilateral yet dependent on American power. Thirty nations sent troops, operating under American command, with United Nations sanction. Dozens more contributed financially. After months of buildup and a month of preparatory bombing, victory — pushing Saddam's forces out of Kuwait — required a mere one hundred hours of ground operations, though with consequences that echoed all the way to the Kremlin. On top of their concessions over Germany, the Gulf War forced Soviet officials to at long last accept that any future partnership with the United States would never entail full parity. They had, in fact, lost. The Gulf War was thus the Cold War's final act. That era climaxed not with a tectonic struggle for Europe, but with Soviet capitulation to the hard realities of American power over a petty despot surrounded by sand. It was Gorbachev — weakened, embarrassed, and increasingly alone — who became the Gulf War's final casualty.

It took Bush the better part of a week after the invasion of Iraq to declare his resolve to remove Saddam Hussein from Kuwait, and another month to fully articulate his reasons. More than merely oil or American national interests were at stake. What mattered most was precedent. "Nothing like this" had so captivated American policy or been of "such moral importance since World War II," Bush argued. By the time the war peaked, so had his rhetoric, moving well beyond the prudence that had marked his first months in office. "What we're doing is going to chart the future of the world for the next hundred years," he declared. "It's that big," and would ultimately lead to nothing less than a "new world order."

These are not cautious words. That phase of his presidency was over, as was Bush's sense that he lacked the means to respond to the international crisis at his doorstep. Events in Beijing and Eastern Europe had seemed largely beyond his control, while negotiations over Germany demanded quiet persuasion rather than full-throated presidential declarations. Iraq provided a comfortable moral clarity and the resulting freedom to employ his bully pulpit to the fullest. One nation had attacked another, violating a clearly marked border. Blood had been shed. Civilians were at risk, including Americans, as was the global economy.

The conflict also seemed familiar, even strangely comfortable. Iraq's invasion recalled the strategic problems faced by American policy makers over the course of his lifetime, featuring a rapacious dictator, wavering allies who threatened appeasement, a conventional military assault, lines on a map, and an old-fashioned struggle over natural resources in a familiar part of the world. It was, in this sense, a very twentieth-century affair, posing daunting problems but little that felt novel. "I know you're aware of the fact that this has all the ingredients that brought down three of the last five presidents," Baker cautioned in the privacy of the Oval Office. Speaking as Bush's chief diplomat, but also as his friend, he presented a daunting list of worries: "a hostage crisis, body bags, and a full-fledged recession caused by forty-dollar oil."

Bush shared Baker's fears, and others besides. Add Iran to the list, he said. "We just cannot have another Tehran situation," in which American pride fell prey to Middle Eastern politics. They couldn't afford another quagmire like Vietnam, either, or a repeat of Lebanon in 1983, when American forces had fled as soon as they were bloodied, leaving civil war and a deeply bruised reputation behind.

Every crisis is different, but the formidable litany of concerns the pair drew from the past suggested commensurate responses as well. Memories of Tehran in 1979 warned against riling up public passions in the Middle East, for example, and anticolonial and religious passions in particular. "The majority of Arabs were horrified at Saddam's invasion of Kuwait, but the vast majority also find the military intervention by Israel's protector—especially with troops in the kingdom that keeps the Holy Places—hard to stomach," Scowcroft's team advised early in the conflict. "For most Arabs it is now a choice between two negatives, and our battle is to keep the second negative the lesser of the two." The Americans could not win if Saddam succeeded in portraying them as modern-day imperialists or crusaders—or, worse yet, Zionist puppets. The most effective remedy appeared to be for Bush to cultivate Arab allies of his own, no matter the cost, while keeping Israel as far from the battlefield as possible.

Vietnam brought two more lessons to the table. Senior officers like Colin Powell, veterans of the conflict, and of the army's soul-searching in its wake, preached the catechism of overwhelming force, believing they'd been robbed of victory in Southeast Asia by limitations set by politicians back home. Bush's generation, largely veterans of the home front during Vietnam, drew a second though related lesson from their collective memory of the conflict: the virtues of speed and international support. When it came to conflict in the gulf, "I'm convinced they'll support us—the Congress—provided it's fast and surgical," he told his diary. "But if it's drawn out and long, well, then you'll have all the handwringers saying 'they shouldn't have done it,' and they'll be after my neck on, perhaps, impeachment." Democracies loathed drawn-out quagmires, his reading of history taught, and the perception of going it alone. Lyndon Johnson never garnered the degree of international support he desired, Senator Alan Cranston reminded Bush, adding that he hoped this time around the United States "won't be the Lone Ranger the way we were in Vietnam." Thirty-eight countries had sent troops or supplies to South Vietnam's government in Saigon, but historical memory mattered more than facts. As American policy makers remembered it, the United States had fought in the jungle largely on its own, fueling popular frustration at home.

The more flags in their coalition, Bush and those around him thus reasoned, the longer American voters would tolerate the deployment of troops, especially if the actual period of combat seemed short and casualties few. Gulf allies and stalwarts like the British and Australians could send troops. Countries such as Germany and Japan, whose constitutions prohibited direct military participation, could send medical teams or supplies or cash. "My bottom line is that when this chapter of history is written," Bush told Japan's prime minister in mid-August, "Japan and the U.S. and a handful of other countries will have stood side-by-side as much as possible." Americans would remember who'd come to their aid, he added. Indeed, what Baker came to call their "tin-cup" tour, seeking funds from coalition partners, paid immediate dividends. The total amount collected for the war nearly equaled Washington's expenses. Some analysts even calculated that the United States turned a small profit.

Memories of Reagan's failed efforts in Lebanon reinforced the value of holding fast to one's commitments, especially in the volatile Middle East, where political coalitions frequently shifted faster than memories faded. The last two times American presidents had sent troops to the region — in 1979 and 1983 — they had been either unarmed or unwilling to stay. "Why should the King not be concerned," Saudi Arabia's ambassador pointedly asked Scowcroft and Cheney in the first hectic hours of the crisis, that "if the going got tough, the United States would behave in the same manner again?" Cheney offered detailed deployment plans but, in the end, only promises. "We're serious this time," he said, though for the time being, Saudi leaders remained uncertain if they yet trusted the Americans with their kingdom, and their lives.

Of all the lessons of history, one drove Bush especially: World War II's reminder that dictators could never be satisfied or appeased, and certainly never trusted. They responded only to force. Saddam was "Hitler revisited," Bush told crowds throughout the summer and fall. It wasn't just the Iraqi's mustache or military garb. "As was the case in the 1930s, we see in Saddam Hussein an aggressive dictator threatening his neighbors." Unlike during the isolationist era of his youth, when the United States stood by while the world plunged into war, this time "America will not stand aside."

The readings on Bush's desk and bedside table kept the analogy alive. Shocked by news from Kuwait of Iraqi atrocities, he ordered copies of an Amnesty International assessment distributed to every Oval Office visitor. He also spent his dwindling free time consuming histories of the war in which he'd fought, searching less for answers than for confirmation. "I'm reading a book, and it's a book of history — great, big, thick history about World War II," he told one crowd in October. "And there's a parallel between what Hitler did to Poland and what Saddam Hussein has done to Kuwait." Men were being imprisoned and women raped, teenagers murdered. Babies had even been tossed from their incubators by Iraqi soldiers. "So, it isn't oil that we're concerned about," he emphasized. "It is aggression. And this aggression is not going to stand."

By November, outraged by the daily drumbeat of reported atrocities in Kuwait, Bush even began suggesting that Saddam was in some ways worse than the Nazi despot. "This morning, right now, over 300 innocent Americans —

civilians — are held against their will in Iraq, denied the freedoms granted all under international law," he charged. "Many of them are reportedly staked out as human shields near possible military targets, something that even Adolf Hitler didn't do."

The assertion prompted immediate pushback. "Saddam Hussein is pretty bad," one prominent columnist retorted, but "I don't think he's in Hitler's class." Bush's staff implored him to tone down his rhetoric, but not to discard the analogy, which resonated too well with the public to reject entirely. "Go back and take a look at your history," he nonetheless lectured reporters later that day, "and you'll see why I'm as concerned as I am" and so keen to meet this dictator's aggression with all necessary force.

The folly of appeasement may be modern America's most widely credited lesson of history. Although Bush had largely dropped Hitler from his wartime stump speeches by year's end, he therefore retained the allusion. "Appeasement — peace at any price — was never the answer," he argued as January's deadline for Iraq's withdrawal from Kuwait neared. "Turning a blind eye to Saddam's aggression would not have avoided war; it would only have delayed the world's day of reckoning, postponing what ultimately would have been a far more dangerous, a far more costly conflict."

The allusion worked. Just as Congress began debating whether Bush should in fact be authorized to go to war with Iraq, one early 1991 poll found that more Americans listed Saddam than Hitler when asked to name the century's most evil leader. If fighting was the right thing to do in the 1940s, and few Americans (and politicians especially) would stand up and argue otherwise, then logic dictated standing fast against Hitler's successor in evil. Bush "remembers that people said Hitler was just a benevolent dictator," noted Senator Simpson. "He doesn't want to repeat history. He's thinking of his grandkids, of the future."

Saddam Hussein was evil, to be sure, yet the intensity of Bush's feelings toward his opponent in time obscured his ability to tell fact from fiction when it came to the Iraqi leader. The Amnesty International report he kept touting proved largely fictitious, little more than Kuwaiti propaganda, its lists of atrocities eerily reminiscent of those ascribed to German troops who ransacked Belgium in 1914. More damning, supposed eyewitnesses to Iraq's conquest and the ensuing rape of Kuwait who testified before Congress were later discovered to have been in Washington the entire time. One was the Kuwaiti ambassador's daughter. Yet by the time these revelations trickled out, the persuasion campaign had already done its work, shaping both the president's opinion and the public's. The question that dominated the airwaves had turned from *if* American forces would fight to *when*. "It's black and white, good vs. evil," Bush told staffers, sounding more like Reagan than himself. "The man [Saddam] has to be stopped."

Bush wasn't the only one duped by exaggerated reporting, but he was the one whose decisions mattered most. Saddam was "evil personified," he told his diary, later adding, "I have Saddam Hussein now as clearly bad and evil as Hitler." The differences between them, moreover, did not diminish the countless number of individual sacrifices required to ensure their defeat, as Bush knew all too well. Televised images of soldiers leaving their families behind revived painful memories. "I remembered just as clearly as it can be when I went off to Chapel Hill" for flight training "and my dad gave me a hug on that platform," he recalled in his diary. "I didn't know one single soul, and I was off for an experience into the unknown. It shaped my life."

Bush knew what it meant to go to war, and he weighed the benefits and risks intently, laying them out in a long letter to his children, composed as 1990 came to an end, which summed up his overwhelming sense of responsibility and his belief that his decisions might save future generations from similar pain. "My mind goes back to history," he wrote. "How many lives would have been saved if appeasement had given way to force earlier on in the late '30s or earliest '40s? How many Jews might have been spared the gas chambers, or how many Polish patriots might be alive today? I look at today's crisis as 'good' vs. 'evil' — yes, it is that clear."

Bush's repeated references to Hitler and to the nation's past wars, both in public and with no one else around, provided more than mere justification for going to war. They also provided a strategy that played to his strengths. Saddam had generated international opposition by violating a well-recognized border, just like Hitler. Twenty years after serving at the United Nations, Bush knew how to coordinate international condemnation. Saddam had tanks, just like the Nazis, with which he overpowered his ill-prepared neighbors. But just as American industry and technology overwhelmed German forces in time, the United States could bring to bear firepower that Iraq's antiquated army could never hope to match. Indeed, just like Franklin Roosevelt in the 1940s, Bush commanded the only force on earth capable of moving vast armies across distant oceans. Proving that the incredible logistical feat — transporting whole di-visions of armored troops from Europe to the Middle East in weeks — was less a miracle than an example of unmatched American military prowess simulta-neously provided justification for the ongoing global mission Bush considered key for long-term peace. "Only the Americans can mount a swift military re-sponse to a challenge of this sort outside the NATO area," *The Independent*, a British newspaper, editorialized a fortnight into the crisis. Having lived under American protection since the Second World War, "Europe simply has to de-cide whether to tag along . . . Its defense remains in American hands."

Indeed, for all Bush had worried over the preceding months about how he might make the case for continued U.S. engagement in Europe and elsewhere, finding unpersuasive the amorphous threats he could only term "uncertainty" or "the unknown," Saddam made it for him. "A lot of people who were warn-ing against rapid disarmament must be very happy this has happened," an un-named NATO official told the *New York Times*. Many of those people worked in Washington, and for Bush. James Baker certainly believed that the crisis in the Persian Gulf offered the silver lining of demonstrating Washington's post–Cold War potential. "We remain the one nation that has the necessary political, military, and economic instruments at our disposal to catalyze a successful col-lective response," he publicly declared.

Bush saw potential domestic advantages in the crisis as well. "I just hope that Iraq and the country's unity can now be parlayed into support for the budget agreement," he told his diary in September. Conservatives still threatened to shut down the government following his concession on taxes, and Democrats and Republicans alike hoped the prospect of peace in Europe would enable de-fense reductions well beyond what he considered prudent. They'd have a hard time pursuing either goal while American troops were preparing for battle in the Middle East. "It is said that I much prefer to work on international affairs," he conceded in his diary, "I think the answer is I do prefer this" type of diplo-matic work, "but I see a budget deficit as something important — something essential to solve."

None of this is to say that Bush either welcomed the war or thought it would be easy. For all the perceived opportunities embedded within Saddam's inva-sion, expelling him from Kuwait posed a tremendous challenge. Iraq boasted more than a million men under arms, thousands of tanks and artillery pieces, and one of the region's largest air forces. It was also battle tested. "In eight years of war with the Iranians," one analyst noted, the Iraqis had learned one thing above all else: "how to bleed an enemy."

Iraqi forces were also led by a man unconstrained by the supposed "rules of war," or even basic morality, who had deployed poison gas and chemical weapons against his enemies and his own people. Bush's assessment of Saddam was largely right in this crucial sense: force was the only currency that truly mattered in the Iraqi leader's universe. Iraq's jails overflowed with those considered enemies of the regime. So too its cemeteries. He did not have to be Hitler to be horrible, or to surround himself with equally depraved minions. "Don't bother me with these questions about chemical weapons," the country's foreign minister had previously remarked, waving away Saddam's critics. They'd employ every weapon at their disposal to retain power. The Americans would quickly lose their appetite for war once bloodied, Saddam reasoned, ordering his units to prepare for a stubborn defense and publicly boasting of his ability to strike deep inside the United States. "Our rockets cannot reach Washington," Saddam conceded. "But we can send a strapped person [a suicide bomber] to Washington . . . and retaliate just like the old days. This is the thing. Strapped with a bomb . . . We might become quiet for twenty days, but after twenty days, you will be surprised to see boom, boom, boom, boom."

"Are We Ready for This?" *Time* magazine asked in an early September cover story, the words interlaced with the image of an American soldier in full chemical weapons gear. Pundits predicted that Kuwait's liberation would cost thousands of American casualties, following what Saddam boasted would be "the mother of all battles" against the Western infidels. The Pentagon's computer-driven scenarios forecast at least five thousand killed and wounded. Cautious planners quietly shipped thirty thousand body bags to the region while simultaneously ramping up training for funeral details back home. The United States should expect "an overflow of casualties," the director of the Veterans Administration warned.

Barely given time to celebrate the fall of the Berlin Wall, Americans were now told to become newly vigilant, prompting one of the largest civil debates since before World War II, the last time the nation had so deeply pondered whether or not to go to war. Senate majority leader George Mitchell warned of "an unknown number of casualties and deaths" and, beyond the stark human toll, "billions of dollars spent, a greatly disrupted oil supply and oil price increases, a war possibly widened to Israel, Turkey, or other allies, the possibly long-term American occupation of Iraq, increased instability in the Persian Gulf region, long-lasting Arab enmity against the United States," and ultimately "a possible return to isolationism at home." Mitchell's fears actually came true under the leadership of another president named Bush but proved unfounded at the time. Still, by year's end, as deadlines for Saddam's withdrawal drew near, fully 40 percent of Americans polled expected the war would lead "to another prolonged situation like the Vietnam conflict."

America's military leaders shared their concern. Merely containing Iraq would require upward of ten times as many troops as the entire Panama operation, Pentagon strategists initially calculated. Defeating Saddam would take far more, demanding activation of National Guard and reserve units. No president had authorized this last step since the 1960s. When he had, the nation's campuses had erupted in anger.

"A week before mid-term elections," Cheney's briefers thus told Bush at a meeting in mid-September, he would be reaching "into every community in America [to] take people away from their homes and their jobs." The message, Robert Gates recalled, taking notes in the back of the room, was not subtle. Surely, military advisers cautioned, no politician wanted fearful voters heading to the polls. Americans typically rallied around the flag in a crisis, but they would not necessarily support their commander in chief if their vote might otherwise keep their own flesh and blood out of harm's way.

It would be neither the first nor the last time the lines between military leaders and their civilian masters blurred. Army planners in particular hoped to temper if not outright deflate Bush's enthusiasm for a military response, forcing him to rely on diplomacy or existing economic sanctions to induce Saddam's reluctant retreat. "Iraq is losing one billion [dollars] in revenues each day the sanctions are in effect," General Norman Schwarzkopf, who commanded American forces in the region, told reporters in early September. "It's going to be interesting to see how much loyalty he [Saddam] has in his armed forces when he's unable to pay their salaries, feed them, and resupply them with fuel and spare parts and ammunition." Little more than a month after the Iraqi invasion, he believed "the next move right now" was to wait.

And he said so publicly. Schwarzkopf did not yet have the firepower he considered necessary to complete his mission, and his unabashed reluctance to enter into battle exemplified that of the Pentagon's high command. The military's first war plan for ousting Saddam from Kuwait lacked originality, calling primarily for a frontal assault against well-fortified Iraqi positions. "Hi diddle diddle, right up the middle," Haass mocked in response, noting that no sensible president would ever willingly endorse such a bloody scheme. That was its point. The planners "didn't want to do it," Scowcroft perceived, and hoped to deter Bush by presenting only unappealing options. Ordered to come up with something less likely to result in a bloodbath, they next demanded more troops than any president had authorized since LBJ, and more than any Republican since Abraham Lincoln. Their minimum requirements alone totaled almost half of all American combat power throughout the world.

"Whew," Scowcroft audibly gasped in response. Powell took that as a good sign: his intended message had gotten across. He more than anyone else in Bush's orbit embodied the Pentagon's post-Vietnam anxieties. Veteran of several combat tours in Southeast Asia, Powell had been the one who ferried casualty reports from Lebanon to Defense Secretary Caspar Weinberger in 1983. "The shattered bodies of Marines at the Beirut Airport were never far from my mind in arguing for caution" when deploying military force, he later admitted. "Eighty bodies pulled out. A hundred. A hundred and fifty." To go to war, he consequently believed, required more than just a national need; it demanded a national commitment as well. "As a midlevel career officer, I had been appalled at the docility of the Joint Chiefs of Staff, fighting the war in Vietnam without ever pressing the political leaders to lay out clear objectives for them," Powell later remembered. He vowed never to repeat their mistake. "If the President opts for this offensive," he cautioned once Scowcroft had finished exhaling, "we'll need a hell of a lot more" than he'd already authorized.

All eyes turned toward the head of the table. "You got it," Bush replied, pushing back his chair and walking out of the room. "Let me know if you need more."

It was effective political theater. He didn't even blink, Powell recalled thinking. Cheney's jaw dropped, as did those of the rest of the assembled military brass. "Does he know what he just authorized?" Cheney incredulously asked Scowcroft.

"He knows perfectly well what he authorized" came the reply. Having had a front-row seat to the political divisions caused by drawn-out conflicts, Bush and those closest to him wanted the Pentagon to feel authorized to win. He'd not be a "Lyndon Johnson, going down to the situation room picking targets as he [Johnson] did in Vietnam," Gates declared, recalling an oft-told trope of civilian overmanagement of a war. But by the same token, Powell and his ilk would have no one back home to blame if they failed. "We gave them [the Pentagon] absolutely everything they asked for," Cheney later stated. There was going to be "no excuse possible for anybody in the military to say that the civilian side of the house had not supported them." Bush's White House did make one key concession to politics, however, withholding announcement of the final bulk deployment — totaling 150,000 troops — until after the midterm election.

Neither Congress, the American people, nor the president's own military planners were keen to go to war. His strategists cautioned that oil alone did not provide sufficient cause, as international markets were likely to replace whatever production disruptions might occur in the short term. Bush considered Saddam evil, yet he'd learned to stomach plenty of evil despots over the course of his long career. What then ultimately drove Bush to act? It was the opportunity to demonstrate American leadership, on American terms. "The entire planet is in this madman's [Saddam's] debt," James Baker later opined. "His brutal invasion of Kuwait provided the unexpected opportunity to write an end to fifty years of Cold War conflict with resounding finality."

To see precisely how Bush employed this opportunity to the fullest, and ultimately to understand its effect on Soviet-American relations and what he would eventually pronounce his "new world order," consider the manner in which he forged the international response. Saddam wanted Kuwait for its oil. Bush and his team knew whom to call to turn off the taps. "I view you as a friend and an important member of NATO," he told Turkey's president, Turgut Özal, within forty-eight hours of Saddam's invasion. "In your geostrategic location you occupy a key position because of the pipeline going to Iran and Iraq. I wanted to know the conditions under which you might close it."

Though phrased as a question, it was closer to a command, albeit with the promise of rewards for compliance. "I cannot emphasize enough my view that there can be no neutrality in these circumstances," he told Özal the next day, "and my hope that we can get the plan, for the Saudis and you to be prepared to shut down the pipeline if Iraq fails to withdraw from Kuwait as virtually the entire world is calling for them to do." Bush would have no hope of forcing Saddam's withdrawal through economic sanctions if he could not make an oil embargo stick.

"I know funds are involved," he continued, referring to the revenues that would be lost if Turkey could no longer reexport Iraqi oil. "They can be made up if the pipeline closure is needed. Others have called Saddam another Hitler . . . I hope you would tell the Iraqi envoy, tell him we'll close the pipeline — and NATO will back us up."

"I feel strongly about Saddam," Özal responded, though he also attempted to demur, fearing domestic resentment if he facilitated the downfall of a neighboring Muslim regime, not to mention the potential for direct Iraqi retribution. Bush later learned that the Saudis were pressing Özal to forestall a decision, fearing any inflammatory Turkish moves before they'd made up their own minds to accept Washington's aid.

"We can't tell our other Allies to crack down while the oil continues to flow," Bush interrupted. He vowed to procure a United Nations sanction to provide Özal political cover, and force if necessary, if Turkey incurred Saddam's wrath. "I know the West would applaud Turkey and there would be no argument from any country — including most of the Middle East," he continued, promising to establish clearer conditions for payment and protection to help Özal make up his mind. "You're entitled to know what the result would be if you go forward," Bush assured him, adding, "This is not another Lebanon." This time the United States would see the job through. "We have a lot at stake here," he told Özal on August 5. It was their third phone call in as many days. "The future of peace of the whole world is at stake." And if the Turks did their part? "Kuwait is ready to compensate you for loss of revenues."

So too were the Saudis, eventually, who stood to make billions in excess profits by picking up their own production to offset Iraqi and Kuwaiti shortfalls. Bush's analysts calculated that Riyadh was likely to gain at least $9 billion in additional revenue if oil prices remained static. A further, and expected, 20 percent rise in cost per barrel would generate $17 billion, more than enough to cover any potential Turkish losses, produce a fund capable of addressing the "longer-term needs of the poorer Arab nations," Scowcroft noted in mid-August, and still allow the Saudis to keep a tidy profit for themselves. "We need to make them [the Saudis] understand that such investments," specifically funds set aside to promote development throughout the region, "are their best protection against Saddam-type Arab radicalism," he urged.

A visit from Jim Baker to Ankara the following week sealed the deal. Washington promised to guarantee $1 billion in Turkish losses, the funds ideally drawn from Kuwaiti and Saudi coffers, but from the United States Treasury if need be. The American offer was generous, Özal replied. But he estimated that closing the cross-border trade would cost $2.5 billion annually. Baker happily responded that he'd already secured additional World Bank funding of $1 billion to $1.5 billion for the next two years.

Turkey cut the pipelines. They were "short of cash," Özal explained to Britain's Margaret Thatcher a fortnight later, and the Western funds more than made up for their potential losses. Coordinating throughout with the White House, she helped secure a general relaxation of textile quotas within the European Community for Turkish goods as a further reward for Özal's cooperation. Having worked for years to gain full acceptance in Europe, the Turks, as one foreign diplomat put it, were "not going to desert the Western camp — no way. It's not an option." Especially not with Baker promising that his government would push for what Özal's government truly wanted: a future invitation to join the European Community. As with the Eastern Europeans and even the Soviets, the lure of joining Europe proved a powerful incentive, one that American policy makers happily exploited now that their own role on the continent appeared more secure.

The negotiations with Özal offer just one example of Bush's telephone diplomacy. "Assembling the coalition was really a remarkable performance," Cheney later recalled, "and it was a very personal kind of performance." Reagan had largely left such matters to underlings, reserving his phone calls for world leaders he knew well, such as Thatcher. In a sense Bush did the same, but his reservoir of friends and associates was deeper. In a critical moment, when American policy makers could do little more than say to Saudi and other gulf leaders "Trust us," personal history mattered. As Cheney recalled:

> I always remember the first weekend of the crisis . . . The President sent me to Saudi Arabia to obtain Saudi permission to deploy forces to the kingdom. When I called him to tell him King Fahd has signed off on that and got his approval to go ahead and deploy the force, he [Bush] said, "I want you to stop in Egypt on the way back and meet with President Mubarak," which I was happy to do, and he'd already called President Mubarak before I'd arrived. We were able to work out some very important understandings and arrangements at that point; got on the airplane, headed back for Washington. The president called me on the plane and said "Stop in Morocco. I want you to see King Hassan." Again, he'd been on the phone, working the phones, and lined up Moroccan support. As

we worked to put the coalition together, it was very much the personal
stature, if you will, of George Bush and his ability to call in chits and work
on relationships that he'd spent a lifetime building.

Of course, foreign leaders did not accept troops, lend support, or otherwise
act for or against their own national interests merely because their old friend
George Bush called. But the hours and years he'd invested in building relationships around the globe made the process of gaining their acquiescence easier
and faster. Saudi Arabia provides not only the most important but also the
clearest example. Arab nationalists and religious fundamentalists alike, veterans of the jihad against the Soviets in Afghanistan, lobbied for the chance to
defend the kingdom without resorting to Washington's aid. King Fahd heard
their pleas but accepted Bush's offer instead, driving the most radical — including a wealthy young zealot named Osama bin Laden — to pledge to cleanse
holy Muslim lands of any Western stain.

As is the case with all counterfactuals, we shall never know the answer with
complete certainty, but the best available evidence suggests Saud's response
might well have been different if another president had made the request. "We
will stay until we are asked to leave," Bush told the king. "You have my solemn
word on this." Within two weeks of this phone call the United States had thirty
thousand troops and accompanying airpower in Saudi Arabia. By November
it was a quarter million. By the time hostilities began in January, it was twice
that number again. As one expert on the region's politics has written, the Saudis
hesitated because of the profound animus against the stationing of non-Mus-
lim troops on their soil, but Bush's "promise decided the matter."

"All countries in the West clearly have to turn to us," Bush in turn recorded
in his diary, including both the Soviets and the Middle East states. There was
no one else. Washington could act unilaterally in this decisive moment, he reasoned. Or it could work for consensus. Even if the immediate results were the
same, the residual sense of inclusion would bear longer-term benefits long after
this particular crisis was resolved. "It is my theory that the more they [potential
allies] are included on take-off, the more we get their opinion, the more we
reach out, no matter what is involved in terms of time involved, the better it is,"
said Bush, repeating a mantra he'd been espousing since the early 1970s. "Everyone is proud. Everyone has his place in the sun — large or small, they should be
consulted, their opinions considered. And then when the United States makes a

move and I make a decision, we are more apt to have solid support." He worked the phones accordingly as the crisis erupted and then evolved, calling foreign leaders nightly for weeks on end and inviting the most prominent to Washington or Kennebunkport for direct consultations. For the most important of all, Gorbachev, he even flew to Europe to make his case in person.

Not everyone could be persuaded. Sitting in a rotating seat on the UN's Security Council, Cuba refused to back Washington's resolution against Iraq, to no one's surprise. Long critical of American power, and frequently prey to it, Fidel Castro's regime could no more endorse a Yankee-led military operation than a snowball could survive a Havana summer. Lobbying and promises of aid appeared unlikely to sway Yemen, either. "I spoke with the president of Yemen," Bush told Canada's Mulroney. "He asked for more time, that we not enforce the UN sanctions. It was kind of a pathetic call. He and [Jordan's] King Hussein are the only two on that wicket. I think it's money." Saddam promised to split the proceeds of his spoils with any Arab state that stood up to the Americans, increasingly casting the struggle as less about his own aggression than a stand of rich against poor, and of Muslims against Western imperialism and those corrupted by it.

Few took the bait. King Hussein refused to break with Iraq, disappointing Bush, who'd hoped his long personal relationship with the king might help pull Amman into his growing coalition. "I've known the guy forever," Bush told his staff, but to no avail. Yemen ultimately balked too, refusing in the end to side with the American-led majority. Its "permanent rep [to the UN] just enjoyed about $200 to $250 million worth of applause for that speech," Baker quipped to an aide after watching Yemen cast its lot with the Iraqis.

Yemen and Cuba held temporary seats. They could vote against American-inspired resolutions but not block them. The Soviet Union and China, by contrast, as permanent members, held all-important vetoes, and had long-standing reasons for opposing any American-led military action. Beijing's regime habitually railed against the way great powers kept lesser states in line, and Deng Xiaoping in particular found Washington's presumption of global leadership distasteful, in time describing the American-led effort in the Persian Gulf as little more than "big hegemonists beating up on small hegemonists." Chinese leaders remained bitter over the international condemnation they'd suffered during the previous year, moreover, and thus made plain their reluctance to participate in any international consensus authorizing military force.

"China can't go for military measures; except in case of traffic jams," Baker said in another quip to an aide, this time with more biting sarcasm. "Like the one in Tiananmen Square in June of 1989."

Technically speaking, Bush did not need China's endorsement to operate in the gulf as he saw fit. He did not need the United Nations' sanction. But he wanted both nonetheless, and the imprimatur the Security Council's vote represented. "As a matter of principle, we will never vote for the use of force resolution," Li Peng advised Ambassador Lilley privately. Bush hoped the Chinese might be persuaded to change their minds, if not out of a sense of international obligation or because of all he'd done to keep Sino-American relations intact after Tiananmen, then at least for their own self-interest. "I told Qian [Qichen, China's foreign minister]," Baker noted, that we "could accept a decision on their part not to contribute forces to the multinational effort, but we would not understand if they stood in the way . . . Qian clearly got the point."

China had short-term reasons for hoping to remain in Bush's good graces. A sharply contested vote in Congress on its continued most favored nation trading status was pending, as was a prospective visit to Washington by its foreign minister, something that had not happened since Tiananmen. Beijing wanted both, but the White House faced significant opposition. Bush warned Scowcroft that their legislative effort for "MFN could fall apart if [China's] Iraq support erodes." He told China's leaders the same thing in a November letter. "A positive vote in the UN security council would help pave the way for a most successful visit by foreign minister Qian and for improved relations between our two countries," Bush wrote. "I cannot overstate the importance of your decision."

The State Department was blunter. "The question of a meeting with the President depends on how the Chinese vote on the resolution," policy makers in Washington informed embassy staff in Beijing, imploring them to pass word of the prospective quid pro quo to their local contacts. This was something Bush had already made clear to Deng, whom he once more contacted personally. "I want to emphasize the priority I attach to the next U.S. step: winning a continuation of China's most-favored nation status. That is, of course, a crucial next step," he wrote Deng. "I am compelled to say, however, that more needs to be done on the Chinese side before I can guarantee a general improvement in our relations."

As the critical UN vote approached, Bush thought he had a deal: American trade and endorsement in exchange for Chinese support. Baker in particular thought he'd explicitly tied any ensuing meeting at the White House for Foreign Minister Qian to a yes vote in the Security Council. Both proved mistaken. The Chinese abstained in the end, denying Washington the full weight of global opinion Bush desired, but nonetheless enabling the resolution to pass. They did not vote yes. But neither did they veto. Enraged, Baker urged cancellation of the White House's invitation to Qian.

Bush demurred, reasoning that both sides had secured what they needed in the Security Council, if not necessarily what they wanted. The Chinese could say they had taken a principled stand, and the Americans had secured United Nations support. They might need Beijing's endorsement, or at least its abstention, again in the future, and Bush felt that granting China's foreign minister his coveted presidential sit-down would increase the likelihood of Beijing's continued support if further resolutions proved necessary. It wasn't the deal or the diplomatic victory Baker desired, but as Bush put it, "We did not need an international crisis in the wake of our UN success."

Bush believed that Beijing did enough in the end, barely, to demonstrate its commitment to international order because he had refused to sever Sino-American ties after Tiananmen. That theory is impossible to test, but he is likely correct. Given their reluctance to back an American president who so clearly stood as the only thing between them and congressional fury, it is difficult to imagine Chinese leaders voting with a president who had more forcefully severed relations only a little more than a year before. Bush's prudent diplomacy in 1989, however emotionally unsatisfying, ultimately paid unexpected dividends. "China remembered the terrible savaging Bush took for them," Lilley concluded. "And when it came time to collect the fee for that, he could do it."

Similar maneuvers secured Gorbachev's cooperation, even though Soviet strategists largely balked at fighting their former ally, or from doing anything that might help their longtime enemy (the Americans) enhance their strategic foothold in the Middle East. Foreign Minister Shevardnadze in particular put his dwindling political capital on the line to persuade Gorbachev to support Washington over Baghdad, even if the crisis resulted in the kind of military deployment that was anathema to *perestroika*.

Soviet leaders were nonetheless taken aback in mid-August when American troops took up defensive positions in Saudi Arabia without warning, and certainly without asking in a serious way if the Kremlin wished to join in the kingdom's defense. "Are you *consulting* us or are you *informing* us?" Shevardnadze asked Baker after the fact, irritation evident in his voice. "We're informing you, I guess," Baker admitted, according to one version of events. Baker recalled the conversation differently, albeit only subtly so. "Well, Eduard, I'm talking with you," he wrote in his memoir, because this type of military deployment "is not something we want to do by ourselves." Both men hoped cooperative action in the gulf might lead to closer Soviet-American relations, but Bush had not wanted to risk a debate over potential Soviet deployments to the region before authorizing his own. The Americans thus spoke glowingly of the new Soviet-American ties that might arise from cooperation in confronting the crisis, but demonstrated with their actions that those ties would grow only if Moscow was willing to do things Washington's way. Bush put it bluntly when speaking privately to congressional leaders. He would accept Moscow's support but not its dictates, and not necessarily even its help. "There are no longer two superpowers in the world," he told them. "There's only one."

Nevertheless, the Soviets retained more than enough power to complicate American efforts if Gorbachev chose. Maintaining his continued acquiescence thus proved one of Bush's principal priorities as the gulf crisis grew into a standoff. Moscow held a UN veto, and if piqued or disillusioned over its treatment in the Middle East could still thwart the all but signed negotiations over Germany. While constantly working the phones in the weeks following Saddam's invasion, calling and cajoling foreign leaders for hours on end, Bush flew to Helsinki on September 9, a major address to Congress and the American people less than forty-eight hours away, solely for the purpose of previewing his speech for Gorbachev, in person, and expressing his hope for continued partnership.

It was less a consultation than a briefing. "I think it is a wonderful signal to the world that the US and the Soviet Union have been proceeding in concert on the Gulf," Bush began. "The closer we can be together today, the closer the new world order, the closer the US-Soviet relationship, the closer to a solution to the Gulf crisis. I want to go to work with you as equal partners in dealing with this," he told his Soviet counterpart. "I want to go to the American people tomorrow night to close the book on the Cold War and offer them the vision of this new world order in which we will cooperate."

Gorbachev immediately perceived the truth, however: that they weren't equal partners. "I think this crisis is a test of the process we are going through in world affairs and of a new US-Soviet relationship," he agreed, "but let me emphasize that in this new world US–Soviet Union cooperation is essential." Moscow's commitment was firm, he said, but not without limits. "It was difficult for us at first because you decided to send forces and then told us. That made it difficult. Perhaps it happened that way because this process is new," he suggested. The Kremlin was not used to finding out about military moves after the fact, and Gorbachev had learned all too well from bitter personal experience that deploying troops as a show of force frequently led to their use.

"I can accept that as constructive criticism," Bush quickly conceded, demonstrating the old adage that it was easier to seek forgiveness than permission. "We weren't trying to do it behind your back, but I can accept I should have called you and told you I was going to do it."

Gorbachev plowed on, pointing out additional ways the United States had taken the lead without inviting Soviet input, most especially in that Bush had framed the conflict in military terms from the start, moving troops and coordinating Saudi Arabia's defense before attempting substantive talks with Saddam, which left the Iraqi no other options than capitulation or war. "I think we all — especially you — are under great pressure to be decisive," Gorbachev observed, "but if that means military action — therein lies the danger." Having preached for years the wisdom of removing force from international affairs, he could not now endorse its immediate application. The new diplomacy demanded new rules. He'd been saying so well before Bush became president. "We are completely with you on this," Gorbachev noted, referring to their joint demand that Saddam abandon Kuwait, adding, "I agree with you that we should go out of these talks showing our joint commitment to a common cause and our determination to resolve this situation." But he wished to resolve it peacefully, and not sink the new world order they both desired down to Saddam's barbaric level. "Saddam Hussein says he hasn't attacked either of our countries and it was British colonialists who separated Kuwait from Iraq. He also says Panama was invaded without the UN taking action. Why are you ganging up on us?" Saddam was effectively asking.

"Bullshit," Bush exploded in response. He'd kept his cool while Gorbachev critiqued his previous month's work, but the Soviet leader's quoting their generation's Hitler to prove his point was too much to bear. "He also says I'm a pork-eater," Bush retorted, hoping to further paint the conflict as one between

Muslims and infidels, and "he's right." Calming himself somewhat, he swatted away Gorbachev's plea for a new round of diplomatic talks with the Iraqis, one that might include as a concession to Saddam the promise of a broad round of talks on the overarching Arab-Israeli dispute. "Any agreement on a plan which left [the] Kuwait issue open would be a major defeat for the collective action which has gotten us so far," Bush insisted, while making Saddam into the very champion of the Arab world he desired to be. To reward him by turning international attention to Israel would legitimize his invasion and leave him just as capable of pursuing conquest whenever the time seemed ripe. "He gets what he wants" if he wasn't forced to retreat, Bush objected, adding later in their conversation that "this aggression is so clear that if he is rewarded it will set a terrible precedent for the future."

"But if he gets nothing at all," Gorbachev countered, the two men clearly speaking past each other, "if he is backed into a corner, it will be more costly for us all. We need to give him some daylight," he implored, introducing a whiff of appeasement. "If we had offered Hitler [a] way out, would it have succeeded?" Bush shot back. The situations were different, Gorbachev retorted with the typical disdain Soviet leaders demonstrated when being lectured on Hitler. "Only in personality," Bush said. Dictators were dictators. They could not be appeased. Ever.

The two men largely agreed to disagree in Helsinki, setting the stage for further tension in the months to come. Both wanted a new world order, but only one was willing to consider military force as part of it. Not surprisingly, it was the one with force to spare, capable of deploying troops across the world even as his counterpart struggled to bring his own forces home from their long occupation of Germany. "Many things were happening at the same time: the Gulf Crisis, the quick finale of the German unification story, and, what was probably Gorbachev's top priority, the maneuverings on radicalizing the economic reform," one of Gorbachev's close advisers noted. "In a subtle way, all these developments were linked," as they "brought closer the day when the old issues would disappear and an era would end, leaving us to face a more uncertain yet promising future."

21

THE NEW WORLD ORDER

CONGRESS FELL HUSHED as the president strode to the podium. Grayer than the last time he had stood before them, he had come to explain why he had ordered American troops into harm's way. Witness to war in his youth, he wasn't eager to be its instrument. Some memories never faded: of maimed men, fire, and tears. The prospect of securing something truly historic, however, beyond mere battlefield success, steeled his mind. "We entered this war because violations of right had occurred which touched us to the quick and made the life of our own people impossible unless they were corrected and the world secure once and for all against their occurrence," he declared. Others might fight for gain or glory. Americans would fight instead to forge "a new world," including a new assembly of nations capable of guaranteeing the safety of "great and small states alike" in this "war to end wars."

It became one of Woodrow Wilson's most famous phrases, praised and critiqued. Chastened by lengthy casualty rolls and reports of European leaders squabbling for advantage, Wilson's political opponents rebuffed his plea to bring the world closer together. It hadn't been easy to gain their consent to fight overseas in the first place. "We have accepted the truth" of our international obligation, he nonetheless pleaded and preached as prospects for American participation in the League of Nations dimmed, "and we are going to be led by it, and it is going to lead us, and, through us, the world, out into pastures of quietness and peace such as the world never dreamed of before."

The Senate saw things differently and refused to join, demanding exceptional distance to go along with American exceptionalism. Wilson, heartbroken, was soon dead. The war he'd hoped to avoid returned within a generation, thanks in part to Germany's unhappiness over the peace treaty Wilson had helped engineer at Versailles. His assistant secretary of the navy became president in his own right, appearing before Congress a quarter century later, while American troops fought and died overseas yet again. The longest-serving president in American history, he had just returned from a fourteen-thousand-mile trip to see Soviet leader Joseph Stalin. Franklin Roosevelt's hands shook and his voice broke by March 1945, though his dream of completing his mentor's final mission remained undimmed. "This time we are not making the mistake of waiting until the end of the war to set up the machinery of peace," he told the overflow crowd. "This time, as we fight together to win the war finally, we work together to keep it from happening again." Americans needed to join the new assembly of nations due to arise from the ashes of war, not merely wish it well from afar, so "the sons and the grandsons of these gallant fighting men do not have to do it all over again in a few years."

This time, Congress listened. Roosevelt was dead six weeks later, yet the United Nations he helped design took life in San Francisco that spring, when representatives from fifty countries—including the United States and the Soviet Union—formed a new collective body. He considered the organization "the crowning act of his career," veteran journalist Anne O'Hare McCormick wrote after conducting what proved to be one of FDR's final interviews. "He dreamed of going down in history as the President who had succeeded where Woodrow Wilson failed in making the United States the great bastion" of a new international system for collective security "while the forge of war was still hot enough to fuse nations together."

The United Nations wasn't perfect. A General Assembly offered all member nations an equal voice, their will enforced by a Security Council comprising the great powers. "It has already been said by many that this is only a first step to a last peace," Roosevelt's successor, Harry Truman, admitted. No one expected nirvana; indeed, the entire body was designed not to prevent conflicts but to defuse them. Plenty of conflict ensued in the decades to follow. But not a great-power war. This was not peace in an absolute sense. Invasions continued, civil wars erupted and spread, and nuclear Armageddon hung over it all. Yet nothing on the scale of either world war returned, at least in part because the Americans had finally moved beyond isolation into a powerful form of engagement. "Roosevelt is remembered for this achievement," Henry Kissinger subsequently noted, "but it was Wilson who shaped American thought," and in particular

Americans' understanding that they needed to engage the world around them. Isolationist tendencies persisted. Yet for successive Cold War presidents, the question was not *if* the United States would lead in international affairs but *how much*.

George Bush strode to that same podium forty-five years later, on September 11, 1990, another commander in chief with troops in harm's way, and, more profoundly, another president facing a hinge moment in history. Born the same year Wilson died, he too faced fierce opposition over his handling of the current conflict. A mere youthful cog in Franklin Roosevelt's war machine, he'd also just returned from meeting with the Soviet Union's top leader. Two days before, in Helsinki, he'd stood side by side with Mikhail Gorbachev in shared condemnation of Iraq's conquest of Kuwait. "Nothing short of the complete implementation of the United Nations Security Council resolutions [demanding Iraq's withdrawal] is acceptable," Bush told reporters at their joint news conference. Gorbachev vigorously nodded his consent. "As soon as Saddam Hussein realizes that, then there certainly will be a peaceful resolution to this question."

Security was tight around the Capitol. Soldiers patrolled street corners. Police patted down even those whose special pins signifying membership in the House or Senate typically excused them from such indignities. The Secret Service made neither exceptions nor excuses. Agents wore their own telltale pins, and more ominous bulges under their jackets. It felt like war, and with good reason, even if Congress had yet to vote to call it such. Thirty thousand American troops stood guard in Saudi Arabia. By month's end, 107,000 deployed for Operation Desert Shield. By year's end it would be half a million, along with nearly 130,000 tanks and vehicles, and 519,000 tons of ammunition and supplies, or slightly more than one ton for every American serving in theater, who also collectively consumed 2.5 billion gallons of fuel, even more of water, and ultimately received enough mail — 38,000 tons' worth — to cover more than twenty-one football fields to a depth of eight feet. Soldiers complained about the heat, and that their eggs were watery and their bacon — a commodity the Pentagon refused to do without, even in the holy Muslim kingdom — crisped too quickly in the dry air. But the fact that troops only weeks before stationed in Europe and the United States had food and roofs at all, and tanks and artillery to go along with four aircraft carrier battle groups offshore and the region's largest air force above, was an astonishing projection of force. "No way they're going to seize the oil fields" of Saudi Arabia now, General Schwarzkopf assured

Washington about the Iraqis. Prying them out of Kuwait, however, would be another thing entirely.

Despite such rapidly amassed American force, this "was not, as Saddam Hussein would have it, the United States against Iraq," Bush told the legislators, a watching nation, and indeed the globe. "It is Iraq against the world." A dozen countries had their own troops en route to the gulf, and the United Nations Security Council had called for Kuwait's unconditional release five times in as many weeks. Bush had deployed troops to the region on his own authority. But as he stressed throughout his congressional address, they operated with global consent, and with a broader mission beyond merely Saudi Arabia's defense or Kuwait's liberation. They were there because the Cold War constant that had dominated his life, and that for so long he'd refused to concede in its final days, was finally over.

"We stand today at a unique and extraordinary moment" in history, Bush declared. "The crisis in the Persian Gulf, as grave as it is, also offers a rare opportunity to move toward an historic period of cooperation." Moscow and Washington stood unified for the first time in generations. This was historic but also portentous. "Out of these troubled times . . . a new world order — can emerge," Bush said. "A new era — freer from the threat of terror, stronger in the pursuit of justice, and more secure in the quest for peace. An era in which the nations of the world, East and West, North and South, can prosper and live in harmony."

The world had changed since he'd taken office, and so had he. Eighteen months before, he'd stood outside the Capitol and declared the day of the dictator over. That September night he described more fully than ever before his vision for a new international system formed not merely because the stream of history continued its long flow in democracy's direction, but because the United States and its allies, which now included a bevy of states once sworn to its destruction, were all finally pulling in the same direction. "Today that new world is struggling to be born, a world quite different from the one we've known," he declared. "A world where the rule of law supplants the rule of the jungle. A world in which nations recognize the shared responsibility for freedom and justice. A world where the strong respect the rights of the weak." It was also a world in which Americans would not retreat from the responsibility of leadership, to an extent beyond what Wilson ever imagined or Roosevelt lived to see. "How we manage this crisis today," Bush told his listeners, "could shape the future for generations to come."

Bush was not promising anything as utopian as Gorbachev had pledged, merely something better than before. "Freer" than in the past, it did not promise universal freedom. "Stronger" in its pursuit of justice, it would not yet be fully just. "More secure" than the world they'd all inhabited since 1945, though Bush did not promise an end to uncertainty or war. Conflict would remain. Criminals and tyrants would forever seek to exploit cracks in the international order, and Bush believed one thing for certain: so long as humanity existed, there would always be another crisis. Even the most harmonious societies required policing. There would always be the need for an international policeman to walk the beat.

The new world order he offered that night instead promised a more equitable mechanism for dealing with future crises as they arose, a mechanism based on international sanction but also on American power. It would be premature to beat the world's swords into plowshares and pruning hooks, Bush effectively said. Indeed, if the current crisis demonstrated anything, it was that swords, and especially American ones, still had their use. "Recent events have surely proven that there is no substitute for American leadership," he pointed out. "This is the first assault on the new world that we seek" and "the first test of our mettle . . . Had we not responded to this first provocation with clarity of purpose, if we do not continue to demonstrate our determination, it would be a signal to actual and potential despots around the world. America and the world must defend common vital interests — and we will. America and the world must support the rule of law — and we will. America and the world must stand up to aggression — and we will."

Listeners could be forgiven for thinking they'd heard all this before. Bush's "new world order" was merely Franklin Roosevelt's finally put into practice. But whereas FDR's hopes had collapsed under the weight of Cold War tensions, that impediment was now gone. For the first time in forty years the Soviets were on board, while the Chinese (another great power Roosevelt had presumed would help maintain global order) proved willing at least to abstain rather than stymie international consent. "In the Cold War days, we'd say this is black and the Soviets would say, hey, that's white," Bush told reporters in the spring of 1991. "And you'd have a veto and nothing would happen. And the peacekeeping dreams of the U.N. were dashed . . . [S]o, part of this new world order has been moved forward by a United Nations that functioned," in part by enabling the great powers to impose order without imposing upon one another. "We might have still been able to stand up and come to the assistance of Kuwait" without the United Nations, he continued. "I might have said to hell with them, it's right and wrong, it's good and evil, he's evil, our cause is right; and without the United Nations, sent a considerable force to help." He had the authority to do it on his own (at least, once Saudi Arabia blessed the arrival of American troops). "But it was enhanced — it is far better to have this collective action where the world, not just the Security Council but the whole General Assembly, stood up and condemned" Saddam's invasion of Kuwait.

The assembled legislators cheered when Bush finished, even those whose reservations would surface over time, cheering as much in support of American troops in the field as for their commander in chief. The response was far different in Moscow. "You are asking the Soviet Union to approve the use of American force against a long-time ally of the Soviet Union," Gorbachev lamented to Baker in November. Washington's request opened a deep rift within the Kremlin between those who believed that their need for Western cooperation and aid outweighed any responsibility to Iraq's despot and those who feared that abandoning Moscow's allies in the developing world so soon after their retreat from Eastern Europe signaled an unrecoverable erosion of Soviet influence. It was a rift, in effect, between those who trusted the Americans, or felt they had no choice but to try, and those who could not stomach placing their fate in Washington's hands.

Never before had the two oft-sparring superpowers spoken so clearly as one, or been so unified in supporting the Security Council's will. Yet tensions lingered. Bush hoped for a peaceful end to the gulf crisis but seemed increasingly resigned to war — and in some ways, the sooner the better. "The more I think about it, the more I would love to have an incident so we could execute a devastating attack against Iraqi armor and Iraqi air," he had confessed in his diary the week before, revealing his desire to demonstrate the full measure of American power in the post–Cold War world. The longer the crisis persisted, the more Bush feared that Saddam might successfully barter withdrawal for some political victory, perhaps a broad conference on Arab-Israeli affairs, making himself a hero in the region while showing others that they too might profit from military aggression. And the longer the crisis, Bush also feared — and Saddam strategized — the more likely American voters would be to tire of the effort.

Gorbachev desired Iraq's retreat as well but saw little to be gained from bloodshed, or from a vivid display of American power through the defeat of a onetime client state armed almost exclusively with Soviet-made weaponry. He rightly feared that Washington's military deployment made fighting nearly inevitable despite the diplomatic options simultaneously being pursued, and worried about the more than five thousand Soviet citizens in Iraq, and the potential sympathies of the large Muslim populations in his turbulent southern regions. Gorbachev also sympathized with Saddam's calls for a broad Middle Eastern peace conference. Having already transformed one part of the world, the father of *perestroika* could not fathom why the Middle East's riddles should prove more difficult to solve than Eastern Europe's. The pursuit of a solution might even further boost his international standing, temporarily taking his citizenry's minds off their problems at home. "We were embarking on a long journey," one of his principal advisers noted after witnessing Gorbachev and Bush's unprecedented collaboration, predicting that a collision could not be far off. "I was sure that at the end of it Saddam Hussein would be out of Kuwait. The rest — what would happen to Saddam, how much blood would be spilled and whether U.S.-Soviet cooperation would remain intact under the pressure of events — was by no means clear."

Their brewing disagreement behind the scenes reflected a broader difference between the two leaders' prospects and the countries they led. Bush spoke optimistically of a better world. Gorbachev continued to struggle in his. Calls for secession multiplied throughout the Soviet Union as summer transitioned into fall and the Soviet economic free fall continued. "It was not easy for him and Shevardnadze to talk with the Americans in Helsinki when people in Moscow were queuing for bread," Gorbachev's interpreter lamented. "There had been another round of panic-buying" before the Soviet delegation departed, "sparked by fresh rumors of an impending price rise." Gorbachev was also due to host a final meeting of the 2 + 4 negotiations over Germany immediately upon his return from Helsinki, to endorse formally what he'd already conceded: Germany's right to unify, within NATO if it so chose, and thus for all practical purposes the passage into history of both the GDR and the Soviet people's greatest triumph. The new Germany was "the first unification of a country in modern history achieved without war, pain, or strife," Chancellor Kohl proudly declared. All of Europe triumphed in its creation.

That boast sounded hollow to Soviet ears. "Mister General Secretary, congratulations on the prize from the imperialists" one telegram to the Kremlin read. Free at last to speak their minds without fear of jail or retribution, a great many did. "Thanks for ruining the Soviet Union, betraying Eastern Europe, destroying the Red Army, [and] giving away our resources to the Americans and the media to the Zionists." Another writer sarcastically congratulated Gorbachev "for betraying Lenin and the October Revolution, for destroying Marxism-Leninism," and "for utterly ruining your country."

Gorbachev's office overflowed with these angry missives, which he compulsively read out of some masochistic sense of duty. "Why waste time on this debased trash?" one of his closest advisers asked incredulously on seeing the pile of complaints covering his desk. His boss merely shrugged in response, unable to avoid critique even if he tried. Raucous crowds routinely marched past the Kremlin in protest. "Down with Gorbachev!" banners read. "Down with the perestroika clique!" Unused to such popular attention, even if a logical consequence of the Kremlin's own democratic reforms, Foreign Minister Shevardnadze could only ask: "What were we meant to do? Fire on them?"

Someone did pull a trigger. A would-be assassin targeted Gorbachev in early November. The first blasts from his shotgun missed. The rest scattered skyward as police tackled him to the ground. "He was probably crazy," the KGB chief told reporters, though he could not deny that others might have good reason to think their country better off with a new man in charge. Gorbachev had been awarded the Nobel Peace Prize in October for his work in transforming Europe. That same work made many of his countrymen wish him dead.

Celebrated abroad, he felt besieged at home. "We must remember this certainly was not the prize for economics," a Foreign Ministry spokesman joked with the press. In Stalin's day the quip would have won the speaker a one-way trip to Siberia. In Gorbachev's, even supporters criticized. Sucked ever deeper into the political quagmire as the fall progressed, the newest Nobel laureate retained dreams of a better world but also perceived himself as the only one capable of keeping true democratic chaos at bay. "The essence of what we're doing is aimed not at some distant utopia, but at a better life for those living in our country today," Gorbachev optimistically proclaimed in advance of the Soviet Union's annual commemoration of its 1917 Bolshevik revolution. "We are witnessing the advent of an epoch that will be marked by the triumph of a truly humanitarian civilization."

Such words once inspired. By now they sounded hollow. Occupying an increasingly depopulated political middle, Gorbachev retained all the trappings of power yet little support, his orders and dictates ignored. Military commanders moved troops and weaponry without his knowledge or consent, complicating arms negotiations with the Americans, who did not know if Soviet withdrawals from Eastern Europe were designed to decrease tensions or merely to shield armaments from would-be inspectors. Still head of the Communist Party, he'd seen its rolls drop by nearly 3 million over the previous year. Some members formally resigned, or were purged for disloyalty. Others simply refused to pay dues to an anachronistic organization. On average eight thousand citizens deserted the party every day during 1990, following Boris Yeltsin's public lead. Several republics had already declared their independence, even as they continued to acknowledge Soviet dominion over them. They were autonomous yet also subservient, sovereign yet subordinate, free yet still part of the union.

Could these contradictions continue? The answer was not for long. Civil war seemed dangerously likely; indeed, for a country so deeply divided, it appeared to many the obvious solution. "The economic and political crisis in the country has come to a head," Yeltsin blustered, all too eager in his new post as president of Russia to profit politically from his onetime mentor's travails. "The people's patience is coming to an end, and an explosion could occur at any time." Gorbachev feared the end was near as winter approached. He secured immediate American food aid, enough to keep the bread lines reasonably short for at least another few weeks. But after that? "It always turns out that they crucify prophets," he told an aide. "So that's why I wonder whether my time has come to be crucified."

"The political situation in the Soviet Union is now almost completely polarized," Condoleezza Rice reported as 1990 came to an end. "The Republics are strong enough to veto the actions of the Center but too weak to act on their own. And while Rome burns, the Supreme Soviet, the Russian Parliament and Gorbachev, himself, engage in an endless tragicomedy of restructuring and renaming feckless institutions — each more powerful on paper and less real in substance than the last." To her mind, "no one is capable of governing the Soviet Union" in its present form, though one thing was certain: "the Soviet people — reformers and conservatives alike — are tired of the parody of representative government that masquerades as 'democracy' in the Soviet Union."

Gorbachev's domestic approval rating dropped to 19 percent in December, Ambassador Matlock noted from Moscow. Yeltsin's support was better, though not much, settling in at 32 percent. Gorbachev's "only hope for political survival and for preservation of some sort of union seemed to depend on his willingness and ability to make common cause with Yeltsin and the democrats" against those who thought only of taking the country back to a time before *perestroika*. At the same time, the two men despised each other too much to ever consider working together. "It was obvious to me that Gorbachev not only failed to see what seemed clear to me but was actually doing the opposite," Matlock concluded, leading to both a political stalemate in Moscow and the sense that an explosion was on the horizon.

"We need help," Gorbachev confessed to Baker. "As we move toward implementing these reforms, there's going to be great dissatisfaction . . . [T]he domestic situation is going to get much worse." The Soviet economy should pick up in six or nine months, he assured Baker, who was polite enough to refrain from noting that Kremlin economists always forecast an uptick in another six months. "We need help now," Gorbachev repeated. "Can you help get some money from the Saudis for us?" Four or five billion dollars in immediate aid would help. As a start. They'd be able to secure that type of long-term financing from the Germans, the French, the Japanese, and the rest only if they proved themselves willing international partners, which Bush's continued support could ensure.

Bush had promised as much in Helsinki, though not from his own coffers. "I have to tell you in advance: we have no extra money," he'd told Gorbachev. "The deficit of [the] federal budget is close to its record high. In this situation, the Congress shall never give the president a 'card-blanche' [*sic*] to spend billions." He would indeed ask America's allies to help. But at the same time, Bush said, keeping the Soviets on the line like a fading marlin, "the increasing trust between us helps cooperation," and so long as cooperation continued, he'd be willing to press the nation's richer friends and American businesses for further aid. "Frankly, our [joint] statement on Iraq shall be a great help to my administration in pushing forward the projects of economic cooperation with the USSR."

Both sides understood the bargain this implied: Moscow's compliance in the gulf, as with its earlier concession on Germany and NATO, in exchange for Washington's continued blessing and the promise of aid. "My own sense is that in the end they will go along with us," Baker privately wrote Bush after a lengthy negotiating session with Gorbachev and Shevardnadze. "They got the point, and I believe their stake in good relations and partnership and desire for partnership with us will lead them in the right direction."

The irony was not lost on the Soviets that the Americans would willingly pay billions to beat back a petty despot but balked at spending anything near that amount when it came to aiding their onetime adversary's transition to democracy. More ironic still, Gorbachev's backing of the UN resolutions against Iraq, made in part to ensure continued American support, further undermined him at home. "The new relations of cooperation and interaction, particularly with the United States, are now seen as one of the chief levers capable of lifting us out of the current [domestic] crisis," one well-placed Soviet commentator noted, but not everyone appreciated being forced to bargain for their survival. Elements of the Soviet military and Foreign Ministry balked at the way Gorbachev allowed the Americans a free hand in the Middle East, noting that Iraq owed more in Soviet debts than Bush was offering in aid, debts Baghdad would never repay if reduced to rubble. Others lamented Washington's construction of a massive military infrastructure a mere two hundred miles from the nearest Soviet border, which the Warsaw Pact's chief of staff openly called a grand American plan to replace lost influence in Europe. He clearly did not realize that Bush intended to retain influence in both regions. Indeed, newly emboldened to report and to speculate, Soviet pundits openly feared the Kremlin might choose to counter American influence by inserting troops of its own, prompting fears of a new Afghanistan, a name as despised in Red Army circles as Vietnam was in the Pentagon, and with far fresher memories. "Gorbachev cannot afford to send troops abroad," either financially or politically, argued one Moscow commentator. "Perhaps we could send two or three ships to the Gulf as a symbolic contribution to a multinational force," he mused, "but that's all." The very notion of putting Red Army units back into a region they'd just fled, which Shevardnadze seemed to imply might happen as a further demonstration of Moscow's solidarity with the West, undermined Gorbachev's domestic credibility even further.

"Do they intend to wait until the people take to the streets to have their say?" Yeltsin wondered aloud. In November, Kazakhstan became the fourteenth of fifteen Soviet republics to formally declare its sovereignty. Separatists in Georgia won 54 percent of the vote in parliamentary elections. The government of Ukraine began distributing food coupons, the first step toward creation of its own currency. Even cities claimed independence, especially those with Russian majorities set amidst a sea of varied ethnicities in the outer republics. Yeltsin's government openly endorsed each secession movement while continuing down its own path toward autonomy, pledging to support native Russians in the republics. He announced economic reforms in early December beyond anything Soviet authorities stood ready to authorize, including private ownership of land. "This is the first time in 73 years we have ever adopted a law on private property," he boasted, noting the central government's refusal to take this fundamental step on the road to a market economy. Russia would do what Soviet authorities would not. "If they want a fight," he promised, "we'll give it to them."

Gorbachev moved instead to solidify his hold on power along two different fronts. Hoping to demonstrate his continued international influence while still supporting global resolve in the gulf, he dispatched Yevgeny Primakov as a personal emissary to Baghdad in search of a peaceful solution to the standoff over Kuwait. Washington objected, but only quietly, being unable to publicly disavow a mission designed to stave off war. No solution emerged. Saddam refused to withdraw despite Primakov's pleas, though he did promise that the thousands of Soviet citizens currently within his country could depart . . . as soon as they finished the work they'd been contracted to complete. They weren't hostages — at least not in a formal sense. Better to think of them as indentured servants whom Saddam would dismiss once he saw fit. Primakov cautioned him not to stall, warning that the Americans weren't bluffing. "If you don't pull your troops out of Kuwait, a strike against you will be inevitable," he said. "If my only options are to fall on my knees and surrender or fight," Saddam responded, more in resignation than defiance, "I'll choose the latter."

Back in Moscow Shevardnadze fumed, confessing to Baker and his staff that Primakov's mission was misguided, even stupid. Longing for closer superpower relations, he saw little to gain from quixotically demonstrating Moscow's independence, or from openly reviving Moscow's alliance with someone as despicable as Saddam Hussein. If Gorbachev had to send anyone in search of peace, it should have been his foreign minister, Shevardnadze added, not some former KGB operative.

"I've tried very hard to discover a grain of reason" in Primakov's proposals, Shevardnadze angrily told his boss. "But I just couldn't do it." He might secure Saddam's withdrawal in time, or the release of Soviet workers, but only at the price of bartering away the Soviet Union's burgeoning relationship with the Americans and thus "the collapse of our whole foreign policy." No American ally, and in particular neither the Saudis nor the Kuwaitis, whose combined promises of future aid and investment neared $5 billion, would come through with the dollars Moscow needed if the Soviets failed to back Bush in the gulf. Moreover, Shevardnadze complained, "I can't be the minister if various other people are going to be involved in affairs that belong to my sphere."

Buffeted by successive crises at home and torn between advisers who urged a show of Soviet independence and those who encouraged closer ties to the West, Gorbachev seemed "truly at a loss" as 1990 drew to a close, one of his aides recorded in his diary. It was "the first time I'd ever seen him in that state. He could see power slipping from his hands." Bush tried to bolster his bargaining partner's confidence, even as he continued to stress that Washington's support demanded Moscow's ongoing cooperation in the gulf. "You are going through a difficult time," he told the Soviet leader when they met in Paris in late November. "I do not want to interfere with your internal affairs, but permit me to say: we will try to do everything that we can to help you in your burdensome task." The Americans didn't need Soviet troops or planes, just Gorbachev's support in New York as the Security Council took up new resolutions authorizing the use of force to liberate Kuwait. "I am asking you to help me send Saddam this signal," Bush told him. "We calculate that it will be enough [to compel] Saddam to do what is demanded of him."

The "new world order" was at stake, Bush once more insisted, and once more Gorbachev balked at its American design. "This is an extraordinarily important moment not only for us both, but also for everything that you and we have begun doing together in the world," the Soviet leader responded. "If we do not prove now that we are in a condition at this new phase of the world's evolution, to cope with similar problems, then it will mean that what we have begun does not count for much." They agreed that Saddam's aggression could not stand, with Gorbachev assuring Bush, "I am convinced, we must be together with you." But that did not necessarily mean the pair saw eye to eye. They must still "try to avoid a military resolution," Gorbachev implored, securing from Bush the concession that hostilities would not commence until early in the new year, after Saddam had been given one more clear deadline by which to withdraw.

But Gorbachev had plans of his own, which caused Bush's circle to question if they'd placed too much faith in him personally. Opening a new domestic front in his own defense in conjunction with Primakov's ongoing peace initiatives, Gorbachev proclaimed new presidential powers in December, abruptly dismissing his presidential council en masse while firing the head of the state television system for showing too much journalistic independence. "I am the president's man," his replacement immediately declared, "and I have come to carry out the president's will," and not necessarily to stand up for freedom of the press as Western journalists might expect. More chilling still, Gorbachev appeared to have cut a deal with conservative forces within the Red Army and the KGB, whose director told television viewers that "at the president's request," the members of "the security forces have made our choice: we stand for the flowering of our socialist homeland" and for the Soviet Union itself. "To be or not to be — that is the choice of our great state." Acting on cue, Gorbachev decreed all local economic measures — especially those ratified by Russia and the republics — null and void. Only his rules mattered. "We have an absolute crisis," Shevardnadze privately warned Baker, sensing Gorbachev's apparent lurch to the right. "We could have a dictatorship very soon in our country."

The move sent shock waves through Bush's White House, which feared that the right-wing coup seemingly always on the horizon had finally occurred, only with Gorbachev at its head. "I have never found anything he ever told me to be untrue," Baker pondered, thinking back on his last conversation with Shevardnadze. It would in fact be the last time the pair spoke as their nations' chief diplomats, though Gorbachev's handpicked foreign minister had far more still to say to the People's Congress in December. "Dictatorship is coming," Shevardnadze told the assembled delegates. The democracy they touted faced an impending crisis. "No one knows what kind of dictatorship this will be and who will come — what kind of dictator — and what the regime would be like."

"What would a crackdown and the use of force mean in this country?" he asked aloud. "It could be worse than Tiananmen." He would not be party to such violence, or to whatever came next. "I am resigning," Shevardnadze dramatically declared to his comrades. Audible gasps ensued. Only his closest aides knew of his plan beforehand. Certainly Gorbachev did not. He stared into space as dumbfounded as the rest. "Do not respond, do not curse me!" Shevardnadze continued as the collective gasp turned to angry cries. "Let this be my contribution, if you like, my protest against the onset of dictatorship. I express profound gratitude to Mikhail Sergeyevich Gorbachev. I am his friend. I am a fellow thinker of his . . . but I think that it is my duty, as a man, as a citizen, as a Communist."

"We would obviously be foolish not to take the warning in Minister Shevardnadze's resignation statement seriously," Baker told reporters, saying much the same to Bush in private. Scowcroft agreed, advising they voice support for Gorbachev's continued rule and for a peaceful resolution to the secessionist crisis, even as they cautiously prepared to deal with his inevitable successor, or with a Gorbachev who'd finally resigned himself to a military crackdown on dissent. Gorbachev was "still trying to find common ground between the reformers and the reactionaries," Scowcroft advised the president, "even though it doesn't exist."

Recovering from their shock, Soviet officials stressed that Shevardnadze's sudden resignation would do nothing to alter the generally positive trajectory of Soviet-American relations. Gorbachev said the same in a personal letter to Bush. But American officials were worried, and moved to consolidate their unprecedented gains of the past few months. Having secured a unified Germany within NATO, democratic transitions in Eastern Europe, and formal UN sanction for action in the gulf, all only recently considered inconceivable and all possible only with an unprecedented degree of Soviet compliance, Scowcroft and those around him increasingly wondered if perhaps they'd wrung from Soviet reformers all they might hope to win. It was certainly a longer list of victories than anyone would have predicted when Bush first came into office. Perhaps it was time to count their blessings and prepare for the day when Soviet-American good feelings once more abated.

BUSH'S NEW WORLD ORDER threatened to prove surprisingly short-lived as 1991 began. Two overlapping potential roadblocks in the new Soviet-American partnership at its heart troubled American policy makers: the Persian Gulf and the Baltics. Gorbachev continued to urge new overtures for peace, but as January began, Bush seemed resigned to war — even if Congress objected, and even if it cost him his job. Neither the House nor the Senate had yet formally endorsed hostilities, but votes loomed. Five times between November and January, historian Jon Meacham has tallied, Bush alluded in his diary to potential impeachment, each time swearing he'd order Saddam's forcible removal from Kuwait even without congressional consent.

He'd sooner sacrifice his office than let the Iraqi despot win. "It is only the United States that can do what needs to be done," he dictated to his diary. "I still hope that Saddam will get the message; but if he doesn't we've got to take this action; and if it works in a few days, and he gives up, or is killed, or gets out, Congress will say 'Attaboy, we did it, wonderful job; wasn't it great we stayed together.' If it drags out and there are high casualties, I will be history; but no problem—sometimes in life you've got to do what you've got to do."

In truth, Bush expected the war to go quickly once it started, and thus reasoned that legislators, no matter their ire, would have little stomach for removing a victorious—and undoubtedly popular—wartime leader. The final vote in Congress on January 12 went his way, albeit only barely, with the resolution passing, 250–183, in the House, and an even closer 52–47 in the Senate. "The big burden, lifted from my shoulders, is this Constitutional burden," Bush wrote in clear relief, including "the threat of impeachment."

Much to Bush's consternation, however, war simultaneously drew closer in the Baltics. On January 10, even as American voters watched nonstop coverage of congressional debate over the Gulf War resolutions in Congress, Gorbachev continued his lurch to the right, calling for Lithuania's separatist regime to formally submit to Moscow's authority or else face military reprisal. Soviet troops swarmed across the border to back up his threat, seizing key buildings and transportation nodes in Vilnius the next day. Gunfire erupted on the thirteenth. Soviet paratroopers and Lithuanian militias tangled over control of a central television tower. Crowds gathered in response. Then more gunfire. Over a dozen Lithuanians soon lay dead, as did a Soviet officer, accidentally felled by his own troops.

Gorbachev denied ordering the bloodshed, yet he could not avoid responsibility for having escalated the crisis by deploying troops into the middle of a maelstrom. Yeltsin, perhaps the one man who could have saved the Soviet Union by this point by closing ranks with Gorbachev, chose instead to deepen the divide. He formally recognized the independence of the Baltic states, and flew to the region to personally lend his support for its separatists and to propose a mutual defense treaty—against what was still technically their own central government! — all the while encouraging Russians to march for their *own* secession from the Kremlin's rule. Gorbachev continued to deny responsibility for the bloodshed in the days that followed, leaving open the possibility either that he condoned the violence or, perhaps, that he was no longer master of his country's security forces. "He seemed confused, not in control, and on the wrong side of the issue," one of his longtime advisers lamented, while Matlock's staff openly debated which was more frightening: "that Gorbachev planned this or that he has lost control of the armed forces?" Back at the White House, Condoleezza Rice expressed the same fear. "There are only two possibilities," she wrote Scowcroft. Gorbachev "cannot control the MVD [Ministry of Internal Affairs] and the Army or he does not want to. Neither is comforting."

Either way his reputation appeared irreparably harmed. After Vilnius he could no longer count on any support from the left, having ceded that side of the political spectrum to Yeltsin and to those who wished to see the Soviet flag torn down. "We must not become captive of the 'any Gorbachev is better than the alternative' logic," Rice advised. By firing on protesters, even if only tacitly, he had crossed the administration's "red line." The United States had to respond. "Our goal must be to use what leverage we have to persuade him not to commit suicide through a Faustian bargain with forces on the right that will never trust him," she continued. Perhaps they could persuade him by "freezing" current trade negotiations, which sounded nicer than cutting them off, and always left open the possibility that they could be thawed. Yes, the move might destabilize Gorbachev, perhaps even ensure his fall. But "if it comes to that there are alternatives other than Boris Yeltsin," Rice mused, knowing that no one in Bush's inner circle trusted the Russian leader, "including some of the men that Gorbachev himself helped to create — Yakovlev, Shevardnadze, Bakatin, and others." It was time to prepare in a serious way for a post-Gorbachev future, because "even if Gorbachev's successor turns out to be the reincarnation of Joseph Stalin, we will have to do business with the Soviet Union."

For Bush the real problem with the Lithuanian crisis lay in its timing. War in the gulf loomed. Saddam had until January 15 to withdraw or face military ouster. The deadline passed without retreat or resolution. American warplanes and missiles began to fly two days later, setting off a bombardment of Iraqi forces that lasted more than a month. "We own the skies," Bush boasted in his diary. He fretted over the pilots he was ordering into harm's way but feared disruption of his carefully built coalition. The Iraqis lobbed missiles at Israel, hoping to spark a retaliatory strike that might open a fissure between Bush and his Muslim allies. And he worried that Gorbachev was not yet done meddling in the gulf as well.

Bush consequently reacted with restraint to news from the Baltics, largely withholding public judgment yet saying plenty to the Soviets in private. American presidents had for decades pledged to support the region's independence, and the same coalition of left-wing Democrats and right-wing conservatives that had coalesced to criticize his Tiananmen policies stood ready to call him out again if his support for democracy appeared to waver. "It would be a sad irony if the price of Soviet support for freeing Kuwait was American acquiescence in Soviet aggression against another illegally annexed country," Democratic senator Bill Bradley argued. Even reliable Republicans voiced concerns. "Unless Gorbachev puts an immediate end to the threats, blackmails, and aggression," Senator Bob Dole said, "the United States should not deal with him in a business as usual manner." Critics were also quick to point out the apparent contradiction between Bush's newfound high rhetoric when it came to America's willingness to stand up to aggression in a region rich with oil, and his seemingly more muted response to news of aggression on Soviet soil. "The New World Order was born in Washington on September 11, 1990," one widely read columnist opined, "and it was shot dead in the streets of Vilnius on January 13, 1991."

Bush turned, as before, to his pen rather than his bully pulpit in response, hoping to employ the economic leverage he held over Gorbachev without publicly undermining him. "Last June, during the Washington Summit, we talked about the effect of Soviet actions in the Baltic States at great length," he wrote Gorbachev on January 22, reminding the Soviet leader that as part of their overall negotiation over Germany, he'd approved their pending trade agreement despite the ongoing economic embargo against Lithuania. In the wake of the recent violence, similar patience no longer seemed possible. "I had hoped to see positive steps toward the peaceful resolution of this conflict with the elected leaders of the Baltic States," Bush wrote. "But in the absence of that and in the absence of a positive change in the situation, I will have no choice but to respond."

Shortly, Bush continued, "I will freeze many elements of our economic relationship, including Export-Import credit guarantees; Commodity Credit

Corporation credit guarantees; support for 'Special Associate Status' for the Soviet Union in the International Monetary Fund and World Bank and most of our technical assistance programs. Furthermore, I will not submit the Bilateral Investment Treaty or Tax Treaty to the United States Senate for consent to ratification when and if they are completed." These were not steps he relished taking, Bush stressed. "We have come too far in US-Soviet relations to return to a confrontational course," especially as "we have both talked of our desire for a new world order." Yet he simply could not reconcile Gorbachev's optimistic words with his more recent actions.

The letter turned increasingly personal. "Mikhail, I cannot help but recall that you, yourself, told me that you personally could not sanction the use of force in the Baltic States because it would mean the end of perestroika." Perhaps *perestroika* was in fact at its end, Bush suggested. Thus with it the new phase of Soviet-American rapprochement, and certainly any hope the Soviets might retain of further American largesse as they continued their difficult transition to capitalism and democracy. "The United States and the Soviet Union have made tremendous progress over the course of the last two years in improving the relationship between our two countries," Baker simultaneously told reporters, employing language Bush prudently kept private. "And the events of the last 10 days to two weeks, I'm afraid to say, could have the effect of putting that progress in jeopardy."

Bush's direct message, supplemented by Baker's public stance, was as plain a threat as one administration could deliver to another. Gorbachev nonetheless wanted to be sure he understood its meaning in its entirety. It was delivered by Matlock even as a crowd of 300,000 gathered in Moscow to protest the violence in Vilnius, one of the largest demonstrations in Soviet history and also one of the most raucous in recent memory. Previous crowds had featured anti-Gorbachev slogans. This was the first time Matlock had seen a crowd explicitly demanding Gorbachev's resignation. Speakers called for him to step down, for the immediate withdrawal of Soviet troops from Lithuania, for dismissal of the USSR Supreme Soviet, and for those responsible for the bloodshed to stand trial. "A regime in its death throes" has committed a criminal act, a leading Moscow journal editorialized. "After Bloody Sunday in Vilnius, what is left of our president's favorite topics of 'human socialism,' 'new thinking' and 'our European home'?" it asked. The answer: "Virtually nothing."

Matlock almost failed to reach the Kremlin. Moscow was awash in unrest, and a crowd of Iraqi and Palestinian demonstrators besieged the American embassy in protest over Bush's Gulf War. Marine guards forcibly dispersed them just long enough for the ambassador's car to escape. Taking time to translate Bush's letter out loud once in Gorbachev's office, Matlock turned for a response once finished. "Did he say he has taken these steps" to freeze economic ties, the Soviet leader pressed, "or that he will take them?" If the former, he was sunk; if the latter, he could at least still hope to salvage the aid he required. "He said he would," Matlock responded, "if . . ."

Gorbachev understood. "Try to help your president understand," he implored as their conversation wound to a close. "We are on the brink of a civil war. As president, my main task is to prevent it." They should expect a period of "zigs and zags" on the long road to democracy, he said.

Or in this case, retreats. Gorbachev pulled Soviet forces back from the Baltics by month's end, opening negotiations over the region's eventual separation and ordering a thorough investigation of the bloodshed, even though he continued to believe it was the fault of "a virtual nighttime, constitutional coup." Similar negotiations began within Russia as well.

"Tell my friend George" that I will stand by my commitments, Gorbachev consequently told Matlock. "Whatever the pressure put on me regarding the Persian Gulf War, the German question, the ratification of the conventional arms treaty, I'll keep our agreements . . . My deepest desire is to keep our radical changes from drowning in blood," he proclaimed. "Whatever happens, I'll never turn away from my fundamental goals," or from the promise of a new age of Soviet-American cooperation they both desired.

It was a revealing list, more so perhaps than Gorbachev intended. Determined to demonstrate his commitment to Soviet-American cooperation, he instead gave Matlock (and Bush) more a litany of Soviet capitulation than of true and equal East-West cooperation. He promised to continue to hold fast on the gulf, on Germany, on their arms talks, and on his promise to keep violence at bay even if it meant the end of the Soviet Union itself. His opponents despised him for that same inventory. Gorbachev had capitulated on Germany and NATO, they said, weakened the Red Army in Europe, and now was playing Bush's lapdog in the Persian Gulf.

Every one of Gorbachev's pledges had been made so Bush would continue to keep open the spigot of American and Western aid. Faced with Bush's clear threat to cut Western credits entirely unless he caved in the Baltics too, Gorbachev capitulated there as well. American threats were not the only thing that prompted Gorbachev's retreat. He realized that by hoping to forcefully stem the tide of secession, he'd actually catalyzed its proponents. Matlock promised that his government would continue to be understanding of the severity of Gorbachev's problems, but only to a point, as both sides agreed that this was no time for the prospective Soviet-American summit tentatively planned for February. Bush had a war to fight, and Gorbachev had enough problems.

The American-led air war continued for a month, hollowing out Saddam's vaunted armor and army in devastating nightly strikes. Overmatched during the day, Iraqi forces were utterly helpless in the face of American technology once the sun went down. Soldiers took to sleeping in the open, knowing their tanks were likely to become coffins before daybreak. Baghdad lost power repeatedly, its bridges wrecked and its water supply and sanitation system compromised. Able to strike both with precision and at will, American war planners targeted infrastructure rather than population centers (though, sadly, plenty of the latter were inadvertently hit as well in the fog of war), reasoning that a miserable Iraqi populace would in time rise up in revolt, whereas displays of mass casualties, as in prior wars, would only strengthen support for the regime.

Ground combat loomed, and with the forceful eviction of Iraqi forces from Kuwait, a potential bloodbath as well. Eager to retain both Saddam and their own international influence, Soviet envoys tried anew to broker a cease-fire. In mid-February, newly appointed foreign minister Aleksandr Bessmertnykh secured James Baker's consent that the fighting would stop "if Iraq gave its unambiguous commitment to pull out of Kuwait," a promise that infuriated Bush, who'd not authorized it. "My emotion is not one of elation," Bush told his diary upon first hearing the news that Saddam had perhaps indeed backed down. "We've got unfinished business. How do we solve it? How do we guarantee the future peace? I don't see how well it will work with Saddam in power, and I am very, very wary."

Baker's overreach mattered little in the end, as Baghdad once more refused to withdraw without Western concessions. The ground war seemed more likely than ever as February passed its midpoint. Gorbachev once more dispatched Primakov to Iraq in search of a deal, even as American analysts predicted that failure to head off the next phase of the war would prove to be the last straw for Gorbachev domestically. The Soviet leader called Bush to plead for a pause in the bombing, or at least for more time before ground combat

commenced. "I expect you have people in your country that suggest you are
too close with the United States," Bush responded sympathetically. "That is
particularly true because of the good relationship with Iraq you have had over
the years and we understand that. Similarly, we have elements in this country
who suggest we are staying too close to the center in the Soviet Union"—that
is, too close to Gorbachev personally. Yet, Bush promised, "we won't depart
from that." He appreciated all Gorbachev had done in the name of peace.
"What you have tried to do [w]as politically courageous given your long relationship [with Iraq], your standing as a superpower and your minority group's
sentiment."

But enough was enough. "I don't want to see us pull apart even though we
have profound differences at this moment on this particular question" of Iraq,
Bush told him. But at the same time, he and the other coalition members were
determined to see the war through to its end, and to see Kuwait forcibly liberated. "I don't trust Saddam Hussein," he continued. Iraqi troops had been
steadily setting fire to Kuwait's oil wells even as Saddam's foreign minister
pleaded for time so their military could organize an orderly withdrawal. "Very
candidly, he is taking advantage of his time and, I think, of your good faith to
destroy Kuwait," Bush said, before putting Gorbachev fully on the spot. "If you
can't support" us, he said, "we would appreciate your not opposing" us as the
war moved into its final phase.

Gorbachev received a far less evenhanded response the next day, when he
once more called Bush to plead for more time for Iraq. It was the third time the
two leaders had spoken in less than thirty-six hours, and both were clearly exhausted. Bush's patience was also at its end. "We have a white flag from Saddam
Hussein," Gorbachev claimed, but it was too late for Bush to listen. There were
still too many conditions in the tentative Iraqi promise to withdraw, and too
much damage being done to Kuwait, to delay further. "Yesterday while you
were trying to work all this out," Bush noted, "their spokesman went out and
called me a liar about burning up those oil wells. And once again, there have
been a bunch more oil wells set on fire overnight. They are continuing to use
a scorched earth policy and stalling and this has made a profound impact
on me and on other coalition partners." The Iraqis now wanted to negotiate
precisely where the line separating Kuwait from their country should be, he
said, in effect arguing over the location of a disputed border when they should

have been scurrying home across it. Saddam was stalling while destroying Kuwait.

"This is where we are," Bush summed up, "and I think the difference we have is that you think they have agreed to unconditional withdrawal and we and others with us do not agree." But larger issues were at stake than merely whether they should trust Saddam to keep his word. "Let's not let this divide the U.S. and the Soviet Union," Bush said. "There are things far bigger than this conflagration which is going to be over very soon." In other words, the new world order was at stake, and would be determined by whatever happened next. More to the point, what was truly at stake was Moscow's role within it.

It is difficult to detect tone from transcripts of conversations. One cannot know for certain if a phrase is offered with sarcasm, a twinkle in the eye, or even real anger. In this instance, however, Bush left no doubt of his fury. "George, let's keep cool," Gorbachev responded, attesting to his interlocutor's ire. He'd done a fair bit of yelling of his own during their conversation but realized now that infuriating Bush wouldn't help his cause. On the contrary, Bush's anger helped Gorbachev see at last that it was time to accept the reality of American power. "Although, of course, all of us are human beings, I think both of us understand that what we need is not Saddam Hussein — his fate has been determined," Gorbachev conceded. "Our concern is to take advantage of the opportunity before us in order to obtain the goal we set together in the UN Security Council and also prevent a tragic phase in the further development of the conflict."

He would recall Primakov from his last mission to Baghdad, Gorbachev promised, and ultimately follow Bush's lead. "But that's it," his close aide recorded in his diary after witnessing Gorbachev's capitulation. "We're doomed to be friends with America. Otherwise we would face isolation and everything would go haywire again. He [Gorbachev] told me last night, when I advised him against responding to Hussein's final message: 'You're right! There's no point now! It's a new era!'"

Bush soon ordered a halt to the fighting on his own, declaring a cease-fire a symbolically satisfying one hundred hours after ground operations began. By that time Iraqi forces had fled Kuwait, and video images of the devastation they had suffered left American strategists fearful that they'd be accused of causing wanton death and destruction even after fulfilling all that the United Nations mandate authorized. They'd been sanctioned only to liberate Kuwait, Bush reasoned, not to drive all the way to Baghdad. His intelligence analysts predicted that Saddam would not last long, no doubt toppled by his own people or executed by his own generals for having led the country into such an embarrassing defeat. Iraqi casualties topped twenty thousand. Another sixty thousand were prisoners of war. Iraq's infrastructure had been reduced to rubble. Saddam's air force was largely gone, much of it flown, in another great irony, to Iran and out of harm's way. More than 90 percent of Iraq's armor lay destroyed. All at a cost of only 148 American troops killed in action, more lost to friendly fire than to the enemy.

Subsequent critics questioned why Bush failed to take Baghdad, or to destroy Saddam's forces more completely, but no one within his inner circle at the time had any appetite for the full-scale occupation and reconstruction effort that undoubtedly would have ensued. He'd gone to war to set a precedent, not to transform a region. Democracy would come to the Middle East in time, he reckoned, believing its residents as subject to the stream of history as any other. But there seemed little reason to think that time had come, or that American force might somehow catalyze a process best grown organically from below. "As events have amply demonstrated," Baker wrote in 2006, years into a subsequent bloody American occupation of Iraq featuring a rise of sectarian violence, religious fundamentalism, and regional instability in the wake of the dictator's demise, "I am no longer asked why we did not remove Saddam in 1991."

Bush's approval rating soared in the aftermath of victory, even as Gorbachev's plumbed new lows. "I know that these euphoric [public approval] ratings, nothing like it since Truman after World War II, are nothing," Bush confessed in his diary in the weeks to follow. "They go away tomorrow." He'd not secured a clean victory as in 1945, and indeed "Hitler is alive . . . Hitler is still in office, and that's the problem. And that's why, though I take great pride in the fact that the shooting has stopped, our military doing great, American people elated, I have no elation."

Exhausted by the stress and worry of the previous months, and suffering from an as yet undiagnosed thyroid condition that sapped his strength, he even fantasized about stunning the press (and his own advisers) by announcing he would not run for reelection. "I'm not in a good frame of mind right now," Bush dictated to his diary in March, albeit after downing two martinis on an evening flight aboard Air Force One and arriving at an empty White House residence with Barbara out of town. "My whole point is, I really don't care, and that's bad — that's bad. But I'll get in there and try." There still were problems to solve in the Soviet Union, with no clear sign that Gorbachev would remain in charge for long, or even if that was a fully desirable outcome. Winning reelection in another eighteen months would not prove easy either, no matter what his approval ratings currently suggested. "The common wisdom is I'll win in a runaway, but I don't believe that. I think it's going to be the economy [that] will make that determination," he noted, though ultimately, "I love the job, and I think we have a real chance to accomplish a lot; I think we have in foreign policy."

22

"DISUNION IS A FACT"

HE WAS SUPPOSED to have ten minutes. He filibustered for forty instead, after having been ten minutes late to a meeting in his own office. "How long are we supposed to wait for His Highness?" Scowcroft sarcastically asked as Bush once more checked his watch. People usually waited for presidents. Not the other way around. But by July 1991 Boris Yeltsin was a president too, and cared less for diplomatic protocol than for demonstrating his own importance.

A populist born into a communist system, Yeltsin was happiest when at the head of a crowd, a trait with little to recommend it within Bush's White House. The man had the "instinct of a demagogic," Scowcroft told reporters in a barely disguised off-the-record interview. In other private conversations he referred to Russia's new head as "an egoist, a demagogue, an opportunist, and a grandstander" who never missed a chance to "upstage Gorbachev." Yeltsin's pathology was also the result of political savvy, however, not just deep-seated insecurity, revealing a canny recognition that precedents mattered most in unprecedented times. The precedent of flying to Washington less than a week after his election, complete with a picture of him sitting in the Oval Office. The precedent of making Gorbachev attend his inauguration, and then standing still so the Soviet leader had to approach him with congratulations, similarly offered an instant image of credibility. Most powerful of all was the invocation Yeltsin received that day from the head of the Russian Orthodox Church. Technically it was not unprecedented. But it did mark the first time Moscow's leading religious authority had blessed a political leader since the time of the czars. In touch with his people in the way Gorbachev was in touch with ideas, Yeltsin knew that the Russians wanted to move past both socialism and *pere-*

stroika, and that they longed to grasp once more what had made them a nation long before Lenin, Marx, Engels, or the Soviet Union itself.

He even selected an office with precedent in mind, choosing a suite mere yards from Gorbachev's. Indeed, the chamber had once been Gorbachev's before he inherited the Kremlin's top spot. Yeltsin assumed it would be only a matter of time before this bit of history repeated itself. "This is the first time ever in the 1000-year history of Russia that there is a democratically-elected President in the Kremlin," he informed Bush and his entourage as July drew to a close, beginning the meeting that he had demanded be part of Bush's Moscow summit. A year after the Persian Gulf crisis suddenly erupted, nine months since Germany's formal reunification, and five months since Kuwait's liberation, Bush was scheduled to spend three days in the Soviet Union. Or what was left of it.

Most of that time was dedicated to Gorbachev, still titular head of the country and thus his formal host. For the first time, leaders from the Soviet republics were also on Bush's itinerary, including Russia's new president, though it was not clear what their collective, newly declared sovereignty really meant. "My view is, you dance with who is on the dance floor," Bush told his diary. "You don't try to influence this succession, and you especially don't do something that would [give the] blatant appearance [of encouraging] destabilization." Therefore, "we meet with the republic leaders," he reasoned, "but we don't overdo it." Gorbachev would continue to be the focus of his administration so long as he remained in charge, Bush told his staff. "Our goal is to keep Gorby in power for as long as possible," said Scowcroft, "while doing what we can to help them [the Soviet leadership] in the right direction." They'd quietly court other leaders but commit to no one else. After all, as Robert Gates dispassionately noted, for all his troubles, "Gorbachev was doing what we wanted done on one major issue after another — from his willingness to let Germany be reunified in NATO to his partnership with us in taking on Iraq. We had no desire to jeopardize that," especially when the alternative to his rule might mean "fragmentation of the Soviet Union," including "dangerous instability in a country with tens of thousands of nuclear warheads."

"There are two flags in the Kremlin," Yeltsin noted as his session with Bush began: "The Russian flag and the Union flag. This is new and unique." No one quite knew what to make of the distinction. The new union treaty negotiated in recent weeks between the Soviet center and the disparate republics, to be formally signed in late August, ceded most of Gorbachev's power but kept his office intact. The Soviet Union still existed. But it was no longer allowed to dictate foreign and military policy or direct a unified economic program. Under the radical new plan, Gorbachev could instead only coordinate common approaches in conjunction with men he'd long considered his subordinates, yet who in large measure also considered themselves his equal, and, in Yeltsin's case, his most likely successor. Indeed, they'd be his paymaster. To save the union, Gorbachev had agreed to shift tax authority to the separate republics, leaving him to operate the center only with whatever funds they allowed.

"This was the decision to dissolve the empire," one economist commented, "raising hopes that it could be transformed into a soft confederation." The power to tax was historically the power to control, and its surrender "ended the history of the USSR as a single state." Gorbachev perceived little choice in the matter, facing not only dissent but also a deadline. It was 2 AM before the negotiations for the new union treaty ended on July 29. Bush was due to arrive at the Kremlin in less than ten hours. "I might as well go home" if unable to control revenues, Gorbachev had declared when talks threatened to break down. Yeltsin called his bluff. "Don't force us to decide the matter without you," he retorted, making plain that the republic chieftains believed they held the authority to do just that. The new "union" they ultimately approved would not be called either "soviet" or "socialist," and its republics now explicitly had the right to negotiate their secession at a time of their choosing. But at least the union remained. "They will now approve almost anything in order to cling to the vestiges of power," one of Gorbachev's senior advisers said of the conservatives within their ranks. He might as well have been speaking of his boss.

Yeltsin refused to accept Gorbachev's authority even when it came to matters of protocol. Asked to dine with Bush alongside other leaders of the more prominent republics, he demanded a private session instead. Invited to dinner at the American ambassador's residence but seated apart from the head table, he spent much of the evening standing behind Bush, who tried in vain to engage his seatmates rather than the hulk towering over him. Purposefully arriving last to the formal receiving line the night before, he'd even offered to escort Barbara Bush into the ballroom. Is that really all right? she wondered aloud, knowing Gorbachev was due the honor. Quickly defusing the situation, she escorted herself, strategically placing Raisa Gorbachev next to Russia's lumbering new president. "Yeltsin's really grandstanding, isn't he," Bush subsequently complained to Scowcroft. "Everyone was dumbfounded except me," Gorbachev told his staff. "I knew Boris too well."

"Tell him there are certain agreed upon norms of gentlemanly behavior," one of Scowcroft's aides implored Ambassador Matlock, but to no avail. In an age when every moment could be caught on film and broadcast, the images produced by such antics mattered. The Kremlin had seen innumerable rivalries over its long centuries. Rarely if ever were two epicenters of power pitted against each other in a tighter struggle for control within its walls. At stake were questions not just of who sat where or escorted whom, but of which power enjoyed international endorsement.

"In view of the fact that the union treaty is no longer a draft — we resolved the final problem yesterday and initialed an agreement," Yeltsin told Bush as they sat in his new office, "there is no longer any obstacle to going ahead." With the questions of taxation and funding solved, he continued, "there should be no difficulty in finalizing new contracts with American firms, and with Russia accepting direct American aid as well." Indeed, Yeltsin pressed, Washington could now freely recognize Russia's formal independence, and Bush could in turn recognize him as the power to negotiate with from here on.

No one on the American side had yet seen the final text. "Is this something that has been resolved by your agreement with the center?" Bush probed, unsure of the specifics of Yeltsin's claim but well aware of its implications. The United States couldn't recognize two sovereign powers occupying the same space, not without undermining Gorbachev. He'd learned to be wary of Yeltsin's claims in any event. "The treaty reached with Gorbachev was a compromise between the two of us," the Russian brusquely replied. "He removed one line and I initialed that page. So there are no more problems on the tax issue. The treaty calls for a fixed Union tax rate to be agreed upon by the republics."

Baker found that answer wholly unsatisfactory. "Is this figure agreed upon or is it a figure to be agreed upon?" he interjected. A former lawyer and treasury secretary, he as much as anyone else knew that such details were the heart of the agreement. "We have only agreed to agree on a fixed rate in the future," Yeltsin answered, unhelpfully. "The rate of taxation will differ from one republic to the next," he continued. "The center will make an estimate of spending for us." Having explained the complex negotiation to his own satisfaction, Yeltsin again pressed for Bush's approval. "Now, do I understand that you support my idea of formalizing the basics of our relationship?" he abruptly asked.

Poland's roundtable discussions had taken months. Czechoslovakia's were concluded in weeks. In each case the results were widely scrutinized before voters or foreign heads of state were asked for comment, let alone endorsement. Bush and Gorbachev were scheduled the next day to sign a treaty limiting each country's respective nuclear arsenals. The Strategic Arms Reduction Treaty (START) was nearly a decade in the making. Reagan had first proposed it in 1982, as an alternative to the Strategic Arms Limitation Talks (SALT), a process itself negotiated over years and signed by his predecessor though never ratified by the Senate. Nearly a decade later, after countless hours of bargaining and tectonic changes to the international landscape that no one could have imagined in 1982, the START treaty was nearly ready. But not quite. Baker, Scowcroft, and their Soviet counterparts had in fact spent considerable time hammering out the treaty's final language the previous month in London, while Bush and Gorbachev met with their own counterparts during a meeting of the G7, the world's leading economic powers. Hours went into discussing single words, each comma and semicolon vetted by a phalanx of lawyers and specialists, with copies faxed to Moscow and Washington for further scrutiny before both sides agreed they finally had a deal. "There was still a lot to do" to make the world safer, Gorbachev declared when he signed the accord. Bush agreed, but optimistically noted that as the culmination of more than a generation of arms talks between the two sides, their pact offered "a significant step forward in dispelling half a century of mistrust."

This single treaty took nearly ten full years to negotiate. Now, only hours had elapsed since a band of bleary-eyed Soviet and republic officials initialed their draft text of the union treaty, and yet Yeltsin was pressing Bush to endorse it unseen. Yeltsin was many things: brash, manic, frequently depressed, prone to great feats of stamina yet also intimately familiar with the interior of a vodka bottle. But he was not known for patience.

"Which relationship?" Bush responded to Yeltsin's urging. "Do you mean the U.S. and Russia or yours with the center? I am unclear what you are asking. We do want to look forward to cooperation with you," he continued, but "please give us a copy of the Union agreement. When it is finally worked out, we will study it and work within but not ahead of it." After all, Bush explained, "I don't want to plough new ground." Everything about negotiating in Moscow seemed new by midsummer, including his administration's grudging realization that they'd soon be required, in some form or fashion, to negotiate with fifteen new entities at once, as Gorbachev announced that the new union treaty would be formally signed on August 20.

Presuming he survived that long. Weeks earlier, bombarded yet again with evidence that hard-liners were planning to seize power in Moscow, Bush had taken the remarkable step of passing a warning to Gorbachev and Yeltsin. "A coup is being organized," Moscow's new mayor wrote to Ambassador Matlock, thrusting a hastily scribbled note into his hands as the pair shared a drink. He dared not speak. The KGB was listening. It even had microphones in Gorbachev's private quarters. "We must get word to Boris Nikolayevich [Yeltsin]," his next scribbled note implored.

It was a well-timed plea. Yeltsin was in Washington for his post-election visit to the White House. The last time he was there he'd nearly walked out when refused a formal session with Bush, who instead popped in to "spontaneously" greet the Russian leader in Scowcroft's office. This was how diplomacy worked. Yeltsin could say he met with the president, whose office could say with a straight face they'd never "scheduled" any such meeting. This time was different. The pair sat not at Scowcroft's cramped desk but in the Oval Office, befitting Yeltsin's newfound status.

"This is only talk," he responded when Bush passed along the warning that Gorbachev's enemies intended to seize power. "People would take to the streets and there would be civil war" if anyone tried such a thing. Of course, Yeltsin wanted Gorbachev removed too, but through legal channels. "Only a violation of the constitution would remove Gorbachev now," which "cannot happen," Yeltsin promised.

That Bush's White House operators were unable to get Gorbachev on the phone to deliver the message directly did nothing to ease American anxieties. He was "unavailable," their Kremlin counterparts replied. Furious and embarrassed that his aides had failed to put Bush's call through — he'd been asleep, not deposed — Gorbachev nonetheless also brushed aside the American warning. Rumors of discontent were nothing new, he assured Matlock later, and he had more control than perhaps the CIA realized. "Tell President Bush I am touched," he told the ambassador. "But tell him not to worry. I have everything well in hand. You'll see tomorrow."

Gorbachev proved true to his word. Chiding the Supreme Soviet for contemplating a "constitutional coup," he quelled a resolution to have his powers transferred to the prime minister. "The putsch is over," he boasted to reporters.

Yet disquiet remained within what was left of the Soviet hierarchy. Radicals sided with Yeltsin and the secessionists. Conservatives could not hope to govern without Gorbachev, yet neither could they employ the full range of their power to defend their crumbling union so long as he remained in charge. Now he planned to cede formal authority to the republics in late August.

No one seemed happy with his rule. "Two Russians were complaining as they stood waiting in line to buy vodka," Gorbachev joked of his waning popularity while sharing a drink with his fellow world leaders in London in mid-July. The line was too long, one said, and it was all Gorbachev's fault; he was off to shoot the man responsible, he declared. A few hours later he returned. "Well, did you shoot him?" his buddy inquired. "No," came the answer. "The line over there was longer than this one!"

Gorbachev's critics, even those charged with executing his commands, openly vented their frustration. Defense Minister Dmitry Yazov "was in a morose mood," Scowcroft noted of his companion at the formal dinner the evening the START treaty was signed, "complaining that everything was going our way while the Soviet military was deteriorating daily. No new equipment was coming in . . . young men were not responding to the draft, there was no housing for troops returning from Europe, and so on." Born in Siberia the same year as Bush, Yazov earned medals for valor during World War II. But the enemies in that war were long since defeated, Scowcroft said, and the East-West tensions that spawned the nuclear age had diminished. "What was the threat?" Why did the Red Army need more troops and new weapons?

"NATO," Yazov bluntly replied. To Russians, its continued existence and eastward expansion were a visible living symbol of Soviet defeat. Its counterpart, the Warsaw Pact, had only weeks before formally voted itself into oblivion. "We do not hide that our goal is the inclusion of Czechoslovakia in the West European integration," Václav Havel proclaimed at the pact's final session. To Soviets like Yazov, it was not simply that their allies were fleeing and their empire dissolving; it was that their enemies were gaining.

Summer also brought little of the promised aid Gorbachev hoped might come from his personal relationship with Bush and other Western leaders and from his ongoing cooperation. Inviting himself to London to plead his case to the G7, he left largely empty-handed. The Soviets were not yet equipped to digest large investments or Western contracts, warned Bush's commerce secretary, Robert Mosbacher. "The business environment is highly unstable . . . The center, the republics, and the localities all argue over the division of authority and control over national assets. Laws contradict each other and constantly change," while "nobody is willing to make decisions and stick by them."

Bush took Mosbacher's advice to heart, since it corresponded with his own belief that Soviet problems remained less economic than political. Like the Poles before them, Soviet officials were calling for a new Marshall Plan to ease their market transition, yet "Gorbachev must understand the political context in which we all operate," Scowcroft wrote in June. So long as the Kremlin continued to spend "at least one-fifth, and probably more, of its GNP on defense" while simultaneously propping up recalcitrant allies like Cuba, there would be little enthusiasm in Congress for a massive bailout, especially if it required Bush to reopen already tense budget negotiations. The United States would thus be best served by offering Gorbachev "strong support without large-scale financial assistance," including backing the Soviet request for special status within the G7 and the International Monetary Fund, providing access to Western expertise and advice but without the financial buy-in typically demanded of full members.

"We didn't give them economic aid because we just couldn't see how to do it without putting money down a rat hole," Scowcroft subsequently explained, noting that Gorbachev seemed a less reliable investment after his lurch to the right earlier in the year, which "undermined any hope of significant financial assistance tied directly to reforms." Western investors wouldn't open their purse strings, Bush's inner circle reasoned, until they knew for certain who within the Soviet Union would uphold their contracts, or if their guarantees were meaningful. Yeltsin promised that Russia would recognize private property. Gorbachev said that the Soviet Union would not. "I had seen no evidence that even basic economic changes were being implemented," Bush later wrote. "In my view, the Soviet Union suffered more from economic inefficiencies and poor priorities than from lack of money."

Bush consequently urged his partners at London's G7 meeting to vote against any hasty bailout. "Either we send an important message or we don't," Helmut Kohl urged, imploring, "We must try to influence events in the USSR." Of course, he added, the Germans refused to pay for Soviet aid alone. "Amazing political changes are taking place," Italy's prime minister agreed. "The Warsaw Pact is gone without notice," but while "there are times one needs to decide with political foresight," this "doesn't mean writing a blank check." The European Union's representative pointed out that private investment alone couldn't handle the task at present. "About $8 billion has been pulled out" of the Soviet Union in the last year alone, he noted. "Investors need assurance of the application of law," which no one could possibly guarantee. Nonetheless, the G7 had to act, if only by formally deciding not to. "Gorbachev is visiting this group," Canada's Brian Mulroney reminded them, even if only by his own invitation. "We must respond."

"There is no economic logic in lending now," Bush concluded, effectively ending the discussion and Soviet hopes of a speedy bailout. So far as he could tell, "economic policy in the Soviet Union isn't yet together" and wouldn't be until the country's political future solidified. Everyone in the room supported *perestroika* and wanted to see Gorbachev's democratic reforms proceed, Bush acknowledged, adding that he wasn't concerned about the potential linkage between Soviet actions in the Baltics and new forms of financial aid. On the contrary, he said, "I'm making this linkage" between Soviet reforms and the prospect for aid "because of my Congress and because no one else is loaning." Banks and corporations, and governments too, simply would not lend to the Soviets until Gorbachev and the republics made their peace. "No one can accuse President Bush of not doing enough to support Soviet reform," Baker added preemptively. "But we need to consider the political aspects" within the Soviet Union, which can only be done "through a bilateral mechanism." In other words, between one Soviet head of state and the outside world, once it was clear who really held power in Moscow. Until then, they'd best hold their aid in reserve.

Gorbachev disliked being told to wait. "What does George Bush want from me?" he asked rhetorically as the two men sat down to lunch. The Soviet leader was due to address the G7's next session that afternoon but knew that the American president's voice resounded loudest of all in their deliberations. "If my colleagues among the 'Seven' tell me, when we meet later, that they like what I am doing and they want to support me but first I have to stew in my own soup for a while, I just tell them that we are all in that soup." Stability within the Soviet Union mattered to postwar security, Gorbachev stressed. He could not help but add, "Isn't it strange: a hundred billion dollars were scraped up to solve a regional conflict." But when "what we have here is a project to transform the Soviet Union, to give it a totally new quality, to bring it into the world economy so that it will not be a disruptive force and a source of threats," Western pockets were suddenly empty.

No American-produced transcript of their lunch meeting exists or had at this writing been released. Indeed, Gorbachev's aide, who produced the best record of the conversation available, recalled their meeting as taking place over breakfast instead. No matter the time or the bill of fare, one thing was clear. "Bush was visibly offended," he wrote. Speaking coldly, in a "tone of rebuke," he responded to Gorbachev's barbs by stressing his desire for the Soviet Union "to become a democratic country, a market economy, dynamically integrated into the Western economy," noting that he wanted "the federative problems between the center and the republics to be successfully resolved" as well. "The decline of the Soviet Union isn't in our interests," Bush said, but when it came to the lengths to which the Americans could go to provide material aid and financial support, "this is what it is."

The exchange captured a critical dynamic of Soviet-American relations as 1991 reached its midpoint, and of the personal relationship at its heart. Having chosen to favor Gorbachev over other, more unpalatable options, Bush believed that he'd gone out of his way to demonstrate his support, and that he'd done nothing else since ending the pause and pronouncing his desire to move "beyond containment" two years before. He'd not danced on the Berlin Wall lest celebration undermine Soviet reformers. He'd not until recently claimed the Cold War over, and had yet to publicly call its demise an American or Western victory. "I think I've done a lot to keep the process with Gorbachev on track," he told the other G7 leaders, noting that he'd endorsed *perestroika* time and again.

He mentioned Gorbachev more often than Yeltsin when speaking to the press, even when the new Russian president was at the White House, and had repeatedly delivered messages of encouragement to Gorbachev in private correspondence and through various diplomatic and informal channels. "The President is greatly concerned at Gorbachev's perception of a weakening of the President's support for him and the reform process," Scowcroft wrote Matlock in May, in a cover letter for a private note of support from Bush. "That perception is wholly incorrect . . . [H]is commitment to Gorbachev and his reforms is undiminished and he would like to lay to rest Gorbachev's misconceptions." Bush said much the same whenever he and Gorbachev spoke throughout the summer. "As I hope your ambassador informed you, in talking to him [Yeltsin] I stressed that on the main political issues you are my man, the person I'm working with," Bush told Gorbachev over the phone on June 24. "It's satisfying to work with you as the head of the Soviet Union." He said the same the next month in London: "I know the Soviet Union is going through difficult times . . . But we have not vacillated, and we won't vacillate, in supporting your efforts." Finally, sitting together in Gorbachev's dacha outside Moscow as July came to a close, Bush offered his personal endorsement yet again: "Our policy, our choice, is to support you. And I want everyone in the Soviet Union and Europe to know it."

Gorbachev wanted more than words, however. Hoping for aid, he received primarily advice instead. "What are our friends waiting for?" he lamented when pleading his case to the G7. A year before, his people would have celebrated his invitation to speak at such a conference. Much had changed in a year, and from the perspective of Soviet finances, hardly any of it for the good. "We need a new kind and new level of cooperation," he urged, "which would really integrate my country into the world economy. I really don't see when such a chance would come up again." His words failed to open Western coffers to the extent he desired. "Gorbachev came to collect what he obviously felt his due," Scowcroft advised Bush as the G7 ended without a new bailout program for the Soviets. He wanted Western aid as a down payment for reform. "I ask once again in the presence of the delegation that the President instruct them to consider [Soviet] membership in the IMF," Gorbachev pleaded at the Moscow summit. "I have big problems in the next 1–2 years. Call us what you like — associate members, half associate members." The name mattered less than the money. "It is important for us to use that fund," he begged. To no avail.

Bush's supportive words carried their own price, for both men. Visiting Kiev on the way home from Moscow, he heard welcoming addresses rendered in Ukrainian and English, and the American and Ukrainian national anthems. A generation before, President Nixon had been welcomed in Kiev by speeches in Russian, and with Soviet songs. Nixon's hosts yearned to demonstrate their solidarity with Moscow. Bush's trip, conversely, was designed so he could support Ukraine's sovereignty without explicitly endorsing its independence. "We want to retain the strongest possible official relationship with the Gorbachev government," he told reporters in Kiev before diplomatically adding, "but we also appreciate the importance of more extensive ties with Ukraine and other Republics, with all the peoples of the Soviet Union."

It was a Hippocratic answer if ever there was one, but this was not the message best remembered from Bush's stopover. Addressing a Ukrainian parliament shorn of any symbol of Soviet rule, Bush told legislators that "some people have urged the United States to choose between supporting President Gorbachev and supporting independence-minded leaders throughout the USSR." Yet, he added, "I consider this a false choice." *Perestroika* had brought newfound freedoms. But "freedom is not the same as independence. Americans will not support those who seek independence in order to replace a far-off tyranny with a local despotism. They will not aid those who promote a suicidal nationalism based upon ethnic hatred." With Bush due to visit Babi Yar later in the day, site of one of the Holocaust's most deadly massacres, his speechwriters expressed little patience for the rise of ethnic antagonism percolating beneath the secessionist movements in each of the republics. "It is a solemn reminder," Bush said, "of what happens when people fail to hold back the horrible tide of intolerance and tyranny." In short, Washington favored democracy, for Ukrainians as for all peoples. Just not yet, and certainly not if the costs were destabilization, danger, and the potential for civil war.

His plea for patience drew immediate condemnation, in Kiev and in Washington, where his remarks were interpreted as a paean to the status quo. "Bush came here as a messenger for Gorbachev," a leader of Ukraine's independence movement complained to reporters. Indeed, "Bush seems to have been hypnotized by Gorbachev" to support the center's agenda over the republican cause. William Safire of the *New York Times* authored the critique that stuck, writing weeks later about the president's "dismaying Chicken Kiev speech," in which, he argued, Bush "lectured Ukrainians against self-determination, foolishly placing Washington on the side of Moscow centralism and against the tide of history." The White House's fear of "disunion, dismemberment, [and] disintegration" was not irrational, Safire conceded. But it was behind the times. "The moral advantage to disunion is overlooked by diplomats hung up on tidiness."

Safire's piercing critique not only stuck but also struck a chord, with Bush's "Chicken Kiev" address typically placed on a par with his ill-conceived campaign pledge forswearing new taxes. Little remembered is its timing, however. Bush spoke to the Ukrainian parliament on August 1. Safire wrote his much-cited column on August 29. "I've been out of touch for a few weeks," he began sardonically. "Anything happen while I was on vacation?"

Much had occurred indeed. "The argument is moot," Safire sagely concluded about the question at the heart of Bush's address, and his entire diplomatic approach to Gorbachev, bent on keeping the Soviet Union intact. "Disunion is a fact."

23

"I HAVE SIGNED IT"

AUGUST IS TYPICALLY for vacations, once summits are done. James Baker went to his Wyoming ranch after Moscow. Dick Cheney took his fly-fishing gear to British Columbia. Bush decamped as always to Maine, where Brent Scowcroft nervously planned to join him. "Experience taught me that crises mysteriously seemed to break out when the President was at Kennebunkport and the national security leadership dispersed," he later joked. Robert Gates at least had the first watch with the president, before his own planned sojourn in the Pacific Northwest.

Gorbachev sought solace on the Black Sea, withdrawing to the Kremlin's expensive new retreat carved into the Crimean cliffs a mere forty miles from where Stalin, Roosevelt, and Churchill had once divided the world at Yalta. Nobody needed the time away more. "I'm tired," he conceded, the months of negotiations with the republics and with potential foreign lenders weighing him down. Raisa continued her fruitless lobbying for him to resign while the choice remained his. Prophets rarely see likely successors in their midst. "Everywhere you look things are in a bad way," he confessed. "But who would I leave it to?"

Duty followed even on vacation. Gorbachev received his daily briefings and wrote while his grandchildren played in the surf, composing his remarks for the following week's critical ceremony marking the new union treaty. Having already choreographed every detail from the music to the seating chart, he planned a valedictory for the Soviet Union of old, but also an inspiring inaugural for its successor. "We would be poor patriots if we renounced our history, severed the life-giving roots of continuity and mutual ties," he planned to say. "But we would also be poor patriots if we held on to what must die, what we cannot take into the future with us, what would interfere with rebuilding our life on a modern, democratic foundation."

The treaty, for all its potential uncertainties, offered to his mind the best chance to preserve what had once made Soviet society great, and *perestroika* too, couched increasingly in terms of democracy rather than as an economic ideal. Nearly everyone favored democracy, at least rhetorically, its virtues endorsed across the political spectrum even by those who secretly yearned for socialism's triumphant return. Only black marketers praised what *perestroika* had done to their finances. Most Soviet citizens understood only that everything they'd ever known about politics was about to change as August 20 approached. But into what, no one could truly say. "We don't even know what our country will be called in a couple of days," complained Moscow's leading television anchorman, erupting on air in mid-August. The official name, Union of Sovereign States, was too bland to be remembered. "The Union treaty has a different name for our country but who will sign it?" Indeed, "what about those who don't?" The commander of Soviet Airborne Forces believed he knew the most meaningful result: if the treaty were signed, he warned his colleagues, "it would be the end of the USSR."

Bush enjoyed his security updates by the sea as well. "The President and I were sitting on the deck of his house at Walker's Point, looking out over the Atlantic and eating pancakes," Gates later recalled, and the last item in the president's brief that morning offered a familiar warning. "There was very likely going to be a coup attempt" in the Soviet Union within the month, the CIA predicted, before Gorbachev could sign his country formally out of existence. Still chewing, Bush looked up from his reading. "Should I take this seriously?" He'd heard this warning before. He'd even passed along earlier warnings to Yeltsin and Gorbachev, each of whom proved thoroughly unimpressed, and hardly grateful. "Yes, and here's why," Gates replied. Gorbachev had always faced the threat of a conservative revolt. His opponents now had a deadline.

Gorbachev also feared a coup, though not at the hands of the recalcitrant Soviet hard-liners identified by American analysts. Having lurched to the right, he took comfort in the fact that most of that ilk realized he was the only thing standing between them and secessionist anarchy. Instead Gorbachev feared Yeltsin, who continued to negotiate terms of the union treaty with other republic leaders, without Gorbachev's consent and most certainly without his participation. "You understand what's going on," Gorbachev snapped at an aide when told the Russian leader had met with his Kazakh counterpart. Yeltsin soon after claimed for Russia supply chains formerly run by Soviet officials. The union treaty had been silent on the issue, which he took as an invitation for the republics to further expand their influence, and his own, especially if it came at Gorbachev's expense. "Independently, ignoring the opinion of the president of the USSR, local leaders are deciding matters of state," Gorbachev explained, fuming. "This is a conspiracy," whose only purpose could be to wrest more control from the center.

Indeed, this was Yeltsin's plan precisely: to support what remained of Soviet authority only until the moment arose when he, and Russia, might take its place. Only Gorbachev wielded the pen that could legally — and, he hoped, peacefully — sign the Soviet Union out of existence. Plus, Gorbachev still retained the West's favor, as well as command of Soviet military forces. Yeltsin was traveling with an increasingly large entourage by August, befitting his newfound stature. Gorbachev traveled with the nuclear codes.

The real conspiracy was already under way. On August 17, three days before the anticipated treaty ceremony and two before Gorbachev was due to return from the Crimea and Yeltsin from Kazakhstan, KGB chairman Vladimir Kryuchkov summoned a select group of colleagues to one of his organization's quiet retreats in Moscow. Having bugged every other part of the city, Soviet intelligence services maintained their own resort facilities, where powerful men could talk, and plot, without fear of being overheard. Prime Minister Valentin Pavlov attended, as did Defense Minister Dmitry Yazov, Gorbachev's chief of staff Valery Boldin, and Central Committee members Oleg Baklanov and Oleg Shenin, who oversaw defense and government personnel, respectively. In traditional Russian fashion, the old men took a steam bath before retiring to a table laden with food and vodka and, in deference to their host, high-end scotch. Talk turned to politics, then to complaints, and finally to a plan. By the time the group disbanded that evening, they'd agreed to confront Gorbachev with the sorry state of national affairs and present him with a simple choice: he could support them in issuing a state of emergency designed to roll back the more disastrous elements of reform, including the plan to sign away the Soviet Union's key powers, or he could step out of the way while, as one member put it, they did "the dirty work" required to set things right again.

Gorbachev had frequently mused on the possibilities of a state of emergency; perhaps, his subordinates reasoned, he'd appreciate their intervention.

Kryuchkov already had a list of military, political, and economic measures in mind, and the consent of elite army units ready to deploy throughout Moscow to discourage dissent. The KGB head ordered 250,000 pairs of handcuffs readied, just in case, and cells in a nearby prison prepared for scores of new visitors.

These were Gorbachev's men. He'd selected most for their current posts, deciding it was better to co-opt potential enemies than to let their anger fester, even as his increasingly conservative appointments further drove away the reformers he'd once considered allies. "I had promoted these people — and now they were betraying me," he later wrote, recounting his reaction when a delegation of the conspirators confronted him in his Crimean office two days later. Told he had unexpected visitors from Moscow, a nearly unforgivable breach of protocol, he tried phoning Kryuchkov for an explanation. Every line was dead, even the one connecting him to the Soviet military command.

It was then that he knew. This was what happened to Khrushchev. Before his Politburo enemies cut him loose in 1964, they first cut him off from the outside world. He summoned Raisa and his daughter and son-in-law into one of the side bedrooms; they agreed to stand firm, and with him, no matter what happened next. "We all knew our history," as Raisa Gorbachev put it, including its "terrible pages." Khrushchev had been allowed to retire, but no Soviet leader before or after him had left office under his own power. Rumors even swirled that those who died in office had been dispatched before their time.

"Whom do you represent, and in whose name are you speaking?" Gorbachev pressed his visitors. The plotters stood mute in response, more sheepish than confident, most likely realizing in that moment why coups were more easily conceived than executed. Neither would they reveal who had authorized their "emergency committee." Gorbachev pointed out that he certainly had not, and neither had the Supreme Soviet. Knowing then that the men were acting without legal authority — and, frankly, that if they'd intended to take his life, they'd not be having this conversation — Gorbachev quickly surmised that their grasp on power might not be as complete as the plotters hoped. This meant he still had supporters, or at least there were some who might still follow his orders. If only he could reach them.

"If they were truly worried about the situation in the country," he began berating his would-be captors, "we should convene the USSR Supreme Soviet and the Congress of People's Deputies. Let's discuss and decide" how to save the country through constitutional means. "Anything else is unacceptable to me."

Harsher words followed. Their entire mission was ill conceived and destined to fail, he argued, calling the men criminals and traitors. "You take a rest," Baklanov suggested, finally mustering the courage to speak. "While you are away we can do the dirty work and you will return to Moscow."

The thought caused Gorbachev to explode, its implication clear and repulsive. Tiananmen seemed to their eyes an example to be followed rather than shunned. The problem within their own republics was not that Soviet troops had shed civilian blood but that they had shed too little. "Go to hell, shit-faces," he responded, the outburst jolting General Valentin Varennikov from his stupor. The only man in the plotters' delegation from outside Gorbachev's personal orbit, the towering figure who directed Soviet land forces had never been a fan of reform. Having stood at the back of the group thus far, Varennikov stepped forward, raising himself up to his full height over his more diminutive captive. "Hand in your resignation," he ordered in a menacing tone. It was not phrased as a suggestion.

Gorbachev stood too, but Boldin stepped between them. He'd been in the hospital only days before, hooked up to an IV drip for more than a week while his liver was failing. He'd been Gorbachev's loyal deputy for fifteen years, and they'd shared more meals and moments than either man could count. He was their "soul mate," Raisa later lamented, to whom "we trusted everything — our most intimate secrets." Genuine concern for his country, and perhaps even a sense that he was still serving Gorbachev by enabling measures his master could never condone, had rousted him from his sickbed. "Mikhail Sergeyevich," Boldin began, "perhaps you don't understand the situation."

Gorbachev cut him off. "You're an asshole." The plotters realized then that any hope of securing Gorbachev's aid, or even his grudging consent, was pointless. So too, perhaps, was their entire project. Driven by a desperate sense of fear for their country, and by group-think, they had convinced themselves in the hours leading up to their fateful flight from Moscow to the Crimea that Gorbachev would actually accept their intervention. He might even relish the opportunity to continue his vacation, preserving his reputation while others rolled back the obvious mistakes of his policies. Without Gorbachev's authority, however, as their hostage plainly stated, they had no legal mandate whatsoever. "Everything went haywire from the start," Boldin later concluded.

Ordering Gorbachev and his family to remain in their quarters, cut off from the outside world, the plotters flew back to Moscow. Vodka eased their jangling nerves. When they arrived back at the Kremlin late that evening, they hastily signed a proclamation declaring six months of martial law. Vacillating wildly with his pen in one hand and a continuous series of cigarettes in the other, Vice President Gennady Yanayev ultimately added his name to the declaration. Called from home to sign, Foreign Minister Bessmertnykh crossed his name off instead. He refused to join their ranks, yet did nothing to halt them, isolating himself from contact but also thereby enabling the plotters to employ his ministry's international communications network. Yazov and Kryuchkov began issuing orders. Troops converged on the city center, the airport, and Moscow's television and radio centers. Boldin returned to the hospital, there to be heavily sedated once more.

The news went out by dawn: Gorbachev was ill and indisposed; martial law was in effect; Vice President Yanayev and an "emergency committee" were in charge. Issuing an "Appeal to the Soviet People," the plotters declared that *perestroika* had reached "a dead end." They would in its stead "restore law and order, end bloodshed, and declare merciless war against the criminal world." To remain safe, citizens need only follow the committee's commands "without deviation," which included a national curfew and suspension of political parties and the rights of assembly, protest, and the press.

IT WAS LATE EVENING in Maine on August 18 when the news first broke. Recently arrived from Washington, Scowcroft was, as always, surrounded by papers, working in his hotel room near the president's compound. CNN droned in the background. An announcement of Gorbachev's apparent resignation seized his attention. A half hour of phone calls produced little new information. Gates, at the White House, knew nothing more. Neither did any other staffers. Scowcroft nonetheless woke Bush with what scant news he had, and the pair agreed to confer again in another five hours, before the morning news cycle began for real on the East Coast.

Unable to sleep, Bush repeatedly called the White House Situation Room for updates, and turned as usual to his diary. "I keep wondering whether I should go back to Washington or not," he dictated, fearing the move might provoke a heightened sense of crisis and anxiety around the world. Complicating the question was the expected arrival of hurricane-force winds and rain later that night, as well as the fact that "there's not much to do there, except look busy." He had far more questions than answers. "The one thing we worried about was a right-wing coup," Bush continued, "although in recent times I've become convinced that the moves towards democracy are irreversible. We'll have to wait now to see if that's true, see what the populace in the Soviet Union does. Will there be general strikes? Will there be resistance? Will the military use so much force and crack down so much that they won't permit any democratic moves to go forward?"

Bush reasoned he could at least avoid making things worse. This meant once more walking a diplomatic tightrope, reminding the world of his support for democracy and for Gorbachev without simultaneously condemning the coup in such harsh terms that he'd be unable to work with the new government if it succeeded. While not a comfortable middle ground, and certainly one that yet again left him open to critique from those who desired a more full-throated defense of democracy, it seemed the prudent choice. The harsh truth was that coups succeeded more often than not, at least at first, while the element of surprise remained on the plotters' side. Put bluntly, the odds were good that Gorbachev was already dead. Soviet television alternated between news of the emergency committee's proclamation and a looped video of the ballet *Swan Lake*, the same programming as when Brezhnev, Andropov, and Chernenko had died. Whoever controlled Moscow's airwaves was signaling a succession. "The president's inclination was to condemn it outright," Scowcroft later recalled of the coup. "But if it turned out to be successful, we would be forced to live with the new leaders, however repulsive their behavior." They thus decided Bush "should be condemnatory without irrevocably burning his bridges."

Might they have done more to keep Gorbachev in power? Was his fall their fault? The questions were as unavoidable in the press as the question of why they'd not seen the coup coming. Of course they had. No reporter knew of Bush's earlier warnings. Then again, no one was truly surprised by the news emanating from Moscow. "There will be a lot of talking heads analyzing the policy, but in my view this totally vindicates our policy of trying to stay with Gorbachev," Bush preemptively defended himself in his diary. Doing anything more to favor Gorbachev, or Yeltsin for that matter, would only have invited a crackdown sooner, he reasoned.

The Canadian prime minister wondered if perhaps the stingy response to Moscow's plea for financial help at the G7 meeting earlier in the summer might not prove a political pothole if indeed it turned out that Gorbachev's reign was no more. "Every time I come to Maine, all hell breaks loose," Bush joked as his friend came on the line to discuss the latest news, though he quickly vented some anger as well. "Our embassy doesn't know a damn thing," Bush complained, and "the press is saying it was an intelligence failure."

Mulroney quickly added to his list of worries. "They may say, well, if you people had been more generous in London, maybe this wouldn't have happened." He'd recently raised this issue with Kohl. "Now, if a month from now, Gorbachev is overthrown and people are complaining that we haven't done enough," Mulroney later recalled asking, "is what we're proposing the kind of thing we should do?" Germany's leader thought their aid adequate. Bush certainly did as well. "He said absolutely," Mulroney added. "There was no second guessing the nature of the decision in London." They always could have done more, but as Bush stressed at the time, there was little Western money readily available, and even less confidence the Soviets would spend it wisely.

It is unlikely that more money would have helped, even as a symbolic vote of confidence in Gorbachev's path toward market reform. Bush and Mulroney believed that their Soviet colleague had apparently lost power as a result of his perceived proximity to the West. "This crowd is pretty hardline," Bush explained, speaking of Yanayev and the rest of the emergency committee. "Any doubt in your mind that he was overthrown because he was too close to us?" Mulroney pressed. "I don't think there's any doubt," Bush replied.

In truth Gorbachev's enemies cared far more about his willingness to negotiate away Soviet authority to the republicans than about any financial relationship with outsiders. Bush and Mulroney and other Western leaders did not have any real influence over how the Soviet center and periphery interacted. In the end it was not money but sovereignty that drove the coup plotters to seize power. The August 20 deadline mattered most of all.

Unsure if he'd ever hear from Gorbachev again — indeed, doubting very much that the Soviet leader was still alive — Bush told his diary that "what we must do" in response to the coup "is see that the progress made under Gorbachev is not turned around." He listed the strategic issues he valued most. "I'm talking about Eastern Europe, I'm talking about the reunification of Germany, I'm talking about getting the troops out of the pact countries, and the Warsaw Pact itself staying out of business." Soviet "cooperation in the Middle East is vital of course," Bush added, referring to the broader peace initiative for the region, "and we may not get it now, who knows?" At the least, Gorbachev had survived long enough to provide help with each of these goals. He might well be gone, Bush conceded, and while he lamented the apparent loss of his friend on a personal level, wishing him well repeatedly in his diary, his first presidential priority was pondering who in Moscow he might have to deal with next. "Our options are few and far between of course," he reflected. "The best thing we can do at a time like this is calmness, firmness, adherence to principle. Do not get stampeded into some flamboyant statement," he reminded himself. "See where we can go."

On the far side of the continent Baker independently reached the same conclusion. "There goes another vacation," his wife, Susan, remarked as the phone call from Washington jolted her husband from bed. They'd been awakened by the State Department Operations Center with word of the coup an hour after Scowcroft heard from CNN. Assuring her, "I'll be back here in no time," though admittedly "more out of hope and bravado than anything else," he immediately set pen to paper to compose his thoughts. "No leverage — certainly minimal," Baker jotted, acknowledging there was little that anyone outside the Soviet Union might effectively do in the short term to influence events in Moscow or the Crimea. Indeed, the more foreigners said and did in response, Americans in particular, the more likely the coup might be to succeed. "Signs of 'external legitimacy' or 'threat'" were what the "new committee will be looking for," he noted. China's leaders had used every word against them, no matter how carefully calibrated, to inflame public support in 1989. Speaking out, and activating NATO, in the tense atmosphere at present would only "heighten [the] sense of [a] security problem" and "strengthen the coup supporters," who'd probably succeed no matter what. It "will be hard to do business w/new guys for awhile," he wrote, though they should also "emphasize the lack of their political legitimacy."

Baker thought it crucial for the administration to keep in touch with democratic reformers, singling out Yeltsin as "the key guy." They also needed to offer quiet reassurance to their Central European allies and newfound democracies in Eastern Europe, "particularly Poland." Ultimately they had little choice but to "see how events unfold."

The events themselves presented a puzzle. Armored personnel carriers and tanks roamed Moscow's streets, Baker's aides reported, describing the same scene he was watching on television. But why was CNN still broadcasting from the city? "They should have arrested Yeltsin and other democrats by now," he thought to himself. "They should have cut off links with the outside world." Coups (and crackdowns) had rules. The first for any plotter is to dispatch opponents quickly before your initiative dissipates. The second: information is power, and power is best delivered brutally — and privately. As James Lilley had reminded his colleagues two years before while bullets flew in Beijing, including at the American embassy, one typically closed the door to beat the dog. Muscovites could look out their windows and see the military's might on their streets. That was a show of force the plotters wanted. The problem was that people sitting on the edge of their beds halfway around the world could see, and count, those troops and tanks as well.

Bush faced reporters that morning with scant and conflicting information, hoping above all else to do no harm. He typically conducted press briefings on the lawn in Kennebunkport, but early bands of the impending hurricane forced everyone into the cramped Secret Service office. Steam rose from the drenched journalists crowded around the makeshift podium. The new government was "extra-constitutional," Bush declared, and "outside of the constitutional provisions for governmental change." The term meant little in a technical sense. Scowcroft suggested it carried the right tone of disapproval without the baggage of a formal condemnation. Bush's subsequent comments displayed his displeasure more clearly. "It's a disturbing development," he elaborated, "there's no question about that. And it could have serious consequences for the Soviet society and in Soviet relations with other countries including the United States."

He'd spoken to other Western leaders, Bush assured the reporters, among them Helmut Kohl, Britain's John Major (who had taken over from Margaret Thatcher the previous November), and François Mitterrand, and they all shared the hope for a return to calm in Moscow. "At this point what to do is simply watch the situation unfold," he said, leaving no doubt which way he hoped events would turn. "Coups can fail," he reminded anyone listening to his words. "They can take over at first, and then they run up against the will of the people." Unprompted, he returned to the thought a few moments later. "I'm inclined to believe that when people understand freedom and taste freedom," Bush said, "and see democracy in action, that they're not going to want to change." It was still early, and the news from Moscow sparse, but "what hasn't been heard from yet are the people of the Soviet Union."

Nor had anyone heard from Gorbachev. Unable to contact the outside world, he was tracking events on a small Sony radio the plotters had neglected to confiscate. It was "a stroke of luck to have brought it along," Raisa wrote in her diary, and another example of the haphazard manner in which their captors operated. Their treatment of Yeltsin was even more incompetent. Arriving from Kazakhstan before news of the emergency committee hit the airwaves, but after the plotters had begun issuing orders from the Kremlin, Russia's leader was left to land safely and then sleep through the night unhindered at his dacha in the city's suburbs. Troops loyal to the coup leaders had him under surveillance the entire time, even before martial law was officially imposed, yet they made no move to arrest him or stop the flow of advisers in and out of his home. Neither did they impede his subsequent journey to Moscow's White House, home to the Russian parliament. Gorbachev had been told that Yeltsin was under arrest, or soon would be, but this was either wishful thinking on the part of his captors, a ruse to encourage his compliance, or, most likely, further proof of their incompetence.

Unlike Bush, who parsed his words lest he offend a potentially hostile new regime, Yeltsin minced none. "On the night of August 18–19, 1991," he announced, dictating a statement to his staff for immediate distribution, "the legally elected president of the country was removed from power." He appealed to all Russians to protest and called for a general strike until Gorbachev's lawful return. As tanks approached the White House, he boldly strode to the nearest of them, which was already surrounded by a small band of Muscovites. Climbing aboard after him, aides unfurled a Russian flag behind Yeltsin. He felt "a surge of energy and an enormous sense of relief" while declaring his people's refusal to forsake democracy, Yeltsin later wrote. His wife had earlier implored him to stay inside, or at least to do more than merely don a bulletproof vest. Russian troops would never fire on their own flag, he assured her, or at anyone who traveled beneath its authority.

Standing on that tank, "coming out to the people," as he described it, became his most inspiring moment as a democrat. According to Eduard Shevardnadze, who'd made his way to Moscow's White House once *Swan Lake* tellingly appeared on his television screen, Yeltsin's defiance immediately called to mind Vladimir Lenin's legendary appeal for revolution atop an armored vehicle outside Petrograd's Finland Station. Every Soviet citizen could recall the image, the Russian version of Washington crossing the Delaware. Of such legendary moments nations are made.

American witnesses described it differently. The troops surrounding the

White House had no orders to fire. The crewmen of the tank Yeltsin boarded were not friendly to his cause, but neither were they hostile. Indeed, before mounting the vehicle, Yeltsin prudently asked its young commander, "Did you come to kill the president of Russia?" Told "Of course not," he climbed aboard, providing any camera pointed his way the image of a fiery democratic demigod delivering an impassioned "Appeal to the People of Russia," or as Matlock termed it, "one of the best photo opportunities of the decade."

Cameras pointed elsewhere on the square told a different story, though these images rarely made the news. To that point few Muscovites had heeded Yeltsin's call for a general strike. More flowed to the scene once CNN broadcast his defiant address around the world, with news of his speech subsequently beamed back into Russia itself via radio. The coup's organizers, trained in earlier days to seize control of state media, had failed to account for its new international reach. Neither had they shut down Moscow's phone lines, which allowed the city's residents to spread word of Yeltsin's defiance. Faxed copies of his proclamation appeared on lampposts and subway signs. The new technology outpaced its implications for those in charge, trained over decades as good Soviet apparatchiks to control development of new ideas more than their distribution. They could control the way people communicated in 1960 but not in 1991. Thousands descended on the White House in response.

Many more arrived after witnessing the plotters' bungled attempt to solidify their own authority at a news conference that they organized but also failed to control. Yanayev's trembling hands and wavering voice gave the appearance of a man more inebriated than in charge. Worse yet, no one on his staff had thought to limit the number of questions, and reporters in the room bombarded him. Having just suspended the rights of a free press, in other words, the plotters now enabled it. As one American diplomat subsequently reported, referring to the journalists, "The evasive answers they received left the impression that the committee actually had little idea what to do with the power they professed to be seizing."

The ill-fated news conference climaxed with a question posed by a twenty-four-year-old reporter, whose entire adult life had been shaped by *perestroika*. Did the committee realize they had "carried out a coup d'état?" she asked, uninterested in the protocols of earlier generations of Soviet journalists. "Which comparison seems more apt to you," she offered as a follow-up, "the comparison with 1917 or 1964?" That is, the ouster of the czar or Khrushchev's removal? Another reporter mockingly asked if they planned to telephone Chilean general Augusto Pinochet for advice. He knew a thing or two about coups.

"What have we gotten ourselves into?" Defense Minister Yazov muttered at the sight. His wife's critique burned in his ears as well. "Dima, this means civil war," she wailed, arriving at his office upon hearing the news. Hobbled by a recent car accident, she nonetheless navigated the Kremlin on crutches to make her appeal. "You have to stop this nightmare. Call Gorbachev." Nothing was that simple, he responded, waving toward the television set in his office that continued to broadcast his comrades' news conference. "Dima, look at who you have got involved with," she cried. "You always laughed at these people. Call Gorbachev."

Their public face cracking, apparently drunk, ridiculed even by their own families, the plotters launched a desperate appeal for international recognition and prepared to crush their opponents. They held the Kremlin, and with it direct lines to troops, and to the epicenter of global power. "There has been a real threat of the country's disintegration, of a break-down of the single economic space, the single civil rights space, the single defense and single foreign policy," Yanayev wrote to Bush. "Under these circumstances we have no other choice but to take resolute measures in order to stop the slide toward catastrophe." They promised to respect every treaty the Gorbachev regime had to date negotiated, and pledged that "the foreign policy course of our country shall be continued" after the restoration of "political stability." Yanayev claimed, hollowly, that "this step is not an abandonment of the reforms started by M. S. Gorbachev. The reforms will be continued . . . We shall proceed on the basis of democracy" and "shall continue with the policy of ensuring civil rights and freedoms." International support was not necessary for their success, he added. They did not need Washington's blessing in particular, though clearly they hoped to have it. "While the main part of the effort can only be made by ourselves, we cherish your good attitude and believe that continued cooperation between our two states has been and remains in our mutual interest."

Bush was not impressed. Indeed, the more his team studied the news emanating from Moscow, Gates reported, the more they "began to think that the coup plotters did not have their act together." Bush issued a second statement later in the day on August 19, employing the far stronger term "unconstitutional" to describe the seizure of power. With Baker still en route back to Washington, he directed Deputy Secretary of State Lawrence Eagleburger to deliver an even sterner — albeit private — response to the Soviet ambassador. "We have no interest in a new Cold War or in the exacerbation of East-West tensions," Eagleburger told the man, even while threatening precisely that. "This misguided and illegitimate effort to by-pass both Soviet law and the will of the Soviet peoples is in no one's interest," Eagleburger added, reading from a prepared text before handing it to his Soviet counterpart. More to the point, it was in fact "unconstitutional."

Ambassador Victor Komplektov understood. He'd been surprised by the news from Moscow as well. Neither did he disagree with Eagleburger's basic points or attempt to defend the plotters' recent deeds. He'd passed along their message, as duty required. Speaking "purely personally," he noted as their meeting came to an end, he appreciated Bush's statement that morning, since "it had not sought to encourage public disturbances," and showed yet again that his administration appreciated the delicate and dangerous circumstances at hand. Whenever crowds and tanks mingled, both men knew, chaos might not be far behind.

Komplektov subsequently met with Gates at the White House, though the American tried his best to let his body language and tone deliver the message the administration was not yet fully ready to state publicly. "I offered no pleasantries or polite conversation and tried to make the atmosphere as cold as possible," he later wrote. By mid-morning, every report out of Moscow further revealed the plotters' incompetence. "Why were all the telephone and fax lines in and out of Moscow still working?" Gates marveled. "Why was daily life so little disrupted?" Why had the regime's opponents not been arrested, and even more puzzling, "how could the regime let the opposition barricade themselves in the Russian Parliament building and then let people come and go?" The questions all added up to a singular answer: this coup was failing.

It could still turn bloody, especially if the plotters became desperate. Buoyed by further analysis from the NSC and CIA, Bush addressed the press later that afternoon, stating, "We will avoid in every possible way actions that would lend legitimacy or support to this coup effort." A statement put out by his press office employed the key diplomatic word: he "condemned" the coup's organizers and their goals.

Bush was doing even more by the second full day of the coup to throw public support behind Yeltsin, who in Gorbachev's continued absence clearly formed the center of opposition. Initially reluctant to phone a man he so little trusted, Bush was ultimately convinced by Scowcroft that the fact of their conversation meant much more than anything the two men might say. "Calling President Yeltsin this morning allows you to show support for him, and through him, for the constitutional process violated by the coup," he advised. "The mere fact of your call will buoy him up," though "it is important not even to inadvertently leave President Yeltsin with the impression that we can give more than general support."

It was a revealing reminder, for a man who did not need one, that while his ability to reverse the coup was minimal, his ability to make the crisis far worse was significant. Here was a clear example of Bush's ability to succeed merely by avoiding doing the wrong thing, whether that be ratcheting up tensions with a new and clearly stressed regime, potentially with shaky control over its nuclear arsenal, or inspiring Yeltsin or others to believe they had Washington's backing no matter how rash their own response. If the Americans had proven themselves unwilling to risk a war by inserting themselves too far into Budapest in 1956 or Prague in 1968, or even Beijing only two summers before, they'd never contemplate doing anything that might lead Soviet forces to perceive themselves under threat of attack, or Russia's democratic opposition to think Bush might support a counterattack of their own.

Bush's new world order approved of great powers fulfilling international will against lesser intransigent states that violated sovereignty and international borders. It had nothing whatsoever to say about great powers confronting one another over affairs inside their own capitals. Indeed, pondering the current crisis, he could not help but compare the clarity of the previous summer with the more opaque present. "What had happened [then] was Iraq had to get out of Kuwait," he dictated to his diary. "Here, I'm not sure what has to happen. What I'd like to see is a return of Gorbachev and a continuous movement for democracy. I'm not quite sure I see how to get there."

To the amazement of White House operators in Washington, their call to Yeltsin at the Russian White House also went through. "Just checking to see how things are going from your end," Bush explained nonchalantly once Yeltsin picked up the line. It was all the manic Russian leader required. "The building of the Supreme Soviet and the office of the President is surrounded and I expect a storming of the building at any moment," he breathlessly informed Bush, beginning a lengthy monologue. "We have been here 24 hours. We won't leave. I have appealed to 100,000 people standing outside to defend the legally elected government."

"You have our full support for the return of Gorbachev and the legitimate government," Bush replied, finally able to break through Yeltsin's verbal deluge. "We will reiterate that early today." Yeltsin's foreign minister had already called associates in Washington to note how much they appreciated Bush's firmer public stance. The two presidents agreed that Bush would not, however, speak to Yanayev, or any of the other coup members, even to attempt to negotiate their surrender. Such a call would potentially legitimize them, if only by allowing them to say Yanayev and Bush were in communication. "We're not hopeful" of reaching Gorbachev directly, Bush admitted, but even the attempt "legitimizes the Gorbachev regime."

Yeltsin was right to fear an assault. Hard-liners met that night in the Kremlin to buoy the new regime's resolve, calling for troops to "liquidate" the democratic opponents congregating in and around the White House. The KGB issued a merciless report detailing the mistakes already made, many of them obvious to even the most inexperienced plotter. They'd failed to seize potential opposition leaders; failed to halt communications; failed to block key transit nodes or shut down public transportation. Worst of all, they'd failed to account for antigovernment media, which by this point meant any citizen with a camera or access to a phone. Realizing their dwindling hopes of success, they dispatched troops to seize control of the White House and arrest its key occupants, or better yet, silence them forever.

In Beijing this moment had led to slaughter. In East Berlin it had led to the regime's capitulation, and in Prague to its resignation. In Romania, blood had flowed even as the regime fled. Moscow teetered on the brink of violence. "Big crowds are gathering," a senior officer from the paratroopers reported after reconnoitering the White House perimeter. "They are erecting barricades. It will be impossible to complete this operation without significant casualties. There are many armed men inside the White House." Plans called for a multi-tiered attack, with simultaneous strikes from troops on the ground, entering through subway tunnels, and even dropped via helicopter on the building's roof. Those commando squads expected to lose half their ranks but to take far more of the defenders with them. In the Crimea, with only the barest information trickling in, but knowing too well the personalities involved, Gorbachev and those around him expected the worst. "This may not end well," he told his fellow captive Chernyaev. "But you know, in this case I have faith in Yeltsin. He won't give in to them, he won't compromise. But that means blood."

Reports of explosions and automatic gunfire at the Russian White House and nearby American embassy reached the State Department in the late afternoon of August 20. It was after midnight in Moscow, the morning of the twenty-first. "I've seldom felt so powerless in my life," Baker remembered later. Tasked with flying to Europe to coordinate with NATO allies, he "kept waiting for the other shoe to drop" and for "news that KGB and Interior Ministry troops had attacked and overrun the barricades, killing Yeltsin in the process."

That call never came. Stung by their perceived betrayal by civilian authorities who'd left them to sink in an Afghan quagmire, and repulsed by the way those same bureaucrats had vacillated between endorsing and then condemning bloodshed in breakaway republics, the Russian commanders of Soviet units had no desire to follow the orders of their new superiors. Like their German counterparts, they could not bring themselves to fire on civilians. They pulled back. General Pavel Grachev, who commanded the airborne troops designated for the assault on the White House, instead told Yeltsin (through a messenger) that "he was Russian and would never allow the army to spill the blood of its own people."

Blood was spilled that night anyway, as troops directed by Varennikov drove forward toward the parliament building, whose defenders opened fire. They had two advantages over Varennikov's men: time to prepare their defenses, and experience. Most of his youthful conscripts had never seen combat. Many of those who volunteered to defend the White House were veterans of Afghanistan. And they were armed. By the time the first skirmish was over, three bodies lay lifeless on the pavement: two from the attacking side and one defender. More were wounded.

Hearing of the casualties, Marshal Yazov had had enough. The KGB's troops refused to move on the building, as did the paratroopers. The regular army, and in particular its Russian-born troops, he vowed, would not be left to bear the burden of firing on Russians alone. "Give the command to stop," he ordered. Yeltsin was already in his armored limousine in the White House parking garage, awakened roughly from sleep and thrust into its backseat by his bodyguards at the first crack of rifle fire. They planned to make a dash for the American embassy, which by prearrangement had kept its gates open to receive them. Rousing himself, Yeltsin too ordered his men to stop. "We don't need any embassy," he told them. Fleeing that night might secure his survival, but it would destroy the grand political future he desired.

Why did the fighting stop? Why did no Chinese-style crackdown occur? It was clearly a near-miss thing, but the differences between Berlin, Beijing, and Moscow are instructive. China's leaders collectively took responsibility for ordering a crackdown against their own people. In East Germany, as in the crumbling Soviet Union, military commanders believed that their civilian heads wanted someone else to take the blame. In Beijing, the worst violence occurred when soldiers from the hinterland engaged civilians from the city, who neither looked, spoke, nor acted like them. They might as well have been fighting a foreign foe. In Leipzig as in Berlin, a German identity connected police and the populace beyond any obvious regional differences. By the same token, Soviet troops drawn largely from Russia, and certainly commanded by Russians, had opened fire in the Baltics but refused to fire on Russia's flag, or its people. Learning from Beijing's first nights of violence two years before, when civilians surrounded and berated troops and ultimately shamed them into withdrawal, Russian civilian leaders organized units to plead with soldiers sent to take back the White House. Largely barred from standing guard, women — especially elderly grandmotherly types, or comelier, younger ones — were particularly encouraged to approach military vehicles with reminders that no one expected the soldiers to fire on their own people. The vast majority did not.

Finally, the troops who at first marched on Russia's White House and then retreated or laid down their arms had an obvious division of loyalties. Actually, they had several. Trained to follow orders, they could not easily decipher which orders mattered. Did they owe their allegiance to Gorbachev, to their uncertain military commanders, or to Yeltsin and the tens of thousands of citizens who ultimately put their bodies between the barricades and their newfound democracy? After five-plus years of *perestroika* and *glasnost,* and the spirit of dissent and questioning each engendered, it should not be surprising that this amorphous thing called democracy had taken root within the minds of soldiers or citizens. The emergency committee claimed to be operating for the sake of democracy, but its actions belied that motive. Yeltsin claimed democracy's mantle as well. "Law and constitutional order will triumph," he preached from the White House. "Despite everything, Russia will be free!" Gorbachev, still imprisoned in his Crimean mansion, pleaded for the same. "An anti-constitutional coup d'état has been carried out," he told his family's videotape recorder. It too remained in their possession. "I don't know whether I shall succeed in getting this out, but I shall try to do everything to see that this tape reaches freedom."

It was not an idle choice of words. The spirit of democracy infused all aspects of Soviet society by 1991, fueling Gorbachev's drive for a new partnership with Europe, then the drive for sovereignty within the republics and Eastern Europe, and now his people's rejection of the last vestiges of their authoritarian state. While crowds chanted for liberty, leaders at every level strained to show their links to the West, and to the United States in particular, whose imprimatur of support seemed all the proof required to demonstrate true democratic credentials. Yeltsin spoke to Bush. Gorbachev did the same as soon as he could. Even members of the failing emergency committee had tried to win Bush's blessing. The spirit of democracy was even diffused throughout the ranks of the Soviet military. Having led the attack on Lithuania's television station in January, one officer took the unusual step of polling his troops before deciding whether to storm the White House as commanded. One by one they told him they'd not fire on their own. "We will not go to the White House to kill people," one of the men insisted. "And we will not lead you there," his commander replied.

Yeltsin survived, and emerged as the Soviet Union's unlikely savior and its final executioner as well. "In the aftermath of the failed coup," the American embassy in Moscow reported a fortnight later, "Boris Yeltsin is the most powerful individual in the USSR. No decision affecting the country as a whole can be taken against his will." Bush and Scowcroft perceived that dynamic with their own eyes. Both watched the television broadcast of Gorbachev's triumphant return to Moscow on August 23, a day after Yeltsin issued arrest warrants for the coup's leaders. The pair stood before the Russian parliament, not the Kremlin. The authoritarian center had been supplanted by its more democratic alternative, and was now its supplicant.

Addressing the crowd of well-wishers, Yeltsin demanded Gorbachev's approval for Russia's latest reforms. Gorbachev, with the world watching, admitted he'd not yet had time to read them. Poking his finger directly in his face, Yeltsin pushed aside any last semblance of subservience to the man who'd started the greatest revolution of the latter half of the twentieth century. "Well, read them!" he berated Gorbachev.

"It's all over," Scowcroft said, shaking his head in disbelief. "Yeltsin's telling him what to do. I don't think Gorbachev understands what's happened." Bush nodded in agreement. "I'm afraid he may have had it."

Within two weeks what little was left of the Soviet Union came apart at the seams. Yeltsin suspended Russia's Communist Party, with Gorbachev unable to do anything more than look on in bewilderment. "Boris Nikolayevhich, Boris Nikolayevhich . . . I don't even know what you are signing there," he stammered. "I have signed it," came Yeltsin's blunt reply.

Television cameras caught it all. Within a week, Ukraine, Belarus, and five other Soviet republics declared their independence. Lithuania's parliament reaffirmed its earlier declaration of sovereignty. Bush initially refused to recognize them, pending their ongoing negotiations with the Kremlin, but on September 2 he publicly acknowledged the Baltic republics' independence. After all, the United States had never formally accepted their integration into the Soviet Union in the first place. "Over the coming weeks and months the situation in the USSR is likely to be characterized by a race between democracy and disintegration," American diplomats in Moscow observed.

The outcome of that race would ultimately be both, which Bush interpreted as a Western victory. "We have just had a tremendous triumph for our values and for our vital interests," he crowed to Britain's John Major immediately following the coup's collapse. "There is little doubt that the path is now clearer for economic reform, restructuring of the Union, expansion of democracy, and removal of political obstacles." The path clear, they could "proceed in an orderly way."

By Christmas Day, the Soviet Union was gone. A new union of confederated republics, negotiated primarily by Yeltsin and his colleagues, took its place. Gorbachev called the American White House one last time before leaving office, and before turning his nuclear keys over to Yeltsin. "The debate in our union on what kind of state to create took a different track from what I thought right," he confessed. "But let me say that I will use my political authority and role to make sure that this new commonwealth will be effective." Gorbachev would always have a special place in his heart, Bush assured him, noting that the horseshoe pit at Camp David where Gorbachev had once thrown a ringer on his first toss remained in good shape, and ever available for a rematch. "I salute you and thank you for what you've done for world peace," Bush said.

And then the man who had for so long refused to gloat at the Cold War's end and democracy's triumph at long last did just that. "During these last months, you and I have witnessed one of the greatest dramas of the 20th century," Bush told the American people that night in a televised address from the Oval Office. They'd witnessed "the historic and revolutionary transformation of totalitarian dictatorship, the Soviet Union, and the liberation of its peoples." For forty years the United States led the struggle against communism, he said. "That confrontation is now over . . . Eastern Europe is free. The Soviet Union is no more. This is a victory for democracy and freedom. It's a victory for the moral force of our values. Every American can take pride in this victory."

Battles still lay ahead. "I want all Americans to know that I am committed to attacking our economic problems at home with the same determination we brought to winning the Cold War," he promised. Polls showed voters cared far more about the economy than anything else, as was usually the case. His own approval numbers, which had topped 90 percent at the start of 1991, dipped in January 1992 below 50 percent for the first time in his presidency. By summer, as fears of a recession grew and fears of international disorder receded, fewer than 30 percent approved of his performance, opening space not only for a primary challenge from within his own party but for a strong Democratic rival and ultimately an independent's bid for the White House as well. The economy and the election would dominate Bush's coming year, but on Christmas night in 1991, he told Americans they could rejoice in their historic victory, with only one thing truly to fear. "We cannot retreat into isolation," he said. "We will only succeed in this interconnected world by continuing to lead."

He had never in his life thought otherwise, or questioned the ultimate wisdom of the American system. Now he could think of nothing more to say as he proclaimed the Cold War over than to offer an invocation that the country stay its course. "For our children, we must offer them the guarantee of a peaceful and prosperous future, a future grounded in a world built on strong democratic principles, free from the specter of global conflict." Because he'd let history run its course, on that Christmas Day, a new and better world seemed possible.

CONCLUSION

GEORGE BUSH LEFT the White House in 1993, leaving behind what might well have been the most powerful state in human history. It was undoubtedly the most powerful in its own era. Its economy was the world's largest. No military could match it. Its universities drew the best minds and its corporations the most creative workers. Global capital looked to New York and entertainment to Hollywood. Technological innovation associated with the nascent Internet soon ushered in a new era of American competitiveness and economic dominance, and with it a newfound sense of optimism rarely seen in a twentieth century that had known so much conflict and war. The democratic revolutions in Eastern Europe, highlighted by the Soviet Union's remarkably peaceful devolution of power, pointed toward a new era of hope. "There is no other place that I want to be," a British pop band sang, their tune climbing the charts on both sides of the Atlantic. Coincidentally released the same day as Bush's "new world order" address in Washington, "Right Here, Right Now" had lyrics that captured the age, "watching the world wake up from history."

American power at the moment of Bush's departure also derived from influence. Washington headed an increasingly large and prosperous group of military allies and allied trading partners as the 1990s began, augmented by the growing wave of democracies who looked to the United States for leadership and guidance. Indeed, policy makers and pundits who had spent 1987 talking of imperial overstretch and 1989 considering the end of history spent much of the ensuing years considering if American hegemony would likely lead to more or less peace. No one doubted the question's central premise. The United States faced no real peer or competitor on the near-term horizon, and while history had taught Bush's generation to fear unbridled allies and historic rivalries, the next generation of American leaders appeared determined to put that historical lesson to good use. "America has and intends to keep military strengths beyond challenge," the president of the United States declared in 2002, "thereby making the destabilizing arms races of other eras pointless and limiting rivalries to trade and other pursuits of peace."

Peace never materialized, however. It did not even prevail in Europe during Bush's presidency. By 1992 war raged in Yugoslavia, tinged with ethnic barbarism unseen since the Nazi Holocaust. Tens of thousands died. More fled. Still overjoyed by the continent's unexpected democratic surge, Europe's new powers buckled before the conflict in their midst. The Balkans seemed distant to Western Europeans, despite their central role in instigating the First World War. Closer to the conflict, the Soviet Union's successor states had neither the will nor the capacity to intervene. Few in Eastern Europe welcomed the prospect of intervention by troops whose uniforms still bore the insignia of the Soviet hammer and sickle. Most important of all, Washington wanted no part of the conflict. As James Baker caustically but accurately pointed out, the terrible dynamics of the ethnic struggle left few clear paths for American intervention. A civil war could never be as simple as when Iraq invaded Kuwait. "We don't have a dog in that fight," Baker said. More to the point, "we don't want to put a dog in this fight."

Outsiders largely just watched, and worried, as the violence and barbarism spread. "Everything is at stake here, if principle is everything," Democratic senator Daniel Patrick Moynihan would argue by the mid-1990s, as the Balkan conflagration continued to flare. Bush had sent American forces halfway across the globe in order to establish a precedent of opposition to direct aggression, and since 1945 global human rights advocates had vowed to "never again" tolerate genocide. But if NATO could not subsequently summon the will to save lives less than a day's drive from its major bases, and if the United Nations was willing to stand by impotently while pictures of massacred civilians flooded the airwaves, Moynihan asked, "what have we gone through the 20th century for?"

Americans had no particular "dog" in the fight in Somalia either, save the cause of rescuing civilians at risk of starvation and threatened by local warlords bent on accumulating power. "I understand the United States alone cannot right the world's wrongs," Bush told the nation in December 1992, explaining his decision to send troops to support the United Nations humanitarian mission in the war-torn East African nation. He'd already lost his reelection bid a month before, unable to rebut charges that he cared more for international affairs than for the domestic bread-and-butter issues voters were most concerned about. He ironically responded by initiating a new mission overseas. Troubled by news reports filled with plaintive cries of victims asking, in effect, "Won't someone help us?" Bush realized he could. Moreover, no one else would. "Some crises in the world cannot be resolved without American involvement," he said. "American action is often necessary as a catalyst for broader involvement in the community of nations."

Taken collectively, actions in Somalia and the Balkans, in combination with the humanitarian mission in northern Iraq that Bush signed on to in late 1991, demonstrated the position of international primacy that American policy makers had desired for generations, and its immediate downside: absent American leadership, the international community was unlikely to muster any will to act, even in the face of genocide or widespread civilian starvation and suffering. American withdrawal, meanwhile, left no one else willing to fill the void. Bloodied in October 1993 in Mogadishu following a citywide firefight that left eighteen Americans and an estimated two thousand Somalis dead, President Bill Clinton pulled the remainder of American forces from Somalia before the conflict could become a quagmire. A general United Nations pullout soon followed. Full-blown civil war ensued, leading both to an estimated 500,000 additional deaths and establishment of safe havens for international terrorist groups that could effectively function only in lawless lands.

Chastened by the political fallout of withdrawal, and wary of incurring casualties again in the midst of an ethnic conflict whose origins seemed to predate the history of his own country, Clinton subsequently refused to intervene when wholesale civilian violence erupted in Rwanda. An estimated 800,000 people died. Another 2 million became refugees. Genocide offered outside great powers "a problem from hell," Harvard theorist Samantha Power concluded, troubling their collective conscience but simultaneously challenging the notions of state sovereignty crucial to the very structure of contemporary international affairs (and central as well to both George Bush's and Mikhail Gorbachev's competing visions for a post–Cold War world). The Cold War, for all its innumerable faults, presented clear geopolitical lines. A Manichaean world is easy to understand, its gains and losses simple to tabulate. Civil wars and ethnic conflicts, especially those unwittingly unleashed by the end of the Cold War, offered less chance of global nuclear apocalypse but also murkier problems, exchanging the specter of mushroom clouds for the proliferation of mass graves.

Clinton eventually overcame his reluctance to deploy force, having by his second term seen enough of the violence in the Balkans. "If someone comes after innocent civilians and tries to kill them en masse because of their race, their ethnic background, or their religion, and it's within our power to stop it," he pledged in 1999, "we will stop it."

Clinton might as well have added the word "eventually." It took him years to muster the political will, within his own administration and within NATO, but this pace is less an indictment than it might appear at first blush. Ethnic conflicts offer no easy exit strategies following outside intervention, and Bush had wanted no part of the conflict either. Moreover, we should praise rather than criticize leaders whose opinions evolve. It took Bush nearly a year in office even to see that Gorbachev presented more an opportunity than a threat. Even then, what really drove Bush to accept the Soviet reformer was less any particular faith in *perestroika* than the realization that its alternatives were likely worse.

One lesson of studying the presidency of George H. W. Bush, if we keep in mind the problems from hell the post–Cold War world presents, is that even the most experienced presidents take time to warm to their awesome new responsibilities. The best learn in office. The steeper the learning curve, however, the greater the danger. Inexperience usually leads not to stalemate but to disaster. As Baker sagely advised his friend upon taking office back in 1989, despite his gathering a national security team arguably as experienced as any in American history, the worst sins for a new administration were typically those "of commission, not omission." Recent history shows that presidents typically face unexpected international crises during their first year in office, a period when foreign foes rightly probe for potential reactions from the new leader in Washington. John Kennedy faced the Bay of Pigs. Bill Clinton confronted "Black Hawk Down" in Somalia. George W. Bush saw a downed reconnaissance plane in East Asia nearly spark a conflict with China even before facing 9/11 a mere eight months into his presidency. And so on. George H. W. Bush faced the most difficult first year of all, whose strategic problems continued throughout most of his term: the complete collapse of the enemy and existential threat that had defined American foreign policy, indeed American society, his entire life. Faced with uncertainty, and unsure of the best response, he paused, considered, and learned.

Iraq too, site of Bush's greatest military victory, proved unsettled even before he left office. It became a quagmire of its own in time, the repetition of Vietnam his administration desperately strove to avoid. Saddam Hussein survived the Gulf War, to Washington's great surprise. Retaining power, he refused as well to hide, flouting first American-dictated no-fly zones in his zeal to crush domestic opponents, and then international sanctions imposed to thwart his supposedly reconstituted weapons program. Undeterred by sporadic airstrikes and inspection regimes launched throughout the 1990s, Saddam even conspired to assassinate the retired president Bush, whose son won the White House by the narrowest of margins in 2000. By the spring of 2003, the younger Bush initiated his own war in Iraq, having consciously planted in the public's mind the false idea that Saddam was somehow responsible for the terrorist attacks of September 11, 2001. Unlike his father, he ordered the country's occupation. "We will be greeted as liberators," the younger Bush's vice president promised, our troops showered with candy and flowers. Democracy will flourish in Iraq under American tutelage, this Bush promised, catalyzing a great democratic wave across the entire Arab world.

Explosions followed instead. In 2013 the last American combat troops left Iraq, with memories of the more than 4,800 comrades whose lives had been lost since 2003. Tens of thousands more were wounded, many with psychological and cognitive injuries that became apparent only years later. Thanks to medical advances, only 13 percent of wounded American soldiers perished in the second Iraq War. The equivalent proportion for Vietnam was 25 percent. The death toll from Iraq thus belies its damage. Medical victories mean that more suffer the pain of their wounds for the remainder of their days, and that the cost of their care compounds over time. The senior Bush fought his Gulf War with a large coalition and an abundance of financial support. Estimates on the second Bush conflict, lacking similar international goodwill and enthusiasm, suggest that long-term costs to the American taxpayer might well reach $6 *trillion*. None of these figures includes the human cost of the estimated 150,000 Iraqi civilians killed over the same period, or the humanitarian crises that swept the destabilized region. Given the complete breakdown of Iraqi civil society after 2003, accurate numbers are nearly impossible to divine. One reputable British medical agency, employing census data, calculated that the likely death toll among Iraqi civilians well exceeded half a million.

The point here is not to litigate the wars of the 1990s or the 2000s but to view them in terms of the world Bush bequeathed his successors. Given the free-dom to act without peer or existential threat, subsequent presidents were less restrained than Bush and those who had come before him. They could move on the basis of their own timeline, concerned more with domestic politics than geopolitics. They assumed that the international goodwill and moral capital the United States inherited at the Cold War's end would always be at their dis-posal, or at least that battlefield victory would force detractors to capitulate be-fore the reality of inevitable American dominance. Believing they understood the new world order they'd inherited, the fulfillment of George H. W. Bush's vision, which was itself the dream of Franklin Roosevelt's generation, they were both imprudent and wrong.

Bush's new world order never promised permanent peace. Neither did it bestow upon American leaders omniscience to match their preponderance of power. It offered instead a mechanism for reaffirming peace and stability, again and again, internationally but with American leadership more than dictates, as ensuing generations faced the unseen and unpredictable challenges Bush never felt capable of naming. Its purpose was to implement the best ideas of 1945, completing the victory Bush's own generation had sacrificed to achieve through a coupling of both American and international ideals he had embod-ied throughout his long political and diplomatic career. Bush believed in the universality of American values and in their eventual acceptance around the world — in time — and he believed that the United States, and only the United States, could safely shepherd the world to that ultimate, more peaceful and prosperous destination. None of these underlying ideas was his alone. Nor were they new. But because Bush believed in the inherent superiority of the Ameri-can ideas on which he'd been raised and never sought to question, and because he believed too that in time others would come to appreciate their virtues, he never considered the need for original thought when it came to global affairs.

Some see this dearth of originality as a flaw. I consider it both an accurate picture of the man as a strategist and indeed a compliment. The new world order was designed to be better than what came before, predicated upon insti-tutions that worked and, more important, whose continued survival afforded subsequent generations of policy makers the tools to choose better outcomes. It gave new voice to international consensus (in particular through the United Nations) and to international collective bodies (in particular NATO and sim-ilar economic instruments such as the International Monetary Fund and the

World Trade Organization). As Brent Scowcroft aptly noted, the ultimate purpose of their "new world order" was neither peace nor a panacea but rather
continuation of an ongoing process of improvement, as more and more of the
world's peoples chose to follow Washington's lead. It would not work if they
were forced. "The world could be a better place," he said, explaining the White
House's general strategic conception at the Cold War's end. "But don't get carried away."

The world Bush passed on to his successors contained everything they
needed to ensure success, if they chose to employ it, and the general approval of
American leadership that proliferated after 1993. The United Nations entered
the 1990s newly relevant after the first Persian Gulf War. NATO survived the
Cold War's end, and with it an overall American commitment to international
security. No one else seemed interested in taking up that role anyway, at least
not from Washington's perspective. "The pivotal responsibility for ensuring the
stability of the international balance remains ours," Bush's first National Security Strategy had stated. By the time he'd left office the world had changed, but
not his conviction on this central point. "We must," he said, "remain the active
leader of the entire world."

But active did not mean reckless. The central flaw of the Iraq War begun
in 2003 was its pace and enthusiasm. It was "a war of choice," Richard Haass
(a veteran of both administrations) subsequently wrote, in comparison to the
"war of necessity" the senior Bush had waged. The second Iraq War was also
designed to catalyze democracy, speeding its triumph throughout the Middle
East. This is the fundamental difference between both Presidents Bush, and
key to appreciating the elder's prudence seen throughout these pages. Both believed freedom and democracy the world's ultimate destination. But whereas
the father was willing to ride the stream of time, avoiding rocky shoals while
flowing with the current, the son paddled hard. Granted, after the most devastating terrorist attack on American soil in history, George W. Bush did not
believe he had the luxury of time his father had enjoyed. Yet the fact remains
that he pushed and pulled to speed history, while the senior man believed history worked best, and worked out best for the United States and its allies, if
American leaders simply followed its natural flow — after of course securing
control of the helm.

Needless to say, not everyone is keen to see democratic values triumph, or
to witness the ongoing expansion of American influence. A great irony of the

George H. W. Bush presidency is that while he bequeathed a world of opportunity to his successors, he also left them simmering quagmires. Iraq was not the only unresolved issue in the Middle East that grew from the Persian Gulf War. Infuriated by his Saudi king's willingness to host infidels in their midst, Osama bin Laden and other Islamic fundamentalists declared a religious war on the United States, though few initially took heed of his pronouncements. Everyone in the world noticed on September 11, 2001. The elder Bush's strategic decision made amidst an international crisis in the late summer of 1990 thus colored his son's entire presidency, leading in time to the thorough hollowing out of the very prestige, influence, and ultimately American power he had striven so hard during his own White House years to preserve.

NATO too became an international sore. Its survival marked Bush's crowning achievement, the embodiment of his entire worldview and the fruit of his most engaged diplomatic effort, even though its subsequent expansion under his successors poisoned Russian relations with the West, and with the United States in particular, all because of a single promise uttered in February 1990. James Baker, as we have seen in these pages, backed by Robert Gates and then by Germany's Helmut Kohl, told Soviet officials that NATO would not expand "one inch to the east" if Germany were allowed to unite with East Germany under its eventual protection. Gorbachev believed them.

Here we must of necessity split hairs. The Americans did not lie. But neither could they foresee the future or bind their successors. Without a formal treaty promising no further NATO expansion, subsequent Western policy makers acceded to Eastern European desires to join the Western security organization. Leaders in Warsaw, Prague, and Budapest — each long subject to Soviet subordination and yet unable to change their geographic fate — leaped at the opportunity to put strategic distance between themselves and Moscow as the 1990s progressed. All three countries joined NATO in 1999. Seven more Eastern European states joined in 2004. By 2009 another two states, once part of Moscow's broad sphere of influence and security cordon, formally entered the alliance as well. Of its twelve new members, seven had been part of the Warsaw Pact. Three more — Lithuania, Estonia, and Latvia — were not only part of the Soviet Union itself but also the scene of its most violent and searing secessionist ruptures.

It is hard for former Soviet citizens and their descendants not to view these defections to their onetime enemy as a strategic loss, especially when compounded by new economic travails encountered when the former Soviet Union's turn toward free markets led to new levels of poverty and prosperity, and once unimaginable inequity within a society formerly predicated (theoretically) on equality. Put another way, the Russian economy, depicted in terms of sickness throughout this book, bottomed out in the mid-1990s. Inflation soared, as did crime. Food became increasingly scarce in urban centers, but plentiful for those with funds, and especially for those willing to claim the spoils of previous Soviet investments. Common citizens could not. Veterans of the Great Patriotic War, their pensions cut and essentially made worthless, took to selling their medals and treasured keepsakes in subway tunnels and on street corners. These were the men and women who had beaten back the Nazis; who had first put a man into space; and who were promised an end to class struggle, and an end to history on communist terms, by 1980. By 2000 they felt defeated, hungry, and angry, ready to turn on the Cold War's victors and those who'd succeeded in its aftermath.

"Be kind to Russia," Boris Yeltsin implored Vladimir Putin upon handing him the keys to power that year. The former KGB colonel who had witnessed democracy-inspired crowds tear down communist symbols in Dresden a decade before, and who had subsequently joined the general Soviet retreat home, knew the power he could derive instead through blame. "They are constantly trying to sweep us into a corner," he told Russians in 2014, to what the *New York Times* reported was "thunderous applause." The "they" in Putin's mind: NATO, and in particular its leader, the United States. "We were promised that after Germany's unification, NATO wouldn't spread eastward," Putin insisted. The man who'd received that apparent promise agreed. Expansion was neither explicitly discussed nor specifically prohibited in those 1990 negotiations, Mikhail Gorbachev conceded in 2014. Yet expansion remained "a violation of the spirit of the statements and assurances made to us in 1990."

This is both true and worthless. The point of the 1990 negotiations over Germany, NATO, and in effect communism's surrender was merely and yet profoundly to ensure the possibility of future cooperation. This was the spirit of both Bush's new world order and Gorbachev's common European home, or at least what they shared: recognition that cooperation, over time, could produce more than jealously hoarded independence. American officials indeed suggested a quid pro quo for Soviet acquiescence in German unification, but NATO expansion was just one of many aspects of the deal, none of which was explicitly negotiated. The West was also supposed to provide aid (it did, though not as much as the Soviets or their successors wanted) and investment (which, again, never proved as much as Yeltsin's cadre desired). So too were Western institutions, as part of the general settlement of the Cold War, to accept a post-*perestroika* Soviet society into their midst. The former Soviet states were in turn to remain open, free, and democratic.

Few vanquished foes are given so much. Western institutions did their part, but it was never enough. In 1994 Russia joined NATO's Partnership for Peace program, designed to improve relations between onetime Cold War adversaries. Three years later Moscow joined the Euro-Atlantic Council, whose goals were the same. Putin's reign saw formation of a new Russia-NATO Council, while Russia finally joined the World Trade Organization in 2012. It had effectively joined the capitalist world far earlier, though for most Russians, and certainly for their political leaders, the intervening years were better told as a story of domestic suffering than as one of Western aid and integration. "At the end of the Cold War" the West "repeated the mistakes of Versailles in 1919," Ilya Ponomarev has argued, "by imposing shock capitalism instead of integrating Russia into a stable world order." The name of this fierce Putin critic may sound familiar to this book's readers. His uncle had been among Gorbachev's sharpest initial critics during his tenure as one of the Soviet Union's leading international policy makers. It was Boris Ponomarev who questioned aloud, back in *perestroika*'s early days, "What are you trying to do to our foreign policy?" Both men found much to criticize within their own government's leadership. But ultimately, each found Western leadership worse.

Like Beijing in the aftermath of Tiananmen, Russian leaders in the twenty-first century discovered there was more profit in stoking anti-Western sentiment than in promoting cooperation. Xenophobia is often a useful tool for would-be dictators, especially those in command of countries whose long history is filled with perceived slights, insults, and a sense of inferiority and exclusion. Ultimately, the question is: Did the Soviets receive a promise that NATO would not expand? The query is better phrased differently: Did Moscow receive an opportunity to join the West? The answer is yes. Its leaders chose instead, for what might well be argued were perfectly legitimate reasons, to see Western expansion not as an opportunity but as a threat. NATO's enlargement "would bring the biggest military grouping in the world, with its colossal offensive potential, directly to the borders of Russia," the head of Moscow's foreign intelligence service warned in 1993. The men who led Russia after 2000 in particular found no reason to see it otherwise.

Bush, of course, was no longer in office in 1993 but was instead enjoying an earlier retirement than he'd hoped. What, then, was his ultimate legacy? It was one of promise. He guided the world through dangerous moments, whose peaceful outcome in hindsight continues to obscure their difficulty. The Cold War need not have ended so well. By the same token, a China isolated after 1989 by a more vigorous American response to Tiananmen might well have turned on itself anew, as it had previously during the years of the Great Leap Forward and, more important, the Cultural Revolution. While Bush does not deserve credit for keeping violence from erupting in Berlin, Moscow, or Leipzig, and while he simultaneously does not deserve blame for its eruption in Beijing, there can be no doubt that a more bellicose president in the White House during such tumultuous times might well have produced a more dangerous result. We cannot know for certain what a triumphalist, a more hawkish, or a or more virulently nationalist president would have meant for global security in 1989 and immediately after. But the world, fortuitously, had a prudent practitioner of Hippocratic diplomacy in office instead. He was neither creative nor innovative, neither a radical nor a revolutionary, but was instead content to follow "what worked." This is what made him a success. This book began with Otto von Bismarck. So too should it end. "The best a statesman can do is to listen to the footsteps of God," the Prussian strategist intoned long before Bush's birth, "get hold of the hem of his cloak, and walk with him a few steps of the way." George H. W. Bush embodied this best statesman-like virtue. He rode the stream of history. And we all survived the Cold War's surprisingly peaceful end.

A NOTE ON SOURCES

Resources and writing on the Cold War's end have flourished in recent years, catalyzed by a plethora of new sources from both sides of the now defunct Iron Curtain. While not intended as an exhaustive list, what follows is a guide to this book's sources, and also a map for subsequent researchers. This volume relies heavily on documents from the George Bush Presidential Library. Once declassified, all are publicly available. Other important resources include the National Security Archive with its bounty of electronic briefing books and its magisterial collections, such as Svetlana Savranksaya, Tom Blanton, and Vladislav Zubok's edited volume *Masterpieces of History: The Peaceful End of the Cold War in Europe* (New York: Central European University Press, 2010); the Woodrow Wilson Center Digital Archive and associated Cold War International History Project; and the Roy Rosenzweig Center for History and New Media, in particular its *Making the History of 1989* databases. The State Department's Virtual Reading Room offers useful access to American sources, while the Margaret Thatcher Archive complements British and widely culled non-British sources.

Memoirs from this period are particularly plentiful. The list must begin with George Bush and Brent Scowcroft, *A World Transformed* (New York: Vintage Books, 1998), which remains at the top of any ranking of post-presidential writings. Bush's *All the Best: My Life in Letters and Other Writings* (New York: Scribner, 2013) sheds light on his personal style, as no doubt will Jon Meacham's forthcoming edited collection of Bush's White House diaries, which are thankfully previewed in his own insightful *Destiny and Power: The American Odyssey of George Herbert Walker Bush* (New York: Random House, 2015). Other useful writings from Bush administration alumni include James Baker, *The Politics of Diplomacy* (New York: Putnam, 1995) and *Work Hard, Study . . . and Keep Out of Politics!* (New York: G. P. Putnam's Sons, 2006); Robert Gates, *From the Shadows: The Ultimate Insider's Story of Five Presidents and How They Won the Cold War* (New York: Simon & Schuster, 1997); G. Jonathan Greenwald, *Berlin Witness: An American Diplomat's Chronicle of East Germany's Revolution* (University Park: Pennsylvania State University Press, 1993); Richard Haass, *War of Necessity, War of Choice: A Memoir of Two Iraq Wars* (New York: Simon & Schuster, 2010); Robert Hutchings, *American Diplomacy and the End of the Cold War* (Washington, D.C.: Woodrow Wilson Center Press, 1998); James Lilley, *The China Hands: Nine Decades of Adventure, Espionage, and Diplomacy in Asia* (New York: PublicAffairs, 2004); Jack Matlock, *Autopsy on an Empire: The American Ambassador's Account of the Collapse of the Soviet Union* (New York: Random House, 1995); Roman Popadiuk, *The Leadership of George Bush* (College Station: Texas A&M University Press, 2009); Colin Powell, *My American Journey* (New York: Ballantine, 1996); Dennis Ross, *The Missing Piece: The Inside Story of the Fight for Middle East Peace* (New York: Farrar, Straus and Giroux, 2004); Louis Sell, *From Washington to Moscow* (Durham: Duke University Press, 2016); John Sununu, *The Quiet Man: The Indispensable Presidency of George H. W. Bush* (New York: Broadside Books, 2015); and last but certainly not least, Chase Untermeyer, *Zenith: In the White House with George H. W. Bush* (College Station: Texas A&M University Press, 2016), which makes one wish all policy makers were as diligent, and elegant, in recording their deeds. This book also benefited greatly from the personal writings generously shared by Ambassador Larry Napper, formerly of the State Department and more recently (and once more) of Texas A&M University.

While not a memoir per se, the work of administration veterans Philip Zelikow and Condoleezza Rice, *Germany Unified and Europe Transformed: A Study in Statecraft* (Cambridge: Harvard University Press, 1997), remains some twenty years after publication the single best study of German reunification. Firsthand American accounts are also to be found in the comprehensive oral history program conducted by the University of Virginia's Miller Center for Public Affairs, and in the underutilized but illuminating oral history records of the Association for Diplomatic Study and Training, located at the George P. Shultz National Foreign Affairs Training Center. Also not a traditional memoir, Michael Beschloss and Strobe Talbott, *At the Highest Levels: The Inside History of the End of the Cold War* (Boston: Little, Brown, 1994), might just as easily be viewed as a firsthand account, given the authors' immediate insider access to contemporary policy makers. Other useful journalistic accounts include David Halberstam, *War in a Time of Peace: Bush, Clinton, and the Generals* (New York: Touchstone, 2001); and Don Oberdorfer, *From the Cold War to a New Era* (Baltimore: Johns Hopkins University Press, 1998), originally published in 1992 as *The Turn*. Neither oral histories nor memoirs should be trusted absent verification, but they offer at once firsthand color typically unavailable merely from government documents and, more important, a sense of what the policy maker valued, at least at the moment of composition. This melding of memory and insight is done particularly well in William C. Wohlforth's edited volume *Cold War Endgame: Oral History, Analysis, Debates* (University Park: Pennsylvania State University Press, 2003).

Non-American voices have also joined the chorus of memoirs from former policy makers. For Soviet (and successor state) officials, see Valery Boldin, *Ten Years That*

Shook the World: The Gorbachev Era as Witnessed by His Chief of Staff (New York: HarperCollins, 1994); Anatoly Chernyaev, *My Six Years with Gorbachev* (University Park: Pennsylvania State University Press, 2000); Mikhail Gorbachev, *Memoirs* (New York: Doubleday, 1995); Mikhail Gorbachev and Zdenek Mlynar, *Conversations with Gorbachev* (New York: Columbia University Press, 2002); Andrei Grachev, *Gorbachev's Gamble: Soviet Foreign Policy and the End of the Cold War* (Cambridge: Polity Press, 2008); Yegor Ligachev, *Inside Gorbachev's Kremlin* (Boulder: Westview Press, 1996); Pavel Palazchenko, *My Years with Gorbachev and Shevardnadze* (University Park: Pennsylvania State University Press, 1997); Yevgeny Primakov, *Russian Crossroads: Toward the New Millennium* (New Haven: Yale University Press, 2004); and Boris Yeltsin, *Midnight Diaries* (New York: PublicAffairs, 2000). Other European memoirs of note include Hans-Dietrich Genscher, *Rebuilding a House Divided* (New York: Broadway Books, 1998); Margaret Thatcher, *The Downing Street Years* (New York: HarperCollins, 2011); George Urban, *Diplomacy and Disillusion at the Court of Margaret Thatcher* (London: I. B. Tauris, 1996); and Lech Walesa, *The Struggle and the Triumph* (New York: Arcade Publishing, 1992). Brian Mulroney's Canadian perspective, *Memoirs* (Toronto: McClelland and Stewart, 2007), offers a window into Bush's diplomacy, while Ji Chaozhu, *The Man on Mao's Right* (New York: Random House, 2008); Qian Quichen, *Ten Episodes in China's Diplomacy* (New York: HarperCollins, 2005); and Zhao Ziyang, *Prisoner of the State: The Secret Journal of Premier Zhao Ziyang* (New York: Simon & Schuster, 2009), provide uncommon access to developments in Beijing.

For George Bush, both biographies and general studies of his presidency, after first consulting Meacham's encyclopedic and insightful work, see the first and still best analysis of Bush's personality, Richard Ben Cramer, *What It Takes: The Way to the White House* (New York: Vintage Books, 1993) and *Being Poppy: A Portrait of George Herbert Walker Bush* (New York: Simon & Schuster, 1992). Then see Ryan Barilleux and Mark J. Rozell, *Power and Prudence: The Presidency of George H. W. Bush* (College Station: Texas A&M University Press, 2004); Meena Bose and Rosanna Perotti, *From Cold War to New World Order: The Foreign Policies of George H. W. Bush* (Westport, Conn.: Greenwood, 2002); Jeffrey A. Engel, *The China Diary of George H. W. Bush* (Princeton: Princeton University Press, 2008); John Robert Greene, *The Presidency of George H. W. Bush* (Lawrence: University Press of Kansas, 2015); Lori Cox Han, *A Presidency Upstaged: The Public Leadership of George H. W. Bush* (College Station: Texas A&M University Press, 2011); Martin Medhurst, ed., *The Rhetorical Presidency of George H. W. Bush* (College Station: Texas A&M University Press, 2006); David Mervin, *George Bush and the Guardianship Presidency* (New York: St. Martin's Press, 1998); Tim Naftali, *George H. W. Bush* (New York: Henry Holt, 2007); Herbert S. Parmet, *George Bush: The Life of a Lone Star Yankee* (New York: Scribner, 1997); Curt Smith, *George H. W. Bush: Character at the Core* (Lincoln: University of Nebraska Press, 2014); and Tom Wicker, *George Herbert Walker Bush* (New York: Viking, 2004). In a class of its own for understanding the man at Bush's right hand is Bartholomew Sparrow, *Brent Scowcroft and the*

Call of National Security (New York: Public Affairs, 2015). See also the more concise but no less insightful David Schmitz, *Brent Scowcroft: Internationalism and Post–Vietnam War American Foreign Policy* (Lanham, Md.: Rowman & Littlefield, 2011).

For those interested in Bush's domestic legacy, and for American society during the broad transition from the 1980s to the 1990s, see Sidney Blumenthal, *Pledging Allegiance: The Last Campaign of the Cold War* (New York: HarperCollins, 1990); Thomas Edsall with Mary D. Edsall, *Chain Reaction: The Impact of Race, Rights, and Taxes on American Politics* (New York: W. W. Norton, 1992); Richard Himelfarb and Rosanna Perotti, eds., *Principle Over Politics? The Domestic Policy of the George H. W. Bush Administration* (Westport, Conn.: Praeger, 2004); Michael Nelson and Barbara Perry, eds., *41: Inside the Presidency of George H. W. Bush* (Ithaca: Cornell University Press, 2014); James Patterson, *Restless Giant: The United States from Watergate to Bush v. Gore* (New York: Oxford University Press, 2005); Michael Schaller, *Right Turn: American Life in the Reagan-Bush Era, 1980–1992* (New York: Oxford University Press, 2007); and Sean Wilentz, *The Age of Reagan: A History, 1974–2008* (New York: HarperCollins, 2007). For the intersection of domestic and foreign politics during this period, see Colin Dueck, *Hard Line: The Republican Party and U.S. Foreign Policy Since World War II* (Princeton: Princeton University Press, 2010); and Julian Zelizer, *Arsenal of Democracy: The Politics of National Security from World War II to the War on Terrorism* (New York: Basic Books, 2010). Future researchers would benefit greatly from following the insights into American military and imperial culture, and its political implications, laid down through investigation of this period by Andrew Bacevich, *The New American Militarism: How Americans Are Seduced by War* (New York: Oxford University Press, 2005) and *American Empire: The Realities and Consequences of U.S. Diplomacy* (Cambridge: Harvard University Press, 2002); Richard Immerman, *Empire for Liberty: A History of American Imperialism from Benjamin Franklin to Paul Wolfowitz* (Princeton: Princeton University Press, 2010); and Michael Sherry, *In the Shadow of War: The United States Since the 1930s* (New Haven: Yale University Press, 1997).

Histories of American foreign policy at the end of the Cold War (and after) abound. The most useful include Hal Brands, *Making the Unipolar Moment* (Ithaca: Cornell University Press, 2016) and *From Berlin to Baghdad: America's Search for Purpose in the Post–Cold War World* (Lexington: University Press of Kentucky, 2008); David Chollet and Jim Goldgeier, *America Between the Wars* (New York: Public Affairs, 2008); Campbell Craig and Fredrik Logevall, *America's Cold War: The Politics of Insecurity* (Cambridge: Harvard University Press, 2009); John Lewis Gaddis, *We Now Know: Rethinking Cold War History* (New York: Oxford University Press, 1997), *The United States and the End of the Cold War* (New York: Oxford University Press, 1992), *The Cold War: A New History* (New York: Penguin, 2005), and *George F. Kennan: An American Life* (New York: Penguin, 2011); Walter LaFeber, *America, Russia, and the Cold War*, 10th ed. (New York: McGraw-Hill, 2006); Melvyn Lefffler, *For the Soul of Mankind: The United States, the Soviet Union, and the Cold War* (New York: Hill and Wang, 2007);

Norman A. Graebner et al., *Reagan, Bush, Gorbachev: Revisiting the End of the Cold War* (Westport, Conn.: Praeger, 2008); David E. Hoffman, *The Dead Hand* (New York: Anchor, 2009); James Mann, *The Rebellion of Ronald Reagan: A History of the End of the Cold War* (New York: Penguin Books, 2010). In addition, one should not overlook the biographical insights to be gleaned from Mann's brilliant *Rise of the Vulcans: The History of Bush's War Cabinet* (New York: Penguin Books, 2004). See also Jack Matlock, *Reagan and Gorbachev: How the Cold War Ended* (New York: Random House, 2005); Christopher Maynard, *Out of the Shadow: George H. W. Bush and the End of the Cold War* (College Station: Texas A&M University Press, 2008); and James Graham Wilson, *The Triumph of Improvisation* (Ithaca: Cornell University Press, 2014), whose author knows as much about American diplomatic sources as anyone alive, and is thankfully generous with that knowledge.

This book devotes special attention to two other authors from the age, and their political impact: Paul Kennedy, *The Rise and Fall of the Great Powers* (New York: Vintage, 1987); and Francis Fukuyama, *The End of History and the Last Man* (New York: Free Press, 1992). For discussion and similarly themed texts from the same period as the former, see David Calleo, *Beyond American Hegemony* (New York: Basic Books, 1987); James Chace, *Solvency: The Price of Survival* (New York: Vintage Books, 1983); Robert Gilpin, *War and Change in World Politics* (Cambridge: Cambridge University Press, 1983); Walter Russell Mead, *Mortal Splendor: The American Empire in Transition* (Boston: Houghton Mifflin, 1987); and Peter G. Peterson, "The Morning After," *Atlantic Monthly* 260, no. 4 (October 1987): 43–69. Fukuyama's influential work dovetailed with the related and equally influential democratic peace theory, that historical quirk that democratic states rarely fought one another — though on average they actually fight more frequently than other states. The idea nevertheless pervaded post–Cold War discussions of American policy. For contemporary and countervailing discussion, see Christopher Layne, "Kant or Cant: The Myth of the Democratic Peace," *International Security* 19, no. 2 (Fall 1994): 5–49; Bruce Russett, *Grasping the Democratic Peace* (Princeton: Princeton University Press, 1994); and the subsequent critiques found in Steven Hook, ed., *Democratic Peace in Theory and Practice* (Kent, Ohio: Kent State University Press, 2010); and Ellen Schrecker, ed., *Cold War Triumphalism: The Misuse of History After the Fall of Communism* (New York: New Press, 2004).

For Soviet politics and policies during this period, with particular attention of course to Mikhail Gorbachev, see Archie Brown and Lilia Shevstova, eds., *Gorbachev, Yeltsin, Putin: Political Leadership in Russia's Transition* (Washington, D.C.: Carnegie Endowment Press, 2001); Archie Brown, *The Fall of Communism* (New York: HarperCollins, 2009), *The Gorbachev Factor* (New York: Oxford University Press, 1996), and *Seven Years That Changed the World: Perestroika in Perspective* (New York: Oxford University Press, 2007); Dick Combs, *Inside the Soviet Alternate Universe* (University Park: Pennsylvania State University Press, 2008); Anthony D'Agostino, *Gorbachev's Revolution* (New York: New York University Press, 1998); Carolyn Ekedahl and Melvin Goodman, *The Wars of Eduard Shevardnadze* (University Park: Pennsylvania State University Press, 1997); Robert D. English, *Russia and the Idea of the West* (New York: Columbia University Press, 2000); Masha Gessen, *The Man Without a Face: The Unlikely Rise of Vladimir Putin* (New York: Penguin Books, 2012); Jonathan Haslam, *Russia's Cold War* (New Haven: Yale University Press, 2011); Steven Kotkin, *Armageddon Averted: The Soviet Collapse, 1970–2000* (New York: Oxford University Press, 2008), and *Uncivil Society: 1989 and the Implosion of the Communist Establishment* (New York: Random House, 2009); Jeffrey Mankoff, *Russian Foreign Policy: The Return of Great Power Politics* (Lanham, Md.: Rowman & Littlefield, 2009); Chris Miller, *The Struggle to Save the Soviet Economy: Mikhail Gorbachev and the Collapse of the USSR* (Chapel Hill: University of North Carolina Press, 2016); William E. Odom, *The Collapse of the Soviet Military* (New Haven: Yale University Press, 2000); Serhii Plokhy, *The Last Empire: The Final Days of the Soviet Union* (New York: Basic Books, 2014); Sergey Radchenko, *Unwanted Visionaries: Soviet Failure in Asia, 1982–1991* (New York: Oxford University Press, 2014); Mark Lawrence Schrad, *Vodka Politics* (New York: Oxford University Press, 2014); Daniel Treisman, *The Return: Russia's Journey from Gorbachev to Medvedev* (New York: Free Press, 2011); and Vladislav Zubock, *A Failed Empire: The Soviet Union in the Cold War from Stalin to Gorbachev* (Chapel Hill: University of North Carolina Press, 2007). Particularly insightful for the post-Soviet transition and continuities of Russian geopolitical thinking is Marvin Kalb, *Imperial Gamble: Putin, Ukraine, and the New Cold War* (Washington, D.C.: Brookings Institution Press, 2015). The historical profession, and myself in particular, are indebted to Thomas J. McCormick's insights into the continuity of American hegemonic aspirations, best laid out in *America's Half-Century* (Baltimore: Johns Hopkins University Press, 1995). Their Cold War origins are best discussed in Melvin P. Leffler, *A Preponderance of Power* (Palo Alto: Stanford University Press, 1992).

For Germany and its contentious unification, see Frédéric Bozo, Andreas Rödder, and Mary Elise Sarotte, eds., *German Reunification: A Multinational History* (New York: Routledge, 2017); Frédéric Bozo, *Mitterrand, the End of the Cold War, and German Unification* (New York: Berghahn Books, 2009); David Childs, *The Fall of the GDR* (Harlow, Essex: Pearson Education, 2001); Frank Costigliola, "An 'Arm Around the Shoulder': The United States, NATO, and German Reunification, 1989–1990," *Contemporary European History* 3, no. 1 (March 1994): 87–110, which insightfully perceived the basic dynamics of the story a generation before its classified documentary record became available; Gareth Dale, *Popular Protest in East Germany, 1945–1989* (New York: Routledge, 2005) and *The East German Revolution of 1989* (Manchester: Manchester University Press, 2006); Richard Gray and Sabine Wilke, *German Unification and Its Discontents* (Seattle: University of Washington Press, 1996); Konrad H. Jarausch, *The Rush to German Unity* (New York: Oxford University Press, 2004); Charles Maier, *Dissolution: The Crisis of Communism and the End of East Germany* (Princeton: Princeton University Press, 1997) and *The Unmasterable Past: History, Holocaust, and German*

National Identity (Cambridge: Harvard University Press, 1988); James McAdams, *Germany Divided: From the Wall to Reunification* (Princeton: Princeton University Press, 1993); Angela Stent, *Russia and Germany Reborn: Unification, the Soviet Collapse, and the New Europe* (Princeton: Princeton University Press, 1999); Stephen F. Szabo, *The Diplomacy of German Unification* (New York: St. Martin's Press, 1992); Frederick Taylor, *The Berlin Wall* (New York: HarperCollins, 2006); and Alexander Von Plato, *The End of the Cold War? Bush, Kohl, Gorbachev, and the Reunification of Germany* (New York: Palgrave Macmillan, 2015). Brilliant as a documentary source is Patrick Salmon et al., eds., *Documents on British Policy Overseas*, series 3, vol. 7, *German Unification, 1989–1990* (New York: Routledge, 2010).

For the end of the Cold War from a European perspective, see Timothy Garton Ash, *The Magic Lantern: The Revolution of '89 Witnessed in Warsaw, Budapest, Berlin, and Prague* (New York: Vintage, 1993), *History of the Present: Essays, Sketches, and Dispatches from Europe in the 1990s* (New York: Vintage, 2001), and *In Europe's Name: Germany and the Divided Continent* (New York: Vintage, 1994); Frédéric Bozo et al., eds., *Europe and the End of the Cold War: A Reappraisal* (New York: Routledge, 2008); Michael Dobbs, *Down with Big Brother: The Fall of the Soviet Empire* (New York: Knopf, 1997); Saki Ruth Dockrill, *The End of the Cold War Era* (London: Hodder Arnold, 2005); and Gregory F. Domber, *Empowering Revolution: America, Poland, and the End of the Cold War* (Chapel Hill: University of North Carolina Press, 2014), the single best work on Poland's diplomacy and democratic transition. See also Misha Glenny, *The Balkans: Nationalism, War, and the Great Powers* (New York: Penguin Books, 2012) and *The Rebirth of History: Eastern Europe in the Age of Democracy* (New York: Penguin, 1990); Richard K. Herrmann and Richard Ned Lebow, eds., *Ending the Cold War* (New York: Palgrave Macmillan, 2004); William I. Hitchcock, *The Struggle for Europe: The Turbulent History of a Divided Continent, 1945–Present* (New York: Anchor, 2003); Tony Judt, *Postwar: A History of Europe Since 1945* (New York: Penguin, 2005); Padraic Kenney, *A Carnival of Revolutions: Central Europe 1989* (Princeton: Princeton University Press, 2002); Mark Kramer, "The Collapse of East European Communism and the Repercussions Within the Soviet Union," parts 1 and 2 and 3, *Journal of Cold War Studies* 5, no. 4 (2003): 178–256, 6, no. 4 (2004): 3–64, and 7, no. 1, 3–96; Michael Meyer, *The Year That Changed the World* (New York: Scribner, 2009); Frances Millard, *The Anatomy of the New Poland* (Cambridge: Edward Elgar Publishing, 1994); John Prados, *How the Cold War Ended: Debating and Doing History* (Washington, D.C.: Potomac Books, 2011); Constantine Pleshakov, *There Is No Freedom Without Bread! 1989 and the Civil War That Brought Down Communism* (New York: Farrar, Straus and Giroux, 2009); Victor Sebestyen, *Revolution 1989: The Fall of the Soviet Empire* (New York: Pantheon, 2009); Robert Service, *The End of the Cold War, 1985–1991* (New York: Public Affairs, 2015); Peter Siani-Davies, *The Romanian Revolution of December 1989* (Ithaca: Cornell University Press, 2005); Gale Stokes, *The Walls Came Tumbling Down: Collapse and Rebirth in Eastern Europe* (New York: Oxford University Press, 2012); A.

Kemp Welch, *Poland Under Communism: A Cold War History* (Cambridge: Cambridge University Press, 2008); and Michael Zantovsky, *Havel: A Life* (New York: Grove Press, 2014).

For a global rather than a European perspective on these events, see Odd Arne Westad, *The Global Cold War* (Cambridge: Cambridge University Press, 2007), which offers a model for how to write international history. See also Jeffrey A. Engel, *The Fall of the Berlin Wall: The Revolutionary Legacy of 1989* (New York: Oxford University Press, 2009); Padraic Kenney, *1989: Democratic Revolutions at the Cold War's End* (New York: Bedford St. Martin's, 2010); and Christopher Marsh, *Unparalleled Reforms: China's Rise, Russia's Fall, and the Interdependence of Transition* (Lanham, Md.: Lexington Books, 2005).

Mary Sarotte's work deserves a category of its own. No one has done more to illuminate the interplay of Germany's internal and international history during this period, or done so more insightfully. See in particular *The Collapse: The Accidental Opening of the Berlin Wall* (New York: Basic Books, 2014) and *1989: The Struggle to Create Post–Cold War Europe* (Princeton: Princeton University Press, 2009); "Perpetuating US Preeminence: The 1990 Deals to 'Bribe the Soviets Out' and Move NATO In," *International Security* (2010), 110–137; "A Broken Promise? What the West Really Told Moscow About NATO Expansion," *Foreign Affairs* (2014): 90–97; "Not One Inch Eastward: Bush, Baker, Kohl, Genscher, Gorbachev, and the Origin of Russian Resentment Toward NATO Enlargement in February 1990," *Diplomatic History* (2010): 119–40; and "In Victory, Magnanimity: US Foreign Policy, 1989–1991, and the Legacy of Prefabricated Multilateralism," *International Politics* (2011): 482–95.

NATO's survival during the Cold War's end, and its ensuing expansion, have generated a pocket industry for contesting historians, who have worked to both uncover and reinterpret new materials with potentially explosive subsequent political impact. In addition to Sarotte, see also Ronald D. Asmus, *Opening NATO's Door* (New York: Columbia University Press, 2004); Daniel Deudney and G. John Ikenberry, "The Unraveling of the Cold War Settlement," *Survival* 51, no. 6 (December 2009): 39–62; James M. Goldgeier, *Not Whether but When: The U.S. Decision to Enlarge NATO* (Washington, D.C.: Brookings Institution Press, 1999); Mark Kramer, "The Myth of a No-NATO Enlargement Pledge to Russia," *Washington Quarterly* 32, no. 2 (April 2009): 39–60; Mark Kramer and Mary Sarotte, letters to the editor, *Foreign Affairs* 93, no. 6 (September–October 2014): 208; Michael McGwire, "NATO Expansion: 'A Policy Error of Historic Importance,'" *Review of International Studies* 24, no. 1 (January 1998): 23–42; Joshua R. Itzkowitz Shifrinson, "Deal or No Deal? The End of the Cold War and the Offer to Limit NATO Expansion," *International Security* 40, no. 4 (Spring 2016): 7–44, and "Put It in Writing: How the West Broke Its Promise to Moscow," *Foreign Affairs Snapshot:* October 29, 2014. Particularly useful when published will be Shifrinson's *Falling Giants: Rising States and the Decline of Great Powers.*

See also Michael McFaul, "Moscow's Choice," *Foreign Affairs* 93, no. 6 (December

2014): 167–71; John Mearsheimer, "Why the Ukraine Crisis Is the West's Fault: The Liberal Delusions That Provoked Putin," *Foreign Affairs* 93, no. 5 (October 2014): 77–89; Walter B. Slocombe, "A Crisis of Opportunity: The Clinton Administration and Russia," in *In Uncertain Times: American Foreign Policy After the Berlin Wall and 9/11*, eds. Melvyn Leffler and Jeffrey Legro (Ithaca: Cornell University Press, 2011); Kristina Spohr, "Precluded or Precedent-Setting? The 'NATO Enlargement Question' in the Triangular Bonn-Washington-Moscow Diplomacy of 1990–1991," *Journal of Cold War Studies* 14, no. 4 (October 2012): 4–54, and "Germany, America, and the Shaping of Post–Cold War Europe," *Cold War History* 15, no. 2 (April 2015): 221–43. Enlightening for NATO's expansion in a broader framework is James Goldgeier and Michael McFaul, *Power and Purpose: U.S. Policy Toward Russia After the Cold War* (Washington, D.C.: Brookings Institution Press, 2003).

Too often overlooked by Cold War scholars because of the cacophony of events in Eastern Europe, China's history of 1989 may prove more meaningful as the twenty-first century unfolds. For U.S.-Chinese relations during the broad Cold War period and after, see Warren I. Cohen, *America's Response to China: A History of Sino-American Relations* (New York: Columbia University Press, 2000); Rosemary Foot, *The Practice of Power: U.S. Relations with China Since 1949* (New York: Oxford University Press, 1997); David M. Lampton, *Same Bed, Different Dreams: Managing U.S.-China Relations, 1989–2000* (Berkeley: University of California Press, 2001); James Mann, *About Face: A History of America's Curious Relationship with China from Nixon to Clinton* (New York: Alfred A. Knopf, 1999) and *The China Fantasy: How Our Leaders Explain Away Chinese Repression* (New York: Viking, 2007); Andrew J. Nathan and Andrew Scowbell, *China's Search for Security* (New York: Columbia University Press, 2012); Richard H. Solomon, *Chinese Negotiating Behavior* (Washington, D.C.: United States Institute of Peace, 1999); Robert L. Suettinger, *Beyond Tiananmen: The Politics of U.S.-China Relations, 1989–2000* (Washington, D.C.: Brookings Institution Press, 2003); Patrick Tyler, *A Great Wall: Six Presidents and China* (New York: Public Affairs, 1999); Dong Wang, *The United States and China* (Lanham, Md.: Rowman & Littlefield, 2013); Odd Arne Westad, *Restless Empire: China and the World Since 1750* (London: Random House, 2013), adds much-appreciated long-term perspective.

Tiananmen looms large in this story. Essential is Zhang Liang, Andrew J. Nathan, and Perry Link, *The Tiananmen Papers: The Chinese Leadership's Decision to Use Force Against Their Own People* (New York: Public Affairs, 2001). The book is not without its critics, which the *China Quarterly* aired, with rejoinders, in 2001 and 2004. For Tiananmen accounts and histories of the Deng period in particular, see Craig Calhoun, *Neither Gods nor Emperors: Students and the Struggle for Democracy in China* (Berkeley: University of California Press, 1997); Eddie Chang, *Standoff at Tiananmen* (Highlands Ranch, Colo.: SensyCorp, 2009); Zhao Dingxin, *The Power of Tiananmen* (Chicago: University of Chicago Press, 2001); Merle Goldman, *Sowing the Seeds of Democracy in China: Political Reform in the Deng Xiaoping Era* (Cambridge: Harvard University

Press, 1994); Steven M. Goldstein, *China at the Crossroads: Reform After Tiananmen* (New York: Foreign Policy Association, 1992); Louisa Lim, *The People's Republic of Amnesia: Tiananmen Revisited* (New York: Oxford University Press, 2007); Richard Madsen, *China and the American Dream: A Moral Inquiry* (Berkeley: University of California Press, 1995); Maurice Meisner, *The Deng Xiaoping Era* (New York: Hill and Wang, 1996); David Shambaugh, *China's Communist Party: Atrophy and Adaptation* (Berkeley: University of California Press, 2008); Jeffrey N. Wasserstrom and Elizabeth J. Perry, eds., *Popular Protest and Political Culture in Modern China* (Boulder: Westview Press, 1994); Wei-Wei Zhang, *Ideology and Economic Reform Under Deng Xiaoping, 1978–1993* (London: Kegan Paul International, 1996). Particularly good on the early years of one of the century's most influential figures, before Tiananmen, are Alexander V. Pantsov with Steven I. Levine, *Deng Xiaoping: A Revolutionary Life* (New York: Oxford University Press, 2015); and Ezra F. Vogel, *Deng Xiaoping and the Transformation of China* (Cambridge: Belknap Press, 2011).

Memories of China's Cultural Revolution played an inordinate role in the events of 1989. See Alexander C. Cook, *The Cultural Revolution on Trial* (Cambridge: Cambridge University Press, 2016); Frank Dikötter, *The Cultural Revolution: A People's History, 1962–1976* (New York: Bloomsbury Press, 2016); Feng Jicai, *Ten Years of Madness: Oral Histories of China's Cultural Revolution* (San Francisco: China Books & Periodicals, 1996); Daniel Leese, *Mao's Cult: Rhetoric and Ritual in China's Cultural Revolution* (Cambridge: Cambridge University Press, 2013); Roderick MacFarquhar and Michael Schoenals, *Mao's Last Revolution* (Cambridge: Belknap Press, 2006); Maurice Meisner, *Mao's China and After: A History of the People's Republic* (New York: Free Press, 1999).

The Middle East mattered to Bush's presidency as well, though military and cultural histories of the Gulf War have outpaced studies of its diplomacy, which are more burdened by issues of declassification. See Rick Atkinson, *Crusade: The Untold Story of the Persian Gulf War* (Boston: Houghton Mifflin, 2003); Michael Gordon and Bernard Trainor, *The Generals' War: The Inside Story of the First Gulf War* (New York: Atlantic Books, 2006); Susan Jeffords and Lauren Rabinovitz, eds., *Seeing Through the Media: The Persian Gulf War* (New Brunswick: Rutgers University Press, 1994); Richard S. Lowry, *The Gulf War Chronicles* (New York: I Universe, 2003); John R. MacArthur, *Second Front: Censorship and Propaganda in the 1991 Gulf War* (Berkeley: University of California Press, 2004); John Andreas Olson, *John Warden and the Renaissance of American Air Power* (Washington, D.C.: Potomac Books, 2007); Keith L. Shimko, *The Iraq Wars and America's Military Revolution* (New York: Cambridge University Press, 2010).

For Gulf War diplomacy and strategic relations, see Michael F. Cairo, *The Gulf: The Bush Presidencies and the Middle East* (Lexington: University Press of Kentucky, 2012); Jeffrey A. Engel, ed., *Into the Desert: Reflections on the Gulf War* (New York: Oxford University Press, 2011); Lawrence Freedman, *A Choice of Enemies: America Confronts the Middle East* (New York: Public Affairs, 2008); Lawrence Freedman and Efraim

Karsh, *The Gulf Conflict, 1990–1991* (Princeton: Princeton University Press, 1995); Dilip Hiro, *Desert Shield to Desert Storm* (New York: Paladin, 1992); Steven A. Yetiv, *Explaining Foreign Policy: U.S. Decision-Making in the Gulf Wars* (Baltimore: Johns Hopkins University Press, 2011).

Much credit is due the remarkable effort by the United States Joint Forces Command to capture, collate, translate, and make available documents from Saddam Hussein's regime and then, more important, to enable scholars under their employ to write thoughtful independent analyses. See, for example, Kevin M. Woods, *The Mother of All Battles: Saddam Hussein's Strategic Plan for the Persian Gulf War* (Annapolis: Naval Institute Press, 2008); and Kevin M. Woods et al., eds., *The Saddam Tapes: The Inner Workings of a Tyrant's Regime, 1978–2001* (Cambridge: Cambridge University Press, 2011).

For U.S. relations with the Middle East, and with Iraq in particular, leading up to and through the Gulf War, see Andrew J. Bacevich, *America's War for the Greater Middle East* (New York: Random House, 2016); Lloyd Gardner, *The Long Road to Baghdad: A History of U.S. Foreign Policy from the 1980s to the Present* (New York: New Press, 2008); Bruce W. Jentleson, *With Friends Like These: Reagan, Bush, and Saddam, 1982–1990* (New York: W. W. Norton, 1994); William B. Quandt, *Peace Process: American Diplomacy and the Arab-Israeli Conflict Since 1967* (Washington, D.C.: Brookings Institution Press, 2005); Ray Takeyh and Steven Simon, *The Pragmatic Superpower: Winning the Cold War in the Middle East* (New York: W. W. Norton, 2016). Always useful when considering the Middle East is its principal (perhaps solitary) geostrategic importance, thus requiring Daniel Yergin, *The Prize: The Epic Quest for Oil, Money, and Power* (New York: Free Press, 2009). Particularly good on the Gulf War's long-term effects for 9/11 and the ensuing cycle of terror and response is Lawrence Wright, *The Looming Tower: Al-Qaeda and the Road to 9/11* (New York: Random House, 2006).

NOTES

INTRODUCTION

page

2   *"hardline statements"*: George H. W. Bush and Brent Scowcroft, *A World Transformed* (New York: Vintage Books, 1998), 5.
    "cannons": Ibid.

3   *"Stop the car"*: Michael Beschloss and Strobe Talbott, *At the Highest Levels: The Inside Story of the End of the Cold War* (Boston: Little, Brown, 1993), 8.
    "do this a lot": Frances FitzGerald, *Way Out There in the Blue: Reagan, Star Wars, and the End of the Cold War* (New York: Simon & Schuster, 2000), 428.
    "Leaders should be equal": Beschloss and Talbott, *At the Highest Levels,* 9.

5   *"They took it"*: William Hitchcock, *The Struggle for Europe: The Turbulent History of a Divided Continent, 1945–Present* (New York: Anchor Books, 2003), 359.

7   *"moving our way"*: *Public Papers of the Presidents of the United States: George Bush* (Washington, D.C.: Government Printing Office, 1989–1993), "Interview with Foreign Journalists," November 21, 1989. Unless otherwise stated, I have relied on the digital version of the public papers available at the American Presidency Project, University of California, Santa Barbara (www.presidency.ucsb.edu). Hereafter cited as *PPP* with identifying information.
    "role of the captain": Daniel Treisman, *The Return: Russia's Journey from Gorbachev to Medvedev* (New York: Free Press, 2011), 391.

9   *"try men's souls"*: John Armor, ed., *These Are the Times That Try Men's Souls: Then and Now in the Words of Tom Paine* (Indianapolis: Dog Ear Publishing, 2015), 1.
    "freer from the threat": *PPP,* "Address Before a Joint Session of the Congress on the Persian Gulf Crisis and the Federal Budget Deficit," September 11, 1990.

10 *"a better place"*: Jeffrey A. Engel, "'A Better World . . . but Don't Get Carried Away': The Foreign Policy of George H. W. Bush Twenty Years On," *Diplomatic History* 34, no. 1 (2010): 34.

1. SWAN SONG AND SURPRISE

11 *"not really working"*: Molly Worthen, *The Man on Whom Nothing Was Lost: The Grand Strategy of Charles Hill* (New York: Houghton Mifflin Harcourt, 2006), 214.
    "can't tell": University of Virginia, Miller Center for Public Affairs Oral History Project, William Webster interview, August 21, 2002, 27. Hereafter Miller Center oral history with identifying information.

A majority believed: For polling, see Robert J. Samuelson, "The Economy Deserts Bush," *Newsweek,* May 30, 1988.

12 *"ready to criticize":* Margaret Thatcher Archive, Margaret Thatcher Foundation, Thatcher to Reagan, December 4, 1986.

13 *"unusual Russian":* Margaret Thatcher Archive, "Memorandum of Conversation Between Thatcher and Reagan," December 28, 1985.

"form of insanity": Kiron Skinner et al., eds., *Reagan, In His Own Hand* (New York: Simon & Schuster, 2001), 12. See also Melvyn Leffler, *For the Soul of Mankind* (New York: Hill and Wang, 2008), 339.

"no Soviet leader": PPP, "The President's News Conference," January 29, 1981.

14 *"days as a passenger":* David Arbel and Ran Edelist, *Western Intelligence and the Collapse of the Soviet Union, 1980–1990* (New York: Routledge, 2006), 146. See also Stansfield Turner, *Burn Before Reading* (New York: Hyperion, 2005).

"spoken world domination": Martin Anderson, *Reagan's Secret War: The Untold Story of His Fight to Save the World from Nuclear Disaster* (New York: Three Rivers Press, 2009), 63; Jason Saltoun-Ebin, *The Reagan Files* (CreateSpace, 2010), 40.

"will invest": Arbel and Edelist, *Western Intelligence and the Collapse of the Soviet Union,* 145.

"struggle": PPP, "Remarks at the Annual Convention of the National Association of Evangelicals in Orlando, Florida," March 8, 1983.

"crusade of freedom": PPP, "Address to Members of the British Parliament," May 8, 1982.

"malice aforethought": James Mann, *The Rebellion of Ronald Reagan* (New York: Penguin Books, 2009), 29.

"enjoined": PPP, "Remarks at the Annual Convention of the National Association of Evangelicals in Orlando, Florida," March 8, 1983.

"godless": PPP, "Remarks at the Centennial Meeting of the Supreme Council of the Knights of Columbus," August 3, 1982.

"beginning of the Reagan presidency": Beatrice Heuser, "The Soviet Response to the Euromissile Crisis, 1982–83," in *The Crisis of Détente in Europe: From Helsinki to Gorbachev, 1975–1985,* ed. Leopoldo Nuti (London: Routledge, 2008), 37–149.

15 *Strategic Defense Initiative:* Steven Weisman, "Reagan Proposes U.S. See New Way to Block Missiles," *New York Times,* March 24, 1983.

'crusade': Vladislav Zubok, *A Failed Empire: The Soviet Union in the Cold War from Stalin to Gorbachev* (Chapel Hill: University of North Carolina Press, 2007), 275.

false alarm: National Security Archive, *The 1983 War Scare Declassified and for Real,* Electronic Briefing Book no. 533, "The Soviet 'War Scare,' President's Foreign Policy Intelligence Advisory Board," February 19, 1990.

how close to the abyss: For nuclear scares, see Ben Fischer, *A Cold War Conundrum: The 1983 Soviet War Scare* (Washington, D.C.: Center for the Study of Intelligence, 1997); and Nate Jones, *Able Archer 83: The Secret History of the NATO Exercise That Almost Triggered Nuclear War* (New York: New Press, 2016).

16 *"confirms our worst fears":* Woodrow Wilson Center Digital Archive, Office of the Federal Commissioner for the Stasi Records, MfS, ZAIG 5382, 1–19, document record 115717.

"Armageddon": Nigel Hey, *The Star Wars Enigma* (Washington, D.C.: Potomac Books, 2007), 83. For Soviet-American mutual sense of duplicity, see Zubok, *A Failed Empire,* 274.

"bleeding": David Hoffman, *The Dead Hand: The Untold Story of the Cold War Arms Race and Its Dangerous Legacy* (New York: Anchor, 2010), 233.

17 *"call me a fool":* National Security Archive, *The Reykjavik File,* Electronic Briefing Book no. 203, Russian transcript of Reagan-Gorbachev summit in Reykjavik, October 12, 1986 (afternoon).

"if you will agree": H. W. Brands, *Reagan: The Life* (New York: Anchor Books, 2015), 603.

"very sorry": David Reynolds, *America, Empire of Liberty: A New History* (London: Allen Lane, 2009), 517.

"another time": PPP, "The President's News Conference," June 1, 1988.

"different from previous": Igor Korchilov, *Translating History: Thirty Years on the Front Lines of Diplomacy* (New York: Scribner, 1997), 183.

"prick": Richard Reeves, *President Reagan: The Triumph of Imagination* (New York: Simon & Schuster, 2005), 205.

"keep dying": Jack F. Matlock, *Reagan and Gorbachev: How the Cold War Ended* (New York: Random House, 2005), 36.

18   *"sluggish"*: Leffler, *Soul of Mankind*, 374.

"History punishes": Steven Saxonberg, *The Fall: A Comparative Study of the End of Communism in Czechoslovakia, East Germany, Hungary and Poland* (New York: Routledge, 2001), 139.

19   *"waxing nostalgic"*: PPP, "Radio Address to the Nation on Soviet-American Relations," December 3, 1988. See also Mann, *Rebellion of Ronald Reagan,* 269 and 305.

20   *"change a nation"*: PPP, "Toasts at the State Dinner for British Prime Minister Margaret Thatcher," November 16, 1988.

"jury is still out": Jeffrey A. Engel, "When George Bush Believed the Cold War Ended and Why That Mattered," in *41: Inside the Presidency of George H. W. Bush,* eds. Michael Nelson and Barbara A. Perry (Ithaca: Cornell University Press, 2014), 109.

"not in a Cold War": "Gorbachev Policy Has Ended the Cold War, Thatcher Says," *New York Times,* November 18, 1988.

21   *must be ready*: Margaret Thatcher Archives, "Press Conference," November 17, 1988.

2. BUSH'S RISE

22   *"never apologize"*: Michael Kingsley, "Rally Round the Flag, Boys," *Time,* September 12, 1988.

"world looks": PPP, "Remarks upon Returning from a Trip to the Far East," February 27, 1989.

23   *"undoubtedly"*: Jeffrey A. Engel, *The China Diary of George H. W. Bush* (Princeton: Princeton University Press, 2008), 362.

"detached and distant": Felix Gilbert, *To the Farewell Address* (Princeton: Princeton University Press, 1970), 145.

"last best": PPP, "Second Annual Message," December 1, 1862.

"shining city": PPP, "Remarks Accepting the Presidential Nomination at the Republican National Convention in Dallas, Texas," August 23, 1984.

"still a beacon": PPP, "Farewell Address to the Nation," January 11, 1989.

"American exceptionalism": The history of American exceptionalism is both broad and contentious. For primers and critiques, see Godfrey Hodgson, *The Myth of American Exceptionalism* (New Haven: Yale University Press, 2009); and Owen Ullmann, "Bush Era Promises to Be Low-Key Practical Affair," *Philadelphia Inquirer,* January 20, 1989.

24   *"do have principles"*: Engel, *The China Diary,* 254–55.

"philosophical way": Author interview with Brent Scowcroft, March 8, 2007.

"Shelley and Kant": Academy of Achievement induction interview, June 2, 1995, http://www.achievement.org/autodoc/printmember/bus0int-1 (accessed June 1, 2007). On Bush's language, see Martin Medhurst, "Why Rhetoric Matters: George H. W. Bush in the White House," in *The Rhetorical Presidency of George H. W. Bush,* ed. Martin Medhurst (College Station: Texas A&M University Press, 2006), 3–18, and other essays in the volume, esp. Catherine L. Langford, "George Bush's Struggle with the 'Vision Thing,'" 19–36.

"silver foot": Sylvia Badger, "Gov. Ann Richards Puts Women in Their Place — Right at the Top," *Baltimore Sun,* June 2, 1996.

"third base": Hugh Sidey, "Dad Says 'I Don't Miss Politics,'" *Time,* June 14, 1999.

26 *"trying to do"*: Herbert Parmet, *George Bush* (New York: Scribner, 1997), 193.

"lots of friends": Tom Wicker, *George Herbert Walker Bush* (New York: Penguin, 2004), 69.

"really stand for": Margaret Warner, "Fighting the Wimp Factor," *Newsweek,* October 19, 1987.

"I like what works": David Mervin, *George Bush and the Guardianship Presidency* (New York: St. Martin's Press, 1998), 33.

"voodoo economics": Jon Meacham, *Destiny and Power: The American Odyssey of George Herbert Walker Bush* (New York: Random House, 2015), 235.

"blindly": Sean Wilentz, *The Age of Reagan* (New York: Harper Perennial, 2009), 265.

"looking over his shoulder": Timothy Naftali, *George H. W. Bush* (New York: Times Books, 2007), 42.

27 *"own man"*: Margaret Warner, "Bush Battles the 'Wimp Factor,'" *Newsweek,* October 19, 1988.

"blind trust": Maureen Dowd, "Washington Talk: White House," *New York Times,* May 12, 1989.

"not a leader": Warner, "Bush Battles the 'Wimp Factor.'"

the Republican primaries: For GOP primary critiques, see Lloyd Grove, "Haig Running to Take Charge," *Washington Post,* January 30, 1988.

"loyalty is a strength": "The Reagan-Bush Accord," *New York Times,* November 1, 1987. Bush employed the term "loyalty" frequently. See Bernard Weinraub, "Dole and Bush: Dramatic Contrast of Styles," *New York Times,* February 7, 1988.

"Where are you from": Richard Ben Cramer, *What It Takes: The Way to the White House* (New York: Vintage, 1993), 15.

"lobster with his chili": Wicker, *George Herbert Walker Bush,* 23.

28 *"wonder if you agree"*: George H. W. Bush, *All the Best: My Life in Letters and Other Writings* (New York: Scribner, 2014), 47–48.

29 *"life in chapters"*: Peter Schweizer, *The Bushes* (New York: Anchor Books, 2004), 87.

"massive fist": Parmet, *George Bush,* 56.

"terribly responsible": Naftali, *George H. W. Bush,* 8.

30 *"those two"*: James Bradley, *Flyboys* (New York: Little, Brown, 2006), 197. Bush gave his most extensive interview on the subject to Bradley, though key moments occurred, by the author's admission, after he had shut off his tape recorder.

"What else ya got": Author interview with George H. W. Bush, February 13, 2009.

"us Episcopalians": Andrew Preston, "The Politics of Realism and Religion: Christian Responses to Bush's New World Order," *Diplomatic History* 34, no. 1 (January 2010): 95–118.

"foxhole Christianity": Bradley, *Flyboys,* 197–98.

"punishment": Maureen Dowd, "Bush the Impish Vetoes the Imperial," *New York Times,* December 5, 1988.

31 *"did not originate"*: Bush, *All the Best,* 69.

"We like Texas": Ibid., 70.

"still full of fun": Ibid., 76, 77.

32 *"had one once"*: Ibid., 81.

33 *"work hard, study"*: James Baker, *Work Hard, Study . . . and Keep Out of Politics!* (New York: G. P. Putnam's Sons, 2006).

"Number one": Public Broadcasting System, "American Experience: George H. W. Bush," transcript, http://www.pbs.org/wgbh/americanexperience/features/transcript/bush-transcript/ (accessed June 13, 2013).

"We can take care": Baker, *Work Hard,* 17.

34   *"carved out"*: Miller Center oral history, James A. Baker III, March 17, 2011, 3.

"so smart": Baker, *Work Hard,* 17.

"killer instinct": Robert Beisner, *Dean Acheson: A Life in the Cold War* (New York: Oxford University Press, 2006), 122.

35   *"all Republicans"*: Bush, *All the Best,* 86, 89.

'*crazies*': Parmet, *George Bush,* 96; Cramer, *What It Takes,* 419.

"Listen, don't talk": Author interview with George H. W. Bush, July 7, 2005.

"restrain yourself": Alan Elms, *Uncovering Lives: The Uneasy Alliance of Biography and Psychology* (New York: Oxford University Press, 1994), 214.

"get the hell out": Cramer, *What It Takes,* 419.

36   *"get rid of"*: Michael Nelson, "George Bush: Texan, Conservative," in Nelson and Perry, *41,* 36.

'*nuts*': Bush, *All the Best,* 87, 89–92.

"I regret it": Naftali, *George H. W. Bush,* 15.

"given a choice": Parmet, *George Bush,* 141.

37   *"Every cabinet"*: Ibid., 148; see also Baker, *Work Hard,* 25.

"spell out": Wicker, *George Herbert Walker Bush,* 26–27.

"takes our line:" Engel, *China Diary,* 411.

"crash course": George Bush and Victor Gold, *Looking Forward* (New York: Doubleday, 1987), 110; Cramer, *What It Takes,* 611.

"loser" and "nothing": Bush and Gold, *Looking Forward,* 107–10.

"do you know": Parmet, *George Bush,* 149.

"ten days": Ibid.

"competitive instincts": Engel, *The China Diary,* 412.

38   *Shea*: Joseph Durso, "Mets Beat Expos in Rain, 4–2," *New York Times,* April 7, 1971.

"Everybody said": Author interview with George H. W. Bush, July 7, 2005.

"interesting to see": Bush, *All the Best,* 133.

"come see you": Engel, *The China Diary,* 414.

"Burundi": Author interview with George H. W. Bush, July 7, 2005.

"like people": Robert Alden, "Bush, Leaving UN Post, Is Fearful of Bloc Voting," *New York Times,* December 23, 1972.

"best policy": Bush, *All the Best,* 137.

"difference in approach": Author interview with George H. W. Bush, July 7, 2005.

39   *"wiener nations"*: Cramer, *What It Takes,* 611.

"profound effect": Author interview with Brent Scowcroft, March 8, 2007.

"the trust": Engel, "When George Bush Believed the Cold War Ended," 105.

40   *"ballgame is over"*: Julian Zelizer, *Arsenal of Democracy: The Politics of National Security—from World War II to the War on Terrorism* (New York: Basic Books, 2010), 241.

"Night and day": Bush, *All the Best,* 149–52, 153–55. See also Engel, *The China Diary,* 418–25.

"more realistic and vital": Meacham, *Destiny and Power,* 159.

41   *"Look, you little bastard"*: Author interview with George H. W. Bush, July 7, 2005.

"eliminate": Robert Dallek, *Nixon and Kissinger* (New York: HarperCollins, 2007), 434. See also Parmet, *George Bush,* 149; and Bush, *All the Best,* 133.

"moving back": Parmet, *George Bush,* 163–64.

"escape": Bush, *All the Best,* 176.

42   *France or the United Kingdom*: "We went back and talked about England," Bush recorded of

his conversation with Ford. "He wondered if it was substantive enough — so did I. We talked about the money. I told him I had lost a lot of money and didn't know if I could afford it." See Bush, *All the Best*, 196; and Barry Werth, *31 Days: The Crisis That Gave Us the Government We Have Today* (New York: Random House, 2007), 166.

"*won't hurt anything*": Don Oberdorfer, "China: Change of Pace," *Washington Post*, December 2, 1974. James Lilley, who worked with Bush at the U.S. Liaison Office and who eventually served as his ambassador to China fifteen years later, likened Bush's choice to "pulling the covers over your head, and getting the hell out of Washington." Author interview with James Lilley, March 9, 2007.

"*like you or not*": Engel, *China Diary*, 147, 431.

43   "*rather weird*": Bush Presidential Library, Personal Papers of George H. W. Bush, China File, Correspondence File, box 1, Bush to Allison, November 15, 1974. Hereafter BPL with identifying information.

"*How awesome*": BPL, Personal Papers of George H. W. Bush, China File, Correspondence File, box 1, Bush to Rhodes, November 28, 1974. See also BPL, Personal Papers of George H. W. Bush, China File, Correspondence File, box 1, Bush to Clements, March 2, 1974.

44   "*reneged on commitments*": Engel, *China Diary*, 243.

domino theory: Ibid., 449.

45   "*I am wondering*": For Bush's reaction to South Vietnam's collapse, see ibid., entries for March 2, April 26, April 30, and May 6, 1975.

"*preserve freedom*": Ibid., 370, 451.

46   "*please don't put me*": Parmet, *George Bush*, 210.

"*building my own*": Richard Allen, "George Herbert Walker Bush: The Accidental Vice President," *New York Times Magazine*, July 30, 2000.

47   "*his shoulder*": Parmet, *George Bush*, 257.

"*such surface qualms*": Warner, "Bush Battles the 'Wimp Factor,'" 28. See also Parmet, *George Bush*, 309.

3. GORBACHEV AT THE UN

48   "*and risk*": "Leadership on the Issues: George Bush Elected 41st President," United States Agency for International Development, Office of Press Relations, n.d., http://pdf.usaid.gov/pdf_docs/PNABB299.pdf (accessed March 23, 2017). See also Maureen Dowd, "Bush Lays Out Foreign Policy Tenets," *New York Times*, August 8, 1988; Timothy McNulty, "Bush Faces Changes with Pragmatism," *Chicago Tribune*, March 3, 1989.

"*most dangerous time*": Alexis de Tocqueville, *The Old Regime and the Revolution* (New York: Harper and Brothers, 1856), 214.

49   "*Fulton in reverse*": National Security Archive, Electronic Briefing Book no. 261, "Gorbachev's Conference with Advisers on Drafting the U.N. Speech," October 31, 1988.

"*Iron Curtain*": Fraser J. Harbutt, *The Iron Curtain: Churchill, America, and the Origins of the Cold War* (New York: Oxford University Press, 1986).

50   "*present our worldview*": Ibid.

"*try to impose*": National Security Archive, Electronic Briefing Book no. 261, "Arbatov Memorandum to Gorbachev," June 1988.

51   "*already correct*": Robert English, *Russia and the Idea of the West* (New York: Columbia University Press, 2000), 211.

"*referring to our thinking*": William E. Odom, *The Collapse of the Soviet Military* (New Haven: Yale University Press, 1998), 96.

"*trying to do*": Vladislav M. Zubok, "Gorbachev's Nuclear Learning," *Boston Review*, April 1, 2000, http://bostonreview.net/archives/BR25.2/zubok.html (accessed March 23, 2017).

"What was the argument": William Wohlforth, ed., *Cold War Endgame: Oral History, Analysis, Debates* (University Park: Pennsylvania State University Press, 2003), 257.

52 *"must become the basis"*: Treisman, *The Return*, 9. Gorbachev employed similar language in formal declarations, in particular where the language of nonviolent struggle might resonate, such as on his 1986 visit to India to sign the Delhi Declaration with Rajiv Gandhi. See K. Natwar Singh, *Walking with Lions: Tales from a Diplomatic Past* (New York: HarperCollins, 2012), chap. 33.

"go to hell": Hoffman, *The Dead Hand*, 314.

"algebra": National Security Archive, Electronic Briefing Book no. 261, "Gorbachev's Conference with Advisers on Drafting the U.N. Speech," October 31, 1988. Also see English, *Russia and the Idea of the West*, 217.

demonstrating beyond doubt: Advocates of this theory abound, most often outside academia. For a similar view in part from within, see, for example, Douglas E. Streusand et al., eds., *The Grand Strategy That Won the Cold War* (New York: Lexington Books, 2016).

53 *George Kennan*: For thoughtful treatments of Kennan, see Frank Costigliola, *The Kennan Diaries* (Princeton: Princeton University Press, 2014); and John Lewis Gaddis, *George F. Kennan* (New York: Penguin, 2011).

fatal cracks: For this discussion of Soviet economic ailments and reforms, I have relied in particular on Archie Brown, *The Rise and Fall of Communism* (New York: HarperCollins, 2009), and *Seven Years That Changed the World: Perestroika in Perspective* (New York: Oxford University Press, 2007); Stephen Kotkin, *Armageddon Averted: The Soviet Collapse, 1970–2000* (New York: Oxford University Press, 2008), and *Uncivil Society* (New York: Random House, 2009); Chris Miller, *The Struggle to Save the Soviet Economy* (Chapel Hill: University of North Carolina Press, 2016); and Zubok, *A Failed Empire*.

54 *"On taking office"*: Hitchcock, *The Struggle for Europe*, 351.

"could not even be analyzed": Leffler, *Soul of Mankind*, 372.

"really that bad": Treisman, *The Return*, 7–8.

55 *"History is on our side"*: William Taubman, *Khrushchev: The Man and His Era* (New York: W. W. Norton, 2003), 427.

"human face": Alexander Chubarov, *Russia's Bitter Path to Modernity: A History of the Soviet and Post-Soviet Eras* (New York: Continuum Publishing, 2001), 141.

by 1980: Alfred B. Evans, *Soviet Marxism-Leninism: The Decline of an Ideology* (Westport, Conn.: Praeger, 1993), 60.

energy exporter: For a succinct discussion of Soviet petroleum policy, see Kotkin, *Armageddon Averted*, 16.

56 *eroded*: For Soviet economic decline in the Reagan era, see Mann, *Rebellion of Ronald Reagan*, 241–52.

57 *food production*: Woodrow Wilson Center Digital Archive, "Stasi Note on Meeting Between Minister Mielke and KGB Chairman Andropov," July 11, 1981.

defense-related expenses: For defense spending, see Leffler, *Soul of Mankind*, 374.

"deficiencies": Hans Bethe, "The Truth About Chernobyl," *New York Times*, May 5, 1991.

58 *"transparent"*: Zubok, *A Failed Empire*, 289.

Mathias Rust: Tom LeCompte, "The Notorious Flight of Mathias Rust," *Air & Space Magazine*, July 2005.

"economy to mentality": Artemy Kalinovsky, *A Long Goodbye: The Soviet Withdrawal from Afghanistan* (Cambridge: Harvard University Press, 2011), 120.

"sources of distrust": Dick Combs, *Inside the Soviet Alternate Universe: The Cold War's End and the Soviet Union's Fall Reappraised* (University Park: Pennsylvania State University Press, 2008), 176.

59 *Americans queried*: For polling data, see BPL, Chief of Staff Files, Bates Files, Debate Prepa-

ration Files, Firing Line Debate. See also Thomas Omestad, "Candidate's Views on Foreign Policy," *Christian Science Monitor*, October 17, 1988. I have also employed the CBS News–New York Times series of monthly surveys to verify these numbers.

surprise best seller: Paul Kennedy, *The Rise and Fall of the Great Powers: Economic Change and Military Conflict from 1500 to 2000* (New York: Random House, 1987). Following an initial print run of 9,000 copies, Kennedy's book sold more than 221,000 by the close of 1988. Kennedy was hardly alone in making this point, and surely not the first. James Chace, for example, well placed to influence debate as managing editor of *Foreign Affairs*, argued in 1981 that American power stood on an economic foundation of sand. "For a time," he wrote, "a great nation can live off its capital and borrow against its future, and the fact that it is no longer as productive as it once was can go unnoticed. This is no longer possible for the United States." James Chace, *Solvency: The Price of Survival* (New York: Vintage Books, 1983), 53.

nation's annual debt: Wilentz, *Age of Reagan*, 275.

trade debt: Ibid. See also Arthur Schlesinger, "Wake Up, Liberals," *Washington Post*, May 1, 1988.

60 *"only other example"*: Kennedy, *Rise and Fall of the Great Powers*, 527. See also Peter Peterson, "The Morning After," *Atlantic Monthly* 260, no. 4 (October 1987): 43. To further illustrate the point that solvency and the future of American power were widely debated in 1987, Peterson's essay received the National Magazine Award for best public interest article of the year.

"being questioned": Walter Mossberg and John Walcott, "US Redefines Policy on Security to Place Less Stress on Soviets," *Wall Street Journal*, August 11, 1988.

"medieval wheels": "Is the US the Latest World Power in Decline?" *PBS NewsHour*, March 2, 2010.

61 *Two thirds*: James Noteware, "What the Japanese Want in U.S. Real Estate," American Bankers Association, *ABA Banking Journal* 81, no. 9 (1989): 30.

Rockefeller Center: Douglas Frantz, "Great Japanese Land Rush," *Los Angeles Times*, March 8, 1989.

"no future": Walter LaFeber, *The Clash: US-Japanese Relations Throughout History* (New York: W. W. Norton, 1998), 379.

62 *"hegemon in decay"*: Nicholas Wade, "The Ascent of Books on Decline of US," *New York Times*, April 10, 1988.

"book out of Boston": PPP, "Remarks to Reagan Administration Political Appointees," August 12, 1988.

"manage the decline": Wicker, *George Herbert Walker Bush*, 51.

"always forward": Miller Center, "Acceptance Speech at the Republican National Convention," August 18, 1988, http://millercenter.org/president/bush/speeches/speech-5526 (accessed December 5, 2016).

63 *"can't take that risk"*: "Tank Ride," http://www.livingroomcandidate.org/commercials/1988 (accessed December 3, 2016).

"find out": Sidney Blumenthal, *Pledging Allegiance: The Last Campaign of the Cold War* (New York: Perennial, 1991), 317.

"grave error": BPL, Chief of Staff Files, Bates Files, Debate Preparation Files, "Atlanta Debate," n.d. The debate in question was held February 28, 1988. See Andrew Rosenthal, "Candidates Facing Series of Debates," *New York Times*, January 8, 1988.

"boutique": Maureen Dowd, "Bush Paints Rival as Elitist, with 'Harvard Yard' Views," *New York Times*, June 10, 1988.

"can they pull": "The Presidential Debate: Transcript," *New York Times*, September 26, 1988.

64 *"advance the entire process"*: National Security Archive, Electronic Briefing Book no. 261, "Politburo," November 3, 1988.

"light up the room": Walter Goodman, "The Gorbachev Visit," *New York Times*, December 8, 1988.

"history of the past": For Gorbachev's speech, see http://isc.temple.edu/hist249/course/Documents/gorbachev_speech_to_UN.htm (accessed June 13, 2013).

"has been achieved": Jacques Delors, "European Integration and Security," *Survival* 33, no. 2 (March 1991): 100. For discussion of Europe's cultural transformation in the wake of unification, see Timothy Garton Ash, *In Europe's Name: Germany and the Divided Continent* (New York: Vintage Books, 1993); James J. Sheehan, "The Transformation of Europe and the End of the Cold War," in Engel, *The Fall of the Berlin Wall*, 26–68, and *Where Have All the Soldiers Gone: The Transformation of Modern Europe* (New York: Houghton Mifflin Harcourt, 2008); Tony Judt, *Postwar: A History of Europe Since 1945* (New York: Penguin Books, 2006); and Hitchcock, *The Struggle for Europe*.

65 *"at another's expense"*: Jeffrey A. Engel, Mark A. Lawrence, and Andrew Preston, eds., *America in the World: A History in Documents from the War with Spain to the War on Terror* (Princeton: Princeton University Press, 2014), 324.

"present realities": Mikhail Gorbachev, *Prophet of Change: From the Cold War to a Sustainable World* (Forest Row, East Sussex: Clairview Books, 2011), 17.

"threat of force": Andrei Grachev, *Gorbachev's Gamble: Soviet Foreign Policy and the End of the Cold War* (Cambridge: Polity, 2008), 164.

"sick of us": Treisman, *The Return*, 11.

"future U.S. administration": John Ehrman and Michael W. Flamm, *Debating the Reagan Presidency* (New York: Rowman & Littlefield, 2009), 223.

"unfurled a blueprint": Mary McGrory, "A Smashing Performance," *Washington Post*, December 8, 1988.

66 *"not since Woodrow Wilson"*: Reeves, *President Reagan*, 482.

"astounding": Svetlana Savranskaya et al., *Masterpieces of History: The Peaceful End of the Cold War in Europe, 1989* (Budapest: Central European University Press, 2010), 311. See also Richard Rhodes, *Arsenals of Folly: The Making of the Nuclear Arms Race* (New York: Vintage Books, 2008), 282.

"In all honesty": National Security Archive, Electronic Briefing Book no. 261, "U.S. Senate, Select Committee on Intelligence, Soviet Task Force," December 7, 1988.

67 *"propaganda advantage"*: PPP, "Informal Exchange with Reporters Prior to a Meeting with Soviet President Mikhail Gorbachev," December 7, 1988.

"contained surprises": National Security Archive, Electronic Briefing Book no. 261, "Memorandum of Conversation: The President's Private Meeting with Gorbachev," December 7, 1988.

"Have I ever told you": "New York (Reagan-Gorbachev Meeting, Governors Island)," from the Reagan Presidential Library, NSC System File Folder 8791367, Margaret Thatcher Archives.

"what the President says": Steven V. Roberts, "The Gorbachev Visit; Table for Three, with Talk of Bygones and Best Hopes," *New York Times*, December 8, 1988.

"could not be reversed": National Security Archive, Electronic Briefing Book no. 261, introduction. See also "New York (Reagan-Gorbachev Meeting, Governors Island)," from the Reagan Presidential Library, NSC System File Folder 8791367, Margaret Thatcher Archives.

68 *"continuity"*: National Security Archive, Electronic Briefing Book no. 261, introduction.

"Have you completed": Bush and Scowcroft, *A World Transformed*, 7.

"What assurance": Norman Graebner et al., *Reagan, Bush, Gorbachev: Revisiting the End of the Cold War* (Westport, Conn.: Praeger, 2008), 115. See also Beschloss and Talbott, *At the Highest Levels*, 10–11.

"Don't misread me": Thatcher Archives, "New York (Reagan-Gorbachev Meeting, Governors Island)."

69 *"commission"*: Beschloss and Talbott, *At the Highest Levels,* 12.

"potentially more dangerous": Ivo H. Daalder and I. M. Destler, *In the Shadow of the Oval Office: Profiles of the National Security Advisers* (New York: Simon & Schuster, 2009), 174.

"kill us with kindness": John Lewis Gaddis, *The Cold War: A New History* (New York: Penguin, 2005), 35.

"heady atmosphere": Bush and Scowcroft, *A World Transformed,* 46.

mutual antagonism: For Bush and Kissinger, see Engel, *The China Diary,* 420–32. See also Maureen Dowd, "Aux Barricades," *New York Times,* January 17, 2007, and "Don't Pass the Salted Peanuts, Henry," *New York Times,* October 4, 2006; and Bob Woodward, *State of Denial* (New York: Simon & Schuster, 2006), 406–10.

70 *"want to create"*: James A. Baker III Papers, Mudd Library, Princeton University (hereafter Baker Papers with identifying information), Monthly Files, box 108, January 1989, "Copy of POTUS Exchange of Notes w USSR President Gorbachev re: January 7, 1989 Meeting Between H. Kissinger and Pres. Gorbachev."

Kissinger further elaborated: For Kissinger's notes, see Baker Papers, box 108, January 1989, "Notes from Henry Kissinger Meeting with Gorbachev, Jan. 17, 1989, 12:00–1:20 PM."

Gorbachev heard Kissinger out: For Gorbachev's formal response, see Baker Papers, box 108, January 1989, "To the Vice President of the United States from M. Gorbachev," January 18, 1989.

"approach the dialogue": Beschloss and Talbott, *At the Highest Levels,* 12–17. See also Mary Sarotte, *1989: The Struggle to Create Post–Cold War Europe* (Princeton: Princeton University Press, 2014), 22–23.

"strange country": "Record of Conversation Between Alexsandr Yakovlev and Henry Kissinger, January 16, 1989," and "Record of Conversation Between Mikhail Gorbachev and Henry Kissinger, January 17, 1989," in Savranskaya, *Masterpieces of History,* 341–46.

4. "WE KNOW WHAT WORKS"

71 *"Americans in"*: Meacham, *Destiny and Power,* 734.

"peace offensive": Odd Arne Westad, *Reviewing the Cold War: Approaches, Interpretations, Theory* (New York: Routledge, 2013), 349.

72 *"regain leadership"*: Robert M. Gates, *From the Shadows: The Ultimate Insider's Story of Five Presidents and How They Won the Cold War* (New York: Simon & Schuster, 1996), 462.

"George to George": Molly Ivins, *Molly Ivins Can't Say That, Can She?* (New York: Vintage Books, 1992), 133.

"moving toward democracy": PPP, "George Bush Inauguration Speech," January 20, 1989.

"day of the dictator": Roman Popadiuk, *The Leadership of George Bush* (College Station: Texas A&M University Press, 2009), 140.

73 *"We know what works"*: PPP, "George Bush Inauguration Speech," January 20, 1989.

"Are We Approaching": James Atlas, "What Is Fukuyama Saying? And to Whom Is He Saying it?" *New York Times Magazine,* October 22, 1989.

"don't understand it myself": Marvin Kalb, *Imperial Gamble: Putin, Ukraine, and the New Cold War* (Washington, D.C.: Brookings Institution Press, 2015), 9.

74 *History had revealed:* Not everyone was immediately impressed. "Only at the University of Chicago, I thought to myself, could intellectuals be so deluded," the noted historian Bruce Cummings recalled thinking. "History just happened to culminate in the reigning orthodoxy of our era, the neoliberalism of Thatcher and Reagan," validating the self-congratulatory

ethos of the time. "I departed after twenty minutes, thinking that his line of argument would go nowhere." Bruce Cummings, "Time of Illusion: Post–Cold War Visions of the World," in *Cold War Triumphalism*, ed. Ellen Schrecker (New York: New Press, 2004), 78–79.

"*first word*": Allan Bloom et al., "Responses to Fukyama," *The National Interest* 16 (Summer 1989): 19.

"*it's wonderful*": Ibid., 28.

"*close this millennium*": Francis Fukuyama, "The End of History?" *The National Interest* 16 (Summer 1989): 3–18.

"*we win*": Atlas, "What Is Fukuyama Saying?"

"*including pornography*": Ibid.

75 "*quest for democracy*": BPL, *NSC Files*, Rice Papers, USSR Subject Files, US-USSR Soviet Relations [3], James Baker, "The Challenge in US-Soviet Relations."

"*resisting the siren song*": BPL, *NSC Files*, Rice Papers, USSR Subject Files, USSR-Gorbachev, "Robert Gates, CSIS Conference," April 1, 1989.

"*stand at the door*": Frances FitzGerald, *Way Out There in the Blue* (New York: Simon & Schuster, 2001), 469.

76 "*Soviet flexibility*": BPL, Scowcroft Papers, USSR Collapse, US-Soviet Relations through 1991 (January–April 1989), Scowcroft to Bush, March 1, 1989.

"*tear itself to pieces*": Robert Bowie and Richard Immerman, *Waging Peace: How Eisenhower Shaped an Enduring Cold War Strategy* (New York: Oxford University Press, 2008), 206.

"*Without the visible assurance*": *Foreign Relations of the United States, 1964–1968*, vol. 13, *Western Europe Region* (Washington, D.C.: U.S. Government Printing Office, 1995), 562.

77 "*contributes to a stable*": Christopher Layne, *The Peace of Illusions: American Grand Strategy from 1940 to the Present* (Ithaca: Cornell University Press, 2006), 108.

"*basic lesson*": BPL, *NSC Papers*, Kanter Files, Subject Files, NATO Summit, May 1989, Scowcroft to Bush, March 20, 1989.

"*we prevented*": "Address by James A. Baker Before the Center for Strategic and International Studies, May 4, 1989," *State Department Bulletin*, June 1989.

"*We must never forget*": PPP, "Remarks at the Boston University Commencement," May 21, 1989.

Crowds numbering: For anti-nuclear rallies, see David Cortright, *Peace* (New York: Cambridge University Press, 2008). I am indebted to Melanie McAlister and David Farber, and the Miller Center for Public Affairs Manuscript Forum organized by Brian Balogh, for deepening my appreciation of Europe's nuclear anxieties during this period.

78 "*deader the German*": This phrase epitomizes ubiquity. Sarotte, *1989*, attributes the phrase to Alfred Dregger (27).

"*defining security issue*": Beschloss and Talbott, *At the Highest Levels*, 36.

79 "*defend our own troops*": Sarotte, *1989*, 27. For nuclear strategy in a broader perspective, see Francis Gavin, *Nuclear Statecraft: History and Strategy in America's Atomic Age* (Ithaca: Cornell University Press, 2012).

"*top priority*": BPL, NSC, Kanter Papers, Subject Files, NATO Summit, May 1989, Scowcroft to Bush, March 20, 1989.

"*Odd as it seems*": Paul Wolfowitz, "Shaping the Future: Planning at the Pentagon, 1989–93," in *In Uncertain Times*, ed. Melvyn Leffler and Jeffrey Legro (Ithaca: Cornell University Press, 2011), 46.

80 "*common European home*": For a contemporary assessment, see Neil Malcolm, "The 'Common European Home' and Soviet European Policy," *International Affairs* 65, no. 4 (Autumn 1989): 659–76.

"*play an essential role*": Grachev, *Gorbachev's Gamble*, 135.

"angered us": David Shumaker, *Gorbachev and the German Question: Soviet–West German Relations, 1985–1990* (Westport, Conn.: Praeger, 1995), 35–36.

81 *"ice has been broken"*: Angela Stent, *Russia and Germany Reborn: Unification, the Soviet Collapse, and the New Europe* (Princeton: Princeton University Press, 1988), 66.

"behave in accordance": Felicity Barringer, "Gorbachev Hails Kohl in Moscow," *New York Times,* October 25, 1988.

"unimaginative, perhaps": Daalder and Destler, *In the Shadow of the Oval Office,* 173.

"forced landing": Ibid., 170. The most authoritative studies of Scowcroft are David Schmitz, *Brent Scowcroft* (Lanham, Md.: Rowman & Littlefield, 2011); and Bartholomew Sparrow, *The Strategist: Brent Scowcroft and the Call of National Security* (New York: Public Affairs, 2015).

82 *"somnolent excellence"*: George Bush and Jim McGrath, *Heartbeat: George Bush in His Own Words* (New York: Citadel Press, 2001), 275.

"became sicker": Miller Center oral history, Robert Gates, July 23–24, 2000, 11.

"very unusual": Miller Center oral history, Brent Scowcroft, November 12–13, 1999, 5, 9.

83 *fuming by some accounts:* Jeffrey Goldberg, "Breaking Ranks: What Turned Brent Scowcroft Against the Bush Administration," *The New Yorker,* October 31, 2005.

"spent the '80s": Miller Center oral history, Brent Scowcroft, 26.

84 *"perils of policy"*: David Rothkopf, *Running the World: The Inside Story of the National Security Council and the Architects of American Power* (New York: Public Affairs, 2006), 251.

"handled too informally": John Tower et al., *Report of the President's Special Review Board* (Washington, D.C.: Government Printing Office, 1987), iv–1.

"incurious president": Schmitz, *Brent Scowcroft,* 85–86.

"atmospherics are not enough": Miller Center oral history, Brent Scowcroft, 16.

85 *"light at the end"*: David Hoffman, "Security Adviser Wary of Gorbachev's Motives," *Washington Post,* January 23, 1989.

"badly needs a period of stability": Jeffrey A. Engel, "1989: An Introduction," in *The Fall of the Berlin Wall: The Revolutionary Legacy of 1989,* ed. Jeffrey A. Engel (New York: Oxford University Press, 2009), 27.

"Cold War is not over": John Broder, "'Cold War Is Not Over,' Bush Adviser Cautions," *Los Angeles Times,* January 23, 1989.

5. THE PAUSE

86 *"Pause"*: Robert Toth, "Bush May Delay Resumption of Strategic Arms Talks," *Los Angeles Times,* December 15, 1988.

"want to be prudent": "Transcript of Bush's Remarks on Domestic and Foreign Issues," *New York Times,* December 5, 1988.

"My advisers": Roy Rosenzweig Center for History and New Media, *Making the History of 1989,* Bush to Gorbachev, January 17, 1989, https://chnm.gmu.edu/1989/items/show/140 (accessed December 3, 2016).

"NOT a friendly takeover": Christopher Maynard, *Out of the Shadow: George H. W. Bush and the End of the Cold War* (College Station: Texas A&M University Press, 2008), 4.

87 *"new man was in charge"*: Beschloss and Talbott, *At the Highest Levels,* 26.

"Don't you think": Lawrence S. Wittner, *The Struggle Against the Bomb,* vol. 3 (Palo Alto: Stanford University Press, 2003), 425.

Ridgway: Association for Diplomatic Studies and Training, Foreign Affairs Oral History Project, "Ambassador Rozanne L. Ridgway," June 4, 1991.

"What are they waiting for?": Mikhail Gorbachev, *Memoirs* (New York: Doubleday, 1996), 496–97.

"Time has its limits": David Remnick, "Gorbachev Criticizes Lack of U.S. Policy," *Washington Post,* April 7, 1989.

"giant hoax": Jack Matlock, *Autopsy on an Empire: The American Ambassador's Account of the Collapse of the Soviet Union* (New York: Random House, 1995), 183.

88   *"Interesting"*: Beschloss and Talbott, *At the Highest Levels,* 34.

"stand there": Maynard, *Out of the Shadow,* 15.

"Maybe it was prudent": Mann, *Rebellion of Ronald Reagan,* 325.

"ad hoc": Beschloss and Talbott, *At the Highest Levels,* 25.

89   *"conscious pause"*: M. J. Heale, *Contemporary America: Power, Dependency, and Globalization Since 1980* (Oxford: Wiley-Blackwell, 2011), 134.

(NSR 3): BPL, National Security Reviews, NSR 3, February 15, 1989, https://bush41library.tamu.edu/files/nsr/nsr3.pdf (accessed December 13, 2016).

"How can the U.S.": Ibid.

90   *"central importance"*: BPL, National Security Reviews, NSR 5, February 15, 1989, https://bush41library.tamu.edu/files/nsr/nsr5.pdf (accessed December 13, 2016).

"real and sustained change": BPL, National Security Reviews, NSR 4, February 15, 1989, https://bush41library.tamu.edu/files/nsr/nsr4.pdf (accessed December 13, 2016).

91   *"ready for it"*: Gale Stokes, *The Walls Came Tumbling Down: Collapse and Rebirth in Eastern Europe* (New York: Oxford University Press, 2012), 6.

92   *"only one nation"*: Gerald Boyd, "Raze Berlin Wall, Reagan Urges Soviets," *New York Times,* June 13, 1987.

"as Tibet": Geoffrey K. Fry, *The Politics of Decline* (New York: Palgrave, 2005), 152.

"just as with Hitler": Evan Thomas, *Ike's Bluff: President Eisenhower's Secret Battle to Save the World* (New York: Little, Brown, 2012), 229. See also National Security Archive, Electronic Briefing Book no. 581, *Eisenhower Concluded Neither U.S. Military Operations Nor Popular Uprisings Were Feasible in Soviet-Controlled Eastern Europe, Despite "Rollback" Rhetoric,* "Almost Successful Recipe: The United States and East European Unrest Prior to the Hungarian Revolution," February 28, 2017.

"Our troops are fighting": Associated Press, "Hungarian Radio Broadcasts Speech Nagy Made in 1956," May 6, 1989.

"should be a lesson": David Pryce-Jones, "What the Hungarians Wrought," *National Review,* October 23, 2006; "How to Help Hungary," *Time,* December 24, 1956.

"real lesson": "Doing It Themselves," *Time,* December 17, 1956.

93   *secured from Kremlin archives*: Andrzej Paczkowski and Malcolm Byrne, eds., *From Solidarity to Martial Law: The Polish Crisis of 1980–1981* (Budapest: Central European University Press, 2007).

"surgery with psychiatry": Tina Rosenberg, *The Haunted Land: Facing Europe's Ghosts After Communism* (New York: Vintage Books, 1996), 179.

"anything we can do": Ronald Reagan Presidential Library, NSC Country File, box 91283, National Security Council Minutes (Poland), December 21, 1981.

94   *Condemn, but don't act*: Ibid.

"breeze of freedom": Lee Davidson, "Bush Takes Nation's Helm in 'New Breeze of Freedom,'" *Deseret News,* January 20, 1989.

"dawning of a new day": David Nichols, *Eisenhower 1956: The President's Year of Crisis* (New York: Simon & Schuster, 2011), 214.

"essential substance": Lyndon Johnson Library, National Security Files, Country File, box 182, Czech Crisis, Central Intelligence Agency Office of National Estimates, "Czechoslovakia: The Dubček Pause," June 13, 1968.

"end zone": Brian Mulroney, *Memoirs* (Toronto: McClelland & Stewart, 2007), 651.

"stir up revolution": Baker Papers, box 108, folder 2, "JAB Personal Notes from 2/10/89 Mtg. with POTUS and Canada PM Mulroney."

"tragedy of Hungary": Bush and Scowcroft, *A World Transformed*, 39.

"time may have come": Hal Brands, *Making the Unipolar Moment* (Ithaca: Cornell University Press, 2016), 283.

95   *"figuring out exactly"*: BPL, National Security Reviews, NSR 4, February 15, 1989, https://bush41library.tamu.edu/files/nsr/nsr4.pdf (accessed December 13, 2016).

"most pressing problems": BPL, Telcon with Prime Minister Thatcher, January 23, 1989.

"right on the mark": Ibid.

96   *"well-armed rabbits"*: Robert L. Hutchings, *American Diplomacy and the End of the Cold War: An Insider's Account* (Washington, D.C.: Woodrow Wilson Center Press, 1997), 14.

"comprehensive concept": BPL, "Memcon with FRG Minister Wolfgang Schaeuble," February 9, 1989.

"Convey to the Chancellor": Ibid.

"stalling": BPL, Telcon with Prime Minister Thatcher, January 23, 1989.

"playing defense": BPL, Telcon with Prime Minister Schlueter of Denmark, January 24, 1989.

"got us figured out": BPL, President's Meeting with Prime Minister Mulroney, February 10, 1989.

97   *"Boston Harbor"*: Baker Papers, box 108, folder 2, "JAB Personal Notes from 2/10/89 Mtg. with POTUS and Canada PM Mulroney." My thanks to Josh Shifrinson for sharing this document.

"but enough": R. W. Apple, "Statesman Bush's Debut," *New York Times*, February 22, 1989.

"grab the ball": PPP, "The President's News Conference," February 21, 1989.

"outrageous hypothesis": Ann Devroy, "Bush Denounces Threat Against Writer," *Washington Post*, February 22, 1989.

"except in Washington": Apple, "Statesman Bush's Debut."

98   *"American leadership"*: James Baker, *The Politics of Diplomacy* (New York: Putnam, 1995), 47–50.

"soft attitude": BPL, John Sununu Papers, White House Office Files, Political Affairs (1989) [2], Wray to Card, May 23, 1989.

"increasingly agitated": BPL, John Sununu Papers, White House Office Files, Communications (1989), Csorba to Wead, February 2, 1989.

"do Gorbachev no favors": Sarotte, *1989*, 23.

"only on the margins": Baker Papers, box 108, folder 2, February 1989, "JAB Notes from February 1989: Re: US-USSR Relations."

"scrambling": Baker, *Politics of Diplomacy*, 67.

99   *"premium on new features"*: Baker Papers, box 108, folder 3, March 1989, "Key Impressions from the Trip."

"take the offensive": Baker Papers, box 108, folder 2, "JAB Personal Notes from 2/10/89 Mtg. with POTUS and Canada PM Mulroney."

"behind the alliance": BPL, President's Meeting with Prime Minister Mulroney, February 10, 1989.

6. "A SPECIAL RELATIONSHIP THERE"

100 *"have stood up"*: Chen Jian, "China's Path Toward 1989," in Engel, *The Fall of the Berlin Wall*, 99. See also Odd Arne Westad, *Restless Empire: China and the World Since 1750* (London: Random House, 2013), 292.

Acheson: Jeffrey A. Engel, "Of Fat and Thin Communists: Diplomacy and Philosophy in

Western Economic Warfare Strategies Toward China (and Tyrants, Broadly)," *Diplomatic History* 29, no. 3 (June 2005): 445–74.

hundreds of millions: Philip Taubman, "Gromyko Says Mao Wanted Soviet A-Bomb Used on G.I.s," *New York Times,* February 22, 1988.

101 *outstretched hand:* Richard Reeves, *President Nixon: Alone in the White House* (New York: Simon & Schuster, 2001), 438. Zhou and Nixon discussed the snub explicitly during the president's visit. See National Archives and Records Administration, Nixon Presidential Materials, President's Office Files, Memoranda for the President, box 87, memorandum of conversation, February 21, 1972.

"week that changed the world": Margaret MacMillan, *Nixon and Mao: The Week That Changed the World* (New York: Random House, 2008).

"Didn't our ancestors": Margaret MacMillan, "Nixon, Kissinger, and the Opening to China," in *Nixon in the World: American Foreign Relations, 1969–1977,* eds. Frederik Logevall and Andrew Preston (New York: Oxford University Press, 2008), 118.

Great Leap Forward: Frank Dikötter, *Mao's Great Famine* (Hong Kong: Walker Books, 2011).

Cultural Revolution: Frank Dikötter, *The Cultural Revolution: A People's History* (New York: Bloomsbury, 2016).

Deng Xiaoping: For Deng, useful primers include Ezra F. Vogel, *Deng Xiaoping and the Transformation of China* (Cambridge: Harvard University Press, 2011); and Benjamin Yang, *Deng: A Political Biography* (London: M. E. Sharpe, 1998).

102 *"personal identity":* Vogel, *Deng Xiaoping,* 22.

"big-power chauvinism": Vladislav Zubok, "'Look What Chaos in the Beautiful Socialist Camp!' Deng Xiaoping and the Sino-Soviet Split, 1956–1963," *Cold War International History Project Bulletin* 10, n.d., 153.

ambitious reforms: For Deng's economic ideas, see Vogel, *Deng Xiaoping,* 217–376; and Maurice Meisner, *The Deng Xiaoping Era: An Inquiry into the Fate of Chinese Socialism* (New York: Hill and Wang, 1996).

103 *"Centralized power":* Vogel, *Deng Xiaoping,* 25. See also Merle Goldman, *Sowing the Seeds of Democracy in China: Political Reform in the Deng Xiaoping Era* (Cambridge: Harvard University Press, 1994).

"short man": Engel, *The China Diary,* 46.

"I lived there": Jeffrey A. Engel and Sergey Radchenko, "Beijing and Malta, 1989," in *Transcending the Cold War: Summits, Statecraft, and the Dissolution of Bipolarity in Europe, 1970–1990,* eds. Kristina Spohr and David Reynolds (New York: Oxford University Press, 2016), 185.

"useful role": James Lilley, *China Hands: Nine Decades of Adventure, Espionage, and Diplomacy in Asia* (New York: PublicAffairs, 2004), 217.

104 *"explain everything":* Ibid., 222.

"hope he'll win": James Mann, *About Face: A History of America's Curious Relationship with China, from Nixon to Clinton* (New York: Vintage Books, 2000), 176. See also Mann's more polemical work *The China Fantasy: How Our Leaders Explain Away Chinese Repression* (New York: Viking, 2007).

new constitution: Lilley, *China Hands,* 279.

105 *"spirit of democracy":* "Bush: 'Our Work Is Not Done; Our Force Is Not Spent,'" *Washington Post,* August 19, 1988.

"special relationship there": BPL, VP Daily Files, Alphabetical Transition Files (November 8, 1988–January 19, 1989), Transition (November 9, 1988–January 19, 1989).

"perverse sort of way": Ibid.

106 *"emperor system should continue"*: John Dower, *Embracing Defeat* (New York: W. W. Norton, 1999), 232–324.

"should have been shot": Susan Chira, "Tokyo Funeral Forces Choices by Old Foes," *New York Times,* January 13, 1989.

"deny the fact": Margaret Shapiro, "Japan's Praise of Hirohito Stirs Rancor," *Washington Post,* January 10, 1989.

"up to the historians": David Sanger, "Takeshita Now Admits World War II Aggression," *New York Times,* March 7, 1989.

107 *"doing the right thing"*: Karl Schoenberger, "World Leaders Pay Respect at Hirohito Rites," *Los Angeles Times,* February 24, 1989.

"what I am symbolizing": "Excerpts from News Session by Bush, Watkins, and Bennett," *New York Times,* January 13, 1989.

"vividly remember": PPP, "The President's News Conference in Tokyo," February 25, 1989.

"heal old wounds": Bush, *All the Best,* 415, also 644.

Americans polled: Andrew Malcolm, "Closer Ties Bind U.S. and Japanese," *New York Times,* February 23, 1989.

"credibility": Miller Center oral history, David Demarest, January 28, 2010, 61.

"key purpose": National Security Archive, Electronic Briefing Book no. 47, "The U.S. Tiananmen Papers," Baker to Bush, February 16, 1989.

108 *"obtain Chinese assurances"*: National Security Archive, Electronic Briefing Book no. 47, U.S. Embassy Beijing to State, February 6, 1989. See also U.S. State Department Virtual Reading Room, U.S. Embassy to State, February 6, 1989.

"oxymoron": Mann, *About Face,* 162.

"since you were here last": Department of State Virtual Reading Room, American Embassy Beijing to White House, February 12, 1989.

109 *"foreign exchange certificates"*: Westad, *Restless Empire,* 379.

"next generation": Department of State Virtual Reading Room, American Embassy Beijing to White House, February 12, 1989.

"brush strokes": Ibid.

110 *"restore our Great Wall"*: Alan A. Lew and Alan Wong, "Existential Tourism and the Homeland: The Overseas Chinese Experience," in *Seductions of Place: Geographical Perspectives on Globalization and Touristed Landscapes,* eds. Carolyn L. Cartier and Alan A. Lew (New York: Routledge, 2005), 274.

"she's still a beacon": PPP, "Farewell Address to the Nation," January 11, 1989.

111 *Lord counseled*: Department of State Virtual Reading Room, American Embassy Beijing to White House, February 12, 1989.

"our intentions": Ibid.

112 *"sage view"*: BPL, Scowcroft Papers, Presidential Correspondence, "Meeting with President Chaim Herzog," February 23, 1989.

"opinion of events": BPL, Memcons and Telcons, "Meeting with King Hussein of Jordan," February 23, 1989.

"looked forward to learning": BPL, Memcons and Telcons, "President's Meeting with President Mario Soares of Portugal," February 23, 1989.

"nothing that adversely": BPL, Scowcroft Papers, Presidential Correspondence, Memorandum of Conversation, "President's Meeting with Prime Minister Noburo Takeshita of Japan," February 23, 1989.

"achieve dominance": BPL, Scowcroft Papers, Presidential Correspondence, Memorandum of Conversation, "Meeting with President Francesco Cossiga of Italy," February 24, 1989.

"propaganda offensive": BPL, Scowcroft Papers, Presidential Correspondence, Memorandum of Conversation, "President's Meeting with President Richard von Weizsaecker of the Federal Republic of Germany," February 24, 1989.

"must stay together": Ibid.

"that of 1917": BPL, Scowcroft Papers, Presidential Correspondence, Memorandum of Conversation, "President's Meeting with Prime Minister Turgut Ozal of Turkey," February 4, 1989.

lao pengyou: Richard Solomon, *Chinese Negotiating Behavior: Pursuing Interests Through 'Old Friends'"* (Washington, D.C.: United States Institute of Peace, 1999).

113 *their "old friends"*: Mann, *About Face*, 135.

"receive the brunt": Solomon, *Chinese Negotiating Behavior*, 102.

"Every time you come": BPL, Scowcroft Papers, Presidential Correspondence, Memorandum of Conversation, "President Bush's Meeting with President Yang Shangkun of the People's Republic of China," February 25, 1989.

"best witness": BPL, Scowcroft Papers, Presidential Correspondence, Memorandum of Conversation, "President Bush's Meeting with General Secretary of the Central Committee of the Communist Party Zhao Ziyang of the People's Republic of China," February 26, 1989.

114 *"there is warmth"*: Author interview with George and Barbara Bush, November 16, 2005.

"out of the question": BPL, Scowcroft Papers, Presidential Correspondence, Memorandum of Conversation, "President Bush's Meeting with General Secretary of the Central Committee of the Communist Party Zhao Ziyang of the People's Republic of China," February 26, 1989.

"taken China that long": BPL, Scowcroft Papers, Presidential Correspondence, Memorandum of Conversation, "Welcoming Banquet in Beijing, China," February 25, 1989.

"suit the taste": Bush and Scowcroft, *A World Transformed*, 92.

Cultural Revolution: useful primers include Roderick MacFarquhar and Michael Schoenhals, *Mao's Last Revolution* (Cambridge: Belknap Press, 2006); and Feng Jicai, *Ten Years of Madness: Oral Histories of China's Cultural Revolution* (San Francisco: China Books & Periodicals, 1996); and Dikötter, *The Cultural Revolution*.

115 *"Hitlerites"*: Westad, *Restless Empire*, 354. As Westad writes, "The Great Proletariat Cultural Revolution was by far the largest and most intense government campaign in Chinese history" (353).

unfettered reform: For political reform and the indigenous reform movement in China, see Craig Calhoun, *Neither Gods nor Emperors: Students and the Struggle for Democracy in China* (Berkeley: University of California Press, 1994); Goldman, *Sowing the Seeds;* Kenneth Lieberthal, *Governing China: From Revolution to Reform* (New York: Norton, 2003); Jeffrey Wasserstrom and Elizabeth Perry, eds., *Popular Protest and Political Culture in Modern China* (Boulder: Westview Press, 1992); and Zhao Dingxin, *The Power of Tiananmen: State-Society Relations and the 1989 Student Movement* (Chicago: University of Chicago Press, 2001).

116 *marches appeared coordinated*: Jim Mann, "Chinese Students Urged to Shun Dissent," *Los Angeles Times*, December 9, 1985.

"lost the initiative": National Security Archive, Electronic Briefing Book no. 6, *Tiananmen Square, 1989: The Declassified History*, "US Embassy Beijing to Department of State, December 23, 1985."

"relaxation of control": Goldman, *Sowing the Seeds*, 198.

"Socialism has failed": Vogel, *Deng Xiaoping*, 577.

117 *"Give Me Liberty"*: "Fang Lizhi, Physicist and Dissident, Died on April 6th, Aged 76," *The Economist*, April 14, 2012.

"Bourgeois liberalization": Goldman, *Sowing the Seeds*, 206. See also Orville Schell and David Shambaugh, *The China Reader: The Reform Era* (New York: Random House, 2010), 184; and

Richard Madsen, *China and the American Dream: A Moral Inquiry* (Berkeley: University of California Press, 1995), 14.

"China would not welcome": BPL, Scowcroft Files, Presidential Correspondence, Memorandum of Conversation, "Welcoming Banquet in Beijing, China," February 25, 1989.

"tremendous beneficial changes": BPL, Scowcroft Files, Presidential Correspondence, Memorandum of Conversation, "President Bush's Meeting with General Secretary of the Central Committee of the Communist Party Zhao Ziyang of the People's Republic of China," February 26, 1989.

"a Western political system": Engel and Radchenko, "Beijing and Malta, 1989," 188.

118 *"chaos will result"*: Sergey Radchenko, *Unwanted Visionaries: The Soviet Failure in Asia at the End of the Cold War* (New York: Oxford University Press, 2014), 159.

"feel warmly": Woodrow Wilson Center Digital Archive, "Memorandum of Conversation Between George H. W. Bush and Zhao Ziyang," February 26, 1989.

"You know well": Ibid.

"overwhelming need": BPL, Scowcroft Files, Presidential Correspondence, Memorandum of Conversation, "President Bush's Meeting with Chairman Deng Xiaoping of the People's Republic of China," February 26, 1989.

"Let's have lunch": Ibid.

119 *"explosion"*: Association for Diplomatic Studies and Training, Foreign Affairs Oral History Project, Ambassador Winston Lord, April 28, 1998.

"Who IS Fang": Mann, *About Face*, 178.

"distinctly unfriendly": Association for Diplomatic Studies and Training, Foreign Affairs Oral History Project, Ambassador Winston Lord, April 28, 1998.

"worst moments": Ibid.

120 *"throwing our hats"*: Ibid.

"Fang not coming": Ibid.

"beyond my wildest dreams": Ibid.

"run aground": Gregory Hywood, "Human Rights Debacle at a Banquet in China Ruffles White House Panache," *Australian Financial Review*, February 28, 1989.

"Human Rights Row": Peter Riddell, "Chinese Human Rights Row Flares During Bush Visit," *Financial Times*, February 27, 1989.

"China Rebukes": Daniel Southerland, "China Rebukes U.S. Over Dissident," *Washington Post*, February 28, 1989.

"Bush's Visit": Gerald F. Seib and Adi Ignatius, "Bush's Visit Is Marred by Flap Over Chinese Dissidents," *Wall Street Journal*, February 27, 1989.

"worst of worlds": Bush, *All the Best*, 415–16. James Mann's account of the banquet affair, and the White House's attempt to pin responsibility solely on Lord and the embassy in Beijing, is particularly illuminating. See Mann, *About Face*, 182–83.

121 *"human rights was not discussed"*: Timothy McNulty, "Rights Flap Mars Bush's China Visit," *Chicago Tribune*, February 27, 1989.

"big, public fanfare": Ibid.

"believes quiet diplomacy": James Gerstenzang, "Poor U.S. Communications Seen in China Dinner Flap," *Los Angeles Times*, March 3, 1989.

"a long way": Bush, *All the Best*, 416.

122 *"China is an exception"*: BPL, Scowcroft Papers, Presidential Correspondence, Memorandum of Conversation, "President Bush's Meeting with Premier Li Peng of the People's Republic of China," February 26, 1989.

"unrest in Eastern Europe": Chen Jian, "Tiananmen and the Fall of the Berlin Wall," in Engel, *Fall of the Berlin Wall*, 110.

"tide was running": Jasper Becker, "No Apologies for Barring Fang from Bush Banquet," *The Guardian*, February 28, 1989.

7. CHENEY RISES AND THE PAUSE ENDS

124 *"resolved through perestroika"*: Savranskaya, *Masterpieces of History*, 423.

Fewer than a third: Christopher Wren, "Breaking Out," *New York Times*, August 4, 1988.

"planners of perestroika": William Doerner and Ann Blackman, "Why the Bear's Cupboards Are Bare," *Time*, January 6, 1989.

"Shortages attack us": Bill Keller, "For Grim Soviet Consumers, the Year of Discontent," *New York Times*, January 1, 1989.

125 *"force is to no avail"*: Zubok, *A Failed Empire*, 319.

"internal political conflicts": Savranskaya et al., *Masterpieces of History*, 446.

"Armed forces were used": Ibid., 448.

"Beyond them is emptiness": Roy Rosenzweig Center for History and New Media, George Mason University, *Making the History of 1989*, "Excerpt from Anatoly Chernyaev's Diary," document 362, https://chnm.gmu.edu/1989/items/show/362 (accessed December 13, 2016).

126 *"Things are moving"*: PPP, "The President's News Conference," April 7, 1989.

unprecedented political defeat: Robert A. Strong, "Character and Consequence: The John Tower Confirmation Battle," in Nelson and Perry, *41*, 122–42. For Tower's personal perspective, see John Tower, *Consequences: A Personal and Political Memoir* (New York: Little, Brown, 1991).

"Let's fight it out": Tower, *Consequences*, 317.

"He is my choice": PPP, "The President's News Conference in Tokyo," February 25, 1989.

"fishy odor": Dorothy Collin, "Hypocrisy Is What You Smell Lingering over the Battlefield," *Chicago Tribune*, March 13, 1989.

127 *"Dick washes"*: Baker, *Politics of Diplomacy*, 22–25.

"closed-door meeting": James Mann, *Rise of the Vulcans: The History of Bush's War Cabinet* (New York: Penguin, 2004), 97.

"Principle is okay": Peter Baker, *Days of Fire: Bush and Cheney in the White House* (New York: Random House, 2013), 29.

"other priorities": Katharine Seelye, "Cheney's Five Draft Deferments During the Vietnam Era Emerge as Campaign Issue," *New York Times*, May 1, 2004.

"dirty word": Mark Danner, "In the Darkness of Dick Cheney," *New York Review of Books*, March 6, 2014.

128 *"I got married"*: Baker, *Days of Fire*, 39.

"raise hell": Bush, *All the Best*, 418.

129 *"yes I can"*: Baker, *Days of Fire*, 35.

"faced with an adversary": "Nominations Before the Senate Armed Services Committee," *Congressional Record*, March 14, 1989.

"caution is in order": PPP, "Remarks at the Swearing-In Ceremony for Richard B. Cheney," March 21, 1989.

"point of view": Savranskaya, *Masterpieces of History*, 73. See also Jeremy Isaacs and Taylor Downing, *Cold War: For Forty-Five Years the World Held Its Breath* (New York: Little, Brown, 2008), 378.

pleading with Gorbachev: National Security Archive, *The Thatcher-Gorbachev Conversations*, Electronic Briefing Book no. 422, "Record of Conversation Between Thatcher and Gorbachev, April 5, 1989."

"a lot of criticism": PPP, "The President's News Conference," March 7, 1989.

130 *"We have to be better"*: BPL, Scowcroft Papers, USSR Collapse, Chronological Files, U.S.-Soviet relations through 1991 (January–April 1989), Scowcroft to Bush, March 1, 1989.

131 *"old curmudgeons"*: FitzGerald, *Way Out There in the Blue*, 469.

"first chapter of Genesis": BPL, NSC Files, Kantor Files, Subject Files, NATO Summit — May 1989, Scowcroft to POTUS, March 20, 1989.

"handsome inheritance": Ibid.

"read this with interest": Ibid. Bush read the document March 26.

132 *"opened up"*: Douglas Martin, "Eduard Shevardnadze, Foreign Minister Under Gorbachev, Dies at 86," *New York Times*, July 7, 2014.

"Who could have expected": Pavel Palazchenko, *My Years with Gorbachev and Shevardnadze: The Memoir of a Soviet Interpreter* (University Park: Pennsylvania State University Press, 1997), 30.

"Civilized person-to-person": Hella Pick, "Eduard Shevardnadze Obituary," *The Guardian*, July 7, 2014.

"Einstein": Baker, *Politics of Diplomacy*, 64.

"can't help but get": Baker Papers, box 108, folder 3, March 1989, "Key Impressions from the Trip." As Baker wrote to Bush, "Same line Varkonyi, Hungarian Foreign Minister, used." See also Baker, *Politics of Diplomacy*, 63–68, for discussion of this trip.

133 *"pace may surprise us"*: Baker, *Politics of Diplomacy*, 67, 70.

status quo plus: Alpo M. Rusi, *After the Cold War: Europe's New Political Architecture* (London: Palgrave Macmillan, 1991), 8.

"mush": Mary Elise Sarotte, "The Wall Comes Down: A Punctuational Moment," in Leffler and Legro, *In Uncertain Times*, 21.

"hope over experience": Gates, *From the Shadows*, 460.

"It's not surprising": Maynard, *Out of the Shadow*, 16–17.

"Some analysts": Benjamin B. Fischer, ed., *At Cold War's End: U.S. Intelligence on the Soviet Union and Eastern Europe, 1989–1991* (Washington, D.C.: Government Printing Office, 1999), NIE 11-4-89, "Soviet Policy Towards the West: The Soviet Challenge," April 4, 1989.

"other analysts believe": James Graham Wilson, *The Triumph of Improvisation: Gorbachev's Adaptability, Reagan's Engagement, and the End of the Cold War* (Ithaca: Cornell University Press, 2014), 152.

134 *"either help or hurt"*: Maynard, *Out of the Shadow*, 17.

"when we are ready": PPP, "The President's News Conference," April 7, 1989.

"environmental briefing": Ibid.

"wasted year": Westad, *Reviewing the Cold War*, 365.

"Washington Fumbles": Matlock, *Autopsy on an Empire*, 177,

"asking an architect": Maynard, *Out of the Shadow*, 16.

135 *"at the creation"*: Dean Acheson, *Present at the Creation: My Years in the State Department* (1969; New York: W. W. Norton, 1987).

"beyond our best expectations": BPL, Scowcroft Files, USSR Chronological Files, NSF Files, Folder: Soviet Power Collapse in Eastern Europe — NSF May 1989 [2], Scowcroft to POTUS, May 25, 1989.

136 *"criticizes its own"*: PPP, "Remarks at the Texas A&M University Commencement Ceremony," May 12, 1989.

"cannot simply be declared": "Transcript of Bush's Remarks on Transforming Soviet-American Relations," *New York Times*, May 13, 1989.

Eisenhower administration: Associated Press, "Bush Revives Ike's 'Open Skies' Proposal," *Los Angeles Times*, May 12, 1989.

"beyond containment": Don Oberdorfer, "Bush Finds Theme of Foreign Policy 'Beyond Containment,'" *Washington Post*, May 28, 1989.

"cautious and prudent": Maynard, *Out of the Shadow*, 25.

137 *"in ferment"*: PPP, "Remarks to the Citizens of Mainz, Federal Republic of Germany," May 31, 1989.

"beyond containment": PPP, "Remarks at the Texas A&M University Commencement Ceremony," May 12, 1989.

"every reason to hope": BPL, Scowcroft Papers, USSR Collapse, US-Soviet Relations Chronological File, US-Soviet Relations through 1991 (May–June 1989), "Remarks by Secretary Cheney," May 10, 1989, and Zelikow to Scowcroft, May 9, 1989.

138 *"may not be over"*: BPL, Scowcroft Papers, USSR Collapse, US-Soviet Relations Chronological File, US-Soviet Relations through 1991 (May–June 1989), "Remarks by Secretary Cheney," May 10, 1989.

"jury": Beschloss and Talbott, *At the Highest Levels*, 10.

"This guy": Wohlforth, *Cold War Endgame*, 236.

"supreme folly": BPL, Scowcroft Papers, USSR Collapse, US-Soviet Relations Chronological File, US-Soviet Relations through 1991 (May–June 1989), "Remarks by Secretary Cheney," May 10, 1989.

"serves too well": Ibid., and Zelikow to Scowcroft, May 9, 1989. "However you dispose of the Cheney speech at breakfast," Scowcroft's aide wrote, "please let me know if it's appropriate afterwards to provide a copy of the [president's] draft speech to DOD. (Incidentally, I'd prefer not to. Just increased chances of leaking the President's message and getting more kibitzers involved)."

Cheney complied: BPL, Scowcroft Files, USSR Collapse, US-Soviet Relations Chronological File, US-Soviet Relations through 1991 (May–June 1989), Snider to Scowcroft, "Secretary Cheney's Speech for Tomorrow Evening," May 10, 1989.

"got to get ahead": P. Edward Haley, *Strategies of Dominance: The Misdirection of U.S. Foreign Policy* (Washington, D.C.: Woodrow Wilson Center Press, 2006), 25.

139 *"on its own"*: Beschloss and Talbott, *At the Highest Levels*, 77.

"how it can be done": Maynard, *Out of the Shadow*, 30–34. See also Meacham, *Destiny and Power*, 370–71.

"intellectual audacity": Meacham, *Destiny and Power*, 372, 371.

140 *"cavalry"*: Maynard, *Out of the Shadows*, 32–33. For Mitterrand and the important calculation of 7,500 total American troops as the realistic net reduction following NATO's negotiations, see Sparrow, *The Strategist*, 305.

"aggressively": Meacham, *Destiny and Power*, 372.

"on the offensive": Beschloss and Talbott, *At the Highest Levels*, 80.

8. FROM A FUNERAL TO A RIOT

141 *"tired but happy"*: Bush and Scowcroft, *A World Transformed*, 85.

"After the euphoria": Bush, *All the Best*, 427.

142 *Hong Kong*: Jan Ackerman, "Andrew Tod Roy: Missionary, Teacher in China," *Pittsburgh Post-Gazette*, May 7, 2004.

older brother: Jennifer Schuessler, "An Old Chinese Novel Is Racy Reading Still," *New York Times*, November 18, 2013.

"reason the Fang issue": BPL, NSC, Paal Papers, China-US January–July 1989 [6], Washington to American Embassy Beijing, "Re: D.A.S. Roy Lunch with PRC Minister Zhao," March 6, 1989.

143 *"disturbance"*: Goldman, *Sowing the Seeds*, 204.

"bourgeois liberalization": Schell and Shambaugh, *The China Reader*, 531–32.

"kill anyone": Roderick MacFarquhar, *The Politics of China: The Eras of Mao and Deng* (Cambridge: Cambridge University Press, 1997), 398. See also Edward Gargan, "Deng's Crushing of Protest Is Described," *New York Times*, January 14, 1987.

144 *"you people"*: Vogel, *Deng Xiaoping*, 585.

"Why are we here?": Dingxin Zhao, *The Power of Tiananmen: State-Society Relations and the 1989 Beijing Student Movement* (Chicago: University of Chicago Press, 2001), 148, 149.

freedom: Vogel, *Deng Xiaoping*, 599.

145 *"the democratic way"*: Martin Walker, "NATO Is a Winner, Says Upbeat Bush," *The Guardian*, May 24, 1989.

"on their lips": Randolph Kluver, "Rhetorical Trajectories of Tiananmen Square," *Diplomatic History* 34, no. 1 (January 2010): 87. Soviet officials, still unclear what their own march toward freedom meant, took a more jaundiced view of Chinese student claims. "It was worrisome," Deputy Foreign Minister Igor Rogachev told Matlock in mid-June, "that students carrying placards calling for democracy and various 'freedoms' were unable to define what these things were. The demonstrators did not seem particularly open to different points of view or different approaches to issues." BPL, NSC Files, White House Situation Room Files, Tiananmen Square Files, Folder: China — Part 4 of 5 Tiananmen Square Crisis, May–June 1989 [3], AmEmbassy Moscow to State, June 8, 1989.

"little time": Vogel, *Deng Xiaoping*, 599.

"Our demand": Dingxin Zhao, *The Power of Tiananmen*, 148.

"political lessons": Ling Shiao to author, May 3, 2015.

146 *"not to mistake"*: Robert L. Suettinger, *Beyond Tiananmen: The Politics of U.S.-China Relations, 1989–2000* (Washington, D.C.: Brookings Institution Press, 2003), 31.

147 *"no god, no czar"*: Goldman, *Sowing the Seeds*, 305.

"turmoil on the campuses": Zhang Liang, Andrew Nathan, and Perry Link, eds., *The Tiananmen Papers* (New York: Public Affairs, 2001), 57. Hereafter *Tiananmen Papers*.

joined the chorus: Ibid., 71–73.

148 *"tiny minority"*: Andrew J. Nathan, "The Tiananmen Papers," in *Foreign Affairs Special Collection: Tiananmen and After*, ed. Gideon Rose (2009): 15.

"got to be explicit": Gerard Lemos, *The End of the Chinese Dream: Why Chinese People Fear the Future* (New Haven: Yale University Press, 2012), 47.

"Clear Stand": Michael Oksenberg, Lawrence R. Sullivan, and Marc Lambert, eds., *Beijing Spring, 1989: Confrontation and Conflict: The Basic Documents* (Armonk, N.Y.: M. E. Sharpe, 1990), 206–8.

free passage: *Tiananmen Papers*, 78.

"innocent and well-meaning": Lemos, *End of the Chinese Dream*, 47.

"If you believe": *Tiananmen Papers*, 83, 90.

149 *"Is this for real"*: Lilley, *China Hands*, 297.

"China I know": Mann, *About Face*, 185.

"did not predict": Association for Diplomatic Studies and Training, Foreign Affairs Oral History Project, Ambassador Winston Lord, April 28, 1998.

"sake of world peace": BPL, NSC, Paal Papers, 1989–1990 China Files, China-U.S. January–July 1989 [6], AmEmbassy Beijing to SecState Washington, April 12, 1989.

dismissed: Patrick Tyler, *A Great Wall: Six Presidents and China* (New York: Public Affairs, 2000), 345–47.

150 *"homework"*: Mann, *About Face*, 185.

"government's desk officer": Baker, *Politics of Diplomacy*, 100.

"really invested": Rothkopf, *Running the World*, 291.

151 *focal points:* Department of State Virtual Reading Room, William Clark to Robert Kimmitt, "Checklist for Your Meeting with Ambassador James R. Lilley, Thursday, April 27, 10:15 A.M," April 26, 1989.

"volcano": Lilley, *China Hands,* 398.

"several coups": Ling Shiao to author, May 3, 2015.

bear of a man: Lilley's memoir, *China Hands,* is extraordinary and worth the read. As for scissors, I had the pleasure of witnessing that myself.

152 *"Henry actually":* Mann, *About Face,* 64. See also Priscilla Roberts, ed., *Window on the World: The Beijing Diaries of David Bruce, 1973–74* (Hong Kong: Centre of Asian Studies, 2001), 271.

"tuck away": Engel, *The China Diary,* 171.

"Demos": Lilley, *China Hands,* 298.

"only three deep": Calhoun, *Neither Gods nor Emperors,* 51.

153 *"ordered his men":* Ibid.

"Students are networking": *Tiananmen Papers,* 106.

"vast numbers": Vogel, *Deng Xiaoping,* 608.

"My children": *Tiananmen Papers,* 138.

"Their motive": Goldman, *Sowing the Seeds,* 312.

"shed a little blood": Tyler, *A Great Wall,* 349.

"worldwide trend": John W. Garver, *China's Quest: The History of the Foreign Relations of the People's Republic of China* (New York: Oxford University Press, 2015), 468.

154 *"doesn't hold up the banner":* Engel, *The Fall of the Berlin Wall,* 13.

"haughty": BBC News, "Witnessing Tiananmen: Student Talks Fail," May 28, 2004.

"create an equal status": Kluver, "Rhetorical Trajectories of Tiananmen Square," 77–78.

"hothouse generation": Perry Anderson, "Sino-Americana," *London Review of Books* 34, no. 9 (February 2002): 20–22.

"month-old infant": Ji Chaozhu, *The Man on Mao's Right: From Harvard Yard to Tiananmen Square, My Life Inside China's Foreign Ministry* (New York: Random House, 2008), 326–27.

"actual physical fear": Miller Center oral history, Robert Gates, 27.

155 *"Old Testament":* Lilley, *China Hands,* 309.

"Government by Old Men": Calhoun, *Neither Gods nor Emperors,* 72.

"Revenge was in his nature": Lilley, *China Hands,* 309.

"lose your youth": Association for Diplomatic Studies and Training, "Moments in U.S. Diplomatic History: The Tiananmen Square Massacre — June 4, 1989," http://adst.org/2013/05/the-june-4-massacre-in-tiananmen-square-1989 (accessed December 14, 2016).

"No government in the world": Lilley, *China Hands,* 300.

"mad dog": Ibid., 303.

"I have words": PPP, "Remarks and a Question-and-Answer Session with Reporters Following a Luncheon with Prime Minister Brian Mulroney of Canada," May 4, 1989.

156 *"time for caution":* PPP, "The President's News Conference with President Mitterrand of France," May 21, 1989.

"lead to bloodshed": Ibid.

"I am old enough": Martin Walker, "The Quiet American," *The Guardian,* May 24, 1989.

157 *"United States is watching":* State Department Virtual Reading Room, "Secretary's Meeting with Wan Li," May 25, 1989.

"required order": BPL, NSC, Situation Room Files, Tiananmen Square Crisis, File: China 2 of 5 Tiananmen Square Crisis (May–June 1989) [2], SecState Washington to American Embassy Beijing, "Secretary's Meeting with Wan Li," May 26, 1989.

"deeply worried": BPL, NSC, Karl Jackson Files, Subject Files, China 1989 [2], "Meeting with

Wan Li," May 23, 1989. See also "Points to Be Made for Photo Op with Chairman Wan Li,"
May 23, 1989; and "The President's Meeting with Wan Li," n.d.

158 *"slowed the pace"*: BPL, NSC, Paal Papers, Subject Files, China, "Wan Li Visit, 1989."
comments in check: BPL, NSC, Karl Jackson Files, Subject Files, China 1989 [2], "Meeting
with Wan Li," May 23, 1989, and "The President's Meeting with Wan Li," n.d.
"I have no desire": BPL, Scowcroft Papers, China 2 of 5 Tiananmen Square Crisis [2], White
House Situation Room Files, Washington to American Embassy Beijing, "Letter from President Bush to Deng Xiaoping," May 27, 1989.

9. CRACKDOWN

160 *"When Gorbachev's here"*: Tiananmen Papers, 143.
"Fang Lizhi incident": BPL, NSC, Paal Series, File: China — Foreign Policy [2], AmEmbassy
Beijing to State, May 5, 1989.
"call this turmoil": Tiananmen Papers, 161.

161 *longer official one*: For the American embassy's view of Gorbachev's visit, see BPL, NSC, Paal
Papers, 1989–1990 China Files, Folder: China-U.S. January–July 1989 [6], AmEmbassy to
State, April 12 and May 20, 1989.
"China's Walesa": Tiananmen Papers, 179. See also American predictions, BPL, NSC, Situa-
tion Room Files, Tiananmen Square Crisis Files, China 1 of 5 Tiananmen Square Crisis [1],
May 16, 1989.
"We do not believe": Qian Qichen, *Ten Episodes in China's Diplomacy* (New York: HarperCol-
lins, 2005), 29–31.
tired and old: Vogel, *Deng Xiaoping*, 613–14.

162 *"still need"*: Suettinger, *Beyond Tiananmen*, 46.
"Deng Xiaoping to the first": Gorbachev, *Memoirs*, 490.
"hotheads, too": Bill Keller, "Soviets and China Resuming Normal Ties After 30 Years; Beijing
Pledges 'Democracy'; the Envy Is Mutual," *New York Times*, May 17, 1989.
"anarchy": Tiananmen Papers, 178.
"guns and cannon": Alexander Pantsov, with Steven Levine, *Deng Xiaoping: A Revolutionary
Life* (New York: Oxford University Press, 2015), 414.
"After thinking": Andrew Nathan, "The Tiananmen Papers," in Rose, *Tiananmen and After*,
24.
"The minority yields": Tiananmen Papers, 190.

163 *"very undisciplined"*: Ibid., 217. See also Jussi Hanhimaki and Odd Arne Westad, *The Cold
War: A History in Documents and Eyewitness Accounts* (New York: Oxford University Press,
2004), 597.
"coup engineered": BPL, Scowcroft Papers, China 2 of 5 Tiananmen Square Crisis [2], White
House Situation Room Files, AmEmbassy Beijing to State, May 24, 1989.

164 *Ji Chaozhu*: David Barboza, "The Man on Mao's Right, at the Center of History," *New York
Times*, February 18, 2012. See also Ji Chaozhu, *The Man on Mao's Right* (New York: Random
House, 2008).
"genuine revolution": BPL, NSC, Situation Room Files, Tiananmen Square Crisis Files, China
1 of 5 T Square Crisis [2], AmConsul Hong Kong to State, May 19, 1989.
"no alternative": Dingxin Zhao, *The Power of Tiananmen*, 183. As Deng supposedly said,
"You carry these things out, and Westerners forget." Lilley, *China Hands*, 309. Lilley made the
same point, employing the same quote, during his 1998 interview with the Association for
Diplomatic Studies and Training, Foreign Affairs Oral History Project, May 21, 1998. Vogel
notes Deng making a similar point (*Deng Xiaoping*, 617).

"official": Madsen, *China and the American Dream*, 25–26.

"Midnight": For Rather's broadcasts that night, see http://eightiesclub.tripod.com/id122.htm (accessed April 22, 2014).

165 *"This is"*: http://www.danrather.com/videos/a-life-in-the-news.html (accessed April 24, 2014).

166 *"Even in the restroom"*: Andrew Jacobs and Chris Buckley, "Tales of Army Discord Show Tiananmen Square in New Light," *New York Times*, June 2, 2014.

"had not expected": Vogel, *Deng Xiaoping*, 620.

"soldiers' hearts": Ibid.

"Forever is the people's day": BPL, NSC, Situation Room Files, Tiananmen Square Crisis Files, China 1 of 5 Tiananmen Square Crisis [3], AmEmbassy to State, May 22, 1989.

"matter of policy": BPL, NSC, Situation Room Files, Tiananmen Square Crisis Files, China 1 of 5 Tiananmen Square Crisis [3], AmEmbassy to State, May 21, 1989.

167 *"permanence about it"*: National Security Archive, Electronic Briefing Book no. 47, "U.S. Tiananmen Papers," AmEmbassy Beijing to SecState, May 20, 1989.

"our principles": BPL, Scowcroft Files, China Files, 2 of 5 Tiananmen Square Crisis [2], AmEmbassy Beijing to SecState, May 24, 1989.

"right side of the winners": Ibid.

168 *"Grab your sticks"*: For Baker's account of this conversation, see his *Politics of Diplomacy*, 103.

"carnage": Ibid.

169 *"History will show"*: Calhoun, *Neither Gods nor Emperors*, 124, 122.

"seemed to wilt": BPL, NSC Files, Situation Room Files, Folder: China — Part 2 of 5 Tiananmen Square Crisis, AmEmbassy to SecState, June 3, 1989. This cable in particular offers a useful snapshot synopsis of the prior fortnight's events from the embassy's perspective.

"passionate statement": BPL, NSC Files, Situation Room Files, Folder: China — Part 2 of 5 Tiananmen Square Crisis, AmEmbassy to SecState, "Suggested Talking Points for CNN Interview," June 3, 1989. See also BPL, NSC, Paal Papers, 1989–1990 China Files, Folder: China-US January–July 1989 [4], Roy to Scowcroft, June 5, 1989, and AmEmbassy Beijing to SecState, June 5, 1989.

170 *"my life and my loyalty"*: *Tiananmen Papers*, 298.

"Tomorrow it won't be there": Lilley, *China Hands*, 377.

"Each American administration": Andrew J. Nathan, "China and International Human Rights: Tiananmen's Paradoxical Impact," in *The Impact of China's 1989 Tiananmen Massacre*, ed. Jean-Philippe Béja (New York: Routledge, 2011), 213.

"to cultivate": Garver, *China's Quest*, 474.

171 *"want to see turmoil"*: *Tiananmen Papers*, 357.

"its entire propaganda machine": *Tiananmen Papers*, 358, 359.

"merciless": "Who Wanted Troops in the Square, Who Didn't and What They Said About It," *New York Times*, June 6, 2001.

"no bloodshed": *Tiananmen Papers*, 370.

172 *"become so dangerous"*: "Interview at Tiananmen Square with Chai Ling," Asia for Educators, http://afe.easia.columbia.edu/special/china_1950_chailing.htm (accessed February 2, 2016). See also Patrick Tyler, "6 Years after Tiananmen Massacre, Survivors Clash Anew on Tactics," *New York Times*, May 30, 1995.

"safeguard their lives": Rana Mitter, *A Bitter Revolution: China's Struggle with the Modern World* (New York: Oxford University Press, 2005), 280.

"live ammunition": Calhoun, *Neither Gods nor Emperors*, 127.

173 *"noise shaking the heavens"*: Timothy Brook, *Quelling the People: The Military Suppression of the Beijing Democracy Movement* (New York: Oxford University Press, 1992), 131.

"*roasted*": BPL, NSC Papers, Situation Room Files, Tiananmen Square Crisis Files, Folder: China — Part 3 of 5 Tiananmen Square Crisis, AmEmbassy to SecState, June 4, 1989.

scenes were repeated: Xinsheng Liu, personal correspondence with author, May 18, 2006. See also Lilley, *China Hands*, 320.

174 "*Chinese don't kill Chinese*": BPL, NSC Papers, Situation Room Files, Tiananmen Square Crisis Files, Folder: China — Part 3 of 5 Tiananmen Square Crisis, AmEmbassy to SecState, June 4, 1989.

"*never dreamed*": Calhoun, *Neither Gods nor Emperors*, 129.

Liu Xiaobo: My thanks to Chen Jian for pointing out Liu's key role at this critical moment.

"*Don't shoot*": *Tiananmen Papers*, 380.

10. UNTYING THE KNOT

175 "*What the hell*": Miller Center oral history, Robert Gates, 28.

"*deeply deplore*": James Gerstenzang and Doyle McManus, "Bush Deplores Troop Assault on China Crowd," *Los Angeles Times*, June 4, 1989.

176 "*going to be difficult*": David M. Lampton, *Same Bed, Different Dreams: Managing U.S.-China Relations, 1989–2000* (Berkeley: University of California Press, 2001), 21.

"*pretty puny reaction*": Thomas Friedman, "Crackdown in Beijing: Administration Ponders Steps on China," *New York Times*, June 5, 1989.

"*folly*": Helen Dewar, "Bush Decries Chinese Decision to Use Force," *Washington Post*, June 4, 1989.

"*do it for him*": Bill McAllister, "Lawmakers Ask Strong U.S. Action; Punish Authorities, White House Told," *Washington Post*, June 5, 1989.

"*murder*": Ibid.

Calls for cessation: Mann, *About Face*, 198.

"*encourage the symbolism*": McAllister, "Lawmakers Ask Strong U.S. Action."

177 "*be able to accommodate*": PPP, "The President's News Conference," June 5, 1989.

"*I'm the president*": Ibid.

"*want us to move*": Bush and Scowcroft, *A World Transformed*, 98. See also McAllister, "Lawmakers Ask Strong U.S. Action."

initial sanctions: Mann, *About Face*, 195.

"*no poetry*": Thomas Friedman, "Taking the Measure of a 'Measured Response,'" *New York Times*, July 2, 1989.

178 "*appalled*": Robert D. McFadden, "The West Condemns the Crackdown," *New York Times*, June 5, 1989.

"*Britain will continue*": Friedman, "Taking the Measure of a 'Measured Response.'"

"*cannot riddle*": Associated Press, "Reagan Urges 'Risk' on Gorbachev: Soviet Leader May Be Only Hope for Change, He Says," *Los Angeles Times*, June 13, 1989.

"*look beyond the moment*": PPP, "The President's News Conference," June 5, 1989.

Rallies: Warren Cohen, *America's Response to China* (New York: Columbia University Press, 2010), 241.

march down Pennsylvania Avenue: For details of these protests, see McFadden, "The West Condemns the Crackdown"; and "World Leaders Outraged by Army Action," *New Straits Times*, June 6, 1989.

"*love affair*": Madsen, *China and the American Dream*, 10.

179 "*not forward*": Walter Goodman, "Many Big Stories to Tell, but the Biggest Is China," *New York Times*, June 5, 1989.

"pushed back her position": Thomas Friedman, "Turmoil in China," *New York Times*, June 10, 1989.

"forcibly": Madsen, *China and the American Dream*, 10.

"What is that guy doing": Kate Pickert, "The Backstory: Tank Man at 25: Behind the Iconic Tiananmen Square Photo," *Time*, June 4, 2014.

180   *"waited for the moment"*: Ibid.

"still believe that": PPP, "The President's News Conference," June 5, 1989. See also Scott Kennedy, ed., *Cross China Talk: The American Debate over China Policy Since Normalization* (Lanham, Md.: Rowman & Littlefield, 2003), 88.

"total break": PPP, "The President's News Conference," June 5, 1989.

"commercial incentive": Ibid.

181   *"long way to go"*: David Hoffman, "China Executions Force Bush to Focus on Future," *Washington Post*, June 25, 1989.

"Don't disrupt": Lampton, *Same Bed, Different Dreams*, 21.

"long haul": Bush and Scowcroft, *A World Transformed*, 98.

"lashing back": Richard M. Nixon, "Revulsion Real; Reprisal Wrong," *Los Angeles Times*, June 25, 1989. Bush received an advance copy of Nixon's essay. BPL, Brent Scowcroft Collection, Special Separate China Notes Files, China Files, China 1989 (sensitive) [2], Office of Richard Nixon to Scowcroft, June 21, 1989.

"practices of Maoism": Henry Kissinger, "The Drama in Beijing," *Washington Post*, June 11, 1989.

"fuzzy-headed": Association for Diplomatic Studies and Training, Moments in U.S. Diplomatic History, "Managing a Crisis: The Ramifications of Tiananmen Square," Richard Solomon interview, http://adst.org/2014/05/managing-a-massacre-the-ramifications-of-tiananmen-square/ (accessed December 15, 2016).

"We believe": BPL, NSC Files, White House Situation Room Files, Tiananmen Square Crisis Files, Folder: China Part 3 of 5 Tiananmen Square Crisis (May–June 1989) [4], AmEmbassy Beijing to SecState, June 6, 1989.

182   *'I did not want'*: National Security Archive, Electronic Briefing Book no. 473, *Tiananmen at 25 Years*, "Cable from DIA Headquarters, June 5, 1989."

"be calm": Bush and Scowcroft, *A World Transformed*, 98.

"great chaos": National Security Archive, Electronic Briefing Book no. 473, *Tiananmen at 25 Years*, "Cable from DIA Headquarters, June 5, 1989."

"reports of clashes": National Security Archive, Electronic Briefing Book no. 16, *Tiananmen Square, 1989: The Declassified History*, "Secretary of State's Morning Summary for June 6, 1989, China: Descent into Chaos."

"blame the foreigners": BPL, NSC Files, White House Situation Room Files, Tiananmen Square Crisis Files, Folder: China Part 3 of 5 Tiananmen Square Crisis (May–June 1989) [4], Lilley to Kimmitt, June 4, 1989.

"flagpole": BPL, NSC Papers, White House Situation Room Files, Folder: China Part 3 of 5 Tiananmen Square Crisis (1989) [1], AmEmbassy Beijing to SecState, June 4, 1989.

183   *"watch our step"*: BPL, NSC Papers, White House Situation Room Files, Folder: China Part 3 of 5 Tiananmen Square Crisis (1989) [1], Lilley to Kimmitt, June 4, 1989.

"Do not go": Roy Morley, *The United Nations at Work in Asia* (Jefferson, N.C.: McFarland, 2014), 149.

"above the second floor": Lilley, *China Hands*, 327.

"warned us": Ibid.

sniper's fire: For an immediate embassy report of the incident, see BPL, NSC Files, File: China Part 3 of 5 Tiananmen Square Crisis 1989 [4], AmEmbassy Beijing to SecState, June 7, 1989.

immediate session: BPL, NSC Files, China, Part 3 of 5 Tiananmen Square Crisis 1989 (May–June 1989) [5], AmEmbassy Beijing to SecState, June 7, 1989.

184 *"one guy on the roof":* Lilley, *China Hands,* 329.

"You have insulted": Ibid. "It was not," Lilley noted, "a blessed exchange."

"close the door": Ibid., 327.

did not want: Chinese efforts to isolate foreigners clearly hampered Washington's ability to gain on-the-ground perspective. "We caution the department that our conclusions are based on an amalgam of informed speculation, rumors that have the ring of truth, and a fairly close reading of the newspapers," Lilley advised on June 1. "This is all we have in a city in which our contacts, out of a justifiable fear for their safety, refuse to see us." BPL, NSC Files, White House Situation Room Files, Tiananmen Square Crisis Files, Folder: China — Part 4 of 5 Tiananmen Square Crisis (May–June 1989) [3], AmEmbassy Beijing to SecState, June 11, 1989.

"are you doing": Mann, *About Face,* 203. For a sanitized version, see Mann, *The Obamians: The Struggle Inside the White House to Redefine American Power* (New York: Penguin, 2013), 80.

"Send somebody": Suettinger, *Beyond Tiananmen,* 70.

"guest": Fang Lizhi, *Bringing Down the Great Wall: Writings on Science, Culture, and Democracy in China* (New York: W. W. Norton, 1992), 281.

185 *"ballistic":* Lilley, *China Hands,* 334.

"storm the compound": BPL, NSC Files, White House Situation Room Files, Tiananmen Square Crisis Files, Folder: China — Part 4 of 5 Tiananmen Square Crisis (May–June 1989) [3], SecState to Beijing, June 12, 1989.

"no choice": Meacham, *Destiny and Power,* 374.

"awful hard": PPP, "The President's News Conference," June 8, 1989.

"pissed off": Meacham, *Destiny and Power,* 374.

"Nobody knows": PPP, "The President's News Conference," June 8, 1989.

186 *"you usually take his call":* Engel, *The China Diary,* 459.

"self-righteous rage": BPL, NSC Files, China, Part 5 of 5 Tiananmen Square Crisis 1989 [1], AmEmbassy Beijing to SecState, June 21, 1989. See also BPL, NSC Files, China, Part 3 of 5 Tiananmen Square Crisis 1989 [1], Briefing from the National Intelligence Council, "Deng's Political Balance Sheet," July 11, 1989.

187 *"situation in China":* National Security Archive, Electronic Briefing Book no. 16, *Tiananmen Square, 1989: The Declassified History,* State Department Bureau of Research and Intelligence, "Status Report of Situation in China as of July 26, 1989."

"transition to a moderate": BPL, NSC Files, China, Part 5 of 5 Tiananmen Square Crisis 1989 [1], DIA to various, June 21, 1989.

"objectively they could be": Miller Center oral history, Brent Scowcroft, 62.

188 *"executing people":* Andrew Bacevich, *American Empire: The Realities and Consequences of U.S. Diplomacy* (Cambridge: Harvard University Press, 2002), 65.

"cannot have a meeting": Bush and Scowcroft, *A World Transformed,* 103.

to reach Deng: Personal ties mattered here as well, facilitating communication even amidst the crisis. Scowcroft had known Han Xu, China's ambassador, since his 1971 trip with Nixon to Beijing, where Han served as vice director of protocol in the Foreign Ministry. "He was literally the first Chinese communist official I met," recalled Scowcroft, and after Han's appointment to Washington for long periods over the ensuing two decades, they "became good friends." Ibid.

"composed it myself": For the complete letter, see ibid., 100–103.

191 *"blasting":* Ibid., 104.

"butchers of Beijing": As Robert Suettinger notes, "Although some of his supporters used

the term 'butchers of Beijing' to describe the Chinese leadership, Clinton used it sparingly" (*Beyond Tiananmen*, 471).

"*very personally*": Bush and Scowcroft, *A World Transformed*, 105.

192 "Had I not met": Engel, *The China Diary*, 461, emphasis added.

"*too soon to tell*": The quote is often misunderstood, at least one diplomat has recently argued. "I distinctly remember the exchange," Chas Freeman, a retired Foreign Service officer recalled, suggesting Zhou referred to 1968 and not 1789, but the "misunderstanding was too delicious to invite correction." See Richard McGregor, "Zhou's Cryptic Caution Lost in Translation," *Financial Times*, June 10, 2011.

194 "*No one thought*": Bush and Scowcroft, *A World Transformed*, 105. Chinese officials recalled events differently. "I am dubious about the perilous experience described by Scowcroft," Foreign Minister Qian Qichen later wrote. "The Chinese government had made careful arrangements," and "the plane entered Chinese territory along a route and at a time designated by China" (*Ten Episodes in China's Diplomacy*, 133).

"*endless time*": Bush and Scowcroft, *A World Transformed*, 106.

195 "*Sino-U.S. relations*": For details of the meeting, see BPL, Brent Scowcroft Collection, Special Separate China Notes Files, China Files, Folder: China 1989 (sensitive) [5], "Meeting with Chairman Deng Xiaoping of People's Republic of China," July 2, 1989.

197 "*kept the door open*": Bush, *All the Best*, 433.

198 "*no global issue*": BPL, Scowcroft Papers, Presidential Correspondence Files, Head of State Message Files, Folder: United Kingdom (outgoing) 1989–1993 [1], White House to Cabinet Office London, July 6, 1989.

11. EASTERN EUROPE ABOIL

199 "*three hours out*": Bush and Scowcroft, *A World Transformed*, 111.

"*external enemies*": Engel, *The Fall of the Berlin Wall*, 85.

"*Chinese solution*": For Honecker's remark, see http://www.chronik-der-mauer.de/index.php/de/Start/Index/id/651992 (accessed December 15, 2016).

Soviets suffered: BPL, NSC, Rice Papers, Soviet Union, USSR Subject Files, Folder: USSR Ethnic, July 19, 1989.

200 "*base of the skull*": Matlock, *Autopsy on an Empire*, 229.

"*levers*": Roy Rosenzweig Center for History and New Media, *Making the History of 1989*, "Excerpt from Anatoly Chernyaev's Diary," no. 362.

"*by force*": Anatoly Chernyaev, *My Six Years with Gorbachev* (University Park: Pennsylvania State University Press, 2000), 226.

"*Can it happen*": BPL, NSC Files, Rice Files, 1989–1990 Subject Files, Folder: China, "Can It Happen in the Soviet Union?" July 20, 1989.

201 "*general rampage*": Matlock, *Autopsy on an Empire*, 225.

"*reinforcing rods*": Ibid., 226.

"*residual doubts*": BPL, Scowcroft Papers, USSR Chronological Files, Soviet Power Collapse in Eastern Europe (July 1989), Rodman to Scowcroft, "Eastern Europe: Why Is Gorbachev Permitting This?" July 28, 1989, and Scowcroft to Bush, with POTUS handwritten comments.

"*pounding*": Mervin, *George Bush and the Guardianship Presidency*, 171.

"*stick in Gorbachev's eye*": Victor Sebestyen, *Revolution 1989: The Fall of the Soviet Empire* (New York: Pantheon, 2009), 301.

202 "*symbolic*": PPP, "Interview with Polish Journalists," June 30, 1989.

"*in a constructive vein*": William Safire, "The George & Gorby Show," *New York Times*, July 10, 1989.

"Soviets feeling comfortable": PPP, "Interview with Polish Journalists," June 30, 1989.

203 *"deliberately do anything"*: Ibid.

"Polish Press Interview": Henry Kamm, "Bush, in Polish Press Interview, Urges a Pullout of Soviet Troops," *New York Times*, July 4, 1989.

"Bush Urges": Janet Cawley, "Bush Urges Soviets to Pull Troops from Poland," *Chicago Tribune*, July 4, 1989. See also David Hoffman, "Bush Suggests Soviets Pull Troops from Poland," *Washington Post*, July 4, 1989.

"bit more considerate": Matlock, *Autopsy on an Empire*, 198–99. For the concert, see "Cliburn Plays in Moscow," *New York Times*, July 3, 1989.

"offended": Matlock, *Autopsy on an Empire*, 198–99.

"hawks": Gorbachev, *Memoirs*, 496–97.

"very strange to me": Savranskaya, *Masterpieces of History*, 497.

"precarious": Woodrow Wilson Center Digital Archive, "Conversation Between M. S. Gorbachev and FRG Chancellor Helmut Kohl," June 14, 1989, document 120811. Gorbachev was most likely referring to the NSC's ongoing evaluations of Soviet prospects, headed by Condoleezza Rice, though for the Soviet leader the distinction between evaluating his prospects and undermining them mattered less than what the former suggested about Bush's lack of faith in his prospects overall.

204 *"an intellectual"*: Rosenzweig Center, *Making the History of 1989*, "First Conversation between M. S. Gorbachev and Chancellor of FRG H. Kohl," no. 373.

"pole into an anthill": Ibid.

"elephant": Savranskaya, *Masterpieces of History*, 497, 491.

205 *"also a revolution"*: James Markham, "Gorbachev Likens Soviets to French," *New York Times*, July 5, 1989.

"was over there": Michael Dobbs, "Soviet Storms Bastille," *Washington Post*, July 5, 1989.

"didn't really say anything": Patrick Marnham, "Gorbachev's Pilgrimage Becomes an Unholy Scrum," *The Independent*, July 5, 1989.

philosophers: Gorbachev, *Memoirs*, 505.

"cannot interfere directly": James Markham, "Gorbachev Says Change Will Sweep Bloc," *New York Times*, July 6, 1989.

206 *"makes it clear"*: Rosenzweig Center, *Making the History of 1989*, "First Conversation Between M. S. Gorbachev and Chancellor of FRG H. Kohl," no. 373.

"not quite getting the message": "Interview with Dr. Condoleezza Rice," December 17, 1997, CNN Cold War, National Security Archive, http://nsarchive.gwu.edu/coldwar/interviews/episode-24/rice1.html (accessed December 15, 2016).

"a common European home": Rosenzweig Center, *Making the History of 1989*, "Europe Is a Common Home," no. 109.

full definition: Marie-Pierre Rey, "'Europe Is Our Common Home': A Study of Gorbachev's Diplomatic Concept," *Cold War History* 4, no. 2 (January 2004): 33–65.

"balance of forces": William Taubman and Svetlana Savranskaya, "If a Wall Fell in Berlin and Moscow Hardly Noticed, Would It Still Make a Noise?" in Engel, *The Fall of the Berlin Wall*, 78–79.

"try to destabilize": Rosenzweig Center, *Making the History of 1989*, "First Conversation Between M. S. Gorbachev and Chancellor of FRG H. Kohl," no. 373.

207 *"understand Moscow much better"*: Ibid.

"destabilizes": Woodrow Wilson Center Digital Archive, "Record of Conversation Between M. S. Gorbachev and Chancellor of FRG H. Kohl," June 12, 1989.

exodus: For the numbers of exiles, see http://www.chronik-der-mauer.de/index.php/de/Start/Index/id/651997 (accessed January 8, 2015).

money they could better spend: Stokes, *The Walls Came Tumbling Down*, 157.

208 *"beginning of a new process"*: Ian Traynor, "Budapest Turns the Iron Curtain into Scrap," *The Guardian*, May 4, 1989.

unlucky enough: Walter Mayr, "Hungary's Peaceful Revolution," *SpiegelOnline*, May 29, 2009.

"German is still a German": Katrin Holtz, "50 Trabants, Heading for Freedom," *Budapest Times*, July 27, 2014.

209 *threatened to resign*: Stokes, *The Walls Came Tumbling Down*, 144.

"negotiate with the devil": John K. Glenn II, *Framing Democracy: Civil Society and Civic Movements in Eastern Europe* (Palo Alto: Stanford University Press, 2001), 76.

"spitting": Mark Gilbert, *Cold War Europe: The Politics of a Contested Continent* (London: Rowman & Littlefield, 2015), 268.

"significant political high point": Charles S. Maier, *Dissolution: The Crisis of Communism and the End of East Germany* (Princeton: Princeton University Press, 1997), 132, 133.

confusing times: Cogent and well researched, the best analysis of Poland's internal politics and Cold War dynamics is Gregory F. Domber, *Empowering Revolution: America, Poland, and the End of the Cold War* (Chapel Hill: University of North Carolina Press, 2014).

210 *"in the depths"*: Stephen Kotkin, *Uncivil Society: 1989 and the Implosion of the Communist Establishment* (New York: Random House, 2009), 129–30.

"sharp defensive reaction": Domber, *Empowering Revolution*, 236. See also Stokes, *The Walls Came Tumbling Down*, 147.

"destabilization": Savranskaya, *Masterpieces of History*, 454.

"disaster": John Tagliabue, "Big Solidarity Victory Seen in Poland," *New York Times*, June 5, 1989.

"too much grain": Stokes, *The Walls Came Tumbling Down*, 147.

"freest country": Milan Svec, "Let's Warm Up Relations with Eastern Europe," *New York Times*, January 23, 1987.

211 *"more authority"*: Woodrow Wilson Center Digital Archive, "Third Conversation Between M. S. Gorbachev and FRG Chancellor Helmut Kohl," June 14, 1989. See also Woodrow Wilson Center Digital Archive, "Report of a Working Visit of Wojciech Jaruzelksi to Moscow," May 9, 1989.

discredit it: Jonathan Bolton, *Worlds of Dissent: Charter 77, the Plastic People of the Universe, and Czech Culture Under Communism* (Cambridge: Harvard University Press, 2014).

"you can go": Mikhail Gorbachev and Zdeněk Mlynář, *Conversations with Gorbachev* (New York: Columbia University Press, 2002), 85. See also Brown, *The Rise and Fall of Communism*, 540.

"already been implemented": Savranskaya, *Masterpieces of History*, 490.

212 *airbrush away*: Larry Napper, "Ceausescu's Gotterdammerung: The Cold War Ends in Romania, Bucharest, 1989–1991," in author's possession.

Actual reality: Stokes, *The Walls Came Tumbling Down*, 159.

"fertilizer": Michael Meyer, *The Year That Changed the World* (New York: Scribner, 2009), 109. See also Kotkin, *Uncivil Society*, 76.

213 *"just like wartime"*: Jill Massino, "From Black Caviar to Blackouts: Gender, Consumption, and Lifestyle in Ceaușescu's Romania," in *Communism Unwrapped: Consumption in Cold War Eastern Europe*, eds. Paulina Bren and Mary Neuburger (New York: Oxford University Press, 2012), 238.

"Breed": Misha Glenny, *The Balkans* (New York: Penguin Books, 2012), 604.

"concentration camp": "Rewiring the Brain: Romania's Orphan Story, 1966–2006," AmericanRadioworks, http://americanradioworks.publicradio.org/features/romania/b1.html (accessed February 27, 2015).

"class of his own": Gorbachev, *Memoirs*, 466; see also 474.

214 *"I began my perestroika"*: Robert English, *Russia and the Idea of the West: Gorbachev, Intellectuals, and the End of the Cold War* (New York: Columbia University Press, 2000), 326.

"running a dictatorship here": Sebestyen, *Revolution 1989*, 274.

"Human dignity": Savranskaya, *Masterpieces of History*, 253.

"primitive phenomenon": Wilson, *The Triumph of Improvisation*, 163.

"wanted to stop": Meyer, *The Year That Changed*, 91.

215 *"destabilize"*: Walter Mayr, Christian Neef, and Jan Puhl, "Winds of Change from the East. Part Six: A Meeting of the Warsaw Pact," *SpiegelOnline* (accessed October 30, 2009).

sat mute: For Jaruzelski's reluctance to attend, see Rosenzweig Center, *Making the History of 1989*, "US Embassy Warsaw to State," June 23, 1989, no. 378. See also Meyer, *The Year That Changed*, 124.

"absolute monarch": Maier, *Dissolution*, 124.

"winked": Michael Meyer, "The Wink That Changed the World," *Slate*, July 6, 2009.

216 *"Graham Greene novel"*: Larry Napper, "Ceausescu's Gotterdammerung: The Cold War Ends in Romania, Bucharest, 1989–1991," manuscript in author's possession.

"constant pressure": BPL, Memcons and Telcons, "Meeting with François Mitterrand, President of France, July 13, 1989."

"some terrible news": James Markham, "Gorbachev Says Change Will Sweep Bloc," *New York Times*, July 6, 1989.

"most are reluctant": National Security Archive, Electronic Briefing Book no. 42, *Solidarity's Coming Victory: Big, or Too Big?* "How to Elect Jaruzelski," June 23, 1989.

217 *"move to overturn"*: A. Kemp-Welch, *Poland Under Communism: A Cold War History* (New York: Cambridge University Press, 2008), 413.

"hand-wringing": National Security Archive, Electronic Briefing Book no. 42, *Solidarity's Coming Victory*, "How to Elect Jaruzelski," June 23, 1989.

"not elected president": Ibid.

"political vacuum": BPL, NSC, Rice Papers, File: President's Trip to Poland/Hungary July 1989, Scowcroft to POTUS, June 29, 1989.

218 *openly lobbied*: BPL, NSC, Rice Papers, File: President's Trip to Poland/Hungary, July 1989, Rice to Scowcroft, "Scope Papers for Poland, Hungary, the Netherlands and Events in Paris," June 28, 1989.

"terrible image": BPL, Memcons and Telcons, "Conversation with Prime Minister Margaret Thatcher of the United Kingdom," June 11, 1989.

asking for money: For the potential for Polish unrest in response to deprivations, see BPL, NSC, Rice Papers, File: Aid to Poland/Hungary [2], "Poland: Impact of Aid on Domestic Food Supplies," July 23, 1989.

"necktie of indebtedness": Domber, *Empowering Revolution*, 230.

promise Bush had made: BPL, NSC, Robert Hutchings Files, Country Files: Poland—Correspondence [2]. See also Scowcroft's response in the same file, which reads in part, "The economic package you will present meets a few of these requests, but the majority are well beyond anything we can support at this time."

"That's nothing": Domber, *Empowering Revolution*, 208.

"exists a real threat": BPL, Blackwill Papers, Chronological Files, Folder: July 1989 [3], Scowcroft to POTUS, July 8, 1989.

219 *"major domestic explosion"*: BPL, NSC, Rice Papers, File: President's Trip to Poland/Hungary, July 1989 [2 of 2], POTUS to Baker and Scowcroft, June 25, 1989.

"pick up the tab": Bush and Scowcroft, *A World Transformed*, 113.

"worst time": Miller Center oral history, Brent Scowcroft, 79. "This encouragement could not take the form of a new Marshall Plan," Bush told Brian Mulroney. "One had to show interest

and concern without making major new financial commitments." BPL, Memcons and Telcons, "Telephone Conversation with Prime Minister Mulroney of Canada, July 6, 1989."

"Polish rat-hole": Sebestyen, *Revolution 1989,* 303.

"squandered it": Miller Center oral history, Brent Scowcroft, 79.

"a fight that never ended": Bush and Scowcroft, *A World Transformed,* 114.

220 *"would have been appalled"*: BPL, NSC, Blackwill, Chronological Files: July 1989 [3], Blackwill to Scowcroft, July 3, 1989.

"gets around": Miller Center oral history, Robert Gates, 21.

"Emotions run high": BPL, Memcons and Telcons, "Telephone Call from the President to Chancellor Helmut Kohl of West Germany, June 23, 1989."

"inadequate": Bush and Scowcroft, *A World Transformed,* 114.

"new Marshall Plan": BPL, Memcons and Telcons, "Telephone Conversation with Prime Minister Mulroney of Canada, July 6, 1989."

221 *"don't have the discipline"*: Associated Press, "Sununu Apologizes for Comparing Poles to Children," July 12, 1989.

"apologize for the metaphor": James McCartney, "Bush Withheld the Sugar from Sweet-Toothed but Diabetic Poland," *Lakeland (Fla.) Ledger,* July 17, 1989.

"painful reminder": Bush and Scowcroft, *A World Transformed,* 116.

"help without interfering": BPL, NSC, Rice Papers, Subject Files: Aid to Poland/Hungary [2 of 2], "Memorandum of Conversation: July 10, 1989."

"workers favor distribution": Ibid.

222 *"delicate period"*: Ibid.

Echoing Thatcher: "The problem might be that there might be nothing concrete to do," Bush told Mitterrand, explaining his reluctance to sit down with Gorbachev. "The two could simply meet as presidents who had not yet met," Mitterrand replied. See BPL, Memcons and Telcons, "Meeting with François Mitterrand, July 13, 1989."

responded cautiously: BPL, NSC, Rice, Subject Files: Aid to Poland/Hungary [2 of 2], "Memorandum of Conversation: July 10, 1989."

"lost cause initially": Ibid.

"opened his heart": Bush and Scowcroft, *A World Transformed,* 117.

official record: PPP, "Statement by Press Secretary Fitzwater," July 10, 1989.

223 *"personal position"*: Domber, *Empowering Revolution,* 242.

"today you have the goodwill": PPP, "Toast at the State Dinner in Warsaw," July 10, 1989.

"Everyone feels uncertain": Maureen Dowd, "For Bush, a Polish Welcome Without Fervor," *New York Times,* July 11, 1989.

"Poland's time of destiny": PPP, "Remarks at the Solidarity Workers Monument in Gdansk," July 11, 1989.

"cheered anything": Bush and Scowcroft, *A World Transformed,* 122.

224 *"somewhat different reception"*: Lech Walesa, *The Struggle and the Triumph* (New York: Arcade Publishing, 1991), 212.

washing it all down: Ibid., 212; "Polish Delicacies for Bush," *New York Times,* July 11, 1989.

"listened attentively": Walesa, *The Struggle and the Triumph,* 212–13.

"I wondered whether": Bush and Scowcroft, *A World Transformed,* 121–22.

225 *"gently asked him"*: Ibid., 121.

"parted like two brothers": Walesa, *The Struggle and the Triumph,* 214.

"Not bad": Bush and Scowcroft, *A World Transformed,* 123.

"beaming": Kemp-Welch, *Poland Under Communism,* 416.

"After Bush's visit": Domber, *Empowering Revolution,* 244.

falling in line: Meyer, *The Year That Changed,* 95.

"For your sake, sir": Walesa, *The Struggle and the Triumph,* 215.

226 *"empathy"*: Constantine Pleshakov, *There Is No Freedom Without Bread!: 1989 and the Civil War That Brought Down Communism* (New York: Farrar, Straus and Giroux, 2009), 171.

"scolded me": Bush and Scowcroft, *A World Transformed*, 125. See also Barbara Bush, *A Memoir* (New York: Scribner, 1994), 302.

"Let Berlin be next": PPP, "Remarks to Students and Faculty at Karl Marx University," July 12, 1989.

227 *"77, 79 countries"*: For the press conference, see PPP, "Interview with Members of the White House Press Corps," July 13, 1989.

"put it in that way": John P. Burke, *Honest Broker? The National Security Adviser and Presidential Decision Making* (College Station: Texas A&M University Press, 2009), 402.

228 *"back of my mind"*: Bush and Scowcroft, *A World Transformed*, 130.

"I am writing": BPL, NSC, Scowcroft Collection, Special Separate China Notes Files, China 1989 (Sensitive) [4], "Bush to Gorbachev." (Bush's letter to Gorbachev appears in this collection because of the presence of a subsequent letter.)

229 *Chairman Deng*: BPL, Scowcroft Collection, Special Separate China Notes Files, China Files, Folder: China — July Letter, Bush to Deng, July 21, 1989.

230 *"sad and embarrassing"*: A. M. Rosenthal, "Inciting Eastern Europe," *New York Times*, July 11, 1989.

12. ANOTHER BORDER OPENS

231 *"cannot escape"*: Baker Papers, box 108, folder 8, Gorbachev to Bush, August 6, 1989.

"elastic formula": Palazchenko, *My Years with Gorbachev*, 143–44.

232 *"serious, substantive, and unstructured"*: BPL, Scowcroft Collection, Special Separate USSR Notes Files, Gorbachev Files, Folder: Gorbachev (Dobrynin) Sensitive 1989–June 1990 [2], Bush to Gorbachev, September 8, 1989.

"no reason for me": BPL, Scowcroft Collection, Special Separate USSR Notes Files, Gorbachev Files, Folder: Gorbachev (Dobrynin) Sensitive 1989–June 1990 [2], "Meeting with Shevardnadze," September 21, 1989.

"really wanted": Bush and Scowcroft, *A World Transformed*, 133.

"restrained, indecisive": Thomas Friedman, "Baker to See Visiting Soviet Politician," *New York Times*, September 12, 1989.

"prepare for a real summit": BPL, Scowcroft Collection, Special Separate USSR Notes Files, Gorbachev Files, Folder: Gorbachev (Dobrynin) Sensitive 1989–June 1990 [2], "Meeting with Shevardnadze," September 21, 1989.

"Just so you hear it": Ibid.

233 *"We heard our allies"*: Ibid.

234 *"inflame Russian nationalist sentiment"*: BPL, NSC, Rice Papers, 1989–90 Subject Files, Master Chronological File Log for USSR — September 1989–December 1989 [1], "Ethnic Tensions in the Soviet Union," September 1, 1989.

underlying absurdity: Padraic Kenney, *A Carnival of Revolution: Central Europe, 1989* (Princeton: Princeton University Press, 2002), 265–69.

"want change too": Meyer, *The Year That Changed*, 139; see also Stokes, *The Walls Came Tumbling Down*, 180.

"people will turn out": Meyer, *The Year That Changed*, 140.

"farther and faster": BPL, NSC, Hutchings, Country Files, Folder: Poland — General [2], Scowcroft to Bush, "Decisions Needed on Additional Assistance to Poland," n.d.

extended vacation: Maier, *Dissolution*, 126.

235 *"not indispensable"*: Frederick Taylor, *The Berlin Wall: A World Divided, 1961–1989* (New York: HarperCollins, 2006), 405.

"kaput": Sebestyen, *Revolution 1989*, 339.

"foster father": Maier, *Dissolution*, 122–23.

"biological solution": Sebestyen, *Revolution 1989*, 339.

"pan-European picnic": Christian Erdi, "The Picnic That Changed European History," August 19, 2014, http://www.dw.com/en/the-picnic-that-changed-european-history/a-4580616 (accessed December 16, 2016).

236 *"Big Bang"*: Meyer, *The Year That Changed*, 104.

"There is no question": G. Jonathan Greenwald, *Berlin Witness: An American Diplomat's Chronicle of East Germany's Revolution* (University Park: Pennsylvania State University Press, 1993), 103.

237 *"tore down the Wall today"*: Meyer, *The Year That Changed*, 118.

"futuristic novels": Condoleezza Rice and Philip Zelikow, *Germany Unified and Europe Transformed: A Study in Statecraft* (Cambridge: Harvard University Press, 1995), 62.

238 *"an inch closer"*: Sheehan, "The Transformation of Europe and the End of the Cold War," 56.

so far-fetched: Rice and Zelikow, *Germany Unified*, 63, 80.

"reckon with us": Serge Schmemann, "Morning After Europe's Vote: Is the Landscape New in London and Bonn?" *New York Times*, June 20, 1989.

Kohl sacked: Ferdinand Protzman, "Westward Tide of East Germans Is a Popular No-Confidence Vote," and "Kohl Replaces Party Official After Losses to the Far Right," *New York Times*, August 22, 1989.

"disregard them": Taubman and Savranskaya, "If a Wall Fell in Berlin," 86, 88.

239 *"late birth"*: Florian Gathmann, "Mercy Is Selective During Wartime," *Chicago Tribune*, September 29, 2006.

"middle ground": Rice and Zelikow, *Germany Unified*, 77.

"future belongs": Rodney Leach, *Europe* (London: Profile Books, 2000), 137.

240 *"slippery man"*: Josef Joffe, "The Secret of Genscher's Staying Power,'" *Foreign Affairs*, January 1, 1998. For contemporary discussion of Genscher, see BPL, NSC, Hutchings, Country Files, FRG — Cables, AmEmbassy Bonn to SecState Washington, August 9, 1989.

"Rarely had I": Meyer, *The Year That Changed*, 105.

"What do you want": National Security Archive, Electronic Briefing Book no. 490, *The Fall of the Berlin Wall, 25th Anniversary*, "Hungarian Discussions with West German Leaders," August 25, 1989.

241 *"can we count"*: Ibid.

"really needed": Tomicah Tillemann, "How to End an Empire" (Ph.D. diss., Johns Hopkins University, 2009), 115.

"choosing Europe": David Reynolds, *One World Divisible* (New York: W. W. Norton, 2000), 555.

"Everything that followed": Sarotte, *1989*, 31.

"treason": Meyer, *The Year That Changed*, 114.

"grave consequences": Rice and Zelikow, *Germany Unified*, 68.

"betraying socialism": National Security Archive, Electronic Briefing Book no. 290, *A Different October Revolution: Dismantling the Iron Curtain in Eastern Europe*, "Transcript of SED Politburo Sessions," September 5, 1989.

"One million": Shevardnadze heard this number from Foreign Minister Horn. See Stent, *Russia and Germany Reborn*, 86.

242 *"still here"*: Sarotte, *1989*, 33.

"freedom and unity": Rice and Zelikow, *Germany Unified*, 73.

"to be deplored": Ibid., 74.

permitting this: BPL, Scowcroft Papers, USSR Collapse Files, Folder: U.S.-Soviet Relations Through 1991 (September 1989), Rodman to Scowcroft, September 20, 1989.

243 *"clash of expectations"*: Ibid.

"expect reform": BPL, NSC, Rice Papers, 1989 — Subject Files, Folder: Aid to Poland/Hungary [1], Rice to Scowcroft, "Strategy Toward Poland and Hungary," September 19, 1989.

"Democrats are already drooling": BPL, NSC, H-Files, DC 046, August 30, 1989, "Meeting on Poland," August 29, 1989. See also "Meeting with the President, Kennebunkport, Maine, August 29, 1989, Talking Points: Poland."

bear the blame: BPL, NSC, Rice Papers, 1989 — Subject Files, Folder: Aid to Poland/Hungary [1], Blackwill to Scowcroft, September 19, 1989.

"stay ahead": BPL, Hutchings, Country Files, Poland — General [2], Scowcroft to Bush, "Decisions Needed on Additional Assistance to Poland." The undated document appears in a series from late September.

proper channels: Greenwald, *Berlin Witness,* 152.

244 *"Struggles of Our Time"*: Ibid., 158.

"abandoned cars": Meyer, *The Year That Changed,* 143.

245 *"catastrophe as in China"*: Ibid., 142.

"crack down": Greenwald, *Berlin Witness,* 162.

"real danger": Gareth Dale, *The East German Revolt of 1989* (Manchester: Manchester University Press, 2007), 15.

246 *"We are the people"*: Andrew Curry, "A Peaceful Revolution in Leipzig," *SpiegelOnline,* October 9, 2009, http://www.spiegel.de/international/germany/we-are-the-people-a-peaceful -revolution-in-leipzig-a-654137.html (accessed December 12, 2016).

"Save us": Engel, *The Fall of the Berlin Wall,* 7. See also Craig R. Whitney, "How the Wall Was Cracked," *New York Times,* November 19, 1989.

"cream of Party": Sebestyen, *Revolution 1989,* 333. See also Whitney, "How the Wall Was Cracked," l and Gale Stokes, *From Stalinism to Pluralism* (New York: Oxford University Press, 1996), 245.

"in a trance": Mann, *Rebellion of Ronald Reagan,* 331.

"Life punishes": Judt, *Postwar,* 613.

247 *"lauded"*: Sebestyen, *Revolution 1989,* 327.

"us or them": Mary Elise Sarotte, *The Collapse: The Accidental Opening of the Berlin Wall* (New York: Basic Books, 2014), 54–55.

13. "IT HAS HAPPENED"

248 *"tear down this wall"*: PPP, "Remarks on East-West Relations at the Brandenburg Gate in West Berlin," June 12, 1987.

"the modern Jericho": Robin Toner, "Democrats Urged to Veer Towards Center," *New York Times,* November 14, 1989.

"Substitute the word": PPP, "Press Conference," October 31, 1989.

"exacerbate your problems": BPL, Memcons and Telcons, "Meeting with Yevgeniy Primakov," October 26, 1989.

"failed to show leadership": PPP, "The President's News Conference," November 7, 1989.

249 *"steps towards democracy"*: Ibid.

"surprised": Beschloss and Talbott, *At the Highest Levels,* 131.

"not meant to bail out": PPP, "The President's News Conference," October 31, 1989.

250 *"I knew exactly"*: Ibid.

"If we mishandle it": Bush and Scowcroft, *A World Transformed,* 148.

251 *"Will they want"*: Liz Sly, "OAS Bid Failed Because Noriega Didn't Want to Give Up Power," *Chicago Tribune,* August 29, 1989.

a hundred times: Tallying United States interventions requires defining an "intervention."

For a thoughtful analysis, see Barbara Torreon, *Instances of Use of United States Armed Forces Abroad, 1798–2015* (Washington, D.C.: Congressional Research Service, 2015).

"fiat is law": Walter LaFeber, *The New Empire: An Interpretation of American Expansion, 1860–1898* (Ithaca: Cornell University Press, 1998), 262.

"ratcheting up the pressure": BPL, NSC, Price, Subject Files, Folder: Panama — May 1989 [4], Hughes to McClure, May 11, 1989.

"any brains at all": Miller Center oral history, Robert Gates, 29.

252 *"reporting was spotty"*: Dick Cheney with Liz Cheney: *In My Time: A Personal and Political Memoir* (New York: Simon & Schuster, 2011), 172.

"inaction": Sparrow, *The Strategist,* 333.

Cheney learned: Cheney, *In My Time,* 173.

"we don't know": Parmet, *George Bush,* 412.

"Amateur Hour": Sparrow, *The Strategist,* 329–30.

Keystone Kops: Robert A. Pastor, "The Delicate Balance Between Coercion and Diplomacy: The Case of Haiti, 1994," in *The United States and Coercive Diplomacy,* eds. Robert J. Art and Patrick M. Cronin (Washington, D.C.: United States Institute of Peace Press, 2003), 147.

"resurgence": "Return of the Wimp Factor," *The Oklahoman,* October 9, 1989.

"don't think it's right": Sparrow, *The Strategist,* 330.

"very candid": Ed Kelley, "U.S. Planned to Seize Noriega," *NewsOK,* October 9, 1989.

253 *"blood on its hands"*: Sparrow, *The Strategist,* 330.

"back here": Ibid.

"something to discuss": Michael Gordon, "Bush and Senators Meet on the Coup That Failed," *New York Times,* October 12, 1989; Sparrow, *The Strategist,* 330.

"very decisively": Sparrow, *The Strategist,* 332.

"vowed never": Parmet, *George Bush,* 414.

"further riots": German History in Documents and Images, vol. 10, "The Triumph of Non-Violence in Leipzig," German Historical Institute, Washington, D.C., http://germanhistory docs.ghi-dc.org/sub_document.cfm?document_id=230 (accessed March 24, 2017).

254 *"all measures"*: Sarotte, *Collapse,* 54.

"main blow": Ibid., 52.

"Three thousand": Woodrow Wilson Center Digital Archive, "Excerpt of a Politburo Meeting," October 4, 1989.

"Leave the Stones": Sarotte, *Collapse,* 55.

"them or us": Dale, *The East German Revolt of 1989,* 23.

"no compromises": Sarotte, *Collapse,* 54.

"unable to confirm": Ibid., 71, 76.

255 *"Turn around"*: Ibid., 74.

"call back": Ibid.

256 *"hundreds of thousands"*: Kenney, *Carnival of Revolution,* 279.

"relieve me of my duties": Sebestyen, *Revolution 1989,* 342.

"it has happened": Taylor, *The Berlin Wall,* 413.

"What is the difference": Ibid., 414.

257 *"earnest political dialogue"*: Serge Schmemann, "East Germany Removes Honecker and His Protege Takes His Place," *New York Times,* October 19, 1989.

"acknowledge paternity": Maier, *Dissolution,* 224.

"explosive": Taubman and Savranskaya, "If a Wall Fell in Berlin," 87.

"accident involving our soldiers": Sebestyen, *Revolution 1989,* 345–46.

258 *"raise the risk"*: BPL, NSC, Blackwill, Subject Files: German Reunification November 1989–June 1990 [1], Blackwill to Scowcroft, November 7, 1989.

"as pressures grow": Ibid.

"exert such leverage": Ibid.

"getting criticism": BPL, Memcons and Telcons, "Telephone Call from Chancellor Helmut Kohl," October 23, 1989.

259 *"Every idea is needed"*: William Tuohy, "'Every Idea Needed,' East German Leader Says," *Los Angeles Times*, October 26, 1989.

"the impression": Ibid.

"Pluralism": Greenwald, *Berlin Witness*, 243.

"We are the party": Ibid., 244, 257.

"cement heads": Savranskaya, *Masterpieces of History*, 582. Kohl employed the term "cement heads" frequently. See Maier, *Dissolution*, 225.

260 *stalling*: Dale, *The East German Revolt of 1989*, 78.

"decided something": Meyer, *The Year That Changed*, 158.

a million participants: BPL, NSC, Hutchings, Country File, Berlin — Correspondence, Hutchings to multiple parties, November 3, 1989.

"bulwark": Sebestyen, *Revolution 1989*, 314.

"pressure": Wendy Tyndale, *Protests in Communist East Germany* (Burlington, Vt.: Ashgate, 2010), 124.

"I can speak German": Sarotte, *Collapse*, 115.

"power of good": Meyer, *The Year That Changed*, 166.

261 *"this tendency"*: Sarotte, *Collapse*, 117.

262 *"also valid for West Berlin"*: Julia Sonnevend, *Stories Without Borders: The Berlin Wall and the Making of a Global Iconic Event* (New York: Oxford University Press, 2016), 69.

"What will happen": Sarotte, *Collapse*, 118.

"Thank you very much": Ibid.

"The gates": Taylor, *The Berlin Wall*, 425.

"has just declared": Hans-Herman Hertle, "The Fall of the Wall: The Unintended Self-Dissolution of East Germany's Ruling Regime," Cold War International History Project, *Bulletin* 12–13 (Fall–Winter 2001): 136–37.

263 *"Nerves are stretched taut"*: Greenwald, *Berlin Witness*, 261.

wild pigs: Sarotte, *Collapse*, 139.

264 *"coward"*: Ibid., 141.

"pushed too far": Ibid.

"If a panic started": "As It Happened Hour by Hour," *The Telegraph*, http://www.telegraph.co.uk/history/11219434/Berlin-Wall-How-the-Wall-came-down-as-it-happened-25-years-ago-live.html?service=artBody (accessed March 24, 2017).

"have to be applied for": Sarotte, *Collapse*, 145.

265 *"largest block party"*: Ibid., 155, 147.

"worldwide catastrophe": Ibid., 158.

"maybe weeks": Savranskaya, *Masterpieces of History*, 580–85.

266 *"long historical processes"*: Ibid.

"wrong party": Sebestyen, *Revolution 1989*, 360.

"horrified by the sight": Patrick Salmon et al., eds., Documents on British Policy Overseas (hereafter DBPO), ser. 3, vol. 7, *German Unification*, "Minute from Sir P. Wright to Mr. Wall," November 10, 1989, 105.

"much too fast": Margaret Thatcher Archives, "Remarks Outside Downing Street," November 10, 1989.

"happy events": Frédéric Bozo, *Mitterrand, the End of the Cold War, and German Unification* (New York: Berghahn, 2010), 113.

"play with world war": Grachev, *Gorbachev's Gamble*, 142.

"The implication": Baker Papers, box 108, folder 11, Roy to Baker, November 9, 1989.

267 *"got to be really serious"*: Sebestyen, *Revolution 1989*, 359.

"hard time": PPP, "Remarks and Question-and-Answer Session with Reporters on the Relaxation of East German Border Controls," November 9, 1989.

268 *"an emotional kind of guy"*: Ibid.

"miss an opportunity": Bush, *All the Best*, 150.

"sitting politely": Andrew Rosenthal, "Bush Sees More Changes Across Czechoslovakia," *New York Times*, November 15, 1989.

"no other goal": Woodrow Wilson Center Digital Archive, "Verbal Message from Mikhail Gorbachev to Helmut Kohl," November 10, 1989.

269 *"happiest people"*: Harold James and Marla Stone, *When the Wall Came Down: Reactions to German Unification* (New York: Routledge, 1992), 46.

"We demand it": Philip Zelikow and Condoleezza Rice, "The Final Soviet Curtain," *Vienna Review*, November 1, 2009.

"German Fatherland lives": Hitchcock, *The Struggle for Europe*, 369.

"an enormous fair": BPL, Memcons and Telcons, "Telephone Conversation with Helmut Kohl," November 10, 1989.

"even more important": Douglas Little, *Us Versus Them: The United States, Radical Islam, and the Rise of the Green Threat* (Chapel Hill: University of North Carolina Press, 2016), 59.

"overall collapse": Taubman and Savranskaya, "If a Wall Fell in Berlin," 70.

270 *"If statements are made"*: Woodrow Wilson Center Digital Archive, "Verbal Message from Mikhail Gorbachev to François Mitterrand, Margaret Thatcher, and George Bush," November 10, 1989. For Bush's formal response, see BPL, NSC, Scowcroft Collection, Gorbachev Files, Folder: Gorbachev (Dobrynin) Sensitive 1989–June 1990 [Copy Set] [5], November 15, 1989.

"countries of Eastern Europe": DBPO, Mr. Powell (No. 10) to Mr. Wall, November 10, 1989, 103.

271 *"being overrun"*: DBPO, Sir R. Braithwaite (Moscow) to Mr. Hurd, November 11, 1989, 106.

"So far no conflicts": BPL, Office of the President, Daily Files, Saturday, November 11, 1989.

"keep the West secure": Bush and Scowcroft, *A World Transformed*, 150.

"Berlin Wall has fallen": Savranskaya, *Masterpieces of History*, 586.

272 *"return to totalitarianism"*: Ibid.

14. GERMANS PAUSE . . . AND ACT

273 *"very carefully"*: Woodrow Wilson Center Digital Archive, "Information About the Content of a Telephone Conversation Between Mikhail Gorbachev and Helmut Kohl," November 11, 1989.

"much too fast": Sheila Rule, "A Sense of Delight, Tempered by Pleas for Caution," *New York Times*, November 11, 1989.

274 *"Summits take on"*: William Safire, "Floating Name Game," *New York Times*, November 26, 1989.

coincidental Malta sessions: Jason DeParle, "The Bitter Legacy of Yalta," *New York Times*, November 26, 1989.

constant pressing: N. Piers Ludlow, "Not a Wholly New Europe: How the Integration Framework Shaped the End of the Cold War in Europe," in *German Reunification: A Multinational History*, eds. Frédéric Bozo, Andreas Rödder, and Mary Elise Sarotte (New York: Routledge, 2017), 135.

275 *"all the fixed points"*: DBPO, Mr. Powell (Strasbourg) to Mr. Wall, December 8, 1989, 165. "My record may sound rather breathless but this actually reflects President Mitterrand's manner and approach," Thatcher's aide wrote.

"already entered history": Elie Wiesel, "I Fear What Lies Beyond the Wall," *Philadelphia Inquirer*, November 27, 1989.

"without hesitation": Günter Grass, *Two States — One Nation* (New York: Harcourt Brace, 1990), 122, and *From Germany to Germany: Journal of the Year 1990* (Boston: Houghton Mifflin Harcourt, 2012), 34.

"too big for Europe": Timothy Garton Ash, *History of the Present* (New York: Vintage, 2001), 287.

"true frontiers": DBPO, Mr. Powell (Strasbourg) to Mr. Wall, December 8, 1989, 165.

"terror in millions": A. M. Rosenthal, "Hidden Words," *New York Times*, February 4, 1990.

276 *"eager to see"*: "One Germany: Not Likely Now," *New York Times*, November 19, 1989.

"almost all Europeans": Michael Gordon, "Kissinger Expects a United Germany," *New York Times*, November 16, 1989.

"panzers": For humor as an insight into history, see Ben Lewis, *Hammer and Tickle* (New York: Pegasus Books, 2009).

"free to go": Charles Krauthammer, "The German Revival," in James and Stone, *When the Wall Came Down*, 175.

"France surrenders": Jeffrey A. Engel, "Bush, Germany, and the Power of Time: How History Makes History," *Diplomatic History* 37, no. 4 (2013): 643.

277 *"spread joy"*: "Why It Won't Be All Reich on the Night," *The Sun*, November 11, 1989.

"wartime children": Archie Brown, "Mikhail Gorbachev," in *Mental Maps in the Era of Détente and the End of the Cold War*, eds. Jonathan Casey and Steven Wright (New York: Palgrave Macmillan, 2015).

"History itself": Rice and Zelikow, *Germany Unified*, 151–52.

"If they want it": David Childs, *The Fall of the GDR* (New York: Routledge, 2001), 98.

"should not disturb it": Savranskaya, *Masterpieces of History*, 593.

278 *"very canny game"*: DBPO, "Minute from Mr. Adams to Sir J. Fretwell, Speaking Note for Use with the Prime Minister," 116.

"know perfectly well": Jonathan Aitken, *Margaret Thatcher: Power and Personality* (London: Bloomsbury, 2013), 593.

"National character": Maynard, *Out of the Shadow*, 56.

"taken a formal position": DBPO, "Extract from Conclusions of a Meeting of the Cabinet Held at 10 Downing Street on 15 November 1989," 124.

"difficult decade ahead": Ibid.; see also Bozo, *Mitterrand*, 196.

"living history": Naftali, *George H. W. Bush*, 86.

"Coventry": BPL, Scowcroft Collection, USSR Chronological, Soviet Power Collapse in Eastern Europe, November 1989, Scowcroft to POTUS, November 22, 1989.

"detrimental": PPP, "The President's News Conference," September 18, 1989.

"I called Brent": For this conversation, see Miller Center oral history, Robert Gates, 30–31. "President Bush's early embrace of German unification silenced those skeptics in the bureaucracy who shared the Anglo-French desire to slow down movement toward unification," Paul Wolfowitz has written; see his "Shaping the Future: Planning at the Pentagon, 1989–93," in Leffler and Legro, *In Uncertain Times*, 49–50. See also Frank Ninkovich, *The Wilsonian Century* (Chicago: University of Chicago Press, 1999), 276.

279 *"we can learn"*: PPP, "Interview with Foreign Journalists," November 21, 1989.

280 *"European squabbling"*: Frank Costigliola, "An 'Arm Around the Shoulder': The United States, NATO, and German Reunification, 1989–90," *Contemporary European History* 3, no. 1 (March 1994): 101. As Thomas J. McCormick has noted, ensuring that there would be no dominant power in Europe has been fundamental to American policy throughout the twentieth century; see his *America's Half-Century* (Baltimore: Johns Hopkins University Press, 1995), 36.

"not in fundamentals": PPP, "President's Thanksgiving Address to the Nation," November 22, 1989.

"all kinds of events": Bush, *All the Best*, 460–61.

281 *"no negative comments"*: Elizabeth Braw, "That Was Not Supposed to Happen," *Newsweek*, November 7, 2014.

"difficult legacy": Konrad H. Jarausch, *The Rush to German Unity* (New York: Oxford University Press, 1994), 82.

"blind or malicious": Richard Gray and Sabine Wilke, *German Unification and Its Discontents* (Seattle: University of Washington Press, 1996), 71.

282 *"Those Who Cried"*: Dale, *The East German Revolt of 1989*, 129, 180.

"Away with": Ibid., 129, 182.

"sell-out": Jarausch, *Rush to German Unity*, 66–67.

283 *"basis for stability"*: Ibid., 66.

"Somebody had to start": Savranskaya, *Masterpieces of History*, 652, 593.

"any major": National Security Archive, Electronic Briefing Book no. 293, cable from U.S. Embassy in Sofia, November 10, 1989.

"utter bemusement": Hutchings, *American Diplomacy*, 81.

"average man": Michael Zantovsky, *Havel: A Life* (New York: Grove Press, 2014), 288.

284 *"Milos, it's over"*: Reuters, "Dubcek Urges Czech Leaders to Step Down," *Los Angeles Times*, November 24, 1989; Dan Stets, "Protesters Jam Square in Prague," *Philadelphia Inquirer*, November 24, 1989.

"to the castle": Zantovsky, *Havel*, 296.

"an anachronism": Judt, *Postwar*, 620.

285 *"Return to Europe"*: Ibid., 619.

"only man dumb enough": Zantovsky, *Havel*, 304.

"might not live": Ibid., 288.

watch in amazement: William Safire, "Floating Name Game," *New York Times*, November 26, 1989.

286 *"coming our way"*: Bush, *All the Best*, 446.

"getting hints": Maynard, *Out of the Shadow*, 47.

"prepared for anything": Bush and Scowcroft, *A World Transformed*, 154.

"bag of tricks": Beschloss and Talbott, *At the Highest Levels*, 149.

"what are the differences": BPL, Scowcroft Collection, Misc. File: Copies from Other Files [2], Bush to Gorbachev, November 22, 1989.

287 *"fate of my own family"*: Ibid.

"the Soviets stress": BPL, NSC, Rodman Papers, Subject Files, Folder: Europe — Eastern, 1989, Matlock to State, November 9, 1989.

"were to appear": Dennis Ross, *Statecraft: And How to Restore America's Standing in the World* (New York: Farrar, Straus and Giroux, 2007), 35.

"no issue of current policy": Hutchings, *American Diplomacy*, 95.

"I have in mind": Woodrow Wilson Center Digital Archive, "Telephone Conversation Between Mikhail Gorbachev and François Mitterrand," November 14, 1989.

"sitting in my chair": Ross, *Statecraft*, 35.

288 *"tightens up"*: Meacham, *Destiny and Power*, 382.

"nothing that will destabilize": National Security Archive, Electronic Briefing Book no. 296, *The Soviet Origins of Helmut Kohl's 10 Points*, "Record of Telephone Conversation Between George H. W. Bush and Helmut Kohl," November 17, 1989.

"comfortable with the way": BPL, Scowcroft Collection, Soviet Power Collapse in Eastern Europe, November 1989 (part 2), Scowcroft to POTUS, November 22, 1989.

"state of peace": Richard Leiby, *The Unification of Germany, 1989–1990* (Westport, Conn.: Greenwood Press, 1999), 149.

"practically unthinkable": Rice and Zelikow, *Germany Unified*, 118.

"incompatible": Jeffrey Gedmin, *The Hidden Hand: Gorbachev and the Collapse of East Germany* (Washington, D.C.: AEI Press, 1992), 53.

secluded: Andrew Moravcsik, *The Choice for Europe* (Ithaca: Cornell University Press, 1998), 397.

289 *"Great speech"*: Bozo, *Mitterrand*, 124.

"on the agenda": Rice and Zelikow, *Germany Unified*, 124.

"destroy the balance": Baker, *Politics of Diplomacy*, 167.

"stand against that tide": DBPO, "Minute from Mr. Adams to Sir J. Fretwell, with Minute by Sir J. Fretwell," November 14, 1989, 115.

290 *"contradict his assurances"*: Chernyaev, *My Six Years*, 237.

"wanted to sour": Ibid., 238.

"respond negatively": BPL, Scowcroft Collection, Soviet Power Collapse in Eastern Europe, November 1989 (part 2), Scowcroft to POTUS, November 22, 1989.

291 *"shared Kohl's reasons"*: Hutchings, *American Diplomacy*, 99.

"within the context": Ibid., 100. See also Baker, *Politics of Diplomacy*, 167–68.

"reason and caution": BPL, Zelikow Papers, November 1989 (German Reunification), "Telephone Conversation with Chancellor Helmut Kohl," November 29, 1989.

"inadvertently to create instability": Ibid.

15. MALTA

292 *"communist doctrine is dead"*: Don Oberdorfer, "Leaders Come to Grips with Post–Cold War Era," *Washington Post*, November 30, 1989.

"Arby's": Tony Kornheiser, "Two Guys and a Summit," *Washington Post*, November 29, 1989.

"victors and vanquished": William Montalbano, "Diplomat: Diplomacy," *Los Angeles Times*, November 30, 1989.

293 *"firmly and irrevocably"*: Don Oberdorfer, *From the Cold War to a New Era* (Baltimore: Johns Hopkins University Press, 1998), 377.

"hotline to God": Woodrow Wilson Center Digital Archive, "Record of Conversation Between Mikhail Gorbachev and Prime Minister of Canada Brian Mulroney," November 21, 1989. Gorbachev employed the same AIDS analogy in 2006 to describe the policies of another President Bush: "Americans have a severe disease — worse than AIDS. It's called the winner's complex. You want an American-style democracy here [in Russia]. That will not work." Claire Shipman, "Gorbachev: 'Americans Have a Severe Disease,'" *ABC News*, July 12, 2006.

"Western values": National Security Archive, Electronic Briefing Book no. 298, *Bush and Gorbachev at Malta*, "Transcript of Gorbachev–John Paul II Meeting," December 1, 1989.

"avalanche of 1989": Beschloss and Talbott, *At the Highest Levels*, 153–54.

294 *"Not in my lifetime"*: Woodrow Wilson Center Digital Archive, "Record of Conversation Between Mikhail Gorbachev and Prime Minister of Canada Brian Mulroney," November 21, 1989. See also BPL, Memcons and Telcons, "Working Dinner with Canadian Prime Minister Brian Mulroney," November 29, 1989.

"lorded it over": BPL, Memcons and Telcons, "Working Dinner with Canadian Prime Minister Brian Mulroney," November 29, 1989.

"Gorbachev understands": Ibid.

295 *"timidity rap"*: Beschloss and Talbott, *At the Highest Levels*, 149–51.

296 *"version of events"*: Miller Center oral history, Brent Scowcroft, 49.

"*tempest in a teapot*": Bush and Scowcroft, *A World Transformed*, 161.

"*shipboard summit*": The best analysis of the Malta transcripts is Joshua R. Itzkowitz Shifrin-son, "The Malta Summit and U.S.-Soviet Relations: Testing the Waters Amidst Stormy Seas," *Cold War International History Project*, e-Dossier no. 40, July 2013, https://www.wilsoncenter.org/publication/the-malta-summit-and-us-soviet-relations-testing-the-waters-amidst-stormy-seas (accessed March 25, 2017). For the complete American texts, see BPL, NSC, Rice Papers, USSR Subject Files, Summit at Malta, December 1989, from which all subsequent quotations from the Malta sessions are drawn unless otherwise specified.

"*seasick summit*": Michael Dobbs, "The Seasick Summit," *Washington Post*, December 3, 1989.

297 "*damnedest weather*": Bush, *All the Best*, 446.

"*kick each other*": Ray Moseley and Timothy McNulty, "Shipboard Talks Underway," *Chicago Tribune*, December 3, 1989.

"*face lit up*": Sparrow, *The Strategist*, 311.

298 "*end of economic warfare*": Don Oberdorfer, "Bush Shifts to Back Moscow," *Washington Post*, December 4, 1989.

"*made a commitment*": Grachev, *Gorbachev's Gamble*, 182.

"*significant political investment*": Gates, *From the Shadows*, 482.

303 "*truly isolated*": Bush and Scowcroft, *A World Transformed*, 169.

"*Hell no*": ABC Evening News, December 3, 1989, http://tvnews.vanderbilt.edu/program.pl?ID=120333 (accessed September 3, 2015).

304 "*virtually no problem*": PPP, "Remarks of the President and Soviet Chairman Gorbachev," December 3, 1989.

305 "*We cannot be asked*": Ibid.

"*firmly convinced*": Beschloss and Talbott, *At the Highest Levels*, 165, 167.

"*situation impossible*": BPL, Memcons and Telcons, "Meeting with Helmut Kohl," December 3, 1989.

306 "*why should there*": See PPP, "The President's News Conference," December 4, 1989, for all quotations from the press conference.

307 "*brutalized*": Meacham, *Destiny and Power*, 388.

"*go in to win*": Colin Powell, *My American Journey* (New York: Ballantine Books, 1995), 192.

capable of drowning out: National Security Archive, Operation Just Cause, "U.S. Southcom Public Affairs After Action Report Supplement," December 20, 1989–January 31, 1990.

"*Death to our Hitler*": Ed Magnusson and John Moody, "The Guest Who Wore Out His Welcome," *Time*, January 15, 1990.

308 "*lives of American kids*": Bush, *All the Best*, 450.

"*build the public support*": Baker, *Politics of Diplomacy*, 194.

"*the incentive*": Beschloss and Talbott, *At the Highest Levels*, 170.

309 "*Freedom*": Pleshakov, *There Is No Freedom Without Bread!*, 217.

"*not just beat them*": Ibid., 218.

all because someone yelled: Peter Siani-Davies, *The Romanian Revolution of December 1989* (Ithaca: Cornell University Press, 2005), 79.

"*Be calm*": Stokes, *From Stalinism to Pluralism*, 190.

310 *casualties*: Stokes notes 1,033 official deaths during the revolution, a terrible number to be sure, but one that belies the fear expressed by international policy makers of a greater conflagration erupting (ibid.).

"*assist the Popular Front*": Thomas Friedman, "Baker Gives U.S. Approval If Soviets Act on Romania," *New York Times*, December 25, 1989.

"France would help": Robert Toth, "Turmoil in Romania," *Los Angeles Times*, December 25, 1989.

Dutch spokesman: Thom Shanker, "Wouldn't Oppose Soviet Invasion of Romania, U.S. Says," *Chicago Tribune*, December 25, 1989.

"die in full glory": Siani-Davies, *Romanian Revolution of December 1989*, 85.

311 *"visualize"*: Savranskaya, *Masterpieces of History*, 91.

"stupid": Siani-Davies, *Romanian Revolution of December 1989*, 176; Carolyn Ekedahl and Melvin Goodman, *The Wars of Eduard Shevardnadze* (University Park: Pennsylvania State University Press, 1997), 124.

"without deviation": Savranskaya, *Masterpieces of History*, 663.

"left a bad impression": Siani-Davies, *Romanian Revolution of December 1989*, 141.

"special case": Palazchenko, *My Years with Gorbachev*, 166.

312 *"explain the situation"*: Ibid., 165–66.

"over to you": Beschloss and Talbott, *At the Highest Levels*, 171. See also Masha Hamilton, "Soviets Scoff as U.S. Explains Its Motives," *Los Angeles Times*, December 22, 1989. Records disagree on precisely how the Soviet official framed this insult though they agree on its intent. See Thomas Blanton, "When Did the Cold War End?" *Cold War International History Project Bulletin* 10 (March 1998): 184.

16. NOT ONE INCH EASTWARD

314 *"don't want our troops"*: BPL, Memcons and Telcons, "Telcon with Prime Minister Thatcher," January 27, 1990.

"exciting changes": Ibid.

"regardless of the decisions": BPL, Scowcroft Collection, Gorbachev Files, Folder: Gorbachev (Dobrynin) Sensitive 1989–June 1990 [3], "Bush to Gorbachev," January 30, 1990, and "Scowcroft to POTUS," January 30, 1990.

"jury is still out": BPL, Memcons and Telcons, "Telcon with Prime Minister Thatcher," January 27, 1990.

"system of alliances": Bozo, *Mitterrand*, 193, 145.

"singling out Germany": Wolforth, *Cold War Endgame*, 59.

315 *"Gorbachev knows this"*: Sarotte, *1989*, 95.

"end-game is over": BPL, NSC, Rice, German Unification, Scowcroft to POTUS, February 14, 1990.

day-to-day bills: DBPO, "Minute from Mr. Hurd to Mrs. Thatcher," January 25, 1990, 224.

"realistic alternative": Jarausch, *Rush to German Unity*, 107.

316 *"reduce the speed"*: Bozo, *Mitterrand*, 148.

"we will go": Sarotte, *1989*, 134.

"ineluctable": Bozo, *Mitterrand*, 167.

strike: Dale, *The East German Revolt of 1989*, 197.

"but for unification": DBPO, "Minute from Mr. Hurd to Mrs. Thatcher," January 25, 1990, 225.

"unmask and paralyze": Dale, *The East German Revolt of 1989*, 197; Sarotte, *1989*, 96.

"law unto themselves": Sarotte, *1989*, 99; Dale, *The East German Revolt of 1989*, 207.

317 *"have no doubt"*: Sarotte, *1989*, 97.

"We were prepared": Vladimir Putin, *First Person* (New York: Public Affairs, 2000), 78.

Stationed in Dresden: Masha Gessen, *The Man Without a Face* (New York: Riverhead Books, 2012), 66–70.

"My comrades": Chris Bowlby, "Vladimir Putin's Formative German Years," BBC News, March 27, 2015, http://www.bbc.com/news/magazine-32066222 (accessed March 25, 2017).

"Moscow is silent": Ibid., 69. See also Celestine Bohlen, "Putin Tells Why He Became a Spy," *New York Times,* March 11, 2000.

"paralysis of power": Fiona Hill and Clifford Gaddy, "How the 1980s Explains Vladimir Putin," *The Atlantic*, February 14, 2013.

318 *"collapse of everything"*: Chris Bowlby, "Vladimir Putin's Formative German Years," BBC News, March 27, 2015. See also Gessen, *The Man Without a Face,* 64–65.

"inevitable": DBPO, Sir C. Mallaby (Bonn) to Mr. Hurd, January 25, 1990, 224.

"only result in chaos": Wilson, *Triumph of Improvisation,* 179.

"difficult to imagine": Zantovsky, *Havel,* 372.

"kept them [Europeans] oppressed": BPL, Scowcroft Collection, Gorbachev Files, File: Gorbachev (Dobrynin) Sensitive 1989–June 1990 [3], Baker to POTUS, February 8, 1990.

"Germany's NATO membership": Rice and Zelikow, *Germany Unified,* 141.

319 *"stood on our heads"*: Sparrow, *The Strategist,* 385.

"Kohl frowned": DBPO, Sir C. Mallaby (Bonn) to Mr. Hurd, January 25, 1990, 223.

"plan B": "James Baker on the Fall of the Wall," *SpiegelOnline,* September 23, 2009, http://www.spiegel.de/international/world/james-baker-on-the-fall-of-the-wall-without-american-leadership-there-would-have-been-no-unification-a-650801.html (accessed March 25, 2017).

"long view": Robert Keatly et al., "Thatcher Sees Eastern European Progress," *Wall Street Journal,* January 26, 1990.

320 *"undoubtedly support"*: BPL, Rice, Subject Files, German Unification, Scowcroft to POTUS, February 14, 1990.

"ineffective brake": DBPO, "Minute by Mr. Hurd," January 27, 1990, 229–30.

"apathy, complacency": BPL, Memcons and Telcons, "Telcon with Prime Minister Thatcher," February 24, 1990.

"stupid and unrealistic": Bozo, *Mitterrand,* 169.

"take advantage": Sarotte, *1989,* 76.

321 *"considerable contingent"*: Rice and Zelikow, *Germany Unified,* 151.

"basis of concessions": Robert Service, *The End of the Cold War, 1985–1991* (New York: PublicAffairs, 2015), 431.

"united Germany will join": Woodrow Wilson Center Digital Archive, "Discussion of the German Question at a Private Meeting in the Office of the CPSU CC General Secretary," January 26, 1990.

"withdrawal of foreign troops": Service, *The End of the Cold War,* 430.

322 *"draw out the process"*: Woodrow Wilson Center Digital Archive, "Discussion of the German Question at a Private Meeting in the Office of the CPSU CC General Secretary," January 26, 1990.

"This is our strategy": Sarotte, *1989,* 102.

"look down on the fight": Woodrow Wilson Center Digital Archive, "Discussion of the German Question at a Private Meeting in the Office of the CPSU CC General Secretary," January 26, 1990.

323 *"force of arms"*: "James Baker on the Fall of the Wall," *SpiegelOnline,* September 23, 2009, http://www.spiegel.de/international/world/james-baker-on-the-fall-of-the-wall-without-american-leadership-there-would-have-been-no-unification-a-650801.html (accessed Mar. 25, 2017).

"no longer exists": Bozo, *Mitterrand,* 171.

"Within six months": Rice and Zelikow, *Germany Unified,* 160.

324 *"judicial"*: Bozo, *Mitterrand*, 174.
"*breakthrough"*: Judt, *Postwar*, 157.

325 *"what new disaster"*: Rice and Zelikow, *Germany Unified*, 161.
seemed plausible: Francis X. Clines, "Wary Gorbachev Sees Momentum Toward German Re-unification," *New York Times*, January 31, 1990.
"certain agreement": Frederick Zilian, *From Confrontation to Cooperation* (Westport, Conn.: Praeger, 1999), 23.
"GIVES IN": Jarausch, *Rush to German Unity*, 108.
"making clear the terms": Rice and Zelikow, *Germany Unified*, 172.
"calculable": UPI, "2 Germanys Disagree on Unity," *Los Angeles Times*, February 1, 1990.

326 *"it is NATO's task"*: Frank Elbe, "The Diplomatic Path to Unity," *Bulletin of the German Historical Institute* 46 (Spring 2010): 33–44.
"special attention": Adam Rotfield and Walther Stutzle, *Germany and Europe in Transition* (New York: Oxford University Press, 1998), 23.
"cooperative security": Dieter Dettke, *Germany Says No* (Baltimore: Johns Hopkins University Press, 2009), 62.
"in practice": Kristina Spohr, "Precluded or Precedent-Setting? The 'NATO Enlargement Question' in the Triangular Bonn-Washington-Moscow Diplomacy of 1990–1991," *Journal of Cold War Studies* 14, no. 4 (Fall 2012): 4–54.

327 *"No reasonable person"*: Rice and Zelikow, *Germany Unified*, 175.
"dumped": Ibid., 168.
"go it alone": Robert Hutchings, "American Diplomacy and the End of the Cold War in Europe," in Robert Hutchings and Jeremi Suri, eds., *Foreign Policy Breakthroughs* (New York: Oxford University Press, 2015), 161.
"pressure cooker": Rice and Zelikow, *Germany Unified*, 179.
"very tempted": DBPO, Sir A. Acland (Washington) to FCO, January 20, 1990, 231.

328 *"no intention of extending"*: Spohr, "Precluded or Precedent-Setting?" 18.

329 *"volcano"*: Eugene Methvin, "Gorbachev Spoils the Party," *National Review*, March 19, 1990.
"drastically restructures": Beschloss and Talbott, *At the Highest Levels*, 179.
"madness": Palazchenko, *My Years with Gorbachev*, 170.
"Welcome to democracy": Baker, *Politics of Diplomacy*, 203.
"steep hill": BPL, Scowcroft Collection, Gorbachev, File: Gorbachev (Dobrynin) Sensitive 1989–June 1990 [3], Baker to POTUS, February 8, 1990.
"lock them into": Ibid.
"no extension": Spohr, "Precluded or Precedent-Setting?" 22.

330 *"End result"*: Sarotte, *1989*, 110; the asterisks are in the original.
"let's think": Joshua Shifrinson, "Deal or No Deal: The End of the Cold War and the U.S. Offer to Limit NATO Expansion," *International Security* 40, no. 4 (Spring 2016): 23.
"moving eastward": Ibid., 24.

331 *"could be capitalist"*: For all quotations from this meeting, see BPL, Scowcroft Collection, Gorbachev Files, Gorbachev (Dobrynin) Sensitive July–December 1990 [1], Memorandum of Conversation, "Robert Gates and V. I. Kryuchkov," February 9, 1990.

332 *"unacceptable"*: Sarotte, *1989*, 111.
"Naturally NATO": Ibid., 112.
"not expand itself": Uwe Klussmann et al., "NATO's Eastward Expansion," *SpiegelOnline Internationale*, November 26, 1990.

333 *"In parallel"*: Mark Kramer, "The Myth of a No-NATO-Enlargement Pledge to Russia," *Washington Quarterly* 32, no. 2 (April 2009): 39–61.

"our breakthrough": Rice and Zelikow, *Germany Unified*, 188.

"matter for Germans themselves": Sarotte, *1989*, 113.

"In no event": BPL, NSC, Wilson Files, Germany — Unification [2], Bush to Kohl, February 9, 1990.

"deeply gratified": Ibid.

334 "critical to assuring stability": Ibid.

"fundamental fact": Ibid.

"great documents": BPL, Memcons and Telcons, "Telcon with Kohl," February 12, 1990.

335 "do what he must": BPL, Rice, Subject Files, German Unification, Scowcroft to Bush, February 14, 1990.

"probable collision course": Ibid.

17. CAMP DAVID

336 "never been so close": William Tuohy, "German Reunification Close at Hand, Kohl Says," *Los Angeles Times*, February 16, 1990.

"substantive issues": Baker, *Politics of Diplomacy*, 216.

"take so long": Palazchenko, *My Years with Gorbachev*, 175.

337 "only complicate things": Vlad Zubock, "With His Back Against the Wall: Gorbachev, Soviet Demise, and German Unification," *Cold War History* 14, no. 4 (Fall 2010): 629.

"crosswise with you": BPL, Memcons and Telcons, "Telcon with Helmut Kohl," February 13, 1990.

"two-by-four": Baker Papers, box 176, folder 14, "Our European Strategy: Next Steps," March 12, 1990.

"roll them": Miller Center oral history, Robert Gates, 88.

338 "surreal": Sarotte, *1989*, 124.

"clear on how that works": For the interview with Baker, see Baker Papers, box 161, folder 22, Department of State (Press), "Interview of the Honorable James Baker on *This Week with David Brinkley* with Guest Interviewers George Will and Sam Donaldson," February 20, 1990. I am indebted to Josh Shifrinson for pointing out this transcript. See his "Deal or No Deal?" 24–27.

339 "special military status": Robert Pear, "Bush and Kohl Try to Allay Fears of a United Germany's Powers," *New York Times*, February 26, 1990.

"categorically rejected": Rice and Zelikow, *Germany Unified*, 190.

"question of questions": Craig Whitney, "Kohl Says Moscow Agrees Unity Issue," *New York Times*, February 11, 1990.

"moving too fast": "James Baker on the Fall of the Wall."

"too late": Rice and Zelikow, *Germany Unified*, 194.

"can't say no": Ibid.

340 "complete clarity": BPL, Rice, Subject Files, File: 2 + 4 Germany no. 1 [2], "Gorbachev on German Unification," n.d.

promote stability: Sparrow, *The Strategist*, 373.

"no formations or institutions": Rice and Zelikow, *Germany Unified*, 214.

341 "de-NATOed": Ibid., 214.

"great faith": Bush and Scowcroft, *A World Transformed*, 242.

"very attractive": Sarotte, *1989*, 121.

"future united Germany": Rice and Zelikow, *Germany Unified*, 203.

"everything is a monument": Zantovsky, *Havel*, 344.

342 *"man on a mission"*: BPL, Scowcroft Collection, Soviet Power Collapse (February–May 1990), Scowcroft to POTUS, February 17, 1990.

cigarette and sausage breaks: Thomas Friedman, "Bush Praises Havel and His 'New Page,'" *New York Times*, February 20, 1990.

"human face of the world": U.S. Congress, *Congressional Record,* 101st Cong., 2nd sess., Representative Dante Fascell, "Extension of Remarks," March 20, 1990.

"consciousness precedes being": Zantovsky, *Havel,* 357.

"pruning hooks": PPP, "Remarks and Question-and-Answer", February 21, 1990.

343 *"who's the enemy"*: For this incident, see *PPP,* "The President's News Conference," February 12, 1990.

"potential instability": Meacham, *Destiny and Power,* 402.

344 *"sell this to the Russians"*: BPL, Memcons and Telcons, "Meeting with Manfred Woerner," February 24, 1990.

345 *"not a player"*: Lally Weymouth, "Being Beastly to Germany," *National Review,* April 16, 1990.

"Nothing about us": Gregory F. Domber, "Pivots in Poland's Response to German Unification," in Bozo, Rödder, and Sarotte, *German Reunification,* 192.

"tantamount to a repetition": BPL, Rice, Subject Files, File: 2 + 4 Germany no. 1 [2], Mazowiecki to Bush, February 21, 1990.

"reluctant": BPL, Memcons and Telcons, "Memorandum of Conversation with Douglas Hurd," January 29, 1990.

"read my history": Aitken, *Margaret Thatcher,* 593.

"brake on everything": DBPO, Mr. Powell to Mr. Wall, February 23, 1990, 305.

346 *"bitter memories"*: Roy Rosenzweig Center for History and New Media, *Making the History of 1989,* "Prime Minister Margaret Thatcher Discusses Jewish Rights in the Soviet Union," February 18, 1990, https://chnm.gmu.edu/1989/archive/files/thatcher-speech-2-19-90_78ce7aa8be.pdf (accessed December 17, 2016).

"get in peace": For this conversation, see BPL, Rice, Subject Files, File: 2 + 4 Germany no. 3 [2], "Bush-Thatcher Telcon," February 24, 1990.

"I tell them apathy": The phrase "trick question" does not appear in the American memcon. See DBPO, "Letter from Mr. Powell (No. 10) to Mr. Wall: Prime Minister's Talk with President Bush," February 24, 1990, 313.

347 *"gives me heartburn"*: BPL, Memcons and Telcons, "Bush-Mulroney Telcon," February 24, 1990.

348 *"not renting our seat"*: Ibid.

"consult with your colleagues": Ibid.

"we will stay": For this conversation, see BPL, Rice, Subject Files, File: 2 + 4 Germany no. 3 [2], "Memorandum of Conversation with Helmut Kohl," February 24 and 25, 1990.

351 *Memorial Hill:* Bush and Scowcroft, *A World Transformed,* 254.

"legitimate security interests": BPL, Blackwill, Subject Files, File: German Reunification November 1989–June 1990 [1], "Statement for Presidential Press Conference."

"without reverting": Bush, *All the Best,* 460–61.

18. CONCESSION

352 *"cutting up the world"*: BPL, Memcons and Telcons, "Telephone Conversation with François Mitterrand of France," February 26, 1990.

353 *"Burundi"*: Engel, *The China Diary,* 416.

"must be consulted": BPL, Memcons and Telcons, "Telephone Conversation with François Mitterrand of France," February 26, 1990.

"particularly reject": BPL, Hutchings Collection, Country Files, File: GDR — Meetings and Visits, Baker to POTUS, June 7, 1990.

"close touch": BPL, Memcons and Telcons, "Meeting with Vaclav Havel, President of Czechoslovakia," February 21, 1990.

"free thinker": BPL, Memcons and Telcons, "Meeting with Acting President Arpad Goncz of Hungary," May 18, 1990.

"divide the world": BPL, Memcons and Telcons, "Meeting with Prime Minister Giulio Andreotti of Italy," March 6, 1990.

354   *"good and close friend"*: Ibid.

"With troops": BPL, Memcons and Telcons, "Meeting with Acting President Arpad Goncz of Hungary," May 18, 1990.

"overlooking the trust factor": Engel, *The China Diary*, 431.

"Openness is the enemy": PPP, "Address to the 44th Session of the United Nations General Assembly," September 25, 1989.

"finest sense of the term": Bush and Scowcroft, *A World Transformed*, 300.

355   *"historic point"*: PPP, "The President's News Conference," May 3, 1990.

curtailment: "Evolution in Europe," *New York Times*, May 4, 1990.

"guaranteeing against": Bush and Scowcroft, *A World Transformed*, 267.

"across the board": BPL, Hutchings Collection, Country Files, File: GDR — Meetings and Visits, Baker to POTUS, June 7, 1990.

"reciprocate": Ibid.

356   *"reliable and calculable"*: BPL, Memcons and Telcons, "Meeting with Acting President Arpad Goncz of Hungary," May 18, 1990.

"American private sector": Ibid.

"We do not hide": Zantovksy, *Havel*, 366.

"could become the seed": Václav Havel, *NATO, Europe, and the Security of Democracy* (Prague: Theo Publishing, 2002), 22–25.

"avoid that linkage": BPL, Memcons and Telcons, "Meeting with Acting President Arpad Goncz of Hungary," May 18, 1990.

357   *"playing no games"*: BPL, Memcons and Telcons, "Meeting with Prime Minister Tadeusz Mazowiecki of Poland," March 21, 1990.

"useful to you": BPL, Memcons and Telcons, "The President's Telephone Call to Chancellor Helmut Kohl of the Federal Republic of Germany," March 20, 1990.

"just what he could expect": Rice and Zelikow, *Germany Unified*, 222.

"identical positions": BPL, Memcons and Telcons, "Memorandum of Conversation Between President Bush and PM Margaret Thatcher," April 13, 1990.

358   *"British choice"*: George Urban, *Diplomacy and Disillusion at the Court of Margaret Thatcher* (London: I. B. Tauris, 1996), 129.

"have to do something": BPL, Memcons and Telcons, "Meeting with President François Mitterrand of France," April 19, 1990.

"unduly disagreeable": Bozo, *Mitterrand*, 248.

"new European dispensation": State Department Virtual Reading Room, "CFE: GDR Rep on Germany, NATO and Bundeswehr Personnel Limits," June 1, 1990.

"without your investments": Walesa, *The Struggle and the Triumph*, 264.

359   *"subject of European architecture"*: BPL, Memcons and Telcons, "Meeting with Prime Minister Brian Mulroney of Canada," April 10, 1990.

"Bush cannot": Mulroney, *Memoirs*, 763.

"no organized center": Palazchenko, *My Years with Gorbachev*, 184.

360   *"unacceptable strategic shift"*: Ross, *Statecraft*, 35.

"we had lost": Beschloss and Talbott, *At the Highest Levels*, 199.

"end of perestroika": BPL, Scowcroft Collection, Special Separate USSR Notes Files, Gorbachev Files, File: Gorbachev (Dobrynin) Sensitive 1989–June 1990 (copy set) [2], "Memcon with Baker and Gorbachev," May 18, 1990.

"will be a dictator": Baker, *The Politics of Diplomacy*, 241.

"nobody to challenge him": Grachev, *Gorbachev's Gamble*, 187.

361 *"difficult times"*: Beschloss and Talbott, *At the Highest Levels*, 215.

"raised me": Serhii Plokhy, *The Last Empire: The Final Days of the Soviet Union* (New York: Basic Books, 2014), 31, 37.

"asking for permission": Bill Keller, "Parliament in Lithuania, 124–0, Declares Nation Independent," *New York Times*, March 12, 1990.

362 *"Lamentable people"*: Kristina Spohr Readman, "Between Political Rhetoric and *Realpolitik* Calculations," in *The Baltic Question During the Cold War*, eds. John Hiden, Vahur Made, and David J. Smith (New York: Routledge, 2008), 166.

"illegal and invalid": Beschloss and Talbott, *At the Highest Levels*, 194.

"Gorbachev in the saddle": Spohr Readman, "Between Political Rhetoric and *Realpolitik* Calculations," 166. Hutchings, "American Diplomacy and the End of the Cold War," names Kohl as author of this metaphorical plea (30).

"Gorbachev is through": Kevin O'Connor, *The History of the Baltic States* (Denver: Greenwood, 2015), 222.

"reasonable coalition": BPL, Memcons and Telcons, "Telephone Call from Chancellor Helmut Kohl of the Federal Republic of Germany," March 15, 1990.

"give up and go home": Rice and Zelikow, *Germany Unified*, 230.

363 *"sensation is perfect"*: Klaus Naumann with Horst Teltschik, "An Unexpected Success: The Rapid Unification of Germany," in *Breakthrough: From Innovation to Impact*, eds. Henk van den Breemen, Douglas Murray, and Maarten Verkerk (Lunteren: The Owls Foundation, 2014), 171.

"hell of a campaigner": Maynard, *Out of the Shadow*, 65.

"Completely unexpected": Rice and Zelikow, *Germany Unified*, 231.

"uphold my principles": Urban, *Diplomacy and Disillusion*, 128.

"new discriminatory restraints": PPP, "News Conference of the President and Prime Minister Margaret Thatcher," April 13, 1990.

"rapid route": Bozo, *Mitterrand*, 214.

"all sorts of consequences": Beschloss and Talbott, *At the Highest Levels*, 197.

364 *"taken note"*: Ibid.

"watching it closely": PPP, "Remarks and a Question-and-Answer Session," March 15, 1990.

"shown restraint": For this conversation, see BPL, Rice, Subject Files, File: Baltics — Other, Matlock to State, March 30, 1990.

365 *"specifically targeted loans"*: BPL, Scowcroft Collection, Special Separate USSR Note Files, File: Gorbachev (Dobrynin) Sensitive 1989–June 1990 [copy set] [2], May 18, 1990.

"lead to suicide": Baker, *The Politics of Diplomacy*, 240.

366 *"piecemeal approach"*: BPL, NSC, Burns Papers, Chronological Files, File: May 1990–June 1990 [2], Brady to Bush, May 24, 1990.

"easing domestic pressures": State Department Virtual Reading Room, "Theme Paper: German Unification," May 22, 1990.

"able to deal": State Department Virtual Reading Room, "Gorbachev's Summit Agenda: Looking Ahead," May 23, 1990.

"moral moments": William Safire, "The Grave Consequence," *New York Times*, April 2, 1990.

"sweet talk": Rowland Evans and Robert Novak, "Bush Sweet Talk on Lithuania," *Chicago Sun-Times*, March 30, 1990.

"beating the drums": Beschloss and Talbott, *At the Highest Levels*, 198.

"toasts": "Drawing the Line: How Far Can Gorbachev Go in Lithuania?" *Philadelphia Inquirer*, April 2, 1990.

367 *"force right now"*: Beschloss and Talbott, *At the Highest Levels*, 200.

"aren't quite as simple": Bush, *All the Best*, 466.

"end in blood": Bozo, *Mitterrand*, 240–41.

"progress stopped": Naftali, *George H. W. Bush*, 94.

368 *"cleanest and most coherent"*: BPL, NSC, NSC0043a — April 23, File: 1990 — Lithuania, Scowcroft to POTUS, April 23, 1990.

"capture the agenda": Ibid.

369 *"left out there all alone"*: PPP, "Remarks at a White House Briefing," April 24, 1990.

"Economic countermeasures": BPL, Memcons and Telcons, "Meeting with President François Mitterrand of France," April 19, 1990.

"plunder of the people": Palazchenko, *My Years with Gorbachev*, 189.

"signs of crisis": Sarotte, *1989*, 155.

"they hate me": Chernyaev, *My Six Years*, 270.

370 *"complicate his life"*: Sarotte, *1989*, 165.

as perestroika *required*: BPL, Scowcroft Collection, Special Separate USSR Notes Files, Gorbachev Files, File: Gorbachev (Dobrynin) Sensitive 1989–June 1990, "Memcon Baker and Gorbachev," May 18, 1990.

"shouldn't we apply": Ibid.

"unusual moves to make": Ibid.

"might be willing": BPL, NSC, Hutchings Collection, Country Files, File: German Reunification, "German Unification — Two Plus Four Process," May 25, 1990.

371 *"never agree to assign"*: Sarotte, *1989*, 165.

"big wild card": BPL, NSC, Hutchings Collection, Country Files, File: German Reunification, "German Unification — Two Plus Four Process," May 25, 1990.

"come out feeling": BPL, Memcons and Telcons, "Telephone Conversation with Chancellor Helmut Kohl of the Federal Republic of Germany," May 30, 1990.

"work together": BPL, NSC, Rice, Subject Files, File: Memcons and Telcons, "Memcon Between POTUS and Gorbachev," May 31, 1990.

"important that we not fail": Ibid.

"humiliate us": Bush, *All the Best*, 472.

372 *"Gorbachev shrugged"*: Bush and Scowcroft, *A World Transformed*, 282.

"I'm gratified": Ibid. See also National Security Archive, Electronic Briefing Book no. 320, *The Washington/Camp David Summit 1990: From the Secret Soviet, American, and German Files*, "Record of Conversation, M. S. Gorbachev and G. Bush," May 31, 1990.

"we will respect that": National Security Archive, Electronic Briefing Book no. 320, "Record of Conversation, M. S. Gorbachev and G. Bush," May 31, 1990.

"unbelievable scene": Bush and Scowcroft, *A World Transformed*, 283.

"politically and psychologically": National Security Archive, Electronic Briefing Book no. 320, *The Washington/Camp David Summit 1990: From the Secret Soviet, American, and German Files*, "On the Directives for Negotiations with the U.S. Secretary of State James Baker in Moscow," May 16, 1990.

373 *"fire off some rounds"*: Steve Goldstein, "Gorbachev, Congressmen, Have at It," *Philadelphia Inquirer*, June 2, 1990.

374 *"further constructive work"*: National Security Archive, Electronic Briefing Book no. 320, "Record of Conversation, M. S. Gorbachev and G. Bush," June 2, 1990.

"interest of both sides": Ibid.

"purely symbolic": Chernyaev, *My Six Years*, 266.

375 *"What a sensation"*: Bozo, *Mitterrand*, 284.
"without a shot": Grachev, *Gorbachev's Gamble*, 190.

19. "THIS WILL NOT STAND"

376 *"Iraq may be"*: Meacham, *Destiny and Power*, 420.
"looks like the next": Hal Brands, *From Berlin to Baghdad: America's Search for Purpose in the Post–Cold War World* (Lexington: University Press of Kentucky, 2008), 49.
377 *"isn't it exciting"*: PPP, "Remarks and Question and Answer Session with the Magazine Publishers of America," July 17, 1990.
"peace-nik[s]": Bush, *All the Best*, 461, 470.
chopping block: Zelizer, *Arsenal of Democracy*, 359.
"loving this job": Bush, *All the Best*, 475.
"No new taxes": *The American Presidency Project*, "Address Accepting the Presidential Nomination at the Republican National Convention in New Orleans," August 18, 1988, http://www.presidency.ucsb.edu/ws/?pid=25955 (accessed March 25, 2017).
378 *"Don't break it"*: Meacham, *Destiny and Power*, 416.
"get something done": Ibid., 413.
"I Lied": Ibid., 409.
"betrayal of Reaganism": Andrew Rosenthal, "3 Little Words: How Bush Dropped His Tax Pledge," *New York Times*, June 29, 1990.
"biggest test": Meacham, *Destiny and Power*, 416.
379 *casualties*: Definitive death tolls for the conflict continue to elude scholars. See http://kurzman.unc.edu/death-tolls-of-the-iran-iraq-war/ (accessed May 26, 2016).
the missiles themselves: Williamson Murray and Kevin Woods, *The Iran-Iraq War* (New York: Cambridge University Press, 2014), 35.
"conspiring bastards": Hal Brands and David Palkki, "'Conspiring Bastards': Saddam Hussein's Strategic View of the United States," *Diplomatic History* 36, no. 3 (June 2012): 626.
380 *"center post"*: Kevin Woods et al., eds., *The Saddam Tapes* (New York: Cambridge University Press, 2011), 131.
"never quite trusted": Joost Hilton, *A Poisonous Affair* (New York: Cambridge University Press, 2007), 77.
"tide of history": Michael Palmer, *Guardians of the Gulf: A History of America's Expanded Role in the Persian Gulf, 1883–1992* (New York: Simon & Schuster, 1992), 156.
"only obstacle": Kevin Woods, *The Mother of All Battles: Saddam Hussein's Strategic Plan for the Persian Gulf War* (Annapolis: Naval Institute Press, 2008), 42.
"much harm": BPL, NSC, Richard Haass Papers, box 43, Working Files, File: Iraq Pre–8/2/90 [2], Glaspie to SecState, April 12, 1990.
"It is not reasonable": For the meeting between Saddam and Ambassador Glaspie, see BPL, NSC, Richard Haass Papers, box 43, Working Files, File: Iraq Pre–8/2/90 [2], Glaspie to SecState, July 25, 1990. See also a complete transcript of the Iraqi version of the discussion at BPL, NSC, David Welch Files, Subject Files, File: Iraq — Key Documents, "The Meeting of the President Commander with Ambassador Glaspie," July 25, 1990.
382 *"effectively what it did"*: Stephen Walt, "WikiLeaks, April Glaspie, and Saddam Hussein," Foreignpolicy.com, January 9, 2011, http://foreignpolicy.com/2011/01/09/wikileaks-april-glaspie-and-saddam-hussein/ (accessed March 25, 2017).
"wrath of God": Frontline, "An Interview with James Akins," http://www.pbs.org/wgbh/pages/frontline/shows/saddam/interviews/akins.html (accessed May 27, 2016).

"any defense treaties": Elaine Sciolino and Michael Gordon, "US Gave Iraq Little Reason Not to Mount Kuwait Assault," *New York Times,* September 23, 1990.

383 *"felt secure"*: BPL, NSC, Richard Haass Papers, box 43, Working Files, File: Iraq Pre–8/2/90 [2], Glaspie to SecState, July 25, 1990.

"best resolved": BPL, NSC, Richard Haass Papers, box 43, Working Files, File: Iraq Pre–8/2/90 [2], SecState (Eagleburger) to AmEmbassy Baghdad, July 28, 1990. For "bilateral disputes," see BPL, NSC, Richard Haass Papers, box 43, Working Files, File: Iraq Pre–8/2/90 [2], SecState to AmEmbassy Baghdad, July 24, 1990.

"take sides": BPL, NSC, Richard Haass Papers, box 43, Working Files, File: Iraq Pre–8/2/90 [2], "Points to Be Made with Iraqi President Saddam Hussein," August 1, 1990.

"summer cloud": Woods, *The Mother of All Battles,* 53.

384 *"talked to the Kuwaitis"*: Miller Center oral history, Thomas Pickering, December 14, 2010, 10.

"really implore you": BPL, Memcons and Telcons, "Telephone Conversation with King Hussein of Jordan," July 31, 1990.

"used to disasters": BPL, Memcons and Telcons, "Telephone Conversation with President Hosni Mubarak of Egypt," August 2, 1990.

"visceral reaction": Sparrow, *The Strategist,* 385.

"generally unloved": Lawrence Freedman, *A Choice of Enemies: America Confronts the Middle East* (New York: Public Affairs, 2008), 219.

385 *"fundamental US interest"*: BPL, NSC, Richard Haass Papers, box 43, Working Files, File: Iraq Pre–8/2/90 [4], Wolfowitz to Gates, Kimmitt, Ross, and Haass, July 26, 1990.

the transcript shows: BPL, NSC, Richard Haass Papers, box 43, Working Files, File: Presidential Meetings Files, "Minutes of NSC Meeting," August 2, 1990.

"extremely vulnerable": BPL, NSC, Richard Haass Papers, box 43, Working Files, File: Iraq Pre–8/2/90 [4], Wolfowitz to Gates, Kimmitt, Ross, and Haass, July 26, 1990.

386 *"intermediate option"*: BPL, NSC, Richard Haass Papers, box 43, Working Files, File: Presidential Meetings Files, "Minutes of NSC Meeting," August 2, 1990.

"interesting opportunity": Ibid.

"no choice": Ibid.

"really care": Sparrow, *The Strategist,* 389.

387 *"the more they shrink"*: BPL, NSC, Richard Haass Papers, box 43, Working Files, File: Presidential Meetings Files, "Minutes of NSC Meeting," August 2, 1990.

"give some spine": Ibid.

"restrain these guys": Beschloss and Talbott, *At the Highest Levels,* 242.

"local dictators": Ekedahl and Goodman, *Wars of Eduard Shevardnadze,* 195. See also Miller Center oral history, Dennis Ross, August 2, 2001, 32.

388 *"serendipity"*: Dennis Ross, *The Missing Piece: The Inside Story of the Fight for Middle East Peace* (New York: Farrar, Straus and Giroux, 2004), 65.

"being defused": Beschloss and Talbott, *At the Highest Levels,* 244.

"no logic to it": Ekedahl and Goodman, *Wars of Eduard Shevardnadze,* 195.

"this opportunity": Ray Takeyh and Steven Simon, *The Pragmatic Superpower: Winning the Cold War in the Middle East* (New York: W. W. Norton, 2016), 309.

"swiftly and affirmatively": Brands, *From Berlin to Baghdad,* 50.

389 *"aggression is inconsistent"*: Wilson, *Triumph of Improvisation,* 188.

"play a good game": Palazchenko, *My Years with Gorbachev,* 208.

"gained nothing": Oberdorfer, *From the Cold War to a New Era,* 434.

390 *"no other way out"*: Alvin Z. Rubinstein, "Moscow and the Gulf War: Decisions and Consequences," *International Journal* 49, no. 2 (Spring 1994): 314.

for them, either: Grachev, *Gorbachev's Gamble,* 192.

"quite obvious then": National Security Archive, *The Cold War*, episode 23, "The Wall Comes Down," interview with James Baker, October 1997, http://nsarchive.gwu.edu/coldwar/inter views/episode-23/baker3.html (accessed December 18, 2016).

"If we were worried": BPL, Memcons and Telcons, "Telephone Conversation with President Brian Mulroney of Canada," August 4, 1990.

391 *"crisis du jour"*: David Schmitz, *Brent Scowcroft*, 141.

"option to be inactive": BPL, NSC, Richard Haass Papers, box 43, Working Files, File: Presidential Meetings Files, "Minutes of NSC Meeting," August 2, 1990.

"No one here": Ibid.

"naked aggression": Sparrow, *The Strategist*, 387.

"Write me a memo": Meacham, *Destiny and Power*, 425. See also Richard Haass, *War of Necessity, War of Choice: A Memoir of Two Iraq Wars* (New York: Simon & Schuster, 2010), 60–64.

"terrible precedent": Little, *Us Versus Them*, 70.

392 *taking the keyboard*: Haass, *War of Necessity*, 68.

"The key": Cheney, *In My Time*, 186.

"keep calm": BPL, Memcons and Telcons, "Telephone Conversation with Jordan's King Hussein and Egypt's Hosni Mubarak," August 2, 1990.

"give us us two days": Ibid.

"following Hitler": BPL, Memcons and Telcons, "Telephone Conversation with King Fahd of Saudi Arabia," August 2, 1990.

393 *"under pressure"*: Meacham, *Destiny and Power*, 427.

"status quo": Ibid., 428.

"Churchill speech": Haass, *War of Necessity*, 62.

"the stakes in this": BPL, NSC, Richard Haass Papers, box 43, Working Files, Presidential Meetings Files, "Minutes of NSC Meeting," August 3, 1990.

"shape of the world": Ibid.

"unacceptable": BPL, Memcons and Telcons, "Telephone Conversation with President François Mitterrand of France," August 3, 1990.

"more urgency": BPL, Memcons and Telcons, "Telephone Conversation with King Fahd of Saudi Arabia," August 4, 1990.

defended or liberated: Haass, *War of Necessity*, 67.

394 *"he lied"*: BPL, Memcons and Telcons, "Telephone Conversation with King Fahd of Saudi Arabia," August 4, 1990.

"will not stand": PPP, "Remarks and an Exchange with Reporters on the Iraqi Invasion of Kuwait," August 5, 1990. "Where'd you get that 'This will not stand'" line? Scowcroft immediately asked. "That's mine," Bush replied. "That's what I feel." Meacham, *Destiny and Power*, 434.

"go to work": Ibid.

20. WITH US, OR NOT AGAINST US

395 *"such moral importance"*: Michael Sherry, *In the Shadow of War: The United States Since the 1930s* (New Haven: Yale University Press, 1997), 464.

396 *"chart the future"*: PPP, "Address to the Nation Announcing Allied Military Action in the Persian Gulf," January 15, 1991.

"new world order": Bob Woodward, *The Commanders* (New York: Simon & Schuster, 2002), 344.

seemed familiar: Christopher T. Jesperson, "Analogies at War: Vietnam, the Bush Administration's War in Iraq, and the Search for a Usable Past," *Pacific Historical Review* 74, no.

3 (August 2005): 411–26. See also Jeffrey Record, *Making War, Thinking History: Munich, Vietnam, and Presidential Uses of Force from Korea to Kosovo* (Annapolis: Naval Institute Press, 2002).

"three of the last five": Philip Taubman, "The Man in the Middle," *New York Times*, October 29, 1995.

"another Tehran situation": Baker Papers, Monthly Files, box 109, folder 4, "Notes, Kennebunkport, Hussein/Saud Visits," August 16, 1990.

397 *"choice between two negatives"*: Maynard, *Out of the Shadow*, 86.

"after my neck": Meacham, *Destiny and Power*, 452.

"Lone Ranger": Duane Tananbaum, "President Bush, Congress, and the War Powers: Panama and the Persian Gulf," in *From Cold War to New World Order: The Foreign Policy of George H. W. Bush*, eds. Meena Bose and Rosanna Perotti (Westport, Conn.: Greenwood Press, 2002), 197.

"this chapter of history": BPL, Memcons and Telcons, "Telephone Conversation with Toshiki Kaifu, Prime Minister of Japan," August 13, 1990.

"tin-cup": Baker, *The Politics of Diplomacy*, 287–90.

398 *small profit*: For a near-time discussion of the accounting complexities involved in any such calculation, see Leonard Silk, "The Broad Impact of the Gulf War," *New York Times*, August 16, 1991.

"serious this time": Meacham, *Destiny and Power*, 429.

"Hitler revisited": PPP, "Remarks to a Fundraising Luncheon," October 15, 1990.

ordered copies: For the Amnesty International report, see BPL, Cabinet Affairs, Will Gunn Files, Folder: Persian Gulf [2].

"a book of history": PPP, "Remarks at a Republican Fundraising Breakfast in Burlington, Vermont," October 23, 1990. Bush most likely refers to Martin Gilbert, *The Second World War: A Complete History* (New York: Henry Holt and Company, 1989). "He read a Martin Gilbert single volume during the presidency," Jon Meacham noted, but "it was more ambient life experience than anything he read" that made World War II the touchstone of his thinking. Jon Meacham to the author, private correspondence, September 16, 2016.

399 *"human shields"*: PPP, "Remarks at a Republican Party Fundraising Breakfast in Burlington, Massachusetts," November 1, 1990.

"Hitler's class": Theo Lippmann Jr., "George Bush Keeps Comparing Saddam Hussein to Hitler," *Baltimore Sun*, November 28, 1990.

"as concerned as I am": PPP, "The President's News Conference in Orlando, Florida," November 1, 1990.

"never the answer": PPP, "Remarks to the Reserve Officers Association," January 23, 1991.

more Americans listed Saddam: Ronald Krebs and Jennifer Lobasz, "The Sounds of Silence: Rhetorical Coercion, Democratic Acquiescence, and the Iraq War," in *American Foreign Policy and the Politics of Fear*, eds. A. Trevor Thall and Jane K. Cramer (New York: Routledge, 2009), 125.

"thinking of his grandkids": Ann Reilly Dowd, "How Bush Decided," *Fortune*, February 11, 1991.

400 *persuasion campaign*: For a thoughtful discussion of this media manipulation and its impact on Congress and the president, see Sparrow, *The Strategist*, 395–403. See also John R. MacArthur, *Second Front: Censorship and Propaganda in the 1991 Gulf War* (Berkeley: University of California Press, 2004), 49; and Michael Massing, "The Way to War," *New York Review of Books*, March 28, 1991.

"has to be stopped": Dowd, "How Bush Decided."

"It shaped my life": Meacham, *Destiny and Power*, 456.

"it is that clear": Bush, *All the Best*, 497.

401 *military prowess:* As Colin Powell subsequently put it, "We could now afford to pull divisions out of Germany that had been there for the past forty years to stop a Soviet offensive that was no longer coming." Powell, *My American Journey,* 487.

"in American hands": Alan Riding, "Allies Reminded of Need for U.S. Shield," *New York Times,* August 12, 1990.

"must be very happy": Ibid.

"to catalyze": Brands, *Making the Unipolar Moment,* 305.

"essential to solve": Meacham, *Destiny and Power,* 442.

402 *"bleed an enemy":* Keith Shimko, *The Iraq Wars and America's Military Revolution* (New York: Cambridge University Press, 2010), 57.

"Don't bother me": Hal Brands, "Inside the Iraqi State Records: Saddam Hussein, 'Irangate,' and the United States," *Journal of Strategic Studies* 34, no. 1 (February 2011): 101.

"boom": Woods, *The Mother of All Battles,* 51.

"overflow of casualties": Meacham, *Destiny and Power,* 545.

"unknown number": Freedman, *A Choice of Enemies,* 228.

403 *"another prolonged situation":* Brands, *From Berlin to Baghdad,* 58.

"week before mid-term elections": Recollections of this meeting abound. See Sparrow, *The Strategist,* 393.

"when he's unable": Theodore H. Draper, "The True History of the Gulf War," *New York Review of Books,* January 30, 1992.

404 *"right up the middle":* Sparrow, *The Strategist,* 393.

"didn't want to do it": Maynard, *Out of the Shadow,* 84.

"bodies pulled out": Mann, *Rise of the Vulcans,* 118.

"I had been appalled": H. W. Brands, "Neither Munich nor Vietnam: The Gulf War of 1991," in *The Power of the Past: History and Statecraft,* eds. Hal Brands and Jeremi Suri (Washington, D.C.: Brookings Institution Press, 2016), 85.

"hell of a lot more": Sparrow, *The Strategist,* 393. See also Maynard, *Out of the Shadow,* 84.

"You got it": Sparrow, *The Strategist,* 393.

"Does he know": Ibid.

"perfectly well": Ibid.

"Lyndon Johnson": PBS, *Frontline: The Gulf War,* oral history with Robert Gates, http://www .pbs.org/wgbh/pages/frontline/gulf/oral/gates/1.html (accessed December 18, 2016).

405 *"no excuse possible":* Andrew Bacevich, *America's War for the Greater Middle East: A Military History* (New York: Random House, 2016), 116.

"unexpected opportunity": Baker, *The Politics of Diplomacy,* 3.

"a key position": BPL, Memcons and Telcons, "Telephone Conversation with President Turgut Ozal of Turkey," August 3, 1990. See also Ekavi Athanassopoulou, *Strategic Relations Between the US and Turkey, 1979–2000* (New York: Routledge, 2014), 97.

"virtually the entire world": For this conversation, see BPL, Memcons and Telcons, "Telephone Conversation with President Turgut Ozal of Turkey," August 4, 1990.

406 *Saudis were pressing:* Meacham, *Destiny and Power,* 430.

"future of peace": BPL, Memcons and Telcons, "Telephone Conversation with President Turgut Ozal of Turkey," August 5, 1990.

"best protection": BPL, NSC, H-Files, NSC0053-August 15, 1990 — Iraqi Invasion of Kuwait, Persian Gulf, "Talking Points."

407 *Baker happily responded:* Baker, *The Politics of Diplomacy,* 284–85.

"short of cash": State Department Virtual Reading Room, AmEmbassy London to SecState, "Meeting with Prime Minister Thatcher," August 23, 1990.

"not an option": Clyde Haberman, "Quick Action by Turkey on Sanctions a Startler," *New York Times,* August 8, 1990.

future invitation: Thomas Friedman, "US Says Turkey Plans to Coöperate on Iraq," *New York Times*, August 9, 1990.

"personal kind of performance": "Discussant: Richard B. Cheney," in Bose and Perotti, *From Cold War to New World Order*, 483.

408 *"solemn word"*: Dennis Ross, *Doomed to Succeed: The U.S.-Israel Relationship from Truman to Obama* (New York: Farrar, Straus and Giroux, 2015), 437.

twice that number: Brands, *Making the Unipolar Moment*, 302.

"decided the matter": Lawrence Wright, *The Looming Tower: Al-Qaeda and the Road to 9/11* (New York: Random House, 2006), 178.

409 *"solid support"*: Bush, *All the Best*, 479.

"I think it's money": BPL, Memcons and Telcons, "Telephone Conversation with Prime Minister Brian Mulroney of Canada," August 7, 1990.

"known the guy forever": Chase Untermeyer, *Zenith: In the White House with George H. W. Bush* (College Station: Texas A&M University Press, 2016), 147.

"permanent rep": Baker Papers, Monthly Files, box 109, folder 7, "Handwritten Note to Bob Kimmitt," UNSC, November 29, 1990.

"big hegemonists": Nicholas Kristof, "Beijing Skeptical of U.S. Gulf Role," *New York Times*, February 20, 1991.

410 *"traffic jams"*: Baker Papers, Monthly Files, box 109, folder 7, handwritten note to Bob Kimmit, November 29, 1990.

"matter of principle": Suettinger, *Beyond Tiananmen*, 113.

"clearly got the point": Maynard, *Out of the Shadow*, 79.

"MFN could fall apart": BPL, Scowcroft Collection, Special Separate China Notes, File: China 1990 (sensitive) [4], POTUS to Scowcroft, September 13, 1990.

"cannot overstate the importance": BPL, Scowcroft Collection, Special Separate China Notes, File: China 1990 (sensitive) [4], POTUS to Deng, August 30, 1990.

"question of a meeting": BPL, Gulf War FOIA, box 44, File: Iraq November 1990 [1], SecState Washington to AmEmbassy Beijing, November 27, 1990.

"emphasize the priority": BPL, Scowcroft Collection, Special Separate China Notes, File: China 1990 (sensitive) [4], POTUS to Deng, August 30, 1990.

411 *Baker in particular*: BPL, Scowcroft Collection, Special Separate China Notes, File: China 1990 (sensitive) [4], Bush to Scowcroft, September 13, 1990.

might need Beijing's endorsement: Mann, *About Face*, 249.

"an international crisis": Suettinger, *Beyond Tiananmen*, 114.

"collect the fee": Sparrow, *The Strategist*, 472.

412 *"We're informing you"*: Beschloss and Talbott, *At the Highest Levels*, 250.

"I'm talking with you": Baker, *The Politics of Diplomacy*, 282.

"no longer two superpowers": Untermeyer, *Zenith*, 149.

"close the book": For this conversation, see BPL, Scowcroft Collection, Special Separate USSR Notes Files, Folder: Gorbachev (Dobrynin) Sensitive July–December 1990 [copy set] [1], "The President's Meeting with President Gorbachev of the Soviet Union," September 9, 1990.

414 *"happening at the same time"*: Palazchenko, *My Years with Gorbachev*, 210, 214–15.

21. THE NEW WORLD ORDER

415 *"a new world"*: Kathleen Hall Jamieson, *Eloquence in an Electronic Age: The Transformation of Political Speechmaking* (New York: Oxford University Press, 1988), 99.

"never dreamed": Thomas J. Knock, *To End All Wars: Woodrow Wilson and the Quest for a New World Order* (Princeton: Princeton University Press, 1992), 275.

416 *"This time"*: PPP, "Address to Congress on the Yalta Conference," March 1, 1945.

"forge of war": Steven Schlesinger, *Act of Creation: The Founding of the United Nations* (Cambridge: Westview Press, 2003), 72.

the great powers: For the UN's initial security structure, see Paul Kennedy, *The Parliament of Man: The Post, Present, and Future of the United Nations* (New York: Vintage, 2006), 51–62.

"only a first step": PPP, "Address in San Francisco at the Closing Session of the United Nations Conference," June 26, 1945.

"shaped American thought": Henry Kissinger, *Diplomacy* (New York: Simon & Schuster, 1994), 44.

417 *"As soon as Saddam"*: PPP, "Joint News Conference of President Bush and Soviet President Mikhail Gorbachev in Helsinki, Finland," June 9, 1990.

Operation Desert Shield: Kenneth Ervin King, "Operation Desert Shield: Thunder Storms of Logistics," *US Army War College Strategy Research Project*, March 30, 2007, handle.dtic .mil/100.2/ADA467240 (accessed March 25, 2017).

their bacon: Associated Press, "The Chow's a Mess, Troops Complain," *Los Angeles Times*, November 6, 1990.

"seize the oil fields": Bacevich, *America's War for the Greater Middle East,* 113.

418 *"Iraq against the world"*: For this speech, see PPP, "Address Before a Joint Session of the Congress on the Persian Gulf Crisis and the Federal Budget Deficit," September 11, 1990.

420 *"United Nations that functioned"*: PPP, "Interview with Middle East Journalists," March 8, 1991.

"stood up and condemned": Roy Joseph, "The New World Order: President Bush and the Post–Cold War Era," in Medhurst, *The Rhetorical Presidency of George H. W. Bush,* 94.

"You are asking": Alexander Thompson, "Screening Power: International Organizations as Information Agents," in *Delegation and Agency in International Organizations,* eds. Darren Hawkins et al. (Cambridge: Cambridge University Press, 2006), 245

"more I think about it": Meacham, *Destiny and Power,* 440.

421 *"by no means clear"*: Palazchenko, *My Years with Gorbachev,* 215.

"not easy for him": Ibid., 213.

422 *"without war, pain, or strife"*: G. John Ikenberry, "German Unification, Western Order, and the Post–Cold War Restructuring of the International System," in *German Unification: Expectations and Outcomes,* eds. Peter Caldwell and Robert Shandley (New York: Palgrave Macmillan, 2011), 15.

"utterly ruining": Chernyaev, *My Six Years,* 301.

"debased trash": Ibid.

"Fire on them": Service, *The End of the Cold War,* 475.

"probably crazy": Bill Keller, "Gunman Fires Twice Close to Gorbachev," *New York Times*, November 9, 1990. See also Elizabeth Shogren, "Gunman Reportedly Wanted to Kill Gorbachev," *Los Angeles Times*, November 16, 1990.

"not the prize": Conor O'Clery, *Moscow, December 25, 1991* (New York: Public Affairs, 2011), 94.

423 *"witnessing the advent"*: Chernyaev, *My Six Years,* 298.

"explosion could occur": G. J. Church, "Depths of Gloom," *Time*, November 26, 1990.

"crucified": Service, *The End of the Cold War,* 485.

424 *"parody of representative government"*: BPL, NSC, Rice Papers, 1989–1990 Subject Files, Folder: Master Chron Log for USSR — January 1990–December 1990 [1], Rice to Scowcroft, November 23, 1990.

"only hope for political survival": Matlock, *Autopsy on an Empire,* 448.

"We need help now": Baker, *The Politics of Diplomacy,* 294.

"spend billions": Pavel Stroilov, *Behind the Desert Storm* (Chicago: Price World Publishing, 2011), 181.

425 *"They got the point"*: Maynard, *Out of the Shadow*, 80.

"chief levers": Carolyn McGiffert Ekedahl, "The Soviet Union and Iraq's Invasion of Kuwait," in *The Diplomatic Record: 1991*, ed. David D. Newson (Boulder: Westview Press, 1992), 89.

Others lamented: Ekedahl and Goodman, *Wars of Eduard Shevardnadze*, 202.

"two or three ships": Carolyn Ekedahl and Melvin Goodman, "The Soviet Union and Iraq's Invasion of Kuwait," in *The Soviets, Their Successors and the Middle East*, ed. Rosemary Hollis (London: St. Martin's, 1993), 99.

"take to the streets": Stephen Handelman, "Russia Passes Law Allowing Private Ownership of Land," *Toronto Star*, December 4, 1990.

426 *"I'll choose the latter"*: Yevgeny Primakov, *Russian Crossroads: Toward the New Millennium* (New Haven: Yale University Press, 2004), 49.

misguided: Bush made a similar point to Gorbachev, albeit less caustically, writing on October 20 that "going to Saddam with a proposal detailing what we are willing to do after he withdraws would violate the basic principles you and I embraced in Helsinki . . . [I]n other words, he would profit from his invasion, and in doing so, he would likely acquire a standing of heroic proportions in the Arab World . . . In addition, it would run counter to the view I know we share that in dealing with this first real crisis of the post–Cold War era, we cannot permit an aggressor to profit in any way from aggression. Anything less would repeat the mistakes of the 1930s." BPL, Rice Papers, USSR Subject Files, File: Gorbachev Correspondence—Outgoing [1], POTUS to Gorbachev, October 20, 1990.

427 *"grain of reason"*: Service, *The End of the Cold War*, 473.

"the collapse": Primakov, *Russian Crossroads*, 51.

"can't be the minister": Service, *The End of the Cold War*, 470–71.

"slipping from his hands": Hoffman, *The Dead Hand*, 353.

"permit me to say": Woodrow Wilson Center Digital Archive, "Record of a Conversation Between M. S. Gorbachev and President of the US George Bush in Paris," November 19, 1990.

428 *"avoid a military resolution"*: Ibid.

"the president's man": O'Clery, *Moscow, December 25, 1991*, 100, 99.

"To be or not to be": Beschloss and Talbott, *At the Highest Levels*, 288.

"Only his rules": O'Clery, *Moscow, December 25, 1991*, 97.

"dictatorship very soon": Beschloss and Talbott, *At the Highest Levels*, 296.

"to be untrue": Ibid.

"Dictatorship is coming": Bernard Gwertzman, "He'd Rather Be Right Than Foreign Minister," *New York Times*, September 22, 1991.

"worse than Tiananmen": Palazchenko, *My Years with Gorbachev*, 241.

429 *"Do not respond"*: Oberdorfer, *From the Cold War to a New Era*, 442.

"obviously be foolish": Thomas Friedman, "Soviet Turmoil Causes Unease in Washington," *New York Times*, December 12, 1990.

"common ground": Beschloss and Talbott, *At the Highest Levels*, 297.

430 *impeachment*: Meacham, *Destiny and Power*, 452.

"I will be history": Ibid.

final vote: Adam Clymer, "Congress Acts to Authorize War in Gulf; Margins Are 5 Votes in Senate, 67 in House," *New York Times*, January 13, 1991.

"big burden": Meacham, *Destiny and Power*, 453. As Richard Haass concluded, "Had he gone ahead in the face of a no vote by one or both chambers of Congress, there would have been bills of impeachment introduced, which then would have gone nowhere given politics and, perhaps just as important, how quickly and successfully the battle unfolded." Haass, *War of Necessity, War of Choice*, 113–14.

431 *"seemed confused"*: Palazchenko, *My Years with Gorbachev*, 245.

"lost control": Matlock, *Autopsy on an Empire*, 456.

"Neither is comforting": BPL, NSC, Rice Papers, USSR Subject File, Folder: Baltics, Rice to Scowcroft, "Responding to Moscow," January 2, 1991.

"the reincarnation": Ibid.

"own the skies": Meacham, *Destiny and Power*, 460.

432   *"sad irony"*: Maureen Dowd, "White House Sticks to Its Subdued Reaction to Baltic Crackdown," *New York Times*, January 15, 1991.

"shot dead in the streets": A. M. Rosenthal, "The New World Order Dies," *New York Times*, January 15, 1991.

"great length": For this letter, see BPL, NSC, Burns Files, Subject Files, GB-Gorbachev Correspondence [3], Rice to Scowcroft, "Letter to Gorbachev Regarding the Baltic Situation," January 22, 1991.

433   *"putting that progress"*: David Binder, "White House's Quandary in Relations with Kremlin," *New York Times*, January 24, 1991.

"After Bloody Sunday": O'Clery, *Moscow, December 25, 1991*, 101.

434   *"has taken these steps"*: Matlock, *Autopsy on an Empire*, 472.

"zigs and zags": Ibid.

"virtual nighttime, constitutional coup": O'Clery, *Moscow, December 25, 1991*, 100.

"I'll never turn away": Chernyaev, *My Six Years*, 327–29.

435   *"unambiguous commitment"*: Primakov, *Russian Crossroads*, 65.

"very wary": Gideon Rose, *How Wars End: Why We Always Fight the Last Battle* (New York: Simon & Schuster, 2010), 221.

analysts predicted: Haass, *War of Necessity, War of Choice*, 127.

436   *"your standing as a superpower"*: BPL, Memcons and Telcons, "Telephone Conversation with President Mikhail Gorbachev of the USSR," February 22, 1991.

"appreciate your not opposing": Ibid.

"scorched earth": For this conversation, see BPL, Memcons and Telcons, "Telephone Conversation with President Mikhail Gorbachev of the USSR," February 23, 1991.

437   *"doomed to be friends"*: Chernyaev, *My Six Years*, 332.

438   *"I am no longer asked"*: Baker, *Work Hard*, 299.

"I have no elation": Meacham, *Destiny and Power*, 466.

439   *"I love the job"*: Ibid., 472.

22. "DISUNION IS A FACT"

440   *"His Highness"*: Plokhy, *The Last Empire*, 45.

"demagogic": Sparrow, *The Strategist*, 435.

441   *"1000-year history"*: BPL, Memcons and Telcons, "Meeting with Boris Yeltsin, President of the Republic of Russia," July 30, 1991.

"on the dance floor": Bush and Scowcroft, *A World Transformed*, 500.

"the right direction": Gates, *From the Shadows*, 502, 528.

"two flags": BPL, Memcons and Telcons, "Meeting with Boris Yeltsin, President of the Republic of Russia," July 30, 1991.

442   *"as a single state"*: O'Clery, *Moscow, December 25, 1991*, 122, 123.

"cling to the vestiges": Palazchenko, *My Years with Gorbachev*, 298.

really all right: Plokhy, *The Last Empire*, 26.

"grandstanding": Beschloss and Talbott, *At the Highest Levels*, 413.

443   *"knew Boris too well"*: O'Clery, *Moscow, December 25, 1991*, 125.

"gentlemanly behavior": Beschloss and Talbott, *At the Highest Levels*, 413.

"should be no difficulty": BPL, Memcons and Telcons, "Meeting with Boris Yeltsin, President of the Republic of Russia," July 30, 1991.

"*I initialed that page*": Ibid.

"*you support my idea*": Ibid.

444 "*half a century*": PPP, "Remarks by President Gorbachev and President Bush at the Signing Ceremony for the Strategic Arms Reduction Talks Treaty in Moscow," July 31, 1991.

"*plough new ground*": BPL, Memcons and Telcons, "Meeting with Boris Yeltsin, President of the Republic of Russia," July 30, 1991.

445 "*must get word*": Matlock, *Autopsy on an Empire*, 541.

"*cannot happen*": Bush and Scowcroft, *A World Transformed*, 505–6.

"*Tell President Bush*": Hoffman, *The Dead Hand*, 363.

"*constitutional coup*": Archie Brown, *The Gorbachev Factor* (New York: Oxford University Press, 1996), 193.

"*putsch*": O'Clery, *Moscow, December 25, 1991,* 122.

446 "*did you shoot him*": Bush and Scowcroft, *A World Transformed*, 509.

"*What was the threat*": Plokhy, *The Last Empire*, 25.

"*NATO*": Ibid.

"*our goal*": Zantovsky, *Havel*, 366.

447 "*nobody is willing*": BPL, NSC, Burns and Hewett Files, USSR Chronological Files, Folder: USSR Chron File: August 1991 [1], Mosbacher to Scowcroft, July 11, 1991.

"*strong support*": Sparrow, *The Strategist*, 444.

"*tied directly to reforms*": Ibid., 446–47.

"*suffered more*": Bush and Scowcroft, *A World Transformed*, 503.

448 "*Gorbachev is visiting*": BPL, Memcons and Telcons, "Second Plenary, London Economic Summit," July 16, 1991.

"*we need to consider*": Ibid.

449 "*disruptive force*": Matlock, *Autopsy on an Empire*, 553. Bush did not take the barb lightly or consider it accurate. "I don't think he understood that the Gulf countries, Japan, and others, their futures threatened by Iraq's aggression and grateful for our leadership role, had dug deeply to help defray the cost of the war." Bush and Scowcroft, *A World Transformed*, 507.

"*is what it is*": Chernyaev, *My Six Years,* 357.

"*I've done a lot*": BPL, Memcons and Telcons, "Second Plenary, London Economic Summit," July 16, 1991.

"*wholly incorrect*": BPL, Scowcroft Collection, Special Separate USSR Notes Files, Gorbachev Files, Folder: Gorbachev — Sensitive January–June 1991 [2], Scowcroft to Matlock via back-channel, May 8, 1991. See also BPL, Scowcroft Collection, Special Separate USSR Notes Files, Gorbachev Files, Folder: Gorbachev — Sensitive January–June 1991 [2], Bush to Gorbachev, May 8, 1991.

450 "*satisfying to work*": Chernyaev, *My Six Years,* 355.

"*vacillated*": Palazchenko, *My Years with Gorbachev*, 293.

"*Our policy*": Chernyaev, *My Six Years,* 361.

"*new kind and new level*": Palazchenko, *My Years with Gorbachev*, 295.

"*Gorbachev came to collect*": James Cronin, *Global Rules: America, Britain, and a Disordered World* (New Haven: Yale University Press, 2014), 220.

"*use that fund*": National Security Archive, Electronic Briefing Book no. 364, "Expanded Bilateral Meeting with Mikhail Gorbachev of the USSR," July 30, 1991.

"*strongest possible official relationship*": PPP, "Remarks at the Arrival Ceremony in Kiev, Soviet Union," August 1, 1991.

451 "*intolerance and tyranny*": PPP, "Remarks to the Supreme Soviet of the Republic of the Ukraine in Kiev, Soviet Union," August 1, 1991.

"*messenger for Gorbachev*": Gaddis, *The Cold War*, 255.

"*hypnotized*": Plokhy, *The Last Empire*, 60.

"*hung up on tidiness*": William Safire, "After the Fall," *New York Times,* August 29, 1991.

452 "*Anything happen*": Ibid.

"*Disunion is a fact*": Ibid.

23. "I HAVE SIGNED IT"

453 "*Experience taught me*": Bush and Scowcroft, *A World Transformed,* 518.

"*who would I leave it*": O'Clery, *Moscow, December 25, 1991,* 124.

"*poor patriots*": Chernyaev, *My Six Years,* 373.

454 "*who will sign it*": Palazchenko, *My Years with Gorbachev,* 308.

"*end of the USSR*": Louis Sell, *From Washington to Moscow: US-Soviet Relations and the Collapse of the USSR* (Durham: Duke University Press, 2016), 296.

"*Yes, and here's why*": Miller Center oral history, Robert Gates, 10.

455 "*conspiracy*": Plokhy, *The Last Empire,* 83.

456 "*betraying me*": Gorbachev, *Memoirs,* 631.

"*terrible pages*": Michael Dobbs, *Down with Big Brother: The Fall of the Soviet Empire* (London: Bloomsbury, 1996), 377.

"*Whom do you represent*": Plokhy, *The Last Empire,* 83.

"*Anything else is unacceptable*": Gorbachev, *Memoirs,* 631.

457 "*dirty work*": Anthony D'Agostino, *Gorbachev's Revolution, 1985–91* (London: Macmillan, 1998), 331.

"*Go to hell*": O'Clery, *Moscow, December 25, 1991,* 142.

"*intimate secrets*": Ibid.

"*Everything went haywire*": Ibid.

458 "*without deviation*": Sell, *From Washington to Moscow,* 297.

459 "*Will the military*": Bush and Scowcroft, *A World Transformed,* 519.

Swan Lake: Amelia Schonbek, "This Portentous Composition: *Swan Lake*'s Place in Soviet Politics," *Hazlitt,* March 26, 2015, http://hazlitt.net/feature/portentous-composition-swan-lakes-place-soviet-politics (accessed November 3, 2016).

"*condemnatory*": Bush and Scowcroft, *A World Transformed,* 520.

"*totally vindicates*": Bruce D. Berkowitz, "U.S. Intelligence Estimates of the Soviet Collapse: Reality and Perception," *International Journal of Intelligence and Counterintelligence* 21, no. 2 (2008): 237–50.

460 "*Every time*": BPL, Memcons and Telcons, "Telephone Conversation with Prime Minister Brian Mulroney of Canada," August 19, 1991.

"*no second guessing*": Ibid.

"*Any doubt*": Ibid.

"*what we must do*": Plokhy, *The Last Empire,* 78.

461 "*Do not get stampeded*": Bush and Scowcroft, *A World Transformed,* 524.

"*There goes another vacation*": Baker, *Politics of Diplomacy,* 517.

"*hard to do business*": Baker Papers, box 110, folder 6, 1991 August, "JAB Notes from Coup." I am indebted, once more, to Josh Shifrinson for providing this document.

"*the key guy*": Ibid.

462 "*cut off links*": Baker, *Politics of Diplomacy,* 516.

"*extra-constitutional*": PPP, "Remarks and an Exchange with Reporters in Kennebunkport, Maine, on the Attempted Coup in the Soviet Union," August 19, 1991.

"*people of the Soviet Union*": Ibid.

463 "*stroke of luck*": Plokhy, *The Last Empire,* 92.

under arrest: "Yeltsin has already been arrested," Baklanov reported, before correcting himself. "He will be arrested." Dobbs, *Down with Big Brother,* 378.

"removed from power": "Yeltsin's Remarks: A 'Reactionary Coup,'" *New York Times*, August 20, 1991.

"surge of energy": "The Struggle for Russia," *Newsweek*, May 1, 1994.

"coming out to the people": Sell, *From Washington to Moscow*, 299. See also Dobbs, *Down with Big Brother*, 388.

464 *"come to kill"*: Sell, *From Washington to Moscow*, 298. See also Matlock, *Autopsy on an Empire*, 586.

"evasive answers": Sell, *From Washington to Moscow*, 301.

"seems more apt": BPL, NSC Files, Burns Files, Subject Files, File: USSR Coup Attempt August 1991 [1], "News Conference with Gennady Yanayev, Acting President of the USSR," August 19, 1991.

465 *"Call Gorbachev"*: Dobbs, *Down with Big Brother*, 395.

"cherish your good attitude": BPL, NSC Files, Burns Files, Subject Files, File: USSR Coup Attempt August 1990 [1], G. Yanayev, USSR, to President Bush, "Re: Comments on Situation in USSR," August 19, 1991.

"began to think": Gates, *From the Shadows*, 522.

466 *"no one's interest"*: BPL, NSC Files, Burns Files, Subject File, File: USSR Coup Attempt August 1991 [1], State to USEmbassy Moscow, August 19, 1991.

"encourage public disturbances": Ibid.

"as cold as possible": Gates, *From the Shadows*, 522. See also BPL, NSC Files, Burns Files, Subject Files, File: USSR Coup Attempt August 1990 [1], "Meeting Between Ambassador Victor Komplektov and Robert Gates."

"let people come and go": Gates, *From the Shadows*, 522.

"lend legitimacy": Beschloss and Talbott, *At the Highest Levels*, 432.

"condemned": "Text of White House Statement," *New York Times*, August 19, 1991. See also BPL, NSC, Burns Files, Subject Files, File: USSR Coup Attempt August 1990 [1], "Statement by the Deputy Press Secretary," August 19 and 20, 1991.

467 *"mere fact of your call"*: Plokhy, *The Last Empire*, 107.

"What had happened": Meacham, *Destiny and Power*, 488.

"Just checking": BPL, Memcons and Telcons, "Telephone Conversation with President Boris Yeltsin of the Republic of Russia," August 20, 1991.

468 *"We will reiterate"*: Ibid.

had already called: Beschloss and Talbott, *At the Highest Levels*, 433.

"We're not hopeful": BPL, Memcons and Telcons, "Telephone Conversation with President Boris Yeltsin of the Republic of Russia," August 20, 1991.

"liquidate": Dmitri Volkogonov, *Autopsy for an Empire: The Seven Leaders Who Built the Soviet Regime* (New York: Free Press, 1998), 515.

seize control: Sell, *From Washington to Moscow*, 301.

"many armed men": Dobbs, *Down with Big Brother*, 397.

469 *"may not end well"*: Chernyaev, *My Six Years*, 376.

"overrun the barricades": Baker, *The Politics of Diplomacy*, 521.

"he was Russian": Plokhy, *The Last Empire*, 118.

"Give the command": Kalb, *Imperial Gamble*, 106.

470 *"We don't need"*: Plokhy, *The Last Empire*, 120–23.

471 *"I don't know"*: "Gorbachev Tape Blasts Coup," *Washington Post*, August 26, 1991. See also Dobbs, *Down with Big Brother*, 402, 393.

"We will not": Dobbs, *Down with Big Brother*, 402.

"aftermath of the failed coup": BPL, NSC, Susan Koch Files, Subject Files, Folder: After the [Soviet] Coup, AmEmbassy Moscow to State, "The USSR Two Weeks After the Failed Coup," September 6, 1991.

"*Well, read them*": Tom Shanker, "Yeltsin Dictates the New Tune," *Chicago Tribune*, August 24, 1991.

472 "*It's all over*": Beschloss and Talbott, *At the Highest Levels*, 438.

"*I have signed it*": O'Clery, *Moscow, December 25, 1991*, 149.

"*characterized by a race*": BPL, NSC, Susan Koch Files, Subject Files, Folder: After the [Soviet] Coup, AmEmbassy Moscow to State, "The USSR Two Weeks After the Failed Coup," September 6, 1991.

"*orderly way*": BPL, NSC, Burns Files, USSR Chronological Files/Folder: USSR Chron File: August 1991 [1], Gompert and Hewett to Scowcroft, "Message from [sic] John Major on the USSR," August 22, 1911 [sic].

"*I salute you*": BPL, Memcons and Telcons, "Telephone Conversation with Mikhail Gorbachev, President of the Soviet Union," December 25, 1991.

473 "*Every American*": PPP, "Address to the Nation on the Commonwealth of Independent States," December 25, 1991.

"*I want all Americans*": Ibid.

"*guarantee*": Ibid.

CONCLUSION

475 "*strengths beyond challenge*": PPP, "Remarks to the United States Military Academy," June 1, 2002.

"*don't have a dog*": Richard Solomon and Nigel Quinney, *American Negotiating Behavior* (Washington, D.C.: United States Institute of Peace Press, 2010), 80.

"*put a dog*": Brands, *From Berlin to Baghdad*, 90.

"*Everything is at stake*": George Will, "A Dog in That Fight?" *Newsweek*, June 11, 1995.

476 "*Some crises*": PPP, "Address to the Nation on the Situation in Somalia," December 4, 1990.

"If there had been no CNN," Robert Gates later noted, there was no way "we would have ever gone into Somalia." Miller Center oral history, Robert Gates, 87.

"*problem from hell*": Samantha Power, *A Problem from Hell: America in the Age of Genocide* (New York: Basic Books, 2002). Power subsequently became Barack Obama's ambassador to the United Nations, assuming in 2013 the same position Bush had held forty years before, there to advocate for humanitarian interventions while also justifying their absence.

477 "*we will stop it*": PPP, "Remarks to Kosovo International Security Force Troops in Skopje," June 22, 1999.

"*commission*": Beschloss and Talbott, *At the Highest Levels*, 12.

first year: This is a key takeaway from the Miller Center, *First Year 2017*, vol. 2, *National Security*," http://firstyear2017.org/issue/national-security (accessed March 25, 2017).

Bay of Pigs: Mark Selverstone, "Epic Misadventure: John Kennedy's First Year Foreign Policy Stumbles Taught Hard-Earned Lessons," ibid.

"*Black Hawk Down*": Jeremi Suri, "It's Not Just the Economy, Stupid: Bill Clinton's Distracted First-Year Foreign Policy," ibid.

downed reconnaissance plane: Melvyn P. Leffler, "Trust but Clarify: George W. Bush's National Security Team Was Beset by Rivalries," ibid.

478 "*greeted as liberators*": E. J. Dionne, "Dick Cheney Reveals His Chutzpah in Iraq Op-Ed," *Washington Post*, June 18, 2014.

$6 trillion: Michael B. Kelley and Geoffrey Ingersoll, "The Staggering Costs of the Last Decade's U.S. War in Iraq — in Numbers," *Business Insider*, June 20, 2014.

half a million: Joshua Keating, "Half a Million Deaths Is a Statistic," *Slate*, October 18, 2013.

480 "*carried away*": Engel, "'A Better World . . . but Don't Get Carried Away,'" 34.

"pivotal responsibility": Sean M. Lynn-Jones and Steven E. Miller, eds., *America's Strategy in a Changing World* (Cambridge: MIT Press, 1993), 32.

"active leader": Brands, *Making the Unipolar Moment*, 331.

"war of necessity": Richard Haass, "The Gulf War: Its Place in History," in *Into the Desert: Reflections on the Gulf War*, ed. Jeffrey A. Engel (New York: Oxford University Press, 2013), 57–83; see also Haass, *War of Necessity, War of Choice*.

482　*"Be kind to Russia"*: Kalb, *Imperial Gamble*, 12, 21.

"We were promised": Shifrinson, "Deal or No Deal," 13.

483　*"Versailles"*: Kalb, *Imperial Gamble*, 239.

"trying to do": Zubok, "Gorbachev's Nuclear Learning."

"borders of Russia": Brands, *Making the Unipolar Moment*, 354.

484　*"hem of his cloak"*: Ibid., 343.

ACKNOWLEDGMENTS

Much life occurred while this book was gestating. More than expected when the project began. Two kids; two cross-country moves; the death of two cats and the acquisition of two more; a new university and the start of a new center. Many are due thanks for their help in bringing this project to fruition, and for their patience.

The book began at Texas A&M University, at the Bush School of Government and Public Service specifically, which provided unexpected access to a former president. Chuck Hermann deserves the first appreciation. With my first book going to press, it was Chuck who suggested I take a gander at a newly released diary at the nearby Bush Library, the namesake's from his days in China. The more I read, the more I realized the end of the Cold War was well worth studying. I thus owe, or blame, Chuck for the last decade of my life. A series of deans and colleagues made that work possible. To Andy Card, Richard Chilcoat, Benton Cocanougher, Ryan Crocker, Michael Desch, Mary Hein, the aforementioned Chuck Hermann, Sam Kirkpatrick, and Arnie Vedlitz, my appreciation indeed for time, support, and in many cases personal memories of the stories that unfold within these pages. Thanks as well to Roman Popadiuk and Fred McClure, who each coupled support for the Bush Library's research mission with personal insights into its namesake. Janeen Wood and Larry Napper deserve particular thanks, the first for arranging my teaching (and so many other aspects of life), the latter not only for his firsthand knowledge of the Cold War's end, but also for his sage recognition that an untenured new center director would benefit from a two-time ambassador publicly calling him "boss." Thanks too are due to Griffin Rozelle, Peggy Holzweiss, Abby Doll,

Aurelia Figueroa, and Sarah Casey for administering the Scowcroft Institute, and my research files. Also from A&M, Terry Anderson, Domonic Bearfield, Angela Bies, Jason Castillo, Joe Cerami, Jonathan Coopersmith, Lorraine Eden, Kishore Gawande, Jim Griffin, Andy Kirkendall, Randy Kluver, Brian Linn, Andrew Natsios, Will Norris, Jim Olson, Jason Parker, Gina Reinhardt, Jim Rosenheim, Adam Seipp, Ron Sievert, Valerie and Eric Simanek, Peter Tarlow, and Lori Taylor made for a collegial and exciting environment in which to share ideas, writings, babysitting, and barbecue.

Archivists are the unsung heroes of this story, in particular those at the George Bush Library in College Station, where Warren Finch and Robert Holzweiss have assembled an exemplary knowledgeable and friendly staff. They are uniformly excellent, energetic, and no doubt underpaid. Zachary Roberts has been my prime contact throughout the long declassification odyssey, one we were both surprised and dismayed to realize dated back more than a decade, and was thus the recipient of many late-night emails that began "Do you know where . . ." He answered each with enthusiasm and patience. Thanks as well to the Bush Library's team of declassification experts who assisted with this project in one form or fashion: Douglass Campbell, Buffie Hollis, Amber Macicek, McKenzie Morse, Kathy Olson, Rebecca Passmore, Christopher Pembleton, Ashley Seely, Elisabeth Staats, Simon Staats, Lisa Trampota, and Deborah Wheeler. Mary Finch and Cody McMillian made sure the photographs for the book were, as always, ideal.

Southern Methodist University provided a new home in 2012, where Provost Paul Ludden's support for the Center for Presidential History proved unfailing. I am especially grateful for his willingness to provide the final unscheduled time required to finish this book. President Gerald Turner and Provost Steve Currall, Deans Bill Tsutsui and Thomas DiPiero, and in particular my department chair Andy Graybill deserve thanks too, the last for his ongoing counsel and friendship. Tom Knock, Brian Franklin, and Ronna Spitz kept our center running when my mind turned to 1989 and always made coming to work a pleasure, the last offering a literal barrier against outside intrusions, the second his earnest camaraderie, the first service as center director in my absence. Though wrong on Reagan, Alan Lowe proved the better half of the best teaching combo in America. My thanks to colleagues from SMU's Department of History for their welcome, and from across campus and town to Mark Chancey, Kim Cobb, Aaron Crawford, Linda Eads, Jim Falk, Jim Hollifield,

Rita Kirk, David Kusin, Evan McCormick, Hervey Priddy, Josh Rovner, Kenny Ryan, Tim Sayle, Harold Stanley, David Stern, and George Yates for opening themselves to perhaps unwanted discussions of the Cold War's end. To our student workers Clara Johnson, Andrew Oh, and Paul Casella fell the unenviable task of trying to decipher and unravel my handwriting, files, and desk. As always, the Mine Hill Road Development Fund offered time and support whenever required.

Several institutions kindly provided a forum for sharing early chapters of this work. First, Fred Logevall, then of Cornell, offered his graduate seminar, and then the terrifying opportunity to lecture in front of my first mentor; second, Brian Balogh and the University of Virginia's Miller Center National Fellows Program provided a large and helpful audience, aided by the close readings of Andy Card, Melanie McAlister, Will Hitchcock, and David Farber. The Lone Star National Security Forum, and in particular the ongoing valued insights of James Goldgeier, offered a third. John Bew and Theo Farrell provided a temporary research post at King's College, London, while Geir Lundestadt and Asle Toje did the same at the Norwegian Nobel Institute. Lunches at the latter in particular fundamentally shaped my sense of "Europe" as a collective aspiration, and I hope each of the aforementioned can see the improvement in the final product their work made possible.

Many friends and colleagues from across the years lent a careful eye or a helpful ear. Sheela Athreya, Frank Costigliola, Ryan Crocker, Alan Dobson, Ben Engel, Josh Engel, Susan Ferber, Geoff and Melissa French, Gregory Gause, Robert Hutchings, Marion Immerman, Marni Karlin, James Mann, Michael McGandy, Andrew Preston, and Thomas Zeiler provided understanding, encouragement, and, most important of all, sympathy. Melvyn Leffler never ceased to encourage me to make the book the one I wanted, with his own story as a model, and took on the heroic task of commenting on the entire finished manuscript. He is both a friend and who I want to be when I grow up. Wise colleagues Chen Jian, Christian Ostermann, and James Graham Wilson offered invaluable correctives, and on a deadline, for which I am eternally grateful. Dr. Toby R. Engel provided a line-by-line deep reading, and the thrill of hearing a request for "more" from the man with a famously cutting red pen provided greater gratification than he can ever know. The aforementioned James Wilson and Josh Shifrinson kindly shared documents at all hours, and the latter and I discussed Germany's unification so many times I don't think either of us can remember whose ideas were originally whose. Jon Meacham most generously provided advice, encouragement, and the unique pleasure of discussing our mutual subject with someone equally interested. Katie Kaufman offered unique insights, and her friendship is a constant reminder of why I went into teaching. Jean Becker's long-term friendship and willingness to open doors for interviews, coupled with her innate appreciation that enthusiasm did not extend into an academic's conclusions, should be a model for all who hold access to power. President Bush, among others, also shared his memories without once trying to influence the story I tell, and serves as a shining example of how those who make history can work with those who write it without losing sight that these are two very different tasks indeed. He remains what Meacham dubbed him: the last gentleman.

Andrew Wylie skillfully put this book in Houghton Mifflin Harcourt's hands and provided calming counsel when needed most. Bruce Nichols's edits and insightful comments undoubtedly made the book better, ably aided by Ben Hyman. Lori Glazer and Taryn Roeder handled publicity, while Ayesha Mirza oversaw the book's marketing. Larry Cooper oversaw the book's production, Deborah Jacobs provided excellent proofreading, and Amanda Heller added a precise and much-appreciated eye as copy editor. I am in her debt.

This book is dedicated to those whose example most shaped my life and career, and to the idea of l'dor va'dor. These debts cannot be repaid save to ensuing generations. But they can at least be acknowledged and remembered. I keep pictures and mementos from each in my office as reminders to show everyone the patience and kindness they afforded me. I will let the reader turn to another page to discover their identities. At home, Marshall and Elaine have cheerfully lived with tales of "the Daddy President Bush" since their earliest memories. Yet they forever said, "Tell me a history story," and meant it, even if often turning to their mother for confirmation. They will one day have books dedicated to them, and will no doubt write more of their own, repairing this world in their own beautiful and creative ways.

Words fail for the final acknowledgment, which is also the most important. For more than twenty years now Kate Carté Engel has provided home, family, partnership, inspiration, and love. Her faith in me never faltered even when my own flagged. These pages are dedicated to others; my life, to her.

INDEX